Expanding the Frontiers of Design

Design Thinking, a method widely used in design business and management, has changed the landscape of contemporary design. Whereas in the past non-designers were called upon to serve as external consultants ad-hoc, in an effort to promote creativity and innovation most design teams now consist of a mix of designers and other professionals. The impact of this development on the design landscape in recent years is so far without thorough investigation and analysis of its various influences. This book comprises an edited collection of selected papers from the 13th Design Thinking Research Symposium (DTRS13) which offers an exploration of Design Thinking from theoretical, practical, and pedagogical perspectives as well as critical analysis of the design process.

The book is arranged in five parts as follows:

Part 1: Thinking about design
Part 2: Design thinking in the studio
Part 3: Design thinking in practice and professional training
Part 4: Design teams of diverse backgrounds, Interdisciplinary projects
Part 5: Design and nature; visual representation

Providing a comprehensive source for new perspectives on design and Design Thinking, *Expanding the Frontiers of Design* is ideal for designers and design academics of all disciplines wishing to strengthen and innovate their practice, as well as industry leaders who seek to consolidate their business strategies and evolve their work.

Gabriela Goldschmidt is a graduate of the Yale School of Architecture. She joined the Technion after years of architectural practice and taught dozens of design studios at the Technion, and as a visitor elsewhere (including the University of Montreal, MIT, Delft University of Technology) Her research interests include design cognition, reasoning, representation and education, visual thinking, sketching, visual analogy, and related topics. She consolidated separate conferences on design thinking into the series *DTRS – Design Thinking Research Symposia*. Gabriela has published extensively, including on design thinking, and is the author of one book (*Linkography: Unfolding the Design Process*, MIT Press 2014), and co-editor of another (*Design Representation*, Springer 2004, with William L. Porter).

Ezri Tarazi is an industrial designer who was affiliated with the Bezalel Department of Industrial Design for many years, including as head of the department, until he joined the Technion. Most of his work is in R&D, particularly for life-support systems of various kinds. He was a partner in a few start-ups and was a co-founder of IDEO Israel until its closure. He participated in many exhibitions including at prestigious museums and galleries and has led many Design Thinking workshops. Currently, he heads Technion's *Design Tech* initiative. He has written two books (Hebrew) including one on Design Thinking.

Expanding the Frontiers of Design

Critical perspectives

Edited by Gabriela Goldschmidt
with Ezri Tarazi

Routledge
Taylor & Francis Group
LONDON AND NEW YORK

Designed cover image: © Gabriela Goldschmidt

First published 2024
by Routledge
4 Park Square, Milton Park, Abingdon, Oxon OX14 4RN

and by Routledge
605 Third Avenue, New York, NY 10158

Routledge is an imprint of the Taylor & Francis Group, an informa business

British Library Cataloguing-in-Publication Data
A catalogue record for this book is available from the British Library

Library of Congress Cataloguing-in-Publication Data
Names: International Design Thinking Research Symposium (13th : 2022 : Haifa, Israel), creator. | Goldschmidt, Gabriela, 1942- editor. | Tarazi, 'Ezri, 1962- editor.
Title: Expanding the frontiers of design : critical perspectives / Gabriela Goldschmidt with Ezri Tarazi.
Description: Abingdon, Oxon : Routledge, 2023. | Includes bibliographical references and index.
Identifiers: LCCN 2023012695 (print) | LCCN 2023012696 (ebook) |
ISBN 9780367772000 (hardback) | ISBN 9780367772079 (paperback) |
ISBN 9781003170242 (ebook)
Subjects: LCSH: Design--Research--Congresses.
Classification: LCC NK1170 .I545 2022 (print) | LCC NK1170 (ebook) | DDC 744--dc23/eng/20230803
LC record available at https://lccn.loc.gov/2023012695
LC ebook record available at https://lccn.loc.gov/2023012696

ISBN: 978-0-367-77200-0 (hbk)
ISBN: 978-0-367-77207-9 (pbk)
ISBN: 978-1-003-17024-2 (ebk)

DOI: 10.1201/b22630

Typeset in Times New Roman
by MPS Limited, Dehradun

Contents

Preface

This book is an outcome of DTRS13 – the 13th Design Thinking Research Symposium. DTRS symposia have been convened in different universities around the world since 1991, when the first symposium took place at the Delft University of Technology, at the initiative of Nigel Cross, Kees Dorst, and Norbert Roozenburg. The purpose was to look at design processes "from within" – to better understand design behavior as it unfolds in real settings.

Every DTRS meeting had its own specific theme, and members of the design research community were invited to submit papers related to that theme. Several meetings featured shared data sets, usually in the form of protocols of design meetings, which researchers were asked to analyze from a perspective of their choice. Most meetings yielded post-conference edited books, and the current volume is no exception.

DTRS13, which was planned for 2021, had to be postponed because of the Covid19 pandemic and took place at the Technion – Israel Institute of Technology in March 2022. under the title: Expanding the frontiers of design: A blessing or a curse? No shared data was distributed. Instead, we accepted papers that related to the theme: Expanding the frontiers of design: A blessing or a curse? which were presented in the conference and stirred lively discussions. The breadth and richness of the material that was presented was refreshing and encouraged us to publish many of the papers in this book, after their authors revised them to take account of reviews and the conference discussions. A few other papers, not included here, are published in a special issue of the journal *Design Studies*. The entire set of papers, unrevised, is assembled in the online proceedings of DTRS at https://dtrs13.net.technion.ac.il/files/2022/05/DTRS13-Proceedings.pdf

DTRS13 was chaired by Gabriela Goldschmidt and Ezri Tarazi, both Technion professors in the Faculty of Architecture and Town Planning, which includes a graduate industrial design program. We are very happy we were able to hold the conference in vivo despite the pandemic (we donned face masks throughout the three days of the event) and thank the brave participants for traveling to Israel from near and far in those difficult times. We are grateful for the generous financial support we received from the Technion-Israel Institute of Technology, DesignTech, Stratasys, and the Azrieli Foundation. Without them, it would not have been possible to hold the conference or create this book.

Gabriela Goldschmidt
with Ezri Tarazi

Introduction

Gabriela Goldschmidt with Ezri Tarazi

Professional design took a long time to become a bona fide discipline in its own right. Seen as a craft, or part of architectural practice, a separate history of the discipline of design is often dated to the industrial revolution (although designed artefacts were of course produced throughout history). In higher education, we see separate design programs as of the 19th century. Academic journals that center on design first appeared quite late in the 20th century (*Design Studies* 1983; *Design Issues* 1984). This was also the time at which a disappointment was felt from beliefs that it would be possible to "scientify" design, that is, to achieve optimal design outcomes by applying scientific methods, largely building on systems theories and advances in computational techniques, as encompassed in the so-called *Design Methods Movement*. Instead, attention turned towards better understanding design processes as they occur in the real world, mainly their cognitive aspects, and as of late, their neurocognitive and neural attributes. Around the turn of the 21st century, the increased interest in design led the business and management communities to adopt Design Thinking, a methodology developed by design and business thinkers, as a springboard for innovation in all sorts of endeavors. The business and management worlds have embraced this concept and Design Thinking is now being taught in business schools and in engineering and other design-oriented programs. This trend parallels the rise of new design fields like service design and social design, which have become very popular in recent years.

Since in industry design is rarely a process led by an individual, teams are formed and, in these teams, we see an increasing presence of team members who do not have a formal design background. For the most part those non-designers are management agents of one or another kind. Teams composed of designers from diverse design disciplines are also very common. Thus, the involvement of various professionals from disciplines other than design, or of a particular design discipline, is today the norm rather than the exception. We can use the evolution of *Design Thinking Research Symposia* (DTRS) meetings to follow the changes in the conceptualization of design practice, especially if we look at some of the meetings that used shared data. DTRS2 in 1994 was the almost mythological Protocol Analysis focused meeting; it used a dataset recording industrial designers at work – an individual, and a team of three industrial designers. DTRS7 in London (2007) used two sets of data, generated in architectural design and mechanical engineering design meetings. DTRS10 (2014) at Purdue University had a wide scope of datasets taken from educational settings – crits (critiques) and reviews – in a variety of disciplines, including industrial design, mechanical engineering, choreography, entrepreneurship and more. DTRS11 (2016) in Copenhagen featured a data set that mainly dealt with strategies for market penetration outside of Europe by a large European corporation. DTRS12 (2018) in Ulsan featured interviews with more managers than designers. We see that the design

DOI: 10.1201/b22630-1

data that were offered for analysis have broadened: from traditional core design disciplines like mechanical engineering, industrial design and architecture they moved towards new horizons that include a variety of disciplines that two decades ago probably would not be considered "design", like choreography (as a representative of an array of additional disciplines) and managerial/business activity like marketing. We also see a gradual increase in the number of studies that deal with social issues rather than tangible artifacts. The participants in the documented design meetings included an ever-growing number of what we call non-designers.

Design Thinking the prescriptive methodology differs from design thinking that denotes the description of how designers think, as observed and analyzed by researchers. Design thinking was used as a generic phrase by researchers in the post "Design Methods" era, when studies of designers in action occupied center place in the design research community. Today, young researchers take it for granted that design thinking refers to the methodology Design Thinking. We maintain that the use of the same terminology for the method that was developed later (even though it is frequently marked by capital initial letters) to discuss two different phenomena poses difficulties that have not yet been sorted out properly (see Goldschmidt, 2022). Furthermore, whereas there seems to be a wide agreement that Design Thinking is beneficial in practice in many disciplines, some observers feel that for designers this represents a loss or the narrowing of territory, and of control over the design process. Attempts to come to terms with this disparity include, for example, a distinction between 'specialist designers' (those with a design education and background), and 'generalist designers' (members of design teams with a professional education and background in another discipline). But such distinctions have not managed to penetrate the discourse mainstream and have thus far not helped to ease the difficulties.

We are thus left with many unresolved questions pertaining to identity and professional responsibilities, to the contents of educational programs and their pedagogical means, and of course to practice and its integration of Design Thinking – how successful are attempts to employ the method, what pitfalls are encountered, where are other methodologies needed in order to be effective? And how can we apply methods that were created mainly for the design of artefacts to the design of services and social settings? We revisit issues related to team composition – by all means not a new topic but one that now centers on the diversity of team members regarding their areas of expertise. One of the central tenets of Design Thinking is that it is human-centered, but now we are aware that this is not enough, and the responsibility of design goes beyond humans to encompass the natural surroundings, including fauna and flora and the environment as a whole, that must be looked after in design plans. How does design widen its scope to do so? Additionally, we also revisit representational means in design: how can they be used to enhance design thinking or Design Thinking?

Accordingly, the chapters of this book are arranged in five parts. Some of the chapters could be classified in more than one way, but we had to choose and hopefully opted for the best classification of each chapter.

Part 1, *Thinking about design*, is concerned with the widest perspective on design and design thinking and offers reflections on what design is and what it is not.

Part 2, *Design thinking in the studio*, takes us into the educational arena and through case studies, exemplifies how design thinking principles are instilled in designers-to-be.

Part 3, *Design thinking in practice and professional training*, includes the largest set of chapters and again through case studies, shows how design thinking, or major values related to such thinking, are being practiced in different settings.

Part 4, *Design teams of diverse backgrounds and interdisciplinary projects*, treats the core question we had posed to participants of this endeavor: what happens when designers do not work in disciplinary isolation but collaborate in different ways with colleagues from other disciplines, close or more remote from their own discipline.

Finally, **Part 5**, *Design and nature; visual representation*, deals with two separate issues. First, how do we think about design when it is seen as serving not just humans but also the natural world around us, including animals and the environment at large. Second, what is the role of visual representation in design and design thinking?

We think the design research community must engage with these and similar questions if we are to remain one coherent community. The chapters of this book do not offer definitive answers; rather, they problematize design, design thinking, and Design Thinking. We hope that this book will contribute to the discourse around these issues and lead readers to shape their own views and ask further questions.

Reference

Goldschmidt, G. (2022). Critical Design and Design Thinking vs. critical design and design thinking. In Eriksson, Y. (ed.), *Different perspectives in design thinking*. CRC Press, pp. 6–20.

Part 1

Thinking about design

DOI: 10.1201/b22630-2

1.1 Designed to write

Christine Toh[a], *Andy Dong*[b], *and Mike Sharples*[c]
[a]*University of Nebraska at Omaha;* [b]*Oregon State University;*
[c]*The Open University*

1.1.1 Introduction

Stories, in the form of news, articles, books, and even tweets, and designs, in the form of objects, environments, and services, are all around us. Together, they represent the bulk of our material culture. Writings and designs share structural similarities. Designs have a function, structure, and behavior (Gero, 1990). Similarly, writings have a function (construe an experience, create a community), structure (genre, grammar, style), and behavior (narrate, report). This leads us to raise the question of whether the *acts* of writing and designing share similarities. Taking a more provocative stance, this chapter raises the possibility that writing is itself a form of designing. Indeed, in writing about writing, Mike Sharples claims that writing is akin to designing: "A writer is not only a creative thinker and problem solver, but also a *designer*" (Sharples, 1999, p. 10). We believe that discovering the extent to which writing is (or can be) a design process can drive a search for cognitive processes, tools, and mediations that underpin both practices. If central concepts such as 'design thinking' and 'primary generator' can be applied to creative writing, then the boundaries of these concepts expand.

If the claim that writing is a form of designing is true, this leads to a broader implication that would explain the expansion of the boundaries of designing, or more popularly 'design thinking' (Brown, 2008). This implication is that writing is an adaptation of the cognitive skills evolved for designing. Perhaps the evolution of cognitive skills for designing (Dong, Collier-Baker, & Suddendorf, 2017) provided the building blocks for language (Bickerton, 1981; Davidson, Noble, Lieberman, Ragir, & Burling, 1993; Deacon, 1997). As early as 1.75 million years ago, long before humans communicated through language, they were engaged in the top-down design of artifacts (Moore, 2020). They were designing hafted (having a handle) tools and symbolic, decorative items (Henshilwood, D'Errico, Marean, Milo, & Yates, 2001; Henshilwood et al., 2002). Moore (2011) suggests that the physical constraints of materials enforced boundaries on design spaces and influenced the emergence of design cognition and practice. A set of cognitive skills have evolved to enable humans to design. To engage in these sorts of activities requires deliberation and planning, both of which are enabled by the cognitive skill of recursion (Hoffecker, 2007). Recursion is considered a building block for language and the creative ability to produce a seemingly infinite variety of combinations from discrete components. It is possible that some cognitive skills involved in design cognition, such as imagination and causal reasoning, are examples of convergent evolution. Both great apes and crows (corvids) have been found to possess these cognitive skills (Emery & Clayton, 2004), which enables them to perform acts such as crafting simple tools. Yet, to

DOI: 10.1201/b22630-3

our knowledge, neither apes nor crows have shown any capability to write – which may simply be the case because there is no need, no natural response necessary, for them to do so even if they could. The capability to design gave humans an inherited difference that we leveraged to improve our species' ability to survive and reproduce. Since humans developed an ability to design before writing, perhaps humans designed language itself. The evolutionary origins of capabilities to envision and reify that world such that it can be shared and communicated to others are more likely than not to be shared between writing and designing.

In short, if there are commonalities in terms of mental structures, processes, and methods between writing and designing, then there are far-reaching implications because the cognitive skills for design predate skills for language.

The central claim of this chapter is that writing is a process of design and therefore writers are ipso facto designers. We have chosen to explore the boundaries of designing through a search for similar cognitive processes, tools, and mediations between writing and designing. We have chosen to investigate creative writing to address the expansion in the boundaries of design thinking because both writing and designing result in an intentional plan for cultural artifacts, one immaterial (writing), one material (design), to express and shape values and beliefs. Researchers have already noted that cognitive processes observed in designing closely resemble those associated with writing including task representing (framing) and planning (Akin & Akin, 1998). Both acts deliver on the goal of communication (Forlizzi & Lebbon, 2002). As stated by Krippendorff (1989, p. 9), designing can be "regarded as meaning that 'the products of design are to be understandable or meaningful to someone'". Similarly, creative writing can be viewed as the process and act of individual expression to share an experience to a specific group (Marksberry, 1963).

The evidence for our claim is a book about the process of writing, *How We Write: Writing as Creative Design* by Mike Sharples (1999), and an interview with the author, provided in the Appendix. Sharples was provided an early draft of this paper and invited to respond to a set of interview questions. The responses to the interview questions have been integrated with the chapter. We chose this book as data for our research because it is the only book we have been able to find that treats the process of writing as both a process and as a cognitive act, mirroring the way that designing has been researched as a process (Braha & Reich, 2003; Jones & Thornley, 1963) and as a cognitive act (Visser, 2004). In addition, while a broader base of literature exists that compares writing with designing, the focus of this chapter is not a comprehensive review of the existing research, but rather a critical analysis of Sharples' claims using recent citations from the design domain to contextualize the authors' claims for our research community. Scholars associated with the research group with whom the author collaborated to produce the research informing this book came from disciplines including psychology, ethnography, computer science, and journalism – which again mirrors the disciplines contributing to design research if we replace journalism with any of the disciplines associated with the practice of design such as architecture, industrial design, or engineering.

We will review the two most intriguing claims made in Sharples' book to test the assertion of the commonalities between writing and designing:

1 The cognition of writing recruits the same cognitive processes used in design.
2 Both writing and design use congruent methods and approaches to structure the process.

1.1.2 Overview of *How We Write: Writing as Creative Design*

How We Write: Writing as Creative Design by Mike Sharples is more an exposé about writing than it is an instruction manual on creative writing. The book is based upon research by the Collaborative Writing Research Group at the University of Sussex, UK. Separated into three parts, each analyzes writing from the following perspectives: i) cognitive psychology; ii) media theory; and, iii) social constructionism. Part 1 deals with creative writing as a mental activity. This part begins with an account of the cognitive development of writing in children and builds toward a general claim of writing as creative thinking guided by internal (i.e., self-imposed) constraints such as topic and style and external constraints such as grammar. Part 2 presents a description of writing as supported by a set of methods and makes the strongest set of claims on the intersection between writing and designing. As Sharples makes clear in the Preface, "My own view is that writing is a process of design, a skill that is grounded in the way we use our intelligence to create and share things of meaning in the world" (Sharples, 1999, p. xii). Finally, Part 3 describes the ways in which an author's social and cultural settings influences their writing.

Across the three parts, Sharples makes four general claims about the intersection of writing and designing:

1. The process of writing is akin if not nearly identical to the process of designing except that writing produces a text and designing can produce everything that is not a text. For example, Sharples states that writing entails activities such as synthesis and analysis and stages such as ideation, analysis, evaluation, and selection, activities and stages encountered in the process of design.
2. The cognitive processes of writing are similar to the cognitive processes in designing because both are mental activities that leverage the mind's capacity for creativity to produce artifacts that are meaningful to others. While it is more likely the case that cognitive processes in designing (if designing is thought of as form-giving) emphasize mental representations that are spatial rather than linguistic (Finke, 1990), designing and writing may entail the same cognitive skills and processes.
3. Both writing and designing use methods and approaches to structure an inherently open-ended and ill-structured task (Goldschmidt, 1997). Even though writing and designing happen in the mind, they rely on methods and external tools to increase productivity and serve as cognitive scaffolds throughout the process.
4. Both writing and designing are mediated by social processes that draw upon resources produced by others. Designing involves stakeholders, technical conversations, and sometimes a client and a brief. Writing involves stakeholders (editors, agents, readers) and sometimes a client (publisher) and a brief (prospectus).

While all four claims outlined above illustrate many similarities that run through the core of both writing and designing, in this chapter, we will focus on claims 2 and 3 for the sake of brevity and since they are both non-obvious assertions that deserve greater explication. The following sections review claims 2 and 3, respectively, discussing the links between writing and designing as both a cognitive process and a set of methods used in practice.

1.1.3 Cognitive aspects of writing and designing

Interest in understanding the mental processes and structures associated with the process of design started from a broader program to develop intelligent software programs to automate the design of specific artifacts, if not entire classes of artifacts (Coyne, Rosenman, Radford, Balachandran, & Gero, 1990). Inspired by foundational concepts from the field of artificial intelligence, research in the mental processes and structures associated with designing came about primarily through an objective to develop computable models of a design and of designing (Gero, 1990). This area of research has come to be known as design cognition (Visser, 2004). The fundamental theory espoused by the design cognition paradigm is that the structures and processes of cognition have causal importance in explaining designers' behaviors and outcomes. Design cognition researchers do not neglect the role of tools, processes, and social context; rather, the interest lies in understanding how cognitive processes intersect with the technical methods of design, such as the intersection between mental imagery and short-term memory while sketching (Bilda & Gero, 2007). Design cognition researchers theorize a set of cognitive processes salient to designing (Farrell & Hooker, 2013) such as abductive reasoning (Dorst, 2011) and evolutionary building blocks such as representation, recursion, and curiosity (Dong et al., 2017).

Sharples (1999) expresses a similar interest in the cognitive aspect of writing. Based on 'think-aloud' research about writing (Hayes & Flower, 1980), a research method widely used in the design cognition community as well (Ericsson & Simon, 1993; Lloyd, Lawson, & Scott, 1995), Sharples focuses on the mental processes associated with creativity in writing. His definition of creativity in writing exactly mirrors the definition of creativity used in design theory. According to Sharples, creativity in writing entails both originality and appropriateness. Appropriateness means that the content and style of the writing suit the author's audience and the purpose of the text. These are the same criteria applied to creative designs – novelty and utility. Novelty in design is usually conceptualized as how unusual or unexpected a design is relative to others within its category. Utility is conceptualized as quality, the extent to which a design objectively meets its intended purpose according to agreed measures of goodness. Originality in writing is achieved by following the rules of grammar or sometimes flexing and extending the expected conventions of style. Originality in design can be achieved by following the rules of a shape grammar (Stiny & Gips, 1972) or algorithms that transform and combine simple representations into original designs (Emdanat, Stiny, & Vakaló, 1999). However, the novelty of certain designs can also come from the deliberate bending of existing paradigms and questioning of established design guidelines, as seen in well-known examples in industrial design, typography, architecture, and the design of musical instruments (Lawson, 2006).

The cognitive pathway to creativity in writing connects with the model of creativity as a psychological phenomenon as explained by C-K theory (Hatchuel & Weil, 2009). C-K theory proposes the existence of two spaces (mathematical sets): the concept space *C* and the knowledge space *K*. The knowledge space contains all true propositions and available knowledge. The concept space contains designs that may be true because it is not yet possible to prove that a particular design is true given the current knowledge space. As such, each design is a proposition. The process of designing transforms undecidable propositions (concepts) into true propositions (designs) in the knowledge space. In other words, the process of designing creates knowledge that makes the undecidable propositions in the concept space true. C-K theory proposes four operators that expand the *C* and *K* spaces until a true proposition (design) can be logically proven.

These C-K spaces appear in Sharples' explanation of a writer satisfying constraints in order to generate an appropriate text. (It should be noted that Sharples was unaware of

C-K theory when writing his book.) Sharples identifies two types of constraints: internal constraints such as the author's self-imposed topic, and external constraints such as the style and structure of writing that would be meaningful to the audience. For example, an author can self-impose an interest in writing a news article on a specific topic, but, in writing for *The New York Times*, the author will adhere to the style stipulated in *The New York Times Manual of Style and Usage*. Sharples claims that writers start with a set of ideas represented in "a tangled web of associated concepts as an organised conceptual space … that can be systematically explored and transformed" (Sharples, 1999, p. 45). The 'organised conceptual space' that Sharples cites are named the same as concept spaces in C-K theory. Ideas in the writer's conceptual space have the same properties as propositions in C-space because ideas have attributes (words) – they originate from a "search through long-term memory for appropriate ideas and appropriate language to express each idea". Similarly, propositions in C-space have attributes that are the properties of a design such as its geometry or material. The K-space is what Sharples describes as internal and external constraints. The constraints, like the K-space, are what are known to be true or false about language, grammar, style, etc.; they guide (constrain) the writer in formulating a meaningful chain of ideas rather than a series of random, disconnected words. Designers retrieve ideas from long-term memory (Bilda & Gero, 2007; Del Missier, Visentini, & Mäntylä, 2015) and connect these ideas "to define, constrain and clarify the problem space of the design task" (Goldschmidt, 1995, p. 208). Likewise, writers "invoke *conceptual spaces* [emphasis added] relevant to the task and then exploring and transforming them to create products that the writer could not have imagined previously" (Sharples, 1999, p. 47) – a behavior that is explicitly called out by the C-K operators and their transformation of propositions in the C-space and knowledge in the K-space. Admittedly, C-K theory applies better to the writers' external constraints (as opposed to internal constraints) since much of the work that goes into the process of writing is concerned with satisficing internal constraints. Nevertheless, both writing and designing involve an exploration of these conceptual spaces to meet internal and external constraints. In sum, the model of thought theorized by C-K theory finds empirical support in Sharples' description of creative thought in writing, most likely since they are both underpinned by Herbert Simon's notions of human problem solving and creativity as explorations within conceptual spaces (Newell & Simon, 1972).

Sharples further elaborates on the fundamental processes of idea generation for writing in hypothesizing 'states of mind' associated with knowledge telling and knowledge transforming. Borrowing a taxonomy proposed by Gelernter (1994), Sharples hypothesizes that knowledge telling and knowledge transforming can either be low-focus or high-focus. High-focus thinking is associated with solving a particular problem, whereas low-focus thinking is associated with inconsequential thinking associated with emotion. Given these dimensions, Sharples hypothesizes four 'states of mind' for idea generation in writing as shown in Table 1.1.1 (Sharples, 1999, p. 48).

Table 1.1.1 States of mind associated with knowledge telling and knowledge transforming

	Knowledge telling	*Knowledge transforming*
High focus	Associational thinking	Problem solving
Low focus	Idea flowing	Daydreaming

While there is no exactly equivalent concept of 'state of mind' in design cognition, these four states could be mapped to the following concepts in design cognition:

1 Associational thinking – Some scholars characterize concept formation as the construction of ideas at increasing levels of complexity (e.g., Dong, 2006; Galle, 1996; Gero, 1998).
2 Idea flowing – A significant body of design cognition research addresses the effects of exposure to far-field, less-common examples during ideation (e.g., Chan et al., 2011; Fu et al., 2013; Tseng, Moss, Cagan, & Kotovsky, 2008).
3 Problem solving – The problem-solving paradigm in design (e.g., Goel & Pirolli, 1992; Simon, 1995) describes the cognitive processes associated with problem-solving in detail.
4 Daydreaming – Other than brainstorming (Osborn, 2001), in which it is shown that brainstorming is most effective when individuals work alone first and then share ideas in a group (Girotra, Terwiesch, & Ulrich, 2010), the most similar research would be avoiding design fixation through incubation and forgetting (Smith & Linsey, 2011). Techniques for de-fixing fixation that involve re-representation and abstraction (Zahner, Nickerson, Tversky, Corter, & Ma, 2010) could be considered low-focus if these activities are undirected externally.

It seems fair to say that categorizing states of mind as a way to distinguish cognitive pathways to creativity in idea generation would not neatly organize research on ideation in design given the very diverse perspectives on the cognitive psychology of creativity (Runco & Chand, 1995). Although Sharples does not review the cognitive psychology behind idea generation in writing, the reasonable overlap between his dimensions of creative thinking in writing and empirical research on idea generation in design allows us to conclude that empirical findings in the design cognition research will likely hold if the research were about writing instead of designing. One important difference stands out, though. Gelernter's high-/low-focus taxonomy and convergent/divergent thinking hypothesizes that low-focus thinking is driven by emotion. In contrast, divergent thinking in design, and creativity, more generally, can be driven by many cognitive processes (Hernandez, Shah, & Smith, 2010) such as incubation, frame of reference shifting, and exposure to novel examples. The effect of emotion on a designer's creativity has received very little attention other than as being affected by emotions about past experiences as a trigger for novel ideas (Solovyova, 2003) and curiosity (Saunders, 2007). However, Sharples (1996) himself has critiqued the notion that emotion is the sole driver of creative imagination, and that other forces may shape creative thought in writing, much like in design.

Sharples explores in depth one topic that has received little attention in the design cognition community: the development of cognitive abilities for writing (or designing) in (neurotypical) children (cf., Dong & Heylighen, 2018). Sharples describes the development of cognitive skills for writing in children as follows. As children acquire the ability to speak, they learn to develop imaginary worlds. Learning indirectly through communication with parents, teachers, and older peers, they learn simple genres such as a book story or a recounted event. For example, a child might be asked to relay a story about a dancing bear or an account of their recent birthday party. The most significant advances in cognitive skills occur when children learn to form 'what next' scenarios and to stop and reflect on their cognitive processes. Children between 10 and 14 develop the ability to talk about their writing. This development suggests that they are thinking about their thinking through "a mixture of engagement and reflection" (Sharples, 1999, p. 24). Sharples calls this cognitive skills knowledge telling, the formation and pursuit of chains of mental associations, and knowledge transforming, the construction of mental spaces to be explored and transformed (Bereiter, 1987). The closest analogous cognitive skills that have been studied in adult designers are mental simulation (Christensen & Schunn, 2009) and reflection-in-action (Schön, 1983; Valkenburg & Dorst, 1998). In a review of the developmental psychology literature, Suddendorf and Dong (2013) conclude that the development of secondary representation and meta-representation skills (Perner, 1991) gives humans the ability to imagine an object that is not present and to see versions of an object in different contexts to imagine how they would perform. At about two years of age, children develop the secondary representation skill to represent the world independent from perception (pretend play). At about four years of age, children develop the skill to represent an object or situation simultaneously in different ways. They may find a TikTok video of a cat swatting away ornaments the cat's guardian recently put away amusing because they can represent the situation as a conflict between the cat's representation of the situation as a game and the guardian's representation of the situation as a chore. The development of these representational skills allows us to imagine objects that are not bound by pre-specified characteristics. These skills should be salient to the ability for writers to imagine the structure of texts in ways that defy conventional understandings.

1.1.4 Methods of writing and designing

One of the most striking similarities between writing and designing lies in the reliance of tools and methods to help navigate an inherently ill-structured (Goldschmidt, 1997) and open-ended process (Buchanan, 1992; Rittel & Webber, 1973). In fact, Sharples describes writing by dictation, a method of freeform writing, as akin to walking in the dark. Similarly, design is often described as involving unclear goals and incomplete information (Voss & Post, 1988) and designers have to resolve conflicting requirements from stakeholders with competing priorities (Agouridas, McKay, & de Pennington, 2004). The uncertainty inherent in the design process shares many properties of walking in the dark, and decades of research and experimentation has led to the use of tools and methods to help designers grapple with this uncertainty. The use of tools to structure the highly unstructured process of writing and designing can be attributed to the many shared properties between the two (written by Bryan Lawson in *How Designers Think* and cited by Sharples): they are both open-ended and cannot be fully specified; the process is endless; there is not an infallibly correct process to follow; the process involves finding and solving problems; it involves subjective judgment; it is a prescriptive activity; and, both designers and writers work in the context of a need for an action (they create change).

For writers, these tools take the form of books, files, notes, word processors, notecards, and many other mediums – all for the purpose of being signifiers of the writers' ideas – external representations that convey meaning to the reader. Ideas live in the head until they can be expressed in the form of these external representations, and their expression is as varied as the type of medium used by each writer. While writers' media of choice are typically constrained by the need to adequately express the written word, designers have a much wider range of external representations and tools at their disposal. Design journals, drafting paper, whiteboards, sticky notes, and even computer-aided design (CAD) software are all used to make external 'marks' on the world during the design process. The act of creating external representations of ideas is so central to design that a wealth of research has explored the relationship between one of the most common methods of expression, sketching, on the design process. Some have even characterized sketching as the 'language' of design itself, a process by which the designer dialogues between internal ideas and external representations of those ideas (Cross, 1999; Goldschmidt, 1991; Shah, Vargas-Hernandez, Summers, & Kulkarni, 2001; Tovey, Porter, & Newman, 2003). Beyond the use of tools for writers to express themselves, Sharples also briefly discusses the use of external resources such as reference books or colleagues to guide the writer in their efforts to effectively express their ideas effectively. One of the most ubiquitous reference books for writing is Strunk's *Elements of Style* (2007). This book has shaped generations of writers and readers, highlighting the power of external resources during the process of writing. In design, the use of external resources has received significant attention in the form of guidelines, heuristics, and principles, which seek to help the designer draw from a rich library of design knowledge. Well-known examples of these external resources include Shneidermanand Plaisant's *8 Golden Rules of Interface Design* (Shneiderman & Plaisant, 2010), Norman's *7 Principles* (2013), and Nielsen's *10 Usability Heuristics* (Nielsen, 1994). So ubiquitous are these external resources that undergraduate design programs often include them as standard training materials for novice designers.

In considering the tools that both designers and writers use to externally represent their ideas, it becomes clear that much of the decision about which tool to use greatly depends on the context and need of the writer or designer. Sharples describes how writers have highly personal preferences for the tools used to capture their ideas, from the saturation of the ink used to write on paper, to the type of word processor used to compose text. So too do designers have highly personal preferences for design tools, albeit with more constraints in place by resources and established platforms used by organizations designers are in. Ferguson (1993) even discusses the different types of sketching used by designers: the thinking sketch, the prescriptive sketch, the talking sketch – all of which serve different purposes during the design process. However, the variety and quantity of tools that exist to help support writers and designers do not fully resolve the central dilemma that Sharples brings into focus in this book: that writing (and designing) are both subject to the tension between the ephemeral nature of ideas as they are formed in the mind and the inflexible nature of the written word (or sketch) once they are put down on paper. In this way, writers and designers both have to struggle to update ideas once they are externally represented since the very act of writing (or sketching) causes the designer to reflect on their idea, interpret it, and modify it. However, not all mediums are created equal in the ability to update and modify ideas easily once they are expressed. Sharples uses the example of a typewriter: it allows a writer to record the flow of words rapidly and easily in a conversational manner. However, revising the text once it is typed on the page is much harder. Similarly, designers sketching rough ideas 'on the back of a napkin' will find that refinements and modifications to the sketches are prohibitively cumbersome. The advent of digital sketching tools, such as Autodesk's Sketchbook, now allow designers to make use of visual layers to make stepwise modifications, and advanced editing tools allow the designer to make substantial changes rapidly and with minimal effort. Despite the reduction in effort for modifications, the use of these digital tools is not without constraints, and designers must work within the bounds of feasible editing actions available within these platforms.

While many frameworks for characterizing the process of both writing and designing exist in the literature, Sharples broadly classifies the activities involved into three main categories: planning, composing, and revising. In planning, the writer is focused on creating an outline and intention for content to be written. Unlike the romantic vision of a lone writer sitting down to compose a novel in a single sitting, Sharples provides compelling evidence that successful planning activities are key to intentional writing and significantly impact the quality of the final outcome. Similarly, designers or teams of designers who proactively perform background research into their users, competitors, and project constraints can plan their design activities more strategically, utilizing the resources available to them to make the best decision in their given design context. In the literature, this planning stage occurs during the conceptual design stage, and is of immense importance to the success of design activities, so much so that it is said that 60–80% of design costs are typically committed to this process (King & Sivaloganathan, 1999). Two types of planning methods are discussed by Sharples as being used by writers. The first is mind-mapping, a process of linking associated concepts together in a visual layout to help organize networks of thoughts and keep writers organized during this stage. While mind-mapping is not a formal tool used in design, the use of associational thinking bears a striking resemblance to the use of computational tools for encouraging idea generation (Hu, Ma, Feng, & Peng, 2017; Siddharth & Chakrabarti, 2018). Another well-established method of planning used in writing is developing outlines, typically in the form of bullet points describing the content to be covered in the writing. While the structure of such outlines is clearly specified for producing written text, designers have their own methods of 'outlining' the design process, or the intended design activities to be engaged with, such as developing a Project Requirements Document (PRD), MoSCoW Prioritization Plans (Waters, 2009), and product specifications (Ulrich & Eppinger, 2016), to name a few. Regardless of the design framework or planning method used, the core purpose of these planning methods is to enable the designer to strategically allocate resources to different stages of the design process to maximize their chances of producing successful design outcomes.

The next main stage of writing described by Sharples is composing, whereby writers generate ideas and transfer them to written language using a medium of their choice. However, this process of idea generation does not happen in a vacuum, as illustrated by Figure 1.1.1 from Sharples's book.

Writers must first translate their own experience that is organized by their own mental schemas (ways of understanding the world) into a written language. The writers' words are then interpreted by the readers' own mental schemas and integrated into their own personal experience of being in the world and reading the written text. In this way, the written artifact is subject to both the experience of the writer and the reader and can never be experienced objectively in both directions. This is not dissimilar to theory of design posited by Norman (2013), which states that the designer's mental model is manifest in the artifacts that they design, and these artifacts in turn are interpreted by users. Users' interpretations are subject to their own mental models of how the world ought to work, as depicted in Figure 1.1.2.

These two mental models (the designers' and the users') do not always align. Therein lies one of the central challenges of design: to design artifacts that can be interpreted

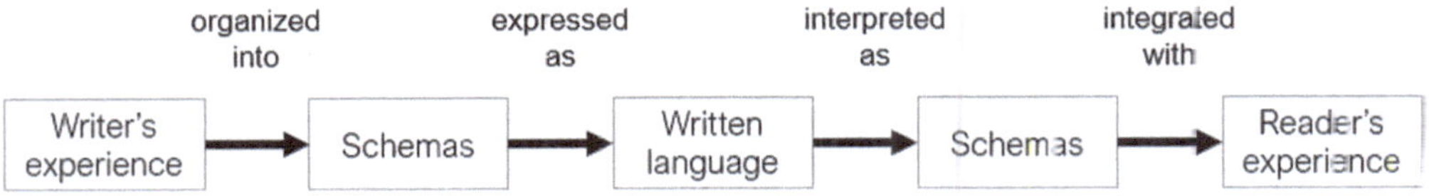

Figure 1.1.1 Process of reflection (Sharples, 1999, p. 91).

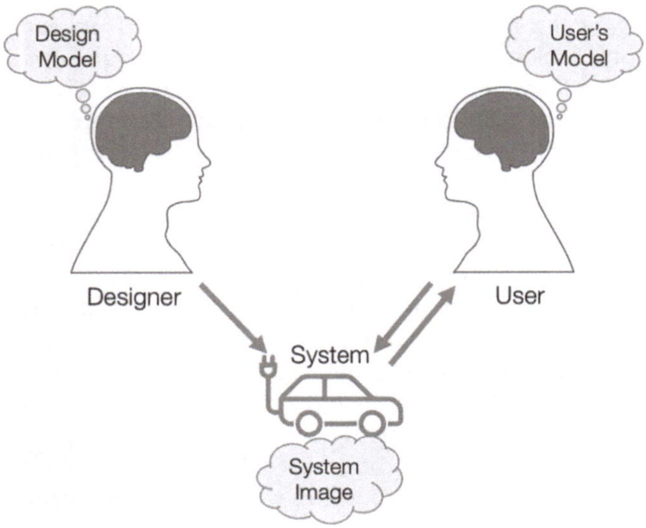

Figure 1.1.2 Mental models (from Norman, 2013, p. 32).

accurately by users who themselves have their own narrative of how the world ought to work to create seamless and effective user experiences. This is an active process on the part of the designer and user, similar to how interpretation of a text sometimes requires mental effort by the reader to relate to the written meaning and the writers' own experience.

Sharples takes this process of expression and interpretation one step further. He asserts that writers themselves interpret their own written word once it is externally represented. A writer's own understanding of their ideas shift once they interpret it on the written page. In this way, the act of composing is essentially captured through cycles of writing, interpreting, and revising to approach a state in which the written word on the page more closely resembles the writers' idea of what it ought to be. This cycle closely resembles the cycle of action and reflection (Schön, 1983; Valkenburg & Dorst, 1998). Descriptions of these cyles characterize designers as using their metacognitive abilities to reflect on design activities while in action. This type of "knowing in practice" has been documented in Schön (1983) work on design as a reflective practice. According to this depiction, designers are especially sensitive to the effects of their actions amidst designing and use this awareness (often in a tacit way) to guide their actions in uncertain and ambiguous situations (as is common in design).

The last stage of writing that Sharples describes is revising. During revision, the composed text is analyzed and modified to fit the goal of the writing task itself. Like designing, there is a psychological barrier inherent in revising text that has been painstakingly composed by the writer. This phenomenon has been studied in design and takes the form of the Sunk Cost Effect (Viswanathan & Linsey, 2013), where designers are reluctant to change their designs substantially, even in the face of evidence to the contrary, because of the amount of resources (time, energy, emotion) sunk into the initial design. Even if certain designs bear smaller costs to change (software code), the emotional investment into the design itself poses significant barriers to changing directions or removing large portions of already developed concepts. This is a challenging phase of the writing process. Sharples notes that experts are more competent at revising than novices:

they revise more, and their revisions are better. Similarly, studies exploring expert designer behavior show that expert designers iterate more and reinterpret the problem and solution concurrently until a good 'fit' is discovered (Dorst, 2019). Expert designers are also able to use more complex reasoning about design at higher levels of abstraction, which helps them transform declarative knowledge into procedural knowledge for recognizing solutions, resulting in better outcomes at the end of the process of iteration (Adelson & Soloway, 1985; Akin, 1990). The value of revising based on interpretation, called recursion by Sharples, is realized through the generation of new ideas that often are better aligned with the writer's goals. The value of critique in design has long been recognized (Costantino, 2015) for its ability to help guide design decisions while they are being made and can enhance design education.

All the steps in the writing process that Sharples described are framed as the foreground for the backdrop of the writer's own personal attributes that can affect this process. Namely, writers' behaviors and attitudes about writing can play a significant role in their process of writing. Factors such as personality traits can affect writers' attitudes, and motivation and apprehension can interact to create problematic behavior in writers, such as procrastination and fear of failure. These behaviors are very familiar to designers who face a similar apprehension about rejection of ideas in which they have invested time and effort. The impact of biases such as Ownership Bias have been studied in design, and research has shown that (Cooper & Lucas, 2006) designers place more value on their ideas over the ideas of others. This can have negative effects on objective decision-making in design (Toh, Strohmetz, & Miller, 2016). Other similar themes are found throughout the engineering design literature, with studies exploring the impacts of personal attributes such as personality traits (Toh & Miller, 2016), cognitive style (Buffinton, Jablokow, & Martin, 2002), and self-efficacy (Starkey, Hunter, & Miller, 2018) on designer behavior and outcomes. Both writing and designing share a core trait of requiring a designer to put substantial investments into endeavors that have inherently uncertain outcomes, risking rejection and failure. These are vulnerable acts of creation that call upon the writer or designer to trust in their ability to navigate the unstructured process with courage. As Sharples wrote, "Being a writer is, above all, having control over how you write and trust in your ability to make progress" (Sharples, 1999, p. 128). So too must designers when faced with the ambiguity of the design process.

1.1.5 Conclusions

This chapter presented a review of a book to claim that writing is a process of design and therefore writers are ipso facto designers. The chapter reviewed the evidence behind this claim by presenting empirical research in the field of design theory. The act of designing, like the act of writing, envisions a future world that does not yet exist and reifies that world in semiotic form.

While this chapter focused on the research on writing as designing from a single book and its author, new research has explored how a conceptualization of writing as designing can be applied in areas as diverse as children's collaborative writing (Vass, Littleton, Miell, & Jones, 2008) and computational creativity (Alvarado & Wiggins, 2018) An important and growing area of research is in multimodal writing – to understand how visual design and imaginative writing can be combined to communicate meaning and achieve effects. New tools and form factors, such as multimedia smartphones, have prompted research into visual and informational design of texts for multiple devices (Ling, Fortunati, Goggin, Lim,

& Li, 2020). All these strengthen a central claim of Sharples' book that writing and designing involve similar approaches, methods, and cognitive skills. Also, regarding creative writing as a design activity can extend perspectives for both fields and inform new ways to write for multiple media.

The chapter's main thesis that writing is a process of design and therefore writers are ipso facto designers leads to two implications. First, there is value in showing and teaching the commonalities of writing and design. Theories from writing, such as narrative cognition, high- and low-focus thinking, engagement and reflection, and the materiality of writing practices can inform design thinking, and theories from design such as primary generators, design spaces, visual balance, and design patterns can inform writing. Skills such as ideation, brainstorming, storytelling, iteration, prototyping, reflection, and revising underpin both practices. The challenge in design education will be to show students how these theories and practices mesh together (or not) to support the activities involved in creative writing, engineering design, product design, and software design.

Second, current research in storytelling in creative writing offers new directions for research in the field of design. Design thinking already embraces storytelling to gain insight into users and to present a prototype artifact. Storytelling in design could be explored further to include shared stories about the design process, storytelling as design, and the rhetoric of storytelling in design (Stevenson, n.d.; Suleman, 2017). One contribution of such research would be to explore whether storytelling in design is a subtractive process (it gets to the essence of the design), an expansive process (it embellishes designs with stories of possible use) or a rhetorical process (it persuades people to 'buy' a design), or all three. Perhaps this heralds a new field of 'narrative design' (already established in videogame design) where powerful emerging tools such as Wu Dao 2.0 (Romero, 2021) allow creative exploration of narrative and visual design.

Through our analysis, we made the case that designing (or design thinking) forms the building blocks of the intentional creation of cultural artifacts. The analysis provides a foundation for understanding how, like mathematics, the design discipline can position itself as a central language for creative and generative disciplines. If all forms of communication modes such as language, mathematics, visual art, and, more broadly, semiotics are all forms of design itself, it becomes possible to cast many disciplines as a form of design. Doing so does not make everything a practice of design. Rather, to view other practices as akin to the practice of design suggests that the thought processes and methods of design should, rightly, be construed as quite valuable. In other words, design thinking is as distinct and valuable as other established modes of thinking such as critical thinking, scientific thinking, and systems thinking, to name a few – and yes, writing. Rather than devaluing design thinking because of the ubiquity of its application, our position elevates design thinking as a distinctive and powerful mode of thinking that can be utilized by many disciplines (such as creative writing) in the same way that mathematics is regarded as a central language in science, engineering, economics, and many other disciplines and practices.

For centuries, the purpose of the academic essay has been to encourage 'writing thinking' – to enable students to expand a thesis, challenge assumptions, clarify arguments, exercise critical judgment, and examine multiple perspectives (see e.g., Fitzgerald (1994)). Writing thinking is at the core of intellectual life in some of the great traditional universities. Perhaps, as 'writing thinking' has become less fashionable, it has been replaced by applying 'design thinking' more widely. In short, design thinking has become core to intellectual life because the cognitive skills and processes associated with it are core to being human.

References

Adelson, B., & Soloway, E. (1985). The role of domain expenence in software design. *IEEE Transactions on Software Engineering*, *SE-11*(11), 1351–1360. doi: 10.1109/TSE.1985.231883

Agouridas, V., McKay, A., & de Pennington, A. (2004). 3.3.1 Consumer product development: A systems engineering approach to the derivation of design requirements from stakeholder needs. *INCOSE International Symposium*, *14*(1), 531–543.

Akin, Ö. (1990). Necessary conditions for design expertise and creativity. *Design Studies*, *11*(2), 107–113. doi: 10.1016/0142-694X(90)90025-8

Akin, Ö., & Akin, C. (1998). On the process of creativity in puzzles, inventions, and designs. *Automation in Construction*, *7*(2), 123–138. doi: 10.1016/S0926-5805(97)00057-5

Alvarado, J., & Wiggins, G. A. (2018). *Exploring the engagement and reflection model with the creative systems framework*. International Conference on Computational Creativity, Salamanca, Spain.

Bereiter, C. A. S., Marlene. (1987). *The psychology of written composition* Hillsdale: L. Erlbaum Associates.

Bickerton, D. (1981). *Roots of language*. Ann Arbor: Karoma.

Bilda, Z., & Gero, J. S. (2007). The impact of working memory limitations on the design process during conceptualization. *Design Studies*, *28*(4), 343–367. doi: 10.1016/j.destud.2007.02.005

Braha, D., & Reich, Y. (2003). Topological structures for modeling engineering design processes. *Research in Engineering Design*, *14*(4), 185–199. doi: 10.1007/s00163-003-0035-3

Brown, T. (2008). Design thinking. *Harvard Business Review*, *86*(6), 84–92.

Buchanan, R. (1992). Wicked problems in design thinking. *Design Issues*, *8*(2), 5–21.

Buffinton, K. W., Jablokow, K. W., & Martin, K. A. (2002). Project team dynamics and cognitive style. *Engineering Management Journal*, *14*(3), 25–33. doi: 10.1080/10429247.2002.11415170

Chan, J., Fu, K., Schunn, C., Cagan, J., Wood, K., & Kotovsky, K. (2011). On the benefits and pitfalls of analogies for innovative design: Ideation performance based on analogical distance, commonness, and modality of examples. *Journal of Mechanical Design*, *133*(8), 081004–081004 doi: 10.1115/1.4004396

Christensen, B. T., & Schunn, C. D. (2009). The role and impact of mental simulation in design. *Applied Cognitive Psychology*, *23*(3), 327–344. doi: 10.1002/acp.1464

Costantino, T. (2015). Lessons from art and design education: The role of in-process critique in the creative inquiry process. *Psychology of Aesthetics, Creativity, and the Arts*, *9*(2), 118–121. doi 10.1037/aca0000013

Cooper, S., & Lucas, W. A. (2006). Developing self-efficacy for innovation and entrepreneurship: an educational approach. *International Journal of Entrepreneurship Education*, *4*, 141–162.

Coyne, R., Rosenman, M. A., Radford, A., Balachandran, M. B., & Gero, J. S (Eds.). (1990). *Knowledge-based design systems*. Reading, MA: Addison-Wesley.

Cross, N. (1999). Design Research: A Disciplined Conversation. *Design Issues*, *15*(2), 5–10.

Davidson, I., Noble, W., Lieberman, P., Ragir, S., & Burling, R. (1993). On the Evolution of Language. *Current Anthropology*, *34*(2), 165–170.

Deacon, T. W. (1997). *The symbolic species: the co-evolution of language and the brain*. New York: W.W. Norton.

Del Missier, F., Visentini, M., & Mäntylä, T. (2015). Option generation in decision making: Ideation beyond memory retrieval. *Frontiers in Psychology*, *5*. doi: 10.3389/fpsyg.2014.01584

Dong, A. (2006). Concept formation as knowledge accumulation: A computational linguistics study. *Artificial Intelligence for Engineering Design, Analysis and Manufacturing*, *20*(1), 35–53 doi: 10.1017/S0890060406060033

Dong, A., Collier-Baker, E., & Suddendorf, T. (2017). Building blocks of human design thinking in animals. *International Journal of Design Creativity and Innovation*, *5*(1-2), 1–15. doi: 10.1080/21650349.2015.1011700

Dong, A., & Heylighen, A. (2018). Central coherence and the shaping of expertise in design: Evidence from designers with autism spectrum conditions. *Artificial Intelligence for Engineering Design, Analysis and Manufacturing*, *32*(3), 321–330. doi: 10.1017/S089006041700066X

Dorst, K. (2011). The core of 'design thinking' and its application. *Design Studies, 32*(6), 521–532. doi: 10.1016/j.destud.2011.07.006

Dorst, K. (2019). Co-evolution and emergence in design. *Design Studies, 65*, 60–77. doi: 10.1016/j.destud.2019.10.005

Emdanat, S., Stiny, G., & Vakaló, E.-G. (1999). Generative systems in design. *Artificial Intelligence Engineering Design Analysis and Manufacturing, 13*(4), 239–240. doi: 10.1017/S0890060499134000

Emery, N. J., & Clayton, N. S. (2004). The mentality of crows: Convergent evolution of intelligence in corvids and apes. *Science, 306*(5703), 1903. doi: 10.1126/science.1098410

Ericsson, K. A., & Simon, H. A. (1993). *Protocol analysis: Verbal reports as data*. Cambridge, MA: MIT Press.

Farrell, R., & Hooker, C. (2013). Design, science and wicked problems. *Design Studies, 34*(6), 681–705. doi: 10.1016/j.destud.2013.05.001

Ferguson, E. S. (1993). *Engineering and the mind's eye*. Cambridge: MIT Press.

Ferretti, F., Adornetti, I., Chiera, A., Nicchiarelli, S., Magni, R., Valeri, G., & Marini, A. (2017). Mental time travel and language evolution: A narrative account of the origins of human communication. *Language Sciences, 63*, 105–118. doi: 10.1016/j.langsci.2017.01.002

Finke, R. A. (1990). *Creative imagery: Discoveries and inventions in visualization*. Hillsdale, NJ: Erlbaum.

Fitzgerald, M. (1994). Why write essays? *Journal of Geography in Higher Education, 18*(3), 379–384. doi: 10.1080/03098269408709282

Forlizzi, J., & Lebbon, C. (2002). From formalism to social significance in communication design. *Design Issues, 18*(4), 3–13.

Fu, K., Chan, J., Cagan, J., Kotovsky, K., Schunn, C., & Wood, K. (2013). The meaning of "near" and "far": The impact of structuring design databases and the effect of distance of analogy on design output. *Journal of Mechanical Design, 135*(2), 021007–021007. doi: 10.1115/1.4023158

Galle, P. (1996). Design rationalization and the logic of design: A case study. *Design Studies, 17*(3), 253–275. doi: 10.1016/0142-694x(96)00004-x

Gelernter, D. (1994). *The muse in the machine: Computers and creative thought*. London: Fourth Estate/Free Press.

Gero, J. S. (1990). Design prototypes: A knowledge representation schema for design. *AI Magazine, 11*(4), 26–36.

Gero, J. S. (1998). Concept formation in design. *Knowledge-Based Systems, 11*(7-8), 429–435. doi: 10.1016/S0950-7051(98)00076-8

Girotra, K., Terwiesch, C., & Ulrich, K. T. (2010). Idea generation and the quality of the best idea. *Management Science, 56*(4), 591–605. doi: 10.1287/mnsc.1090.1144

Goel, V., & Pirolli, P. (1992). The structure of design problem spaces. *Cognitive Science, 16*(3), 395–429. doi: 10.1207/s15516709cog1603_3

Goldschmidt, G. (1991). The dialectics of sketching. *Creativity Research Journal, 4*(2), 123–143. doi: 10.1080/10400419109534381

Goldschmidt, G. (1995). The designer as a team of one. *Design Studies, 16*(2), 189–209. doi: 10.1016/0142-694X(94)00009-3

Goldschmidt, G. (1997). Capturing indeterminism: Representation in the design problem space. *Design Studies, 18*(4), 441–455. doi: 10.1016/S0142-694X(97)00011-2

Hatchuel, A., & Weil, B. (2009). C-K design theory: An advanced formulation. *Research in Engineering Design, 19*(4), 181–192. doi: 10.1007/s00163-008-0043-4

Hayes, J. R., & Flower, L. (1980). Identifying the organization of writing processes. In L. W. Gregg & E. R. Steinberg (Eds.), *Cognitive processes in writing* (pp. 3–30). Hillsdale, NJ: L. Erlbaum Associates.

Henshilwood, C. S., D'Errico, F., Marean, C. W., Milo, R. G., & Yates, R. (2001). An early bone tool industry from the Middle Stone Age at Blombos Cave, South Africa: Implications for the origins of modern human behaviour, symbolism and language. *Journal of Human Evolution, 41*(6), 631–678. doi: 10.1006/jhev.2001.0515

Henshilwood, C. S., d'Errico, F., Yates, R., Jacobs, Z., Tribolo, C., Duller, G. A. T., …Wintle, A. G. (2002). Emergence of modern human behavior: Middle stone age engravings from South Africa. *Science*, *295*(5558), 1278–1280. doi: 10.1126/science.1067575
Hernandez, N. V., Shah, J. J., & Smith, S. M. (2010). Understanding design ideation mechanisms through multilevel aligned empirical studies. *Design Studies*, *31*(4), 382–410. doi: 10.1016/j.destud.2010.04.001
Hoffecker, J. F. (2007). Representation and recursion in the Archaeological Record. *Journal of Archaeological Method and Theory*, *14*(4), 359–387. doi: 10.1007/s10816-007-9041-5
Hu, J., Ma, J., Feng, J.-F., & Peng, Y.-H. (2017). Research on new creative conceptual design system using adapted case-based reasoning technique. *Artificial Intelligence for Engineering Design, Analysis and Manufacturing*, *31*(1), 16–29. doi: 10.1017/S0890060416000159
Jones, J. C., & Thornley, D. G. (Eds.). (1963). *Conference on design methods*. New York: Macmillan.
King, A. M., & Sivaloganathan, S. (1999). Development of a methodology for concept selection in flexible design strategies. *Journal of Engineering Design*, *10*(4), 329–349. doi: 10.1080/095448299261236
Krippendorff, K. (1989). On the essential contexts of artifacts or on the proposition that "design is making sense (of things)". *Design Issues*, *5*(2), 9–39.
Lawson, B. (2006). *How designers think: The design process demystified* (4th ed.). Oxford: Elsevier.
Ling, R., Fortunati, L., Goggin, G., Lim, S. S., & Li, Y. (Eds.). (2020). *The Oxford handbook of mobile communication and society*. New York: Oxford University Press.
Lloyd, P., Lawson, B., & Scott, P. (1995). Can concurrent verbalization reveal design cognition? *Design Studies*, *16*(2), 237–259.
Marksberry, M. L. (1963). *Foundation of creativity*. New York: Harper & Row.
Moore, M. W. (2011). The design space of stone flaking: Implications for cognitive evolution. *World Archaeology*, *43*(4), 702–715. doi: 10.1080/00438243.2011.624778
Moore, M. W. (2020). Hominin stone flaking and the emergence of 'top-down' design in human evolution. *Cambridge Archaeological Journal*, *30*(4), 647–664. doi: 10.1017/S0959774320000190
Newell, A., & Simon, H. A. (1972). *Human problem solving*. Englewood Cliffs, NJ: Prentice-Hall.
Nielsen, J. (1994). *Enhancing the explanatory power of usability heuristics*. Proceedings of the SIGCHI Conference on Human Factors in Computing Systems, Boston, Massachusetts, USA. doi: 10.1145/191666.191729
Norman, D. A. (2013). *The design of everyday things* (Revised and expanded edition ed.) New York: Basic Books.
Osborn, A. F. (2001). *Applied imagination: Principles and procedures of creative problem solving* (3rd ed.). Boston, MA: Creative Education Foundation Press.
Perner, J. (1991). *Understanding the representational mind*. Cambridge: The MIT Press.
Rittel, H. W. J., & Webber, M. M. (1973). Dilemmas in a general theory of planning. *Policy Sciences*, *4*(2), 155–169. doi: 10.1007/BF01405730
Romero, A. (2021). GPT-3 scared you? Meet Wu Dao 2.0: A monster of 1.75 trillion parameters Retrieved from https://towardsdatascience.com/gpt-3-scared-you-meet-wu-dao-2-0-a-monster-of-1-75-trillion-parameters-832cd83db484
Runco, M. A., & Chand, I. (1995). Cognition and creativity. *Educational Psychology Review*, *7*(3), 243–267. doi: 10.1007/BF02213373
Saunders, R. (2007). Towards a computational model of creative societies using curious design agents. In G. O'Hare, A. Ricci, M. O'Grady, & O. Dikenelli (Eds.), *Engineering societies in the agents world VII* (Vol. 4457, pp. 340–353). Berlin: Springer-Verlag.
Schön, D. A. (1983). *The reflective practitioner: How professionals think in action*. New York: Basic Books.
Shah, J. J., Vargas-Hernandez, N., Summers, J. D., & Kulkarni, S. (2001). Collaborative sketching (C-Sketch)—An idea generation technique for engineering design. *The Journal of Creative Behavior*, *35*(3), 168–198. doi: 10.1002/j.2162-6057.2001.tb01045.x

Sharples, M. (1996). An account of writing as creative design. In C. M. Levy & S. Ransdell (Eds.), *The science of writing: Theories, methods, individual differences and applications* (pp. 127–148). New York: Routledge.

Sharples, M. (1999). *How we write: Writing as creative design*. London: Routledge.

Shneiderman, B., & Plaisant, C. (2010). *Designing the user interface: Strategies for effective human-computer interaction* (5th ed.). Boston: Addison-Wesley.

Siddharth, L., & Chakrabarti, A. (2018). Evaluating the impact of Idea-Inspire 4.0 on analogical transfer of concepts. *Artificial Intelligence for Engineering Design, Analysis and Manufacturing*, *32*(4), 431–448. doi: 10.1017/S0890060418000136

Simon, H. A. (1995). Problem forming, problem finding, and problem solving in design. In A. Collen & W. W. Gasparski (Eds.), *Design and systems: General applications of methodology* (Vol. 3, pp. 245–257). New Brunswick, NJ: Transaction Publishers.

Smith, S. M., & Linsey, J. (2011). A three-pronged approach for overcoming design fixation. *The Journal of Creative Behavior*, *45*(2), 83–91. doi: 10.1002/j.2162-6057.2011.tb01087.x

Solovyova, I. (2003). *Conjecture and emotion: An investigation of the relationship between design thinking and emotional content*. Expertise in Design: Design Thinking Research Symposium 6, Sydney.

Starkey, E. M., Hunter, S. T., & Miller, S. R. (2018). Are creativity and self-efficacy at odds? An exploration in variations of product dissection in engineering education. *Journal of Mechanical Design*, *141*(1). doi: 10.1115/1.4041173

Stevenson, N. (n.d.). Establishing a Culture of Storytelling. Retrieved from https://www.ideou.com/blogs/inspiration/establishing-a-culture-of-storytelling

Stiny, G., & Gips, J. (1972). *Shape grammars and the generative specification of painting and sculpture*. IFIP Congress, Information Processing 71.

Storr, W. (2020). *The science of storytelling: Why stories make us human and how to tell them better*. New York: Abrams Press.

Strunk, W. (2007). *The elements of style*. New York: Penguin.

Suddendorf, T., & Dong, A. (2013). On the evolution of imagination and design. In M. Taylor (Ed.), *The Oxford handbook of the development of imagination* (pp. 453–467). Oxford: Oxford University Press.

Suleman, B. (2017). The role of storytelling in design: Separating fact from fiction.Retrieved from https://uxdesign.cc/https-medium-com-babarsuleman-the-role-of-storytelling-in-design-separating-fact-from-fiction-6e43a42161a7

Toh, C. A., & Miller, S. R. (2016). Creativity in design teams: The influence of personality traits and risk attitudes on creative concept selection. *Research in Engineering Design*, *27*(1), 73–89. doi: 10.1007/s00163-015-0207-y

Toh, C. A., Strohmetz, A. A., & Miller, S. R. (2016). The effects of gender and idea goodness on ownership bias in engineering design education. *Journal of Mechanical Design*, *138*(10). doi: 10.1115/1.4034107

Tovey, M., Porter, S., & Newman, R. (2003). Sketching, concept development and automotive design. *Design Studies*, *24*(2), 135–153. doi: 10.1016/S0142-694X(02)00035-2

Tseng, I., Moss, J., Cagan, J., & Kotovsky, K. (2008). The role of timing and analogical similarity in the stimulation of idea generation in design. *Design Studies*, *29*(3), 203–221. doi: 10.1016/j.destud.2008.01.003

Ulrich, K. T., & Eppinger, S. D. (2016). *Product design and development* (6th ed.). New York: McGraw-Hill Education.

Valkenburg, R. C., & Dorst, K. (1998). The reflective practice of design teams. *Design Studies*, *19*(3), 249–271. doi: 10.1016/S0142-694X(98)00011-8

Vass, E., Littleton, K., Miell, D., & Jones, A. (2008). The discourse of collaborative creative writing: Peer collaboration as a context for mutual inspiration. *Thinking Skills and Creativity*, *3*(3), 192–202. doi: 10.1016/j.tsc.2008.09.001

Visser, W. (2004). *Dynamic aspects of design cognition* (pp. RR–5144). Paris: Institut National de Recherche en Informatique et en Automatique (INRIA).

Viswanathan, V., & Linsey, J. (2013). Examining design fixation in engineering idea generation: The role of example modality. *International Journal of Design Creativity and Innovation*, *1*(2), 109–129. doi: 10.1080/21650349.2013.774689

Voss, J. F., & Post, T. A. (1988). On the solving of ill-structured problems. In M. T. H. Chi, R. Glaser, & M. J. Farr (Eds.), *The nature of expertise* (pp. 261–285). Hillsdale: L. Erlbaum Associates.

Waters, K. (2009). Prioritization using MoSCoW. *Agile Planning*, *12*, 31.

Zahner, D., Nickerson, J. V., Tversky, B., Corter, J. E., & Ma, J. (2010). A fix for fixation? Rerepresenting and abstracting as creative processes in the design of information systems. *Artificial Intelligence for Engineering Design, Analysis and Manufacturing*, *24*(Special Issue 02), 231–244. doi: 10.1017/S0890060410000077

Appendix: Interview with Mike Sharples

We provided Mike Sharples with a draft of this paper in preparation for this interview. The following are questions posed to Sharples and his typewritten responses.

"Since the publication of your book *How We Write*, what new research strengthens or weakens the claims you made about writing as creative design?"

> Since publication of *How We Write*, new research has explored how a conceptualization of writing as design can be applied in areas as diverse as children's collaborative writing (Vass et al., 2008) and computational creativity (Alvarado & Wiggins, 2018). An important and growing area of research is in multimodal writing – to understand how visual design and imaginative writing can be combined to communicate meaning and achieve effects. New tools and form factors, such as multimedia smartphones, have prompted research into visual and informational design of texts for multiple devices (Ling et al., 2020). All these strengthen a central claim of the book that writing and visual design involve similar cognitive processes. Also, regarding creative writing as a design activity can extend perspectives for both fields and inform new ways to write for multiple media.

"The key question raised by this Design Thinking Research Symposium is that 'it is no longer clear what exactly design means and how well it is served by the expansion of its boundaries'. Do you think that making the claim that writing is (like) designing expands or sharpens the boundaries of what it means to be a designer or the practice of design?"

> I think it illuminates design practice from a different angle, that of creative writing. It shows how some central concepts such as 'design thinking' and 'primary generator' and can be applied in creative writing and so expands the boundaries of these concepts. The claim that writing is (or can be) a design process may help to clarify what writing and visual design have in common, and drive a search for cognitive processes, tools, and mediations that underpin both practices.

"The field of design has grown tremendously in popularity especially with the coining of the term 'design thinking' – which is stated in the name of the conference at which this paper will be presented. Why do you think it is that there is no equivalent concept of 'writing thinking'?"

> I suggest there is the construct of 'writing thinking', it's just not described as such. For centuries, the purpose of the academic essay has been to encourage 'writing thinking' – to enable students to expand a thesis, challenge assumptions, clarify arguments, exercise critical judgment, and examine multiple perspectives (see e.g., Fitzgerald (1994)). Writing thinking is at the core of intellectual life in some of the great traditional universities. Perhaps, as 'writing thinking' has become less fashionable, it has been replaced by applying 'design thinking' more widely.

"Given that the core proposition of your book is that writing is like designing, what research would you like the design thinking community to perform that would explore this relationship deeper, and what would the contributions of such research be?"

I would love to see the design community explore storytelling in more depth. If I were authoring *How We Write* now, I would cover recent research into storytelling (e.g., Storr (2020)), including neuro-cognitive processes of storytelling, and the suggestion that language evolved from storytelling, not the reverse (Ferretti et al., 2017). If storytelling pre-dates language, then it can be seen as an early design practice (the first storytellers were mime artists and visual designers). Design thinking already embraces storytelling to gain insight into users and to present a prototype artefact. I suggest this could be explored further to include shared stories about the design process, storytelling as design, and the rhetoric of storytelling in design. Good starting points are Stevenson (n.d.) and Suleman (2017). One contribution of such research would be to explore whether storytelling in design is a subtractive process (it gets to the essence of the design), an expansive process (it embellishes designs with stories of possible use) or a rhetorical process (it persuades people to 'buy' a design), or all three. Perhaps this heralds a new field of "narrative design" (already established in videogame design) where powerful emerging tools such as Wu Dao 2.0 (Romero, 2021) allow creative exploration of narrative and visual design.

"There has been a tremendous amount of debate in evolutionary psychology on the origins of the human abilities for language, which, as you state in the book, precede writing. Why do you think that there has not been an equivalent debate about the origins of the cognitive skills needed to design?"

There is some research into the evolution of tool design in early societies, e.g., Moore (2011). One obvious issue is that we can only study the evolution of design through the surviving artefacts, such as arrow heads. I like the way that Moore has studied the experiences of modern experimental knappers to understand the design space and make assumptions about tool design by hominin stoneworkers. Moore suggests that the physical constraints of materials enforced boundaries on design spaces and influenced the emergence of design cognition and practice.

"Extrapolating from your central claim that a writer is a designer, do you think we could educate better writers by teaching design skills? And, conversely, do you think we could educate better designers by teaching writing skills?"

I think it is valuable to show and teach the commonalities of writing and design. Theories from writing, such as narrative cognition, high and low focus thinking, engagement and reflection, and the materiality of writing practices can inform design thinking, and theories from design such as primary generators, design spaces, visual balance, and design patterns can inform writing. Skills such as ideation, brainstorming, storytelling, iteration, prototyping, reflection and revising underpin both practices. The challenge is to show how these theories and practices mesh together (or not) to support the differing fields of creative writing, engineering design, product design, and software design.

1.2 What counts as design? No one right answer

Claudia Eckert[a], *Chris McMahon*[b], *and Martin Stacey*[c]
[a]*The Open University, Milton Keynes, UK;* [b]*University of Bristol, Bristol, UK;* [c]*De Montfort University, Leicester, UK*

1.2.1 Introduction: The word 'design' and the goals of design research

The meanings of the terms 'design' and 'designing' and 'designer' – so important to so many people's senses of who they are – have been the subject of decades of repetitive and seemingly endless debate (see Blackler et al., 2021). Can we end it, or focus it on a productive direction? We argue that 'design' has no one right definition, and consider what follows from that.

In design research what is included in the term 'design' is approached from two fundamentally different angles. The first is "Everything that is associated to the design of an artefact". According to this perspective, design is the process of creating plans for new artefacts or systems and may involve activities that in themselves would not be considered as design. The alternative view is to consider design as a distinct kind of human activity, restricted to the uniquely designerly activities involved in the process of designing (see Cross, 2006). This chapter began as a contribution to the Thirteenth Design Thinking Research Symposium, exploring how the frontiers of designing might extend beyond the fields traditionally associated with the term. As the call for the DTRS symposium states, there has been an "expansion of what is understood as design (and Design thinking), whereby many activities by professionals from different disciplines are seen as included in the term design". This reflects the recognition that many fields not traditionally considered design disciplines address problems that resemble design problems and require methods used by designers; regarding them as design requires a broader view of what might constitute a designed artefact. Another view would be that the scope of design is defined by the distinctive knowledge possessed by designers and evidenced by their qualifications and participation in professional communities. Participatory design approaches have increasingly involved users and other lay people in the design process – for some, this is as much an attempt to democratise the act of creation as an effort to get the user requirements right (see Luck, 2018). This would imply that design cannot be characterised by the knowledge a designer possesses.

Over the last 60 years or so, design as a field of research and education has worked on establishing itself as an academic discipline. This has included attempting to define key characteristics and properties that all design processes share. However, this has proved difficult. To illustrate some of the difficulties, this chapter draws on our studies of two industries: diesel engine design as a representative of traditional incremental engineering design, and sustainable fashion as example of an artistic craft-based domain. Both illustrate that design processes comprise many activities that would not intuitively be thought of as design, but, depending on the scale of the process, are carried out by

DOI: 10.1201/b22630-4

designers. This raises the question whether these activities count as design and need to be included in the characterisations of design and be supported by generally applicable design methods. This chapter argues that instead of trying to characterise design by what is common across all design activities, we should embrace difference between design processes to understand the rich phenomenon of design. It might therefore be more fruitful to think of design as a family resemblance concept, in the Wittgensteinian sense, where two instances do not necessarily share common features, but collectively all the instances covered by the term provide a rich and coherent picture. Only by taking a rich and inclusive view of design in this way can design research develop the nuanced support for designing in different situations that designers need in practice.

In recent years, design research has pursued different goals. One aim of design research is to develop ways to support designers in their activities. To do this, we have to acknowledge that designers may spend only a fraction of their time designing, while also engaging in other artefact-development-related activities or working with people who engage in other activities related to the design of artefacts. Another aim of design research over recent years has been to find what is common across different traditional fields of design, despite huge variations between them. Other researchers see design as a means to transforming our society and focus on what design should achieve, rather than what design is.

Regardless of the perspective in trying to understand design it is necessary to look both at the varied practices of design and at many different design processes. The practice might well be different from the rhetoric by which different domains present themselves; in this regard, the rhetoric may not be helpful. Design domains with an origin in applied arts and crafts, such as graphic design or fashion design, often emphasise the creative aspect of their practice; the designers often refer to themselves as 'creatives'. By contrast, the discussion of complex technical products often centres on performance, cost, quality, safety, and reliability combined with focused innovation. Designers with a strong user involvement, such as in architecture or service design, often focus on user-centred aspects of design.

In section 1.2.2, we note the multiplicity of definitions of design and designing that have been proposed, and then introduce in broad terms some perspectives on design that illustrate the rich insights that have been arrived at by a wide range of design researchers. After a brief discussion of our empirical methodology in section 1.2.3, section 1.2.4 argues, drawing on our own case studies and experience of design, that synthesis, which is often thought of as the essence of design, is in practice remarkably elusive and that many designers carry out many important activities that have very little to do with creating new things, although they are an essential part of their work. Section 1.2.5 looks at how the boundaries of what we consider as design varies in different contexts. Section 1.2.6 argues that design is in fact a *family resemblance concept* and that what matters for something to count as design is often whether those who undertake the activity think of it as design. Section 1.2.7 draws out some implications for how we understand and use the concept of design. Section 1.2.8 concludes with a recommendation to action: to recognise and advertise the centrality to designing of analysis and knowledge generation activities and the people who do them; and to focus comparative design research on the patterns of similarities and differences between design in different fields.

1.2.2 Perspectives on design

Design – as the word is used in ordinary English – has fuzzy boundaries, which raises the question what should or should not be included in design. Typically, solving well-defined problems, for example through the use of mathematical equations, would not be thought of as design. However, mathematics plays a very important role in many design processes and unsolved mathematical problems can be approached in a designerly way. Many people associate design with creating something new and thereby focus on the synthesis aspect of designing. Throughout the history of design research, the academic literature on design also offered a multitude of different complementary perspectives on design, which all capture important aspects of many design situations and problems, but do not fit others. Each perspective brings out facets of design that are extremely important in some fields, but much less so in others. The variety of perspectives is such that we hesitate to try to summarise them in a single short paper, so what follows here are only a few illustrative viewpoints; however, they are influential, and strongly flavour views of what design *is*. Although we observe that while research about the generic aspects of design processes across all different domains still continues (e.g., Daly, 2008; Reymen et al., 2006), the differences between design processes in different domains are however much less researched (Stacey & Eckert, 2010).

Herbert Simon called his well-known book on human-made systems “Sciences of the Artificial”. He distinguishes science, as concerned with what is, from technology, which is concerned with how things ought to be. He defined design as “to devise courses of action aimed at changing existing situations into preferred ones” (Simon, 1969/1996). Even after over 50 years of design research, the concept of design remains remarkable slippery. Fifty years ago, J. Christopher Jones’ influential book *Design Methods* (1970) listed a remarkably diverse collection of definitions; he wrote “There seems to be as many kinds of design process as there are writers about it” and we can say the same today. Writing about discussions on definitions of design over 20 years in a large email forum, Blackler et al. (2021) wrote: “Despite robust discourse around relevant perspectives on design, the list discussions are and have been repetitive, with no significant progress made towards a consolidated definition of design”. Auernhammer and Ford (this volume) present a computational analysis of 77 different definitions of design that collectively point to a core meaning with fuzzy boundaries, plus a number of sources of difficulty for agreeing on a clear meaning for the term (see Section 1.2.6). The multiplicity of views is also intensified by the widespread use of the term ‘design’ in everyday language. As noted by the systems theorist Peter Checkland (discussing the related difficulties in defining the term ‘system’), “there is no language available for serious discussion which is separate from everyday language … terms are fuzzy as a result of their unreflective use in everyday chat” (Checkland, 1999).

1.2.2.1 The emerging field of design

With the recognition that design as a professional activity shapes our world, it has been a subject of study for centuries, but it was in the 1950s and 1960s that research into a theoretical understanding of the topic, and into methods and tools and other supports for design activities, really began to take off. From these early days, design was studied from a wide range of perspectives, with researchers coming from architecture, engineering, the applied arts, planning, computer science, and other disciplines, and this has led to a multiplicity of views on the nature of design. The researchers came from different backgrounds and worked with different types of artefacts. For example, the highly influential work by Pahl and Beitz (1977) came from huge experience with building specialist manufacturing equipment, that transforms inputs using a series of machines and machine elements, but is less well suited to highly integrated, complex and incrementally developed products like jet engines. Early theoretical work on design has been strongly influenced by architectural design, for example in the work of Jones (1970), Schön (1983), or Akin (1986).

1.2.2.2 The rhetoric of design

The early research spawned a new emphasis on design in higher education as evidenced by the UK's Moulton Report on engineering design education, advocating that design "should be a thread running through" courses (Moulton, 1976), and by new teaching programmes, such as the Open University's Design and Innovation programme in the 1980s, supported by readers (such as Roy & Weald, 1986).

A view of design that has emerged quite strongly in the 21st century, especially in public discourse, has been the association of design with creativity, and creativity with cultural products. The UK's Cox Review of Creativity in Business (UK Government, 2005) proposed that "Design is what links creativity and innovation. It shapes ideas to become practical and attractive propositions for users or customers. Design may be described as creativity deployed to a specific end". In recent years, the UK's Design Council has highlighted the 'creative industries' as a driver for the UK economy (Design Council, 2018). The term was borrowed from the economist Richard E. Caves (2000), who used it to describe the subject matter of a study of the social and economic organisation of the production and marketing of art, focusing on the contractual relationships between artists and the commercial operations that act as facilitators and gatekeepers. Thus defined, the 'creative industries' included traditional design fields such as architecture and artistic design domains such as product design, graphic design, or fashion, as well as the creation of cultural artefacts such as advertisements, media, music, visual arts, and books. This excluded engineering and much of software design because they have different economic models.

1.2.2.3 Design as/versus structured problem solving

In engineering, there was a significant emphasis on the search for systematic, generally applicable approaches. Early engineering design theories, like the Theory of Technical Systems (Hubka & Eder, 1988) or Domain Theory (Andreasen, 1980), focussed on structuring products, relating design elements and their mapping to physical parts, and on the activities associated with their creation. Later theories like Axiomatic Design (Suh, 1990) also emphasised relationships between function and structure, advocating an alternation between the functional domain (derived from the customer needs) and the physical domain, but also placed a strong emphasis on the identification of axioms that were universally applicable such as the need to maintain the independence of functional principles. These approaches embody a view of design focusing on how decisions about what a product should do affect its structure and decisions about the embodiment affect what the product can and can't do.

George Hazelrigg (1998) put forward a view of engineering design as an application of mathematical decision theory, comprising a series of choices between alternatives that can be rationally elicited. The sequence of decisions may affect the result, and some decisions may need to be treated concurrently (Mistree et al., 1993). However, finding an optimal solution is rarely possible. In *The Sciences of the Artificial*, Herbert Simon (1969/1996) argued for seeing *design as a rational problem-solving process* applied to ill-defined problems, where designers act with bounded rationality (Simon, 1957). That is, the designers act rationally within the resource limitations of time and mental capacity that restrict their ability to gather information, explore problems and define optimal solutions, so they satisfice – settle for a satisfactory solution.

Design is not necessarily a structured and pre-planned process but is often an interactive engagement with the emerging design. Many analyses of design at different scales have drawn on Asimow's (1962) view of engineering design as a cycle of problem formulation, synthesis, and evaluation. Dorst and Cross (2001) point to the co-evolution of problem and solution.

In *The Reflective Practitioner,* Donald Schön (1983) called this process "*a conversation with the situation,* where designers engage in two different kinds of seeing, "seeing that", i.e., recognising something, and "seeing as", i.e., interpreting what they saw" (Schön & Wiggins, 1992). Schön views designers as engaging in a process involving *Naming*, in which elements of the task are identified, *Framing*, in which elements of the situation are picked out and connected to construct the problem to be solved, *Moving*, in which the design is advanced within the current frame, and *Reflecting*, in which the design is evaluated before the next cycle begins. (The view is quite widespread that the Schön view is radically in conflict with the design-as-problem-solving view of Simon and followers, stemming from Schön's own critique of Simon. We think it is not: this reflects a misinterpretation of Simon stemming from a lack of understanding of *The Sciences of the Artificial*'s radical abstraction away from a sophisticated understanding of the psychology of problem-solving; see Chua, 2009, who points out that the third edition of 1996 emphasises more the fluidity of design reasoning.)

Horst Rittel introduced the idea of a 'wicked problem' in the 1960s. Rittel and Webber (1973, p. 161) state: "The formulation of a wicked problem is the problem! The process of formulating the problem and of conceiving a solution (or re-solution) are identical, since every specification of the problem is a specification of the direction in which a treatment is considered." Rittel and Webber were thinking of town planning, where problems are unique and have no definitive formulation or right answer or even clear endpoints, solutions cannot be tested and are irreversible, so the designers have one chance to get it right, and any change has serious consequences. Many later researchers and practitioners have viewed all design problems as 'indeterminate' and 'wicked'; this is reasonable for town planning and systems engineering, not true of much of engineering, and absurd for fashion or knitwear or most of the 'creative industries'. Buchanan (1992) influentially made this claim without connecting his arguments to any particular industry or type of designing. He sees design as thus in fundamental contrast to science, which doesn't deal with radically indeterminate problems. The idea that design problems can't be clearly formulated and subjected to rational problem-solving procedures has gained a lot of currency.

1.2.2.4 User-centred views of designing

Entirely different views of what designing involves emerge from fields and situations where the hard challenges come from understanding and meeting complex and non-obvious human needs. The term 'Design Thinking' has come to refer to a particular view and set of techniques for tackling design problems, especially when applied in fields not traditionally thought of as design. Design Thinking (with capital letters) puts the interaction with users and figuring out user needs at the centre of design activities, and thereby de-emphasises the technical challenges in building complex systems. In his article with this title, Brown (2008) describes designing as an iteration between three pillars: *Inspiration*, *Ideation* and

Implementation, with an emphasis on finding needs and opportunities and engaging with normal and extreme users throughout the process. The d.school at Stanford developed this into five phases (Plattner et al., 2009): *Empathize, Define, Ideate, Prototype, Test*.

In software development, user-centred design, organising the design process around procedures and techniques for understanding user needs, predates the Design Thinking movement. Participatory design (Sanders et al., 2010), which is also closely related to cooperative design and co-design (Sanders & Stappers, 2008), goes back to the 1970s. It advocates bringing the user as an active participant into the design process including the generation of new ideas. Being part of the creation process also increases the willingness of users to accept the eventual result. Participatory design deemphasises the technical aspects of design, which designers carry out almost behind the scenes.

Products however rarely act in isolation but are embedded in user processes with other products. The design of Product-Service Systems (Baines et al., 2007) emphasises that the service provided depends on the characteristics of the product or system that enable service provision and involves the concurrent design of products and services. According to Mont (2004), "PSS is a system of products, services, supporting networks and supporting infrastructure that is designed to be competitive, satisfy customers' needs and have a lower environmental impact than traditional business models".

1.2.2.5 The granularity of views on design

Many of the differences in the perspectives on design arise from differences in the granularity in which design is studied, ranging from individual design activities, such as an episode in which a textile designers would sketch out the pattern for a fabric, to the design of systems of systems, such as new urban transport systems. This is a not only a matter of the granularity of the tasks in focus, but also of their description. As Wynn and Clarkson (2018) point out, models of design processes, which characterise design or are used to prescribe how designing is carried out, fall under three categories: "*Micro-level models*, which focus on individual process steps and their immediate contexts; *Meso-level models*, which focus on end-to-end flows of tasks as the design is progressed; *Macro-level models*, which focus on project structures and/or the design process in context".

Of course, designers often operate at different levels of detail within one activity. For instance, Buchanan (1992) argues that design operates on two levels: a general level, on which the designer forms ideas and hypotheses about the nature of products or the artificial word; and a practical level, being embedded in specific circumstances.

1.2.2.6 Disciplinary perspectives on design

Analyses of designing taking a cognitive perspective typically consider how designing works at a micro-level, whereas sociological studies and product or organisational views typically consider activities at a meso-level or a macro-level. The cognitive perspective sees design as a sequence of mental activities or types of reasoning steps; much research has focused on identifying mental actions and the structure of design reasoning (e.g., Akin, 1986). The idea that design is a unique mode of thinking is widespread and influential. Cross (2006) argued, based on interviews with expert designers, that designerly ways of thinking are exploratory, emergent, intuitive, reflective and embracing of ambiguity. This perspective fits with concerns for creativity in design, where creativity is related to novelty of the end results or the process of generation (e.g., Boden, 1990). Theoretical approaches to understanding creative thinking such as conceptual blending theory (Fauconnier & Turner, 2002) or the reinterpretation of ambiguous information (Schön & Wiggins, 1992) place synthesis at the heart of creativity. These influence the studies of design that draw on them.

1.2.3 Empirical methodology

Our own empirical research is grounded in a cognitive perspective but has focused on meso-level analyses of design processes. The chapter draws primarily on two very different sets of studies: diesel engine design (Eckert et al., 2004; Flanagan et al., 2007; Eckert et al., 2019); and recent studies of designer-led micro businesses in the fashion industry (Black et al., 2021). Diesel engines are an example of highly technical and incremental products, that have been optimised for efficiency and reliability, but also have been subject to significant innovations over the last 30 years to decrease their environmental impacts. Diesel engines are designed by large teams often involving hundreds of specialist engineers. By contrast the designer-led fashion businesses were very small and the designers not only designed the garments, but also ran the businesses and organised the production.

For these studies we conducted interviews with designers and followed their processes over several months. We have studied both sectors in multiple companies for 20 years and therefore can also observe long-term changes in industrial practice. We also draw on our much earlier studies of the relationship between artistic and technical design in the knitwear industry (Eckert, 2001; Eckert & Stacey, 2003), on our studies of information use in design processes (Lowe et al., 2004; McMahon, 2004; McMahon, 2021) as well as observations of parts of the construction industry (Eckert et al., 2012; Jones et al., 2018) and the second author's early experience of working in diesel engine design.

1.2.4 Finding the core and the boundaries of design

Design practice is extremely varied within different design domains and across domains. This section draws on our empirical work to bring out some of the differences between design practice in different contexts. Many insights into designing have been gained through detailed study of design episodes involving individual designers or small teams of designers, as exemplified by the original Delft protocol studies (Cross et al., 1996). In these studies, designers have been given specific design assignments and have been studied up to the point where they have generated a concept. This experimental set up also favours the study of synthesis tasks, as designers have little time to explore the problem, usually no access to users or clients; and little time to evaluate their solutions in detail. This can give a very misleading picture of where the effort and significant contributions are in developing new artefacts.

In this section, we will argue that many designers spend only a very small fraction of their time creating something new or different, and rather more on making sure they can meet new requirements with minimal changes. The contrast between the different design examples also illustrates that design happens at different scales, from a large team working together, so that each individual has a specific role, to small processes where a single designer carries out most of the activities.

1.2.4.1 The elusive act of synthesis

At a sufficiently abstract level, professional design processes can be mapped onto a pattern of analysis, synthesis and evaluation (see Asimow, 1962). However, when studying professional practice, the moments of synthesis can be difficult to track down. The product or system architecture defines the mapping between functions and components and the interfaces between components (Ulrich & Eppinger, 1995); and thereby incorporates many of the fundamental design decisions. This makes system architecture design a prime candidate for study of synthesis in engineering design. We (Wyatt et al., 2009) analysed how the system architecture of a diesel engine came into being. In the early 2000s, the first author was invited to a large meeting, where the conceptual design of a new engine generation was presented by the conceptual design team in a diesel engine company. Based on this experience, the first author and her colleagues assumed that the company would go through a distinct activity where the system architect would design the system architecture. While this had been the case in the company for earlier generation engines in the 2000s, system architecture design was no longer a recognisable process. Through a series of interviews, it become clear what information was required to develop a system architecture and at which point the system architecture was established. However, when and how the system architecture was designed was strangely elusive. Through the analysis of the interviews, it become clear that the system architecture emerged as the by-product of a requirement cascade, where the company first assured that they could meet regulatory performance requirements, and then focussed on a prioritised list of bundled requirements (see Wyatt et al., 2009). Diesel engines are highly incremental products in which the fundamental solution principle remains largely invariant, but developments are made in embodiment – for example in part dimensions and materials – and in components to deliver secondary functions such as fuel injection equipment and emissions control. They are often designed through the minimal amount of modification necessary from past designs but have to meet very stringent emissions regulations in each product generation. The company starts with tests of existing engines, and of modifications to them, to establish whether they can meet legally binding emission targets (Tahera et al., 2019). At the same time, they also try to minimise the 'newness' of the engine, in terms of newly designed components or existing components deployed in a new context. Very often constraints will be imposed by the production tooling required to make the engines, so things like bolting patterns or key dimensions may be retained. For existing components, the company also knows that the components are working reliably. Parameters of the reused components are entered into the requirement cascade as constraints on the performance calculations in the cascade. Through this process the system architecture is largely inherited from the previous product generation.

Synthesis can also prove difficult to pinpoint in the work of designer-led fashion businesses. The "Rethinking Fashion Design Entrepreneurship: Fostering Sustainable Practices" project studied over 25 such fashion businesses in the UK (Black et al., 2021). The 'fast fashion' culture of recent decades has become highly unsustainable and many fashion designers and other industry experts see the industry as fundamentally broken for this reason among several others (Black, 2012). The fashion businesses that the project studied were largely set up by people with fashion design degrees, who wanted to make a living in their chosen profession in a way they could combine with their consciences. They aimed to produce garments that were environmentally sustainable and produced in an ethical way, while also being financially viable. In fashion there is also an element of cultural sustainability, where businesses want to maintain or connect to traditional garment types or production methods. The majority of the businesses studied were micro-enterprises where the designers worked alone or with a small team of full-time or part-time staff. The survival of many small businesses is based on finding their niche in the fashion system. In some cases, this is centred around a specific product or small range of products, where the original designs are rarely updated.

Designing garments only took a small fraction of their time for most of the businesses. Most designers were occupied with running their businesses and carrying out a multitude of business functions. A lot of time is taken up by organising production for the garments and dealing with sales and other customers facing activities.

Unlike large fashion brands, small fashion businesses tend not to renew their collections regularly. Rather they try to follow customer demand and keep designs that sell well in their ranges and replace designs that have either sold out or are not selling well. The successful businesses proved to be agile and adaptable. For example, several designers produced face masks in the COVID-19 crisis using stock fabric. Others engaged in short-term collaboration with designers with complementary expertise as a way of broadening their sources of inspiration and customer bases. Much of their creative energy goes into finding business models that enable them to operate in the fashion industry while being sustainable, finding ways to engage customers and transforming the fashion system. These fashion designers see themselves as designers and as 'creatives', but the generation of new product designs only takes a tiny fraction of their time.

1.2.4.2 Design at different scales

The two design processes in the previous section are happening at very different scales. While the design of an individual garment takes a single designer a matter of days, including the creation of samples, the design of a diesel engine occupies many people across a supply chain for some years. A diesel engine is also only a component of a product, such as a tractor or a truck. The products themselves operate as parts of systems and of systems of systems. A truck is part of the wider transport system and has to interact with other products or systems that are also part of the transport system, for example when it is unloaded or uses a ferry.

A garment is also not designed in isolation. The designer creates multiple designs either as parts of a coherent collection, as typical high street chains would do, or when they see a gap in their offering. This depends on the garments offered by competing companies and on the wider development of fashions. In fashion design, an individual designer would need to understand this broader product context, whereas in engineering this would be split into many different activities carried out by different people. Whether all of these activities are included in the design is a matter of where the boundaries are drawn. Maybe in fashion one would intuitively include the research and understanding of the wider product context and market in design. While engineers need to understand the state of the art and the competition, many of these activities would be carried out by a marketing department.

The development of products involves a wide variety of actions that we would not intuitively include in design, carried out by experts from different disciplinary backgrounds. To illustrate, Figure 1.2.1 shows some of the disciplines involved in the course of the design of a complex engineering product. For example, the market research and trend analysis would be carried out in a sales or marketing department. Basic science, for example for new materials, would be carried out by scientists. The diesel engine company studied by Wyatt et al. (2009) typically centralises the R&D for their engines across the parent company and works with universities on the underlying science, so that only technology at a fairly high technology readiness level (see Mankins, 1995), i.e., technology that has been tested on engines, would be considered in the product development process.

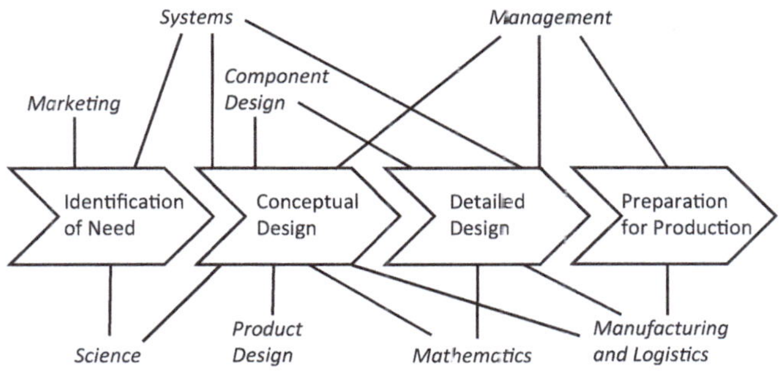

Figure 1.2.1 Disciplines throughout the design process. Redrawn from Eckert & Clarkson (2004).

Would R&D be included in the design process? We might argue that there is design occurring in R&D, but R&D is typically not focussed at the development of a specific engine, so maybe it is not part of any specific design process. However, the research of the fashion context and many activities in the early stages of any fashion design process are not garment-specific either, yet they are recognised as a vital part of designing (Eckert and Stacey, 2003).

What designers include in the term design is also fluid depending on the context. For example, when fashion designers are asked to talk about the design process of a specific garment, they often start with sketching a specific design or looking for a specific source of inspiration. However, when they talk about their activities in general, it becomes clear that researching fashion trends or materials is a very important part of their process, and that they come up with many ideas while they are carrying out general research.

In all these cases, we find that the designers are engaged in many activities that do not fall under the cognitive definition of design. Fashion designers in high street brands usually spend a few days to design their collection and then spend many months sampling and refining their products and setting up the production for them. Small independent designers spend a lot of time running their businesses, finding premises, getting grants, finding suppliers, writing invoices, marketing their product, to name a few activities. They interact with their customers and suppliers and spend an increasing amount of time on social media informing their customers about their businesses and values.

The range of activities carried out by the fashion designers, versus the high degree of specialisation of the engineering design process is not an inherent difference between the domains, but between working in organisations at a different scale. In small and often specialised engineering firms the engineers also carry out a wide range of different activities including design and business tasks. Conversely large fashion companies also have highly specialised designers who largely work on specific tasks, and consequently do more designing.

1.2.4.3 Different roles of a designer

Even when the focus is placed on the development of a specific product, there are many activities that that might or might not be counted as designing.

When the second author worked as a Design Engineer in a different engine company in the early 1980s, there were four groups of employees in the Design Office:

1 Designers, who did the actual scheme development for the design work of the company – layout drawings when he first worked there, then developing into CAD models of the artefact.
2 Design Analysts, who carried out the calculations on the emerging designs – largely stress and vibration analysis, some thermal.
3 Design Engineers, who did a variety of work in the office – they could do some simple design work, and design analysis, but mainly were collecting data, writing design guides, meeting with clients, managing information resources etc. He was a Design Engineer.
4 Design Draughtsmen (and it was all men then), who prepared the detail drawings of parts and assembly drawings, from the layouts prepared by the Designers.

In addition, the company had Development Engineers who advised the Designers on parameters to choose, especially for combustion systems, and did the development work on prototypes – including combustion systems development (injection settings, etc.) and durability testing.

This distinction still largely holds. A small group of (conceptual) designers oversees the development of the system architecture, where the functions the product needs to carry out are mapped to specific components and the interfaces between the components are defined. This was one of the teams we studied (Wyatt et al., 2009). They really understand the users and, maybe more importantly, what different types of users would wish to do with their engines, such as repair them with equipment and clothing that impose the need for access space around parts. The modern equivalent of the diesel engine company's Design Engineers are component specialists, who become experts in particular subsystems, such as pistons, and design these for multiple product generations, usually by making small but significant tweaks to existing designs. Very often such specialists also interact with specialist parts manufacturers e.g., for pistons, fuel injection equipment, and bearings. The majority of their time is spent understanding how their component needs to behave under particular circumstances and how it is likely to behave. They analyse their components, their interfaces to other components, and their role in other products through the product platform. For each product generation the time they spend modifying the CAD models might only be a matter of hours. The other role the Design Engineers carried out was the interaction with customers; the modern company has given this role to application engineers, who handle the direct customer interaction and understand their customers' products, needs and preferences. These are typically engineers by background who understand the significance of elements of the product, can explain it to customers, and can negotiate with customers what can reasonably be accommodated for the engine.

At the diesel engine company where the second author was a Design Engineer, only the Designers (and to some extent the Draughtsmen) were actually doing the designing in the sense of creating the plans for an artefact, but all of the four professional groups in the Design Office, and Development Engineers and more, were involved in the design process. The role of the Design Draughtsman has largely disappeared as engineers make their own CAD models. However, the detailed design of many components, from bearings to pistons, fuel injection equipment and turbochargers, will be outsourced to suppliers, and design consultancies are also active in this field. The detail design of the product is also often changed by manufacturing engineers to suit the production process.

The group that has become much more influential, in the meantime, is the Design Analysts. Modern diesel engine design often starts with an analysis of the previous design and virtual or physical tests to see whether the desired performance can be achieved

(Tahera et al., 2019). Through the entire product development process, the emerging design is analysed and simulated to understand its properties, typically making use of very sophisticated computational analysis tools. These engineers focus on all aspects of the behaviour of the engine, including thermal, stress and vibration analysis. Physical test and development are nevertheless still important, especially for such properties as durability and emissions performance, because until the product can be demonstrated to meet certain requirements, concerning emissions for example, the company cannot legally sell it.

In most engineering companies, the analysis of the market, the logistics, and the sales operations are carried out by the business functions in the enterprise. By contrast, the fashion designers in the small sustainable businesses also had to take these roles on themselves. Even if they had staff who carried out the activities, these staff members also often had a design background.

1.2.5 Examples of changing boundaries of design activities

The previous section illustrated the range of activities carried out by designers, which further raises the question what a designer does. We can go back to J. Christopher Jones' (1970) distinction between designing as the production of representations of the artefact and designing as contribution to the change in man-made things. In the narrower sense, designing is a small part of the designers' activities. In the broader sense, they are designing much of the time. However, even if we are including everything a designer does under design, the boundaries are still fuzzy and constantly shifting. The above example of the diesel engine design also illustrates the changing boundaries of engineering design tasks. In the 25 or so years between the two diesel cases, the emphasis in design practice had shifted to some extent from designing the components to analysing the behaviour of the product mathematically and computationally. This section looks at two examples of these shifts, where other areas of experience, mathematics and customer relations, are brought closer into the design process than ever before.

1.2.5.1 Engineering and mathematics

The relationship between engineering and mathematics is changing as engineering becomes more driven by analysis. To minimise risk and maximise the reuse of existing components and subsystems, engineering companies design their products around key performance equations or models so that parameter values are cascaded through different levels of design, such that the form of the product is to a large extent the response to performance equations and input constraints (see Eckert et al., 2012). This changes the skills profile of the engineers from many engineers having a holistic understanding of the product to engineers with a very high degree of numeracy and good computational skills but often a less well-developed ability to imagine and anticipate the behaviour of the design.

Another shift away from traditional engineering to data analysis has already begun (see Isaksson & Eckert, 2020). Engineering companies collect a huge amount of data about their products. By the beginning of the century, diesel engine companies (for example) had already accumulated large quantities of warranty data from all of their past engines, which they tried to analyse to understand failure patterns. At the time this could be frustrating. It also took a long time to accumulate data, as faults in highly durable and robust products take a while to occur. This left the companies with a lot of data on old and therefore not particularly relevant products and not enough data on recent products. However, developments are taking place in two respects: analysis and simulation software and hardware is improving, and better, smaller and cheaper sensors enable the companies to gather much more data. The former means that trade-offs, design margins and uncertainties can be better understood and accounted for, while the latter means that the performance of real engines may be better understood, and feedback may be gained on the capabilities of analysis and simulation models (although what it is possible to sense may not always be what it is important to try to measure). The increased data collection will also mean that data analysts, who may have little understanding of the engineered products, will need to be employed, and data analysis will increasingly be a central part of design.

1.2.5.2 Fashion design and customer relations

The small fashion businesses studied by Black et al. (2021) have more and different interactions with their customers. Traditionally designers have studied the products of other designers and positioned their products in relation to these. The customers select the products that they liked out of the wide offering on the market. As concern for the sustainability of textiles is growing, some customers want to find out more about the companies they buy from, so that the designers find themselves spending time generating blogs or posting on social media about their designs, their values and their ideas.

Designers see reaching out to the customers and educating them about sustainability as an integral part of their design activities and their contribution to changing the fashion system they want to change. They aim to build a relationship with their customers and respond to the customers wishes and desires to increase the time that customers wear garments and also the customer's bond to the fashion brand. This becomes part of understanding needs and constraints on the designs. Interaction with customers on the design of specific garments has always existed (think of a traditional tailor), but the deeper interaction around the values and processes of brands is new. The customers are engaged in tasks that would traditionally be done by the designers, but does this then make them designers?

1.2.5.3 The boundaries of design

The scope of the words 'design' and 'designing' and 'designer' are increasingly in conflict. Our preferred definition of 'designing' is the creation of a plan for an artefact. Our everyday linguistic intuitions tell us that the synthesis steps of making moves in design space by envisioning new design elements is central to this, but that does not reflect the reality of design practice, where the development of understanding of the design problem is often crucial and sometimes the central challenge, and the development of understanding of the behaviour of proposed designs is frequently the most difficult and time-consuming part of the process. Making actual design moves in terms of specifying geometry through drawing or modelling may be comparatively trivial. For example, the diesel engineers, who aim to carry as many parts across from one product generation to the next as possible (see Eckert et al., 2012), spend most of their time analysing whether the component could withstand altered conditions in the engine, e.g., higher temperatures or increased vibration. They also need to check all interfaces the components have to assure that modifications to the connected or adjacent components do not affect their component. In addition, they need to make sure that the component will still work under altered use conditions. After these evaluations they might make minor changes to features of the component or a particular interface, or just sign the component off to go to testing. For another example, the fashion and knitwear designers we have studied put a lot of effort into understanding the space of acceptable fashions and how to position their designs within it, and finding sources of inspiration, to construct the constraints that guide their designing (Eckert & Stacey, 2003; Stacey & Eckert, 2010.) As Vincenti (1990) pointed out, knowledge generation activities are central to how engineers produce new products; rather different knowledge generation activities are crucial for fashion and knitwear.

1.2.6 Design as a family resemblance concept

Throughout this chapter, we have highlighted different aspects of design and brought in different conceptualisations of design. The readers of this paper probably have their own examples they draw on when they think about design, which are very different from ours; and will agree with some of the definitions we cited from the literature and not others. Everybody who has studied design or is interested in design sees it through his or her own lens, and therefore sees different patterns and characteristics in design. In this section, we will argue that these different personal notions of design can be entirely valid, because they all add to the rich picture of design and agreeing on a definition of design in terms of necessary and sufficient conditions has proved an infeasible enterprise.

Wittgenstein points out in *Philosophical Investigations* (1953) that we frequently use single terms to denote a range of concepts that overlap in their uses. In some cases, no single characteristic is shared by all usages. He points to the example of "game", which we often think of in terms of winning or losing, but also covers a child playing with a ball. As in a family, some characteristics are shared with some members but not others. Yet putting all together we can see the resemblance across a family. In Wittgenstein's conception of communities of language users, the concept is determined by usage with no standard of correctness over and beyond this. It follows that a requirement of understanding a family resemblance concept is that there must be intersubjective knowledge of its exemplars.

Rosch and Mervis (1975) argued from experimental evidence that the prototypicality of exemplars of category concepts depends on the total cue validity of their attributes, and from this that family resemblance offers a good account of prototypicality as well as how people form concepts, and a logical alternative to criterial definitions of concepts in terms of necessary and sufficient conditions.

Many ill-defined concepts, such as power (Haugaard, 2010) or partnering between companies (Nyström, 2005) are defined slightly differently by different writers. Putting all of these definitions together provides a rich picture of the phenomenon. Ward and Scott (1987) point out that we learn family resemblance concepts both in an analytic mode where properties are analysed and extended and in a holistic mode according to global relations of overall similarity. Similarity, which is ultimately reducible to shared features, plays an important role in learning and establishing categories, but we are flexible in our similarity assessment and allow exceptions to our expectations (Goldstone, 1994).

Design is also such a family resemblance concept, that we learn from a sequence of experiences of uses of the word, where everybody has a slightly different understanding of what is meant by design. The examples have shown that there is very little that is shared across the activities that designers at the extreme ends of the spectrum carry out. Some people who think of themselves as designers might spend their time in highly mathematical analysis tasks, whereas others might be mainly engaged in customer interaction. Many are occupied with managing other designers. However, every designer shares many tasks and attitudes with many other designers, just not necessarily with all. We also have to accept that the activities that people called 'designers' engage in might not have one defining characteristic making what they do designing. Designers in a wide range of fields perform very similar activities – at least when viewed from a sufficiently abstract perspective. But designers at the extreme points may not have activities in common, making 'designer' a chain concept – where each use shares many features with to its neighbours and fewer and fewer with others. However, the notion of design as a family resemblance concept allows for this.

Auernhammer and Ford (this volume) present a computational analysis of 77 English-language definitions of design. Drawing on Wittgenstein (1953) and the work of Rosch and Mervis (1975), they argue that collectively these contain bundles of inter-predictable attributes that indicate the core meaning of the concept of design, but that defining design clearly is prevented by five dilemmas. The *complexity dilemma* is that design is too complex and multifaceted for easy definition. The *Zeitgeist dilemma* is that what design involves is context-dependent and influenced by technology and cultural values. The *expertise dilemma* is that design activities vary between ones performed by amateurs and ones requiring a high level of professional training and skill. The *identity dilemma* is whether design is art, science, both or neither. The *disciplinary perspective dilemma* is that different types of design look different.

Different disciplines describe design in different ways and emphasise different elements and activities, for example fashion designers would emphasise placing their work in the context of other designs, expressing the Zeitgeist and increasingly the need for sustainability, whereas product designers might stress user needs and desires, and engineers might focus on performance and safety. There is an element of rhetoric in how different design domains represent themselves, whereby artistic design domains tend to emphasise the novelty and creativity involved in their designs, whereas engineering designers try to emphasise their trustworthiness and rigour. Of course, rhetoric is inherently slippery, and people don't say the same things all the time.

The fact that so many different people call themselves designers is significant. Being a designer is an important element of their self-perception. They studied to become designers, now they work in the area and now they are designers. This is particularly strong with the fashion designers in our case studies who actually did very little designing, as opposed to producing garments or organising the production of garments. They referred to themselves as 'creatives' and looked for the company of other 'creatives', either other designers, crafts people or musicians, with little reflection on whether these people were either designers or were in fact creative by any accepted definition of creativity.

1.2.7 Implications

The search for the essence of design or a description of design that covers all scenarios looks like a futile quest. However, if we accept that design is a family resemblance concept, then we can also see different conceptualisations of design as being valid, complementary, and enriching. This would also free us as a community from the desire and drive of particular subgroups to "own" the concept, define design, or speak on its behalf.

Rather it points towards accepting different takes on design as equally valid. This points to a different approach to looking for a comprehensive, cross-domain analysis of design: developing it incrementally by looking for what is similar and different in different design domains and what causes these patterns of similarity and difference (Stacey et al., 2002). Design research has for a long time focussed on what is common and in consequence has developed very high-level theories and conceptualisations of design. These achieve generality by being highly abstract; often, they are nothing like as general as they are claimed to be. Over the years many important insights into the behaviour of designers have been discovered, but they are not universal. For example, it is widely recognised that designers are "reflective practitioners", going back to the work of Schön (1983), however this characterisation is not normative. Many designers engage in reflected action some of the time or occasionally, but not all do, and many people are also reflective in other everyday situations.

While these abstract conceptualisations of design are undeniably useful for design research or design teaching, their utility for design practice has often been questioned. Bryan Lawson (2006) put this very succinctly apropos of process maps as: " ... about as much help in navigating a designer through his task as a diagram showing how to walk would be to a one-year-old child. Knowing that design consists of analysis, synthesis and evaluation will no more enable you to design than knowing the movements of breast-stroke will prevent you from sinking in a swimming pool".

What designers in practice want from design researchers and to some extent from each other is actionable insights and advice. Practically useful generalisations sit at a meso-level between the very detailed and domain specific accounts and the general abstract descriptions of design. Much of this advice is formulated as methods or tools for design. However, these are only usable if the designers understand how they can be applied and deployed in a specific context. This is often a challenge for two reasons: (a) how to use the method is not well explained, and (b) a scope of applicability is not explained, which in some cases implies that the methods should be universally applicable (Gericke et al., 2020, 2022).

However, it is important to find ways in which designers in different contexts can learn from each other. Different areas of design have particular strengths that are often a direct response to the greatest risks or challenges of the field (Eckert et al., 2005). However, designers in other areas are often not aware of this and design subcultures may have a "not invented here" culture. The failure to look across different fields of design has also been a barrier to understanding the drivers behind different aspects of design behaviour. For example, all design processes generating safety-critical products have many common characteristics, as do processes that interact with users directly (Eckert & Stacey, 2014).

Design in all fields is deeply embedded in the context in which it is operating. Many design problems are highly constrained, and designers need to make many trade-offs, as is the case in diesel engine design. The fashion designers are also constrained, by what they can do in a given time and at a given price point. They often have to make tough decisions about whether what they can practically achieve can be combined with their consciences. These concerns primarily affect professional designers. Amateur designers, such as hobby knitters and dressmakers, who also engage in designerly behaviour, might have a far greater flexibility. They can shift their own goalposts in a way that professional designers cannot.

Understanding the differences between different types of design and the drivers for differentiation is important. Bringing designers from different fields together also requires an appreciation of where the hard problems are in individual design fields and what they are good at.

1.2.8 Conclusion

The field of design research has gained a great deal from a very wide variety of attempts to work out what design or designing *is*. But what emerges from all these attempts is that design, and what people mean by the term, is far too diverse for any one definition to fit

Many characterisations of what design is have claimed to cover all design but are implicitly grounded in particular types of designing and just don't fit many others. Design is best regarded as a family resemblance concept, where the many different instances of it have many features in common with many others, but there may be no characteristic that they all share. A better approach is to look at how and why particular types of designing are similar to and different from other types of designing and be very clear about the scope of any claims about the nature of design.

Whether an activity counts as design depends not so much on the nature of the activity itself but also on the context in which it is carried out. A mathematical problem solved in a class might be just that – a maths problem – but in an engineering context it can become a design task. A making task can be just following instructions or a vital part of prototyping.

One characteristic shared across many very different types of designing is the crucial importance to the development of new artefacts of the analysis and knowledge generation activities that contribute to framing design problems and constraining the form of their solutions. How much these are regarded as part of designing depends on the field and on the self-understanding of the people who do them. They tend to get neglected in rhetoric about design and creativity.

One actionable conclusion we draw is that in the view of design as a cycle of analysis, synthesis, and evaluation, all of these are part of design and are important and deserving of respect. For those of us concerned with what is or should be design, the challenge is to get problem framing, analysis and evaluation activities given due recognition as central to designing. This requires increasing awareness among ourselves, our students, and the non-designing public that the knowledge generated about the problem the system is produced to solve, and the performance of the product, is an integral part of what is produced by a design process. This is true to some extent for under-constrained artistic design fields as well as tightly constrained performance-driven engineering design fields. In the latter, we can call the generators of knowledge as well as the makers of moves in design space 'design engineers'; in other fields the right terms giving appropriate status might be harder to find.

After 60 years it is time that design research embraces the richness and diversity of design and looks broadly at all design activities to address the imbalance in the research effort devoted to understanding and improving different aspects of design. For example, much research effort is invested in creativity and idea generation, but very little into the role of testing and validation in design, even though this takes up many more resources in practice.

References

Akin, Ö. (1986). *The psychology of architectural design*. Pion.

Andreasen M. M. (1980). *Syntesemetoder på systemgrundlag* (Machine design methods based on a systemic approach). Doctoral Thesis, Lund University.

Asimow, M. (1962). *Introduction to design*. Prentice-Hall.

Auernhammer, J., & Ford, C. (this volume). What is design? The semantic core and periphery of design definitions. In G. Goldschmidt & E. Tarazi (eds.), *Expanding the frontiers of design: Critical perspectives*. Boca Raton, FL: CRC Press.

Baines, T. S. et al. (2007) State-of-the-art in product-service systems. *Proceedings of the Institution of Mechanical Engineers, Part B: Journal of Engineering Manufacture*, 221, 1543–1552.

Black, S. (2012). *The sustainable fashion handbook*. Thames and Hudson.

Black, S., Burcikova, M., Williams, D., Rocamora, A., Lyon, F., Elf, P., Werner, A., Vickers, I., Eckert, C. M., & Crommentuijn-Marsh, P. (2021). Fostering sustainable practices: Case studies of micro and small designer fashion enterprises building back better. *Sustainable Innovation 2021*.

Blackler, A., Swann, L., Chamorro-Koc, M., Mohotti, W. A., Balasubramaniam, T., & Nayak, R. (2021). Can we define design? Analyzing twenty years of debate on a large email discussion list. *She Ji: The Journal of Design, Economics, and Innovation*, 7, 41–70.
Brown, T. (2008). Design thinking. *Harvard Business Review*, 86(6), 84–92.
Boden, M. A. (1990). *The creative mind: Myths and mechanisms*. Abacus.
Buchanan, R. (1992). Wicked problems in design thinking. *Design Issues*, 8(2), 5–21.
Caves, R. E. (2000). *Creative industries: Contracts between art and commerce*. Harvard University Press.
Checkland, P. B. (1999). Systems thinking. In W. L. Currie & B. Galliers (eds.), *Rethinking management information systems: An interdisciplinary perspective*. Oxford University Press.
Chua, S. M. J. (2009). Donald Schön, Herbert A. Simon and the sciences of the artificial. *Design Studies*, 30, 60–68.
Cross, N. G. (2006). *Designerly ways of knowing*. Springer.
Cross, N. G., Christiaans, H. H. C. M., & Dorst, K. (eds.). (1996). *Analysing design activity*. Wiley.
Daly, S. R. (2008). *Design across disciplines*. PhD dissertation, Purdue University.
Design Council (2018). *The design economy 2018 – The state of design in the UK*. https://www.designcouncil.org.uk/resources/report/design-economy-2018
Dorst, K., & Cross, N. G. (2001). Creativity in the design process: Co-evolution of problem-solution. *Design Studies*, 22, 425–437.
Eckert, C. M. (2001). The communication bottleneck in knitwear design: Analysis and computing solutions. *Computer Supported Cooperative Work*, 10, 29–74.
Eckert, C. M., & Clarkson, P. J. (2004). The reality of design. In Clarkson P. J., & Eckert, C. M. (eds.), *Design process improvement—A review of current practice*, pp. 1–29. Springer.
Eckert, C. M., Earl, C. F., Stacey, M. K., Bucciarelli, L. L., & Clarkson, P. J. (2005). Risk across design domains. *Proceedings of the 15th International Conference on Engineering Design*. Melbourne.
Eckert, C. M., Isaksson, O., & Earl, C. F (2019). Design margins: A hidden issue in industry. *Design Science*, 5.
Eckert, C. M., & Stacey, M. K. (2003). Sources of inspiration in industrial practice. The case of knitwear design. *Journal of Design Research*, 3(1).
Eckert, C. M., & Stacey, M. K. (2014). Constraints and conditions: drivers for design processes. In A. Chakrabarti & L. T. M. Blessing (eds.), *An anthology of theories and models of design*, pp. 395–415. Springer.
Eckert, C. M., Stacey, M. K., Wyatt, D., & Garthwaite, P. (2012). Change as little as possible: Creativity in design by modification. *Journal of Engineering Design*, 23, 337–360.
Fauconnier, G., & Turner, M. (2002). *The way we think*. Basic Books.
Flanagan, T., Eckert, C. M., & Clarkson, P. J. (2007). Externalising tacit overview knowledge: A model-based approach to supporting design teams. *Artificial Intelligence for Engineering Design, Analysis and Manufacturing*, 21, 227–242.
Gericke, K., Eckert, C. M., Campean, F., Clarkson, P. J., Flening, E., Isaksson, O., Kipouros, T., Kokkolaras, M., Köhler, C., Panarotto, M., & Wilmsen, M. (2020). Supporting designers moving from method menagerie to method ecosystem. *Design Science*, 6.
Gericke, K., Eckert, C., & Stacey, M. (2022). Elements of a design method – a basis for describing and evaluating design methods. *Design Science*, 8, e29.10.1017/dsj.2022.23.
Goldstone, R. L. (1994). The role of similarity in categorization: Providing a groundwork. *Cognition*, 52, 125–157.
Haugaard, M. (2010). Power: A 'family resemblance' concept. *European Journal of Cultural Studies*, 13, 419–438.
Hazelrigg, G. A. (1998). A framework for decision-based engineering design. *Journal of Mechanical Design*, 653–658.
Hubka V., & Eder W. E. (1988). *Theory of technical systems: A total concept theory for engineering design*. Springer.
Isaksson, O., & Eckert, C. M. (2020). *Product development 2040*. 10.35199/report.pd2040

Jones, D. A., Eckert, C. M., & Gericke, K. (2018). Margins leading to over-capacity. In *DS 92: Proceedings of the DESIGN 2018 15th International Design Conference*, pp. 781–792. Dubrovnik, Croatia.
Jones, J. C. (1970). *Design methods: Seeds of human futures*. Wiley.
Lawson, B. (2006). *How designers think: The design process demystified* (2nd ed.). Elsevier.
Lowe, A., McMahon, C., & Culley, S. (2004). Characterising the requirements of engineering information systems. *International Journal of Information Management*, 24, 401–422.
Luck, R. (2018). What is it that makes participation in design participatory design? *Design Studies*, 59, 1–8.
Mankins, J. C. (1995). *Technology readiness levels, a white paper*. Washington, DC: NASA.
McMahon, C. A. (2004). Empirical design research: Fuzzy glimpses of practice. In E. Subrahmanian, R. D. Sriram, P. Herder, H. H. C. M. Christiaans & R. Schneider (eds.), *Workshop Proceedings: The Role of Empirical Studies in Understanding and Supporting Engineering Design*. Delft University Press.
McMahon, C. A. (2021). Situation, patterns, exploration, and exploitation in engineering design. *She Ji: The Journal of Design, Economics, and Innovation*, 7, 71–94.
Mistree, F., Smith, W. F., & Bras, B. (1993). A decision-based approach to concurrent design. In H. R. Paresai & W. Sullivan (eds.), *Handbook of concurrent engineering* (pp. 127–158). Chapman & Hall.
Mont, O. (2004). *Product-service systems: Panacea or myth?* Doctoral thesis, Lund University.
Moulton, A. (1976). *Engineering design education* (Moulton Report). Design Council.
Nyström, J. (2005). The definition of partnering as a Wittgenstein family-resemblance concept. *Construction Management and Economics*, 23, 473–481.
Pahl, G., & Beitz, W. (1977). *Konstruktionslehre*. Springer. (Current edition Bender, B. & Gericke, K. (eds.) (2021). *Pahl/Beitz Konstruktionslehre*, 9th ed. Springer.)
Plattner, H., Meinel, C., & Weinberg, U. (2009). *Design-thinking*. Mi-Fachverlag.
Reymen, I. M. M. J., Hammer, D. K., Kroes, P. A., van Aken, J. E., Dorst, K., Bax, M. F. T., & Basten, T. (2006). A domain-independent descriptive design model and its application to structured reflection on design processes. *Research in Engineering Design*, 16, 147–173.
Rittel, H. W., & Webber, M. M. (1973). Dilemmas in a general theory of planning. *Policy Sciences*, 4, 155–169.
Rosch, E., & Mervis, C. B. (1975). Family Resemblances: Studies in the internal structure of categories. *Cognitive Psychology*, 7, 573–605.
Roy, R., & Wield, D. (1986). *Product design and technological innovation*. Open University Press.
Sanders, E. B. N., & Stappers, P. J. (2008). Co-creation and the new landscapes of design. *CoDesign*, 4, 5–18.
Sanders, E. B. N., Brandt, E., & Binder, T. (2010). A framework for organizing the tools and techniques of participatory design. In *Proceedings of the 11th Biennial Participatory Design Conference* (pp. 195–198).
Schön, D. A. (1983). *The reflective practitioner: How professionals think in action*. Basic Books.
Schön, D. A., & Wiggins, G. (1992). Kinds of seeing and their functions in designing. *Design Studies*, 13, 135–156.
Simon, H. A. (1957). A behavioral model of rational choice. In *Models of man, social and rational: mathematical essays on rational human behavior in a social setting*. Wiley.
Simon, H. A. (1969). *The sciences of the artificial*, MIT Press. (3rd ed. 1996).
Stacey, M. K., & Eckert, C. M. (2010). Reshaping the box: Creative designing as constraint management. *International Journal of Product Development*, 11, 241–255.
Stacey, M. K., Eckert, C. M., Earl, C. F., Bucciarelli L. L., & Clarkson, P. J. (2002). A comparative programme for design research. In *Proceedings of the Design Research Society 2002 International Conference: Common Ground*. Brunel University, Runnymede, London, 2002.
Suh, N. P. (1990). *The principles of design*. Oxford University Press.
Tahera, K., Wynn, D. C., Earl, C. F., & Eckert, C. M. (2019). Testing in the incremental design and development of complex products. *Research in Engineering Design*, 30, 291–316.

UK Government (2005). *Cox review of creativity in business: Building on the UK's strengths.* http://www.hm-treasury.gov.uk/d/Cox_review-foreword-definition-terms-exec-summary.pdf

Ulrich, K. T., & Eppinger, S. D. (1995). *Product design and development.* McGraw-Hill.

Vincenti, W. G. (1990). *What engineers know and how they know it.* Johns Hopkins University Press.

Ward, T. B., & Scott, J. (1987). Analytic and holistic modes of learning family-resemblance concepts. *Memory & Cognition*, 15, 42–54.

Wittgenstein, L. (1953). *Philosophical investigations* (G. E. Anscombe, Trans.). Blackwell.

Wyatt, D. F., Eckert, C. M., & Clarkson, P. J. (2009). Design of product architectures in incrementally developed complex products. In *Proceedings of the 17th International Conference on Engineering Design*, Vol. 4, *Product and Systems Design* (pp. 167–178). Palo Alto, CA.

Wynn, D. C., & Clarkson, P. J. (2018). Process models in design and development. *Research in Engineering Design*, 29, 161–202.

1.3 What is Design?

A semantic analysis of definitions of Design

Jan Auernhammer[a] *and Chris Ford*[b]
[a]*University of Technology Sydney;* [b]*Stanford University*

1.3.1 Introduction

In 1965, Tomas Maldonado (1965) expressed his concern about the meaning of the word *Design* and that it "has been strangely corrupted by general usage." One explanation for today's conflation surrounding the definition of design thinking is a proportional amount of conflation surrounding contemporary definitions of Design. A definitional consensus of the terms *design thinking* and *Design* loses fidelity through semantic creep hastened further with each invocation. The graphic designer questions when an engineer characterizes themselves as a designer. The design engineer then disputes when the organizational designers make similar claims. Tension ensues. Any meaningful expansion of Design inclusiveness ought to first ascertain those forces causing expansions and contractions of its boundaries. A valuable first step is redirecting attention currently focused on the dynamism inherent in the expansion of Design to instead refocus on its shared family resemblance (Wittgenstein, 1953). Shared characteristics among different definitions from diverse disciplines may provide increased clarity into the universal characteristics of all forms of Design. However, the semantic categorization of the term: Design may not have a common definition among different Design disciplines. They instead resemble each other as they are related to one another through overlapping similarities and connected in different ways (Wittgenstein, 1953). The semantic categorization of a specific way of acting and thinking in "Design" is used in diverse disciplines as it corresponds to bundles of inter-predictable attributes (Auernhammer & Roth, 2022; Rosch & Mervis, 1975). Importantly, the consideration of what is Design must also examine the fuzzy boundaries that make it difficult to have a shared definition. This research examines: What are the resemblances inherent in diverse Design definitions and what are the inherent dilemmas in Design descriptions that make it difficult to have a clear Design definition?

The primary data set for this investigation is compiled from published definitions of Design authored by distinguished contributors representing diverse disciplinary backgrounds. They are practitioners, educators, historians, scholars, researchers, and in some cases, are non-designers, such as journalists and psychologists. These 77 definitions are based Buchanan's (1992) on all four orders of design, including graphic design (first-order: Communication and symbols), architecture and engineering (second-order: Objects and artifacts), interaction design (third-order: Interactions and experiences), and organizational and business design (fourth-order: Systems). Their published definitions have been compiled as a dataset for analysis to locate semantic emphasis between definitions of Design and determine whether aspects of these definitions present family resemblances

DOI: 10.1201/b22630-5

forming a core, or some amount of idiosyncrasy forming a periphery. This chapter contributes by providing insights on the (1) resemblances among diverse Design definitions and (2) the dilemmas preventing clear boundaries for defining Design. This research contributes by explicating shared concepts and terms among diverse Design definitions and discusses the various dilemmas of why establishing boundaries for a universal definition of Design is difficult. This chapter closes with insights and implications for design thinking, design research, and design practice, emphasizing the need for clarifying the actual studied phenomena beyond the use of the overarching term: Design.

1.3.2 Background

Since humans first fashioned stones into tools, Design has evolved in ways guaranteeing both the shaping of its body of knowledge and its practices remain in flux. In sociology, Thomas Kuhn (2012) examined how disciplines structure shared principles, norms, and understanding into a paradigm. However, Design evolves at a constant pre-paradigmatic phase resisting the fixing of definitional boundaries (Verganti, 2009). Design contains many connotations and has been corrupted by general usage (Maldonado, 1965; Moholy-Nagy, 1969). It describes the primary characteristics of humans, human attitudes, everyday activity, and professional practice (Lawson & Dorst, 2009; Moholy-Nagy, 1969; Protzen & Harris, 2010). Design practices are continuously evolving and expanding, and with them, the definition and understanding of Design. Bruce Archer (1965) expressed, "There has been a world-wide shift in emphasis from the sculptural to the technological. Ways have had to be found to incorporate knowledge of ergonomics, cybernetics, marketing, and management science into design thinking." The design of human interactions with interactive systems emerged with personal computing and information and communication technologies (Card, Moran, & Newell, 1983; Moggridge, 2007; Norman & Draper, 1986). In management, establishing organizations based on specific design qualities resulted in innovation management approaches (Auernhammer & Roth, 2021; Brown, 2009; Martin, 2009). However, each claim by a "non-traditional" discipline presents a new dilemma for "traditional" Design disciplines, which collectively question this expansion of inclusionary understandings of Design when compared against disciplinary definitions.

László Moholy-Nagy (1969) expressed "the designer must see the periphery as well as the core, the immediate and the ultimate, at least in the biological sense. She/he must anchor her/his special job in the complex whole." The same applies to design research. Design researchers must see the core of Design and its fuzzy periphery to place their findings and insights within the complex whole. Therefore, the investigation presented in this chapter examines diverse definitions of Design to identify valuable shared characteristics forming Design's semantic core and its fuzzy periphery.

1.3.3 Methodology

The research is designed to identify shared characteristics among different definitions of Design from different designers, design scholars, researchers, and historians. We employed a systematic search to identify and capture diverse Design definitions.

1.3.3.1 Data collection

We searched for "design is" and "[FIRST NAME] [LAST NAME]" on search engines including Google and Google Scholar, to capture general, practitioner, and academic perspectives on the definitions of Design. We furthermore examined several books by design authors to identify Design definitions to collect diverse perspectives from different fields. We collected over two hundred thirty statements, examined each statement, and categorized them as either a definition or a description. While discerning the textual dataset, definitions were found to speak directly to requirements and attributes of Design operations, whereas descriptions instead spoke to only a particular attribute, property, effect, or result. Therefore, we included the definitions and excluded the more detailed descriptions in the examination of family resemblances among diverse definitions. The resulting 77 definitions, included in this analysis, coming from individuals with diverse disciplinary backgrounds, representing the four orders of Ddesign (Buchanan, 1992).

1.3.3.1.1 Sample bias and limitation

A limitation of this research is the focus on definitions published in the English language, excluding a large amount of diverse cultural perspectives and semantic meaning. Furthermore, the data sample incorporates selection bias as definitions were searched by the researchers selecting specific authors. The data sample also incorporates a gender, cultural, and ethnicity bias of which there is no self-identified data available. Table 1.3.1 outlines the range of disciplinary backgrounds represented by the authors, including a categorization of the four orders of design as outlined by Buchanan (1992). The limitation of this chapter is that it includes a limited number of definitions.

1.3.3.2 Data analysis

The data was analyzed through three different analysis techniques. Firstly, we employed the software Leximancer to identify inherent concepts in the definitions. Leximancer allowed investigating the qualitative data through unsupervised semantic mapping of natural language to map concepts inherent in the dataset (Smith & Humphreys, 2006). The visualization of the findings is presented in Figure 1.3.1, and the identified main concepts are listed in Table 1.3.2. Secondly, we analyzed the definitions through a word count analysis to reveal shared words among the different definitions. Table 1.3.3 shows the shared word groups with the highest frequency (word count of six and higher). Lastly, we analyzed the definitions and descriptions to identify emergent themes that make the boundaries of Design challenging to define. Table 1.3.4 outlines the five emergent themes that are discussed controversially in the various Design definitions.

Table 1.3.1 Categorization of the disciplinary background of definitions authors

Disciplinary background	*Count (person)*	*Buchanan's order of design*
Design Education, Methodology, Research	25	General
Engineering, Industrial, Product Design	18	Second
Architecture, Interior, Urban Design	10	Second
Business Design, Economics	8	Fourth
Communication, Graphic Design	7	First
Electrical, Structural Engineering	3	Second
Computer Science, Interaction Design	4	Third
Textile, Fashion Design	1	Second
Others (Journalism, Psychology)	3	n/a

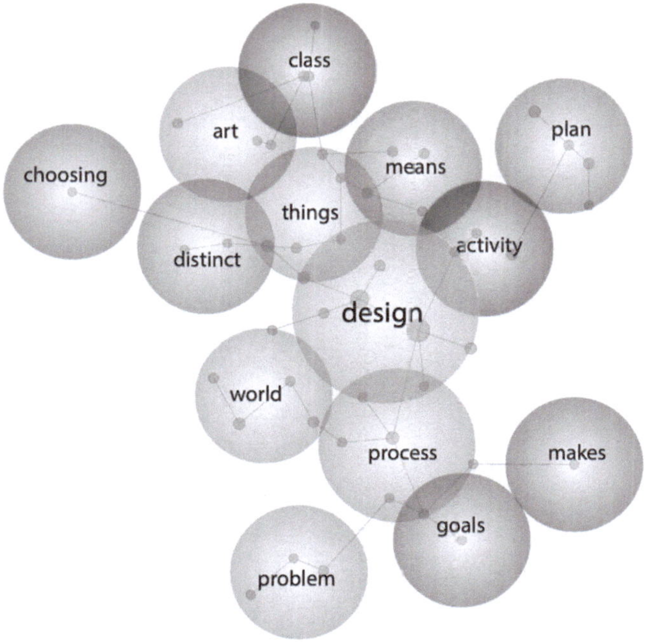

Figure 1.3.1 Visual representation of concepts inherent in the Design definitions. The figure shows a visual representation of the themes as a conceptual map from the Leximancer analysis of the design definitions. The map shows three distinct main patterns: (1) process, problem, goals, makes, and world; (2) things, means, distinct, class, and art; (3) activity and plan.

1.3.4 Findings

The data analysis revealed several (1) concepts, (2) shared terms, and (3) themes of dilemmas inherent in the Design definitions. This section first discusses the revealed concepts identified through unsupervised semantic mapping of natural language with Leximancer. Secondly, the top shared terms are outlined and discussed to provide a semantic core of the meaning of Design. Lastly, the identified themes of dilemmas are summarized, explaining why the definition of Design has various fuzzy boundaries.

1.3.4.1 Concepts inherent in the design definitions

The semantic mapping of the natural language of the Design definitions revealed the inherent concepts. Figure 1.3.1 shows the interconnected concepts within the diverse Design definitions.

A more detailed analysis revealed that design is conceptualized as outlined in Table 1.3.2. The inherent themes in the design definitions are "Changing World" and "Changing Situations." Unsurprisingly, thirteen definitions conceptualize Design as changing situations and the world. Furthermore, Design is conceptualized as "Plan Function," "Analysis Process," and "Making Meaning." Design is planning for a specific purpose, analysis process of a goal, problem, and solution, and making sense and giving meaning to things. The themes in the design definitions also include "Activity" of

Table 1.3.2 Main concepts inherent in the definitions of Design (unsupervised semantic analysis – Leximancer)

Topic	*Freq.*	*Concepts*	*Example definition*
Changing World	7	**World**: change, changing, materials, shape, thinking,	"Design is really a way of looking at the world with an eye toward changing it." (Warren Berger)
Plan Function	7	**Function**: plan, purpose, specific **Plan**: purpose, specific **Purpose**: specific	"Design [i]s a plan for arranging elements to accomplish a particular purpose." (Ray and Charles Eames)
Changing Situations	5	**Changing:** class, situation, situations, environment, means **Class:** situation, situations, environment, means **Situation(s):** situations, environment, means	"The whole class of situations means that the experienced environment is constantly changing, but nevertheless the object must always be just as good a solution, or fit, to the situation." (Horst Rittel)
Making Meaning	5	**Meaning:** making, sense, things, giving, means **Making:** sense, things, giving **Sense:** things, means **Things:** giving	"The etymology of design goes back to the Latin de + signare and means making something, distinguishing it by a sign, giving it significance, designating its relation to other things, owners, users, or gods. Based on this original meaning, one could say: design is making sense [of things]." (Klaus Krippendorff)
Analysis Process	4	**Process:** analysis, goals, problem, solutions	"The premise is that designing is a combinatorial problem-solving process in which analysis of a body of available information (which must first be collected) is used to generate candidate solutions to the problem, or to subproblems. Candidate solutions are compared and assessed, and the most appropriate solution is then selected." (Gabriella Goldschmidt)
Activity	4	**Activity:** bring, plan, specific, change	"Design' is a fundamental human activity, as well as the name of a number of quite specific professions." (Lawson & Dorst)
Planning, Making	4	**Planning:** making, meaning, means	"a particular purpose held in view by an individual or group; deliberate, purposive planning; a mental project or scheme in which means to an end are laid down." (Friedman & Stolterman)
Nature Design	4		"Design is a circumstantial act. It is a battle with the nature of man, with the nature of nature, with the laws of nature, with the rules of [hu]man, and with principles." (Louis Kahn)
Solution Used	4	**Solutions:** used, analysis, problem **Used:** analysis, problem	"Design is a multi-answer situation, and analysis is used to reach an end, not for its own sake." (James Adams)

Table 1.3.3 Word frequency analysis of the definitions of Design

Word	*Count*	*Similar words*
Design	153	Design, designating, designed, designer, designers, designing, designs
Making	22	Make, makes, making
Plan	21	Plan, planning, plans
Process	19	Process, processes
Things	15	Phing, things, things'
Using	15	Use, used, useful, usefulness, using
Arts	15	Art, arts
New	14	New
Purpose	14	Purpose, purposeful, purposes, purposive
Human	14	Human, humanities
Problem	14	Problem, problems
Products	13	Product, production, productive, products
Action	12	Action, actions, act, acts
Industrial	11	Industrial, industrially, industry
Situations	11	Situate, situation, situations
Forms	11	Form, forming, forms
Activity	10	Activities, activity
Change	10	Change, changed, changes, changing
Function	10	Function, functional, functionality, functions
Way	10	Way, ways
World	10	World
Materials	9	Material, materials
Solutions	9	Solution, solutions
Means	9	Meaning, means
Objects	9	Object, objects
Nature	8	Naturally, nature
Specific	8	Specific, specification
Creating	7	Create, creates, creating
Needs	7	Need, needs
Technology	7	Technological, technologies, technology
Language	6	Language
Order	6	Order, orderly
Sense	6	Sense, sense'
Solving	6	Solve, solving
Towards	6	Toward, towards
Living	6	Live, lives, living
Creative	6	Creative
People	6	People
Something	6	Something

planning for bringing about a specific change, "Planning Making" of making meaning, and "Nature Design" and "Solution Used."

The analysis revealed the different themes and main conceptions inherent in the Design definitions. Table 1.3.2 shows that only seven and below definitions share semantic meaning, revealing that there is not a dominant conceptualization of Design inherent in the diverse definitions. We further employed a word group frequency analysis to reveal the most used words in the definitions of Design.

1.3.4.2 Shared words among the design definitions

The word frequency analysis revealed shared terms (word groups) among the different definitions. Table 1.3.3 shows the word groups with a word count of six or higher.

Table 1.3.4 Identified main themes are preventing a clear definition of Design

Theme	*Codes*	*Example quote*
Complexity	Mystification Complex	[industrial design] is the only profession that became a myth before it reached maturity (Nelson, 1965).
Zeitgeist	Historical context Cultural context	Design is reflective of its social, historical, and regional culture of production. This is manifest in specific cultural influences on the designer (Micklethwaite, 2002).
Expertise	Everyday design Attitude Profession	'Design' is a fundamental human activity, as well as the name of a number of quite specific professions (Lawson & Dorst, 2009).
Identity	Not Art Not Science Combination of Art and Science Distinct from Art and Science	Design is a new liberal art of technological culture that has the capacity "to connect and integrate useful knowledge from the arts and sciences alike, but in ways that are suited to the problems and purposes of the present (Buchanan, 1992).
Disciplinary Perspective	Focus on function/fit Focus on aesthetics Focus on communication Design as search Design as exploration	Designing is not a search for the optimum solution to the given problem, but that it is an exploratory process. The creative designer interprets the design brief not as a specification for a solution, but as a starting point for a journey of exploration (Cross, 2011).

The most shared terms are "making," "plan," and "process." Making emphasizes that design incorporates the activity making sense and things. The plan stresses that designing is about the consideration of pre-action and making representations of the designed (situated) artifact, while the process highlights the semantic meaning that Design is a series of actions taken to achieve a particular end. However, there is not a clear conceptualization and shared terms among the design definitions. For example, "making" is shared among eighteen definitions, representing around 23.38%. For this reason, establishing the definitional boundaries of defining Design remains difficult.

1.3.4.3 Themes of boundary dilemmas

The thematic analysis of the dilemmas inherent in the Design definitions and descriptions revealed five themes, as shown in Table 1.3.4. These dilemmas explain why Design has fuzzy boundaries, representing the periphery of Design definitions.

The first identified dilemma is the *Complexity* dilemma. A clear conceptualization of Design prevents the flexibility, adaptability, and evolution inherent in the creative practice of Design. Designers and design scholars avoid the ridged definition, which leads to the mythification of Design. Similarly, the *Zeitgeist* dilemma produces fuzzy boundaries as Design depends on historical and cultural context and is constantly evolving. Different technological possibilities and cultural values at specific points in time and place influence and determine the design practices. Further complexity is that Design is both an everyday activity and a professional practice, making the boundaries to define Design fuzzy. This complexity is the *Expertise dilemma* of who designs. Another theme is the Identity dilemma. Designers and design scholars debate that Design is neither Art nor Science or a combination of both. Related to this theme is the *Disciplinary Perspective* dilemma. The rational technical disciplines emphasize process, function, and goal orientation, while the humanistic artistic disciplines emphasize exploration, culture, aesthetics, and needs. These different dilemmas produce a fuzzy boundary for defining Design.

1.3.5 Discussion: What is design?

The investigation of the family resemblance and fuzzy boundaries of diverse Design definitions revealed diverse conceptualizations within the definitions. Therefore, we are using a metaphor for visualizing the findings through a method used to measure peripheral vision to illustrate the semantic core and peripheral of Design definitions, as illustrated in Figure 1.3.2.

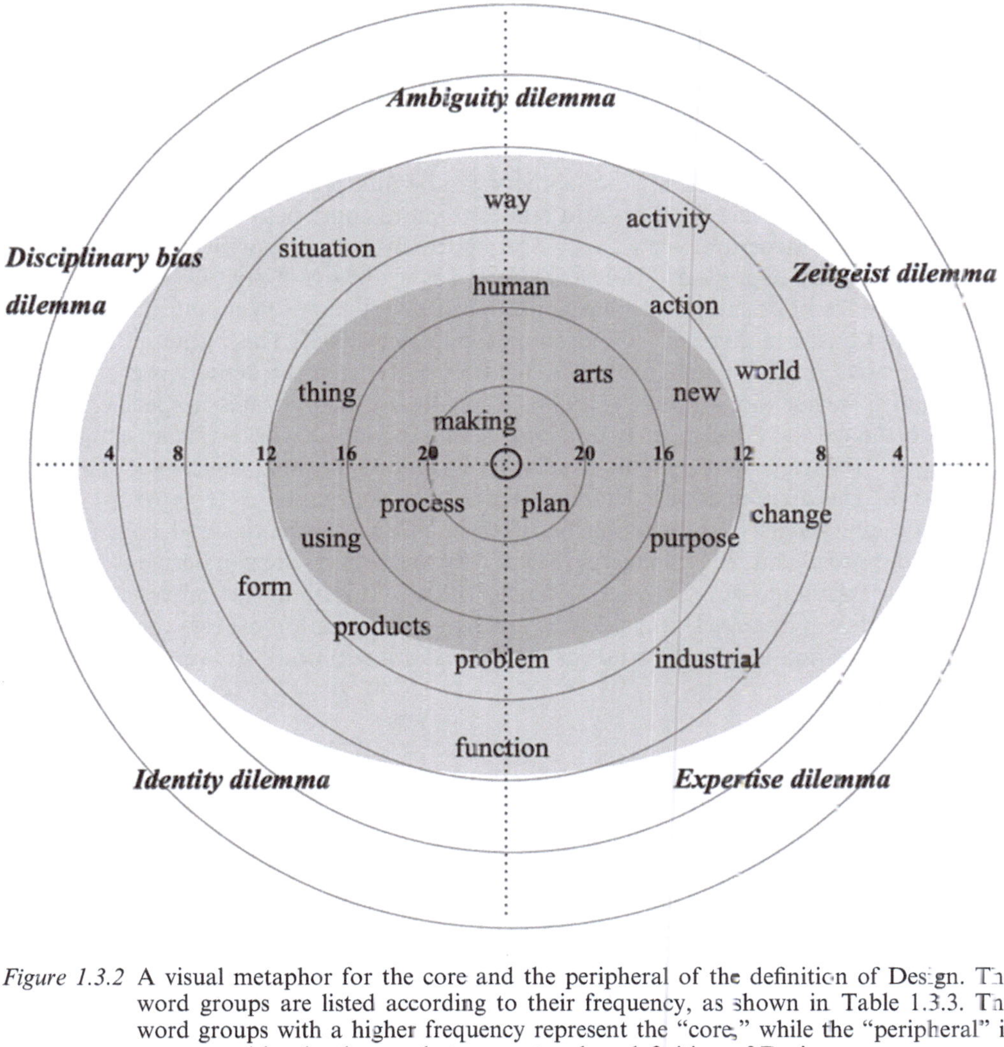

Figure 1.3.2 A visual metaphor for the core and the peripheral of the definition of Design. The word groups are listed according to their frequency, as shown in Table 1.3.3. The word groups with a higher frequency represent the "core," while the "peripheral" is represented by the themes that prevent a clear definition of Design.

The semantic core incorporates the shared words, such as making, plan, and process, and the peripheral is represented by the dilemmas that make it challenging to define a clear boundary for defining design.

1.3.5.1 The core: Family resemblance of design

We interpreted the central conceptualization and shared keywords, which allowed explicating a family resemblance of Design. The first resemblance indicates that Design is a set of specific "activities," a "way," and a "process." The second resemblance is that designing incorporates various embodied cognitive activities, such as "crafting," "meaning making," and "problem-solving." These activities have the aim to establish "new" and "purposeful" "change." The last resemblance is that Design is not an "end" by itself but a "means to an end" to satisfy "needs," and nurture "nature." These word groups do not provide a general definition. However, they provide the corresponding bundles of inter-predictable attributes (Rosch & Mervis, 1975). These attributes span epistemological (ways of knowing) boundaries and capture the fundamental embodied human cognitive activity and practice of design.

1.3.5.1.1 Activity, way, and process

At the core, Design is an activity, way, and process, representing different ways of designerly knowing (Cross, 1982). An activity can be interpreted as a pragmatist epistemology, while way of viewing the world relates to more constructivist view. A process is associated with a structural view that is grounded in a more positivist epistemology. Tomás Maldonado expressed, "Industrial design is a creative activity whose aim is to determine the formal qualities of objects produced by industry" (Verganti, 2009). It is a "generative, transformative act" (Goldschmidt & Porter, 2004). This "generative activity is about creating order" (Sara Turnbull). This view of design as an activity is grounded in a pragmatist epistemology (Dewey, 1938). Design also incorporates a specific way of looking at the world. "Design is really a way of looking at the world with an eye toward changing it" (Warren Berger). "It is a way of creating art that functions" (Michael Vanderbyl). "Design is all about 'defining a sensible way of building'" (Ove Arup). Such a view of design, such as sensemaking, is inherent in an experiential and constructivist epistemology (Arnheim, 2004; Piaget, 1981). Lastly, the process view of design is based on a structuralist epistemology (Selz, 1922; Wallas, 1926). "It is about the process of turning ideas into reality" (Gorb & Dumas, 1987). "Design is the creation process through which we employ tools and language to invent artifacts and institutions" (Owen, 1998).

1.3.5.1.2 Diverse activities

Design incorporates creativity, planning, forming, crafting, making meaning, sense-making, creating order, analyzing, and solving problems. "Design is the creative thinking process of delivering innovation" (Sohrab Vossoughi). It is "deliberate, purposive planning" (Ken Friedman & Erik Stolterman). It "is a matter of composition, of the integrated relation of all constituent parts in forming a whole" (Dewey, Boydston, & McDermott, 2008). "Where craftsmanship joined with technology through the use of design" (Florence & Hans Knoll). "Design is making sense [of things]" (Krippendorff, 2006). It "is not about an aesthetic style, but problem-solving" (Robby Johnston). Design incorporates many complex embodied cognitive activities, often referred to as design thinking and actions, the actual craft and ability of designers in generating tangible change.

1.3.5.1.3 New and purposeful change

This change is new and purposeful. "It's about creating new choices, not just choosing from things that already exist" (Tim Brown). "The central concern of Design is 'the conception and realization of new things'" (Cross, 1982). It is "a plan for arranging elements to accomplish a particular purpose" (Ray and Charles Eames). Milner Gray states that "design is an intention, purpose, plan and that good design is therefore by inference, where such plan has been well conceived, well executed, and of benefit to someone" (Moggridge, 2007). Such new and purposeful change is an end accomplished by a means, a tangible design.

1.3.5.1.4 A means to an end

Design is a means, including technology, objects, and things, to establish an end for the living world. "To design is to plan and to organize, to order, to relate and to control. In short, it embraces all means of opposing disorder and accident. Therefore, it signifies a human need and qualifies [hu]man's thinking and doing" (Josef Albers). "Design is the response to a human need" (McKim, 1959). It is also "a mental project or scheme in which means to an end are laid down" (Collegiate Dictionary as cited in Goldschmidt, 2014). Underlying any design outcome (means) is a purpose motivating and leading designers, which can be the designers' own needs (self-centered) or tensions in the environment or others' needs felt by the designers (self-actualized) (Arnheim, 2009; Auernhammer & Roth, 2022).

1.3.5.2 The periphery: Fuzzy boundaries of defining design

Clear boundaries of defining Design are complex because of various dilemmas. Designers are reluctant to set clear boundaries to avoid limiting the creative practice, design is context-dependent and constantly evolving, and design is a fundamental human activity enacted by all people. This primary activity is inherent in both Science and Art, making it hard to distinguish it from both. Furthermore, different disciplines have cultivated their defined design paradigm. These dilemmas make it difficult to define a clear boundary of what design is.

1.3.5.2.1 Complexity dilemma

The first dilemma of establishing a universal definition for Design is the complexity dilemma. Designers and design scholars avoid providing clarity on the definition of Design to avoid limiting designers' creativity and its evolution. However, this ambiguity led to the mythification of Design. Nelson (1965) expressed that "[industrial design] is the only profession that became a myth before it reached maturity." However, "designing is not a mysterious process, [...] but it is complex" (Cross, 2011). "Designing is a complex and intricate task. It is the integration of technological, social and economic requirements, biological necessities, and the psychophysical effects of materials, shape, color, volume, and space: thinking in relationships" (Moholy-Nagy, 1969). Because designing requires considering these diverse aspects, no single explanation and process incorporate all aspects. Reductionists representation of simplified process models do not represent the complexity involved in the actual practice of design.

1.3.5.2.2 Zeitgeist dilemma

Design is also temporal context-dependent, and the artificial and cultural context is constantly evolving. New technology and cultural values produce new design activities and possibilities. Designers and design scholars constantly integrate new approaches to expand their creative practice. It changes with new technological possibilities. Owen (1998) expressed this as follows: "Design is the creation process through which we employ tools and language to invent artifacts and institutions. As society has evolved, so has our ability to design." It also evolves through changing cultural values and emerging needs based on changing environments. "Design is reflective of its social, historical, and regional culture of production. This is manifest in specific cultural influences on the designer" (Micklethwaite, 2002). Similarly, Horst Rittel emphasized that "Design is directed towards a particular situation which is historically unique" (Protzen & Harris, 2010). Thus, design is a continuously evolving practice as newly designed technologies provide new possibilities and produce new emergent needs within the societal progress of ever-changing human values and beliefs.

1.3.5.2.3 Expertise dilemma

Design is a profession and everyday activity and attitude (Lawson, 2006; Lawson & Dorst, 2009; Moholy-Nagy, 1969). This expertise dilemma exists because of the fuzziness of who designs. Lawson (2006) expressed this paradox of design as follows: "Design is now clearly a highly professional activity for some people, [...] And yet design is also an everyday activity that we all do." The discourse on who designs makes it problematic to establish a clear boundary. For example, Moholy-Nagy (1969) expressed that "Designing is not a profession but an attitude." "[D]esign is basic to all human activity" (Papanek, 1973). Such an attitude can be learned and enacted by everyone. Another argument is that "Design is too important to be left to designers alone" (Gorb & Dumas, 1987). However, Cross (2011) expressed that "everyone can design and expert designers exercise very developed forms of certain tacit, deep-seated cognitive skills." This discourse implies a spectrum of design from everyday designing to professional design for specific domains. Like Design, creativity is defined within the spectrum of everyday creativity and culture-changing creativity, the capital C creativity (Kaufman & Beghetto, 2009). Underlining professional design is the development of specific cognitive skills and attitudes, i.e., creativity, which is combined with domain-specific knowledge of a specific domain, such as architecture (environment and buildings) or industrial design and engineering (products). Amabile (1996) discusses the same attributes of cognitive processes, domain-specific knowledge, and motivation in general creativity. From this perspective, creative or productive thinking of converting parts to a coherent whole is the shared denominator among everyday design activities and diverse professional design disciplines with domain-specific know-how (Auernhammer & Roth, 2022; Maldonado, 1965).

1.3.5.2.4 Identity dilemma

Designers and design scholars distinguish Design from Art and Science as either a combination of both, or a distinct paradigm. This dilemma generates fuzziness in defining Design. Scholars expressed that "Design is the marriage of art and science" (Micklethwaite, 2002). While others emphasized that "Design should not try to find a spot between engineering and art, but should be a separate discipline other than engineering and art" (Katz, 2015). Similarly, Lawson and Dorst (2009) expressed that "Design is always to some extent grounded by functionality, [...] art is not functional; it does not need to make 'sense.'" Like Art, Design is distinguished from Science. "Design is the opposite of the traditional scientific explanation" (Gero, 1990). However, Design is inherent in Science and Art. The scientist designs the scientific experiment to generate new knowledge that brings about changes in the scientific discourse, while the artist generates and conveys new meaning through "designed" artifacts. From an evolutionary perspective, humans designed tools to satisfy needs before they produced art or developed rigorous knowledge systems. "Design is, if we look at it in this broad sense, one of the primary characteristics of the human race. It is the foundation of our adaptivity that we can learn about situations and plan for them in the future. It is the foundation of civilization, of art, of science" (Protzen & Harris, 2010). From this perspective, Art and Science are the offspring of Design as both inherent the human activity of designing. Design is fundamental in all human activity. This identity dilemma makes it challenging to define the clear boundaries of Design.

1.3.5.2.5 Disciplinary-perspective dilemma

Design is "to 'contrive for a purpose,' makes it equally the property of engineers, architects, advertising [hu]men, artists, and dressmakers" (Dreyfuss, 1955). Design includes material objects, verbal and visual communication, organized activities and services, and complex systems or environments for living, playing, working, and learning (Buchanan, 1990). Each discipline produced its predisposed perspective on Design. The disciplinary perspective inherent in visual and graphic design is the focus on communication. "[Visual D]esign is concerned with finding the representation best suited to the communication of some specific information" (Mullet & Sano, 1995). Service design focuses on "intangible experiences that reach people through many different touch-points, and that happens over time" (Moggridge, 2007). The perspective inherent in more technical rational disciplines emphasizes function and procedure. "Design is meant to enhance some critical procedure or function" (Norman, 2007). In this perspective, aesthetics serves as a support for the performance of the procedure and function. In contrast, the humanistic perspective focuses on human values, such as needs. "Design is the response to a human need" (Auernhammer & Roth, 2021; McKim, 1959). "It is the making of plans to shape the world to our needs" (Protzen & Harris, 2010). "Design is the central factor of innovative humanization of technologies and the crucial factor of cultural and economic exchange" (ICSID as cited in Verganti, 2009). Another perspective is market orientation. Carlotta des Bevilacqua expressed that "Nowadays, every market-oriented company has understood that design is an advantage. As a result, all companies can use it. Design is not only a way to give a nice form, but it should rather anticipate a need, proposing a vision" (Verganti, 2009). In organizational design, design is seen as "the diagnosis of misfits and the action to fix them" (Burton, Obel, & Håkonsson, 2015). In this perspective, design is about "matching" internal to external environments. Each discipline has its own perspective that prevents clear boundaries of the meaning of design.

1.3.6 Conclusion and implications

The research presented in this chapter explicated that the definition of design is viewed from diverse perspectives. While shared words provide a family resemblance among different epistemological and disciplinary perspectives, it does not clarify what constitutes design. The identified shared conceptualizations and boundary dilemmas have several implications for design thinking, research, and practice.

1.3.6.1 Implications for design thinking

The debate of definitions of Design is describing essentially *productive thinking*, the embodied cognitive activity of resolving and satisfying a felt tension, need, or problem situation through a tangible design, such as an artifact and collaborative action (Dewey, 1938; Duncker, 1945; Eastman, 1970; McKim, 1980; Schön, 1963; Selz, 1922; Wertheimer, 1945). Lawson (1972) expressed that design thinking is productive and structural thinking, based on, e.g., Wertheimer (1945) and Bruner et al. (1956). Similarly, McKim (1968) outlined that design thinking is productive and visual thinking as discussed by e.g., Wertheimer (1945) and Arnheim (1965). Simon (1969) and Archer (1965) refer to design and design thinking to productive thinking in mathematics (see Pólya (1957) and de Groot (1965) for earlier developments of this perspective). Design thinking incorporates productive thought processes to make sense, meaning, and order of ill-defined and ill-structured design problems and situations (Auernhammer & Roth, 2022; Goldschmidt, 2014; Wertheimer, 1945). It requires many embodied productive thought processes incorporating various perceptual, cognitive, and physical activities, including analyzing situations and reorganizing parts into a new whole (Faste, 1994; Goldschmidt & Porter, 2004; McKim, 1980; Purcell & Gero, 1998; Rowe, 1987; Suwa & Tversky, 1996; Tversky, 2009). Such productive thinking can be an everyday activity as well as employed to solve open and ambiguous problem situations, which requires diverse domain-specific knowledge. For example, productive thinking is part of designing scientific experiments to collect and analyze data to advance the scientific body of knowledge and generate an artwork for the sole purpose of conveying meaning (Duncker, 1945; Wertheimer, 1945). It is inherent in designing illustrations, technology, spaces, and organizations that empower people in their lives. This fundamental embodied cognitive activity of productive or creative thinking allows making sense of the world by enacting it.

1.3.6.2 Implications for design research

This semantic analysis identified and discussed several boundary dilemmas of defining Design. These dilemmas have several implications for research. Research requires clarity as otherwise examination and comparability are unfruitful. Referring to "Design" without expressing the underlying epistemological and disciplinary perspective will result in semantic ambiguity and unwelcome imprecision. We recommend design researchers employ a strategy we call "un-naming." Un-naming describes the activities, processes, and ways of looking at the world without referring just to an overarching term, such as Design. For example, researchers can refer to a specific professional practice to avoid the expert dilemma, outline their epistemological perspective to overcome the debate on the disciplinary perspective dilemma, and outline the specific context in which they examine "Design" to overcome the Zeitgeist dilemma. Such a strategy will help to demystify Design. Furthermore, future research should advance this preliminary study by investigating a broader spectrum of Design definitions to incorporate a greater diversity of diverse perspectives from different epochs, cultures, and situational contexts. Such investigation can identify additional boundary dilemmas and a more comprehensive understanding of the essential semantic meaning and family resemblance of Design, providing further clarity on the multifaceted meanings.

1.3.6.3 Implications for design practice

The research outlined in this chapter presents several implications for practice Design is a fundamental human activity in which professional designers bring unique abilities and domain-specific knowledge relevant for finding and solving specific problems and making sense in specific situations. Such practices will continuously advance with new emerging needs and challenges, new technological possibilities, and evolving cultural values. The disciplinary design practices developed over the last centuries may not serve as the necessary collaborative actions to address the grand challenges of our time, including environmental sustainability, social equity, and models of governance. Therefore, there is the need to continuously "design" new design practices and education programs to enable a culture of Design in which many people contribute to bringing about new and purposeful change. Otl Aicher (1991) expressed this evolution best: "Design transcends theory and practice and not only opens up a new reality but new insights." With each new reality and new insights, new ways can be found to enact, professionalize, and advance the practices of Design and its semantic meaning.

Acknowledgment

The authors would like to thank the wider design community at Stanford University and the valuable feedback from the reviewers.

References

Aicher, O. (1991). *Die Welt als Entwurf*. Berlin: Ernst & Sohn.

Amabile, T. M. (1996). *Creativity in Context: Update to the Social Psychology of Creativity*. Avalon Publishing.

Archer, B. L. (1965). *Systematic Method for Designers*. Boulder: Council of Industrial Design.

Arnheim, R. (1965). Visual Thinking. In G. Kepes (Ed.), *Education of Vision* (Vol. Vision + Value Series). New York: George Braziller.

Arnheim, R. (2004). *Visual Thinking*. Berkeley, CA: University of California Press.

Arnheim, R. (2009). *The Power of the Center – A Study of Composition in the Visual Art*. Berkeley, CA: University of California Press.

Auernhammer, J. M., & Roth, B. (2021). The Origin and Evolution of Stanford University's Design Thinking: From Product Design to Design Thinking in Innovation Management. *Journal of Product Innovation Management*, *38*, 623- 644. 10.1111/jpim.12594

Auernhammer, J. M., & Roth, B. (2022). Different types of Productive Thinking in Design: From Rational to Social Design Thinking. In C. Meinel & L. Leifer (Eds.), *Design Thinking Research: Achieving Real Innovation* (Vol. 13). Cham: Springer Nature.

Brown, T. (2009). *Change by Design: How Design Thinking Transforms Organizations and Inspires Innovation*. New York: HarperCollins.

Bruner, J. S., Goodnow, J. J., & Austin, G. A. (1956). *A Study of Thinking*. Oxford, UK: John Wiley and Sons.

Buchanan, R. (1990). Myth and Maturity: Toward a New Order in the Decade of Design. *Design Issues*, *6*(2), 70–80. doi:10.2307/1511439

Buchanan, R. (1992). Wicked Problems in Design Thinking. *Design Issues*, *8*(2), 5–21. doi:10.2307/1511637

Burton, R. M., Obel, B., & Håkonsson, D. D. (2015). *Organizational Design: A Step-by-Step Approach*. Cambridge University Press.

Card, S. K., Moran, T. P., & Newell, A. (1983). *The Psychology of Human-Computer Interaction*. New York: L. Erlbaum Associates.

Cross, N. (1982). Designerly ways of knowing. *Design Studies*, *3*(4), 221–227. doi:10.1016/0142-694X(82)90040-0
Cross, N. (2011). *Design Thinking: Understanding How Designers Think and Work*. Oxford, UK: Bloomsbury Publishing.
de Groot, A. D. (1965). *Thought and Choice in Chess*. The Hague: Mouton.
Dewey, J. (1938). *Logic The Theory of Inquiry*. New York: Henry Holt and Company.
Dewey, J., & Boydston, J. A. (2008). *The Later Works of John Dewey, 1925-1953: Essays, Reviews, Trotsky Inquiry, Miscellany, and Liberalism and Social Action*. Carbondale: Southern Illinois University Press.
Dreyfuss, H. (1955). *Designing for People*. New York: Simon and Schuster.
Duncker, K. (1945). *On Problem-solving*. Westport, CT: American Psychological Association.
Eastman, C. (1970). On the analysis of intuitive design processes. In G. T. Moore (Ed.), *Emerging methods in environmental design and planning*. Cambridge, MA: MIT Press.
Faste, R. A. (1994). Ambidextrous thinking. In *Innovations in Mechanical Engineering Curricula for the 1990s*. American Society of Mechanical Engineers.
Gero, J. S. (1990). Design Prototypes: A Knowledge Representation Schema for Design. *AI Magazine*, *11*, 26–36.
Goldschmidt, G. (2014). *Linkography Unfolding the Design Process*. Cambridge, MA: The MIT Press.
Goldschmidt, G., & Porter, W. L. (Eds.). (2004). *Design Representation*. London: Springer Verlag.
Gorb, P., & Dumas, A. (1987). Silent Design. *Design Studies*, *8*(3), 150–156. doi:10.1016/0142-694X(87)90037-8
Katz, B. M. (2015). *Make It New: A History of Silicon Valley Design*. Cambridge, MA: MIT Press.
Kaufman, J. C., & Beghetto, R. A. (2009). Beyond Big and Little: The Four C Model of Creativity. *Review of General Psychology*, *13*(1), 1–12. doi:10.1037/a0013688
Krippendorff, K. (2006). *The Semantic Turn: A New Foundation for Design*. Boca Raton: Taylor & Francis.
Kuhn, T. S. (2012). *The Structure of Scientific Revolutions: 50th Anniversary Edition*. Chicago: University of Chicago Press.
Lawson, B. (1972). *Problem Solving in Architectural Design*. (Ph.D.). Aston University, Birmingham.
Lawson, B. (2006). *How Designers Think: The Design Process Demystified*. Westfield, NJ: Elsevier/Architectural.
Lawson, B., & Dorst, K. (2009). *Design Expertise*. New York: Architectural Press.
Maldonado, T. (1965). Design Education. In G. Kepes (Ed.), *Education of Vision* (Vol. Vision + Value Series). New York: George Braziller.
Martin, R. L. (2009). *Design of Business: Why Design Thinking Is the Next Competitive Advantage*. Boston, MA: Harvard Business Review Press.
McKim, R. H. (1959). Designing for the Whole Man. In J. E. Arnold (Ed.), *Creative Engineering Seminar, 1959*. Stanford, CA: Stanford University.
McKim, R. H. (1980). *Experiences in Visual Thinking*. Boston, MA: PWS Engineering.
McKim, R. H.(1968). *Visual Thinking and the Design Process*. Mechanical Engineering. Stanford, CA: Stanford University.
Micklethwaite, P. H. (2002). *What Is Design?: An Empirical Investigation into Conceptions of Design in the Community of Design Stakeholders*. (Doctoral thesis). University of Huddersfield.
Moggridge, B. (2007). *Designing Interactions*. Cambridge, MA: MIT Press.
Moholy-Nagy, L. (1969). *Vision in Motion*. New York: Winterborn and Co.
Mullet, K., & Sano, D. (1995). *Designing Visual Interfaces: Communication Oriented Techniques*. SunSoft Press.
Nelson, G. (1965). *Problems of Design*. New York: Whitney Library of Design.
Norman, D. A. (2007). *Emotional Design: Why We Love (or Hate) Everyday Things*. Cambridge, MA: Basic Books.
Norman, D. A., & Draper, S. W. (1986). *User Centered System Design: New Perspectives on Human-computer Interaction*. Hilsdale, NJ: Taylor & Francis.

Owen, C. L. (1998). Design Research: Building the Knowledge Base. *Design Studies, 19*(1), 9–20. doi:10.1016/S0142-694X(97)00030-6

Papanek, V. (1973). *Design for the Real World*. Bantam Books.

Piaget, J. (1981). *Intelligence and Affectivity: Their Relationship during Child Development*. Oxford, UK: Annual Reviews.

Protzen, J. P., & Harris, D. J. (2010). *The Universe of Design: Horst Rittel's Theories of Design and Planning*. London, UK: Taylor & Francis.

Purcell, A. T., & Gero, J. S. (1998). Drawings and the design process: A review of protocol studies in design and other disciplines and related research in cognitive psychology. *Design Studies, 19*(4), 389–430. doi:10.1016/S0142-694X(98)00015-5

Pólya, G. (1957). *How to Solve it: A New Aspect of Mathematical Method*. Doubleday.

Rosch, E., & Mervis, C. B. (1975). Family Resemblances: Studies in the Internal Structure of Categories. *Cognitive Psychology, 7*(4), 573–605. doi:10.1016/0010-0285(75)90024-9

Rowe, P. G. (1987). *Design Thinking*. Cambridge, MA: MIT Press.

Schön, D. A. (1963). *Displacement of Concepts*. London: Tavistock Publications.

Selz, O. (1922). *Über die Gesetze des geordneten Denkverlaufs: Zur Psychologie des Produktiven Denkens und des Irrtums*. Bonn: F. Cohen.

Simon, H. A. (1969). *The sciences of the artificial*. Cambridge, MA: MIT Press.

Smith, A. E., & Humphreys, M. S. (2006). Evaluation of Unsupervised Semantic Mapping of Natural Language with Leximancer Concept Mapping. *Behavior Research Methods, 38*(2), 262–279. doi:10.3758/BF03192778

Suwa, M., & Tversky, B. (1996). *What Architects See in Their Sketches: Implications for Design Tools*. Paper presented at the Conference Companion on Human Factors in Computing Systems, Vancouver, Canada. 10.1145/257089.257255

Tversky, B. (2009). Spatial Cognition: Embodied and Situated. In *The Cambridge Handbook of Situated Cognition* (pp. 201–216). New York: Cambridge University Press.

Verganti, R. (2009). *Design-Driven Innovation: Changing the Rules of Competition by Radically Innovating What Things Mean*. Boston, MA: Harvard Business Press.

Wallas, G. (1926). *The Art of Thought*. London: J. Cape.

Wertheimer, M. (1945). *Productive Thinking*. New York: Harper.

Wittgenstein, L. (1953). *Philosophical Investigations*. Oxford, UK: Basil Blackwell.

1.4 What design is not

Newton D'souza[a] *and Tilanka Chandrasekera*[b]
[a]*Florida International University;* [b]*Oklahoma State University*

1.4.1 Introduction

The design discipline has experienced brisk changes in the past 35 years, evolving from a singular, technology-aided, trial and error–based esoteric discipline to a more collaborative, technology-intense, evidence-based, and publicly accessible one. These changes have blurred the disciplinary boundaries of design. In critically examining such blurring, it might be prudent to pose the question, "What is not design?" in the hopes that such a negative question might elicit disciplinary clarity to traditional questions of "What is design?" or "What ought design be?"

Such an approach, however, is daunting given that design discipline has been shaped by different perspectives and philosophies emerging out of an essentially multidisciplinary domain. Additionally, *design* has been referred to as a transitive verb (to create or plan), a noun (a scheme or an idea), a process (activities of design), and a discipline. Traditionally, design activity has been described in numerous ways, ranging from design as creative problem solving (Chen & Liu, 2002), problem framing (Cross, 2001), puzzle making (Archea, 1987), abductive reasoning (March, 1984), holographic thinking (D'souza, 2007), and reflection-in-action (Schon, 1991) to a very complicated act of faith (Jones, 1970). Design behavior is described as a form of unique inquiry that is characterized by the production of novel and unexpected solutions, tolerance of uncertainty, access to incomplete information, application of imagination, constructive forethought to practical problems, and the use of drawings and modeling media for problem-solving (Cross, 2001). As a unique form of intelligence, design has been regarded as different from scientific or scholarly thinking styles but just as powerful, encompassing "designerly ways of knowing, thinking, and acting" (Cross, 1982). Design is hence distinguished from other ways of thinking such as in accounting, finance, statistics, software engineering, and medicine (Chan, 2015).

Design problems are also considered unique. They are often referred to as "wicked" (i.e., vicious, tricky, and aggressive) as opposed to "tame" problems consisting of straightforward solutions. Several factors contribute to design's wicked quality, including indefinite formulation of the problem, absence of a stopping rule, and lack of true or false solutions (Rittel & Webber, 1973). The co-evolution of the design problem and solution in design activity may also factor in, making the problem and solution interdependent (Dorst & Cross, 2001).

While the argument toward design uniqueness has been prevalent in design research, the notion that design is a special kind of knowledge accessible only to people trained in design has been challenged. Rather, design is considered a generalized process that even

DOI: 10.1201/b22630-6

non-designers are capable of performing to some extent. For example, designs that are both highly practical and very beautiful have been attributed to the common folk of craft-based societies (Cross, 1990). Design is hence considered a generalized body of knowledge capable of bridging diverse disciplines. The multidisciplinary view of design has gained currency in fields such as architecture, in that building design and construction have increased in complexity, involving multiple consultants, and designers are often expected to accomplish their work in shorter time frames. The challenge of integrative project leadership has left designers to don the role of the traditional master-builder or an "integrator" of various skills and knowledge (Barrow, 2000).

The lead-up to discussions on what is not design is also a time to reflect on what constitutes "design proper." Instead of the "hard dry ground" of an architect's job, design is viewed as ambiguous, a "swamp" in which boundaries are less delineated (Farbstein & Kantrowitz, 1991). Others have described the design as a continuum of activities starting with pre-design (e.g., scenario planning) and ending with post-design activities (e.g., post-occupancy evaluation; Schermer, 2001).

1.4.2 Methodology

To elicit more clarity in the disciplinary boundaries of design, a semantic analysis was conducted to investigate the question "What is not design?" Initially, three prominent design journals from the past 35 years (1985–2020) were considered a subject of the study. However, as the study progressed, the majority of research yield was found in one specific source: the *Journal of Design Studies*. Hence, the inquiry was limited to that journal, which covers a relatively long history on design cognition and addresses design activity across multiple disciplines including engineering and product design, architectural and urban design, computer artifacts, and systems design. Therefore, it provides an interdisciplinary forum for analysis, development, and discussion of fundamental aspects of design activity that are both applied and philosophical. The semantic analysis procedure is as follows:

- Initial data were elicited from the journal database using a keyword search. Keywords consisted of declarative clauses that imply negative connotations of design (e.g., *not design, non-design, design is not, isn't design, how not to design, design can't do*); phrases or idioms that imply how design is expected not to perform (*short of design, unlike design, lack of design, design is more than*); and prefixes that imply activities that occur before or after design proper (*pre-design, post-design*).
- A total of 603 papers were retrieved from the initial search. The papers were then manually coded for relevance. One of the challenges in this process was to distinguish between the conventional use of negative words ("*hope not*") and the use of negative words in a design context ("*design is not*").
- Once the keywords were sorted out for relevance, the individual papers were open coded to reflect the context in which the negative words were used. Open codes consisted of categorization such as "*design is not problem solving*." The idea was to capture *design* as the subject and the corresponding activity *problem solving* as the object, modified by a negative *not*.
- Further thematic coding was conducted on the basis of commonality between the open codes. For example, papers that were coded as "design is not problem solving" and "design is not merely craft skills but also problem framing" were grouped under a common thematic code such as "problem/solution."

- This first cycle of coding resulted in 188 open codes and 21 thematic codes.
- They were independently reviewed by two coders to check intercoder reliability. The initial intercoder agreement was about 80%.
- The second cycle of coding resulted in 52 open codes and 10 thematic codes. The 10 thematic codes are listed below and address the nature, product, and design activities.
- Design is *not science* (although they share similar values).
- Design is *more than pure artistic creativity* (it should address a problem).
- Design is m*ore than problem solving* (and a unique one).
- Design is m*ore than a mechanistic process* (it is also a social one).
- Design is n*ot merely a technical process* (it cannot be pinned down).

Products of design

- The goal of design is *not to produce knowledge* (although knowledge is produced in some form).
- The product of design is *not the artifact itself.*

Design activities

- Non-design backgrounds
- Non-design contexts
- Non-design, pre-design and post-design Tasks

1.4.3 Disciplinary and chronological trends

Further analysis was conducted by organizing the codes in chronological and disciplinary categories. The question of "what design is not" is evident in seven different design and allied disciplines: architecture, computing, product/industrial design, engineering, philosophy, psychology, and technology. In the context of this chapter, disciplines were identified on the basis of the subject focus of the paper and the authors' expertise/affiliation. In cases of authors representing multiple disciplines, the codes were designated to the affiliation of the first author.

Figure 1.4.1 shows thematic discussions related to discipline on what is not considered design. The year of publication is plotted on the y-axis, while the thematic codes are plotted on the x-axis. The majority of disciplinary discussions seemed to be focused on why design is not problem solving or of technical nature. Specifically, discussions on why design should not be considered a problem-solving activity were most prevalent in the discipline of computing. This was also demonstrated through statistical significance conducted via a one-way analysis of variance, $F(10, 51) = 2.17$, $p = .035$, when the computing discipline was compared with all other disciplines. These findings imply an internal debate within the computing discipline to question its traditional affiliation with problem solving. Figure 1.4.1 also shows an evolution from disciplines' historical debates about science, art, and problem solving to more current debates on design artifacts, products, and social role.

Figure 1.4.2 shows disciplinary discussions of specific themes related to specific disciplines. The discussions on what is not considered design mostly occurred in three disciplines (architecture, product/industrial design, and technology). Although no statistical

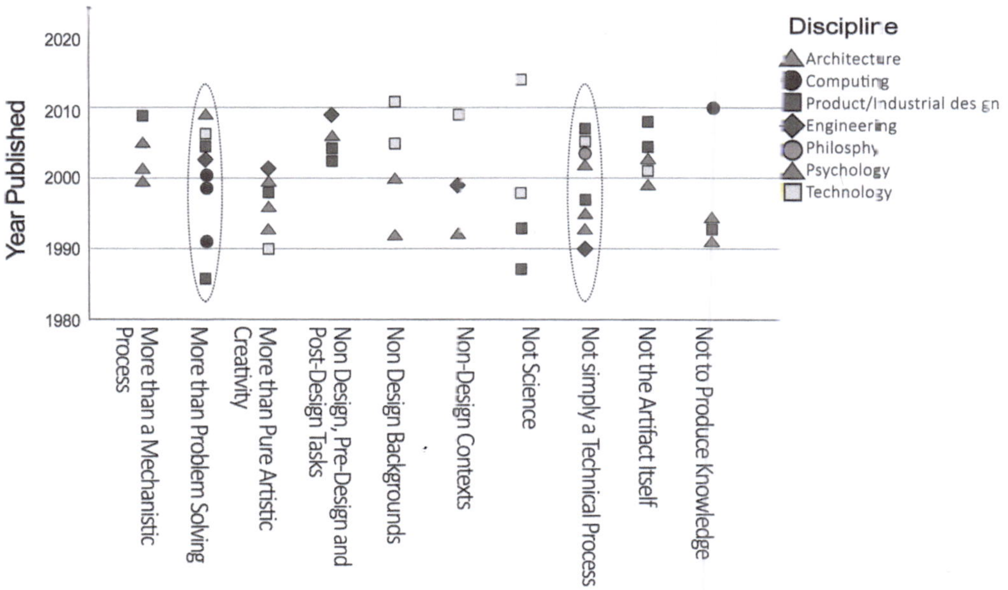

Figure 1.4.1 Thematic discussions within discipline.

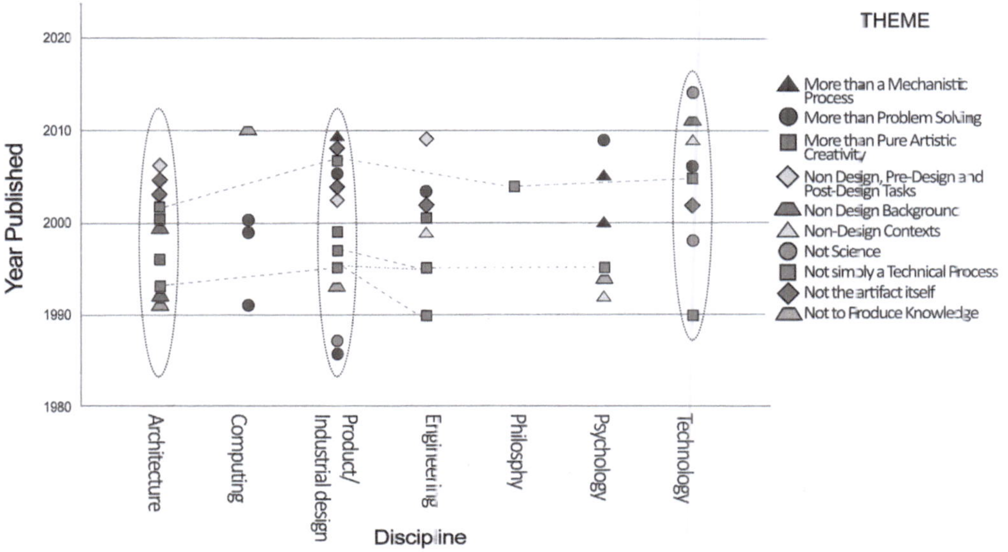

Figure 1.4.2 Disciplinary discussions within themes.

significance was found between the year of publication and themes, the papers show more robust discussions in recent years. The findings also show there was no cross-disciplinary discussions on any themes except for why design should not be considered merely a technical process. This was a little surprising since the design research discourse has been attempting to generate more cross-disciplinary discussions on the nature of design, as evident in the Design Thinking Research Symposium forums.

1.4.4 Nature of design

1.4.4.1 Design is not science (although they share similar values)

A number of authors have advanced the view that design is not the same as science although similarities exist. Science is considered analytic, whereas design is considered constructive. In the discipline of mechanical engineering, for example, engineering science is located on the analytic side, whereas engineering design is placed on the synthetic side (Owen, 1998). The scientific method is considered problem-solving behavior used to determine the nature of what exists, whereas design behavior is thought to invent things of value that do not yet exist (Cross et al., 1981).

Whereas scientists strive to isolate the superior option, architects narrow down their options by satisfying the design goals as much as they can (Akin, 2001). For scientists to know truth, scientific theory would be to put to the most stringent tests, requiring scientists to vigorously become an antagonist to their own theories. On the other hand, the truth the architects seek is temporal-context specific.

In advancing the science versus design distinction, the mental activities of a scientist are characterized by a "mind-to-world" direction of fit; in contrast, a designer's mental activities are dominated by a "world-to-mind" direction of fit (Heylighen et al., 2009). In general, it might be sufficient for a designer to propose a novel idea without actually designing the artifact, whereas the scientist must produce an artifact such as a scientific theory (Galle & Kroes, 2014). Design is considered an action process aimed at predetermined goals, whereas science is considered to create knowledge about something. In this sense, design is considered a service discipline, unlike science (Jonas, 1993).

However, more conciliatory views on design and science can also be found. Much like science, design as a discipline can be studied on its own terms and within its own rigorous culture. A science of design is possible based on the reflective practice of design. The design process is therefore inherently fluid, because it is not blinkered by the quest to achieve one design goal; rather, it is open to new goals discovered accidentally and unintentionally (Meng, 2009).

1.4.4.2 Design is more than pure artistic creativity (it should address a problem)

While the previous section considers design different from science, design is sometimes distanced from pure art. In this view, design is intentional, hypothetical, and predictive, transcending creative intuition, and includes an explicit political dimension. In other words, design activities are considered not merely impulsive, habitual, or coincidental, but rather conscious, selective, and intelligent (Uluoğlu, 2000). Design involves choosing a range of options to allocate material resources despite possible conflicts between the short- and long-term requirements (Lawrence, 1993).

Distinct from art, creative design is not necessarily considered "good" design in terms of quality. A designer aims to achieve a high-quality design, with creativity as only one aspect. In fact, creativity is not a condition unique to design because scientific and artistic processes are considered creative in nature (Alexiou et al., 2009).

Although the creation of design is considered simple, it is not considered a simplistic act. For example, in the case of wayfinding design, logic and efficiency are as important as making design products architecturally interesting (Passini, 1996). Creative design involves problem-framing ability (Dorst & Cross, 2001), and creative designers attempt more problem-finding segments (Liu, 2000). Contrary to the view that beginning designers are more creative, some researchers have observed that expert designers solve design problems in a more creative manner (Hanna & Barber, 2001).

Others consider design as not just art, engineering, or science, but as integrative discipline. It is this "generalizm" that is considered to make a good designer valuable (Owen, 1991). There is also the debate within the design disciplines about which disciplines are more artistically inclined than others. Some refer to architecture and industrial design as more artistic than engineering design, the latter being relatively constrained by the needs of customers and company shareholders (Eder, 1991).

1.4.4.3 Design is more than problem solving (and a unique one)

Beyond mere problem solving, in design there is an "affective" requirement characterized by a leap of faith placing the designer in a creative situation and thereby the designer deriving vicarious satisfaction about the creation (Louridas, 1999). Additionally, design is not just about the creation of useful artifacts. It is equally involved with the creation of beautiful artifacts that intertwine utility and aesthetics (Louridas, 1999).

Beyond mere problem solving, some authors suggest that the job of a designer is to give the client not simply what they want, but what they never dreamed they wanted; and when they get it, they recognize it as something they wanted all the time. In other words, clients want designers to transcend the obvious and the mundane and to produce proposals that are exciting and stimulating. What this means is that design is not only a search for the optimum solution to the given problem, but also exploratory (Cross, 1999) Similarly, others consider that evaluation of the solution is equally important to analysis of the problem. Hence, a designer is considered to be solution-led, not problem-led. Designers formulate and solve problems differently from other professions. In some cases, they generate conjectures about possible (partial) solutions and use these conjectures to explore and define the problem and the solution together (Restrepo & Christiaans, 2004). A problem-solving paradigm is apt when the problem is fairly clear-cut, but that reflection-in-action is better suited to conceptual design. The process is one of a designer interacting with an external representation and performing a series of experiments, the results of which influence further experiments (Smith & Gero, 2005). Design processes are considered not orthodox problem solving in which problem statements reveal solution paths; rather, they are a way to resolve conflicting criteria to validate results and multiple solutions (Bijl, 1985).

Creative design is not a matter of first fixing the problem and then searching for a satisfactory solution. It is a matter of concurrently developing and refining both the formulation of a problem and ideas for a solution, with constant iteration of analysis, synthesis, and evaluation through a "co-evolution" of the problem and the solution (Dorst, 2004). All designing involves some conjecture. But like medical research, the texture of design is not literally trial and error, nor is it merely unremitting guesswork (Bamford, 2002).

While design problems are generally considered ill structured, further distinctions from traditional problem-solving can be observed. The first difference is that in design, imagination needs to be applied right at the fundamental level of concept interpretation. A second difference is that design requires the use of experiments and simulation techniques. Third, designing is fundamentally a social process. The stakeholders need to develop a way of reaching a solution, and this cannot exist before the design situation arises (Dorst & Royakkers, 2006).

The distinction between design and problem solving is also made in studies of cognition and brain. Compared with problem-solving, studying design tasks require a more extensive network of brain areas. Although a problem-solving task may require creative thinking and hypothesis formation, the problem itself is well defined. In design, there is no predetermined final state or criterion for deciding the termination of a task (i.e., it is open-ended). The task requires defining the problem as well as the solution space: It requires creation and interpretation, as well as the creation of a function for evaluating the solution (Alexiou et al., 2009).

1.4.4.4 Design is more than a mechanistic process (it is a social one)

Some authors believe that design is not a mechanistic process that can be fully described in a manual, but a complex and elaborate socially mediated activity (Baird et al., 2000). Even when engineers conduct design, it is not mechanical or rational in the manager's or economist's sense. Instead, design is a social process (Ball & Ormerod, 2000). Design is not only individual action but has to be implemented on an organizational level to be effective. Good design practice requires more than individual strategies and methodology (Lauche, 2005). The conception of a design is not simply a representation in visual form of predetermined values, but a creative, catalytic process in which external factors interact with the beliefs, talents, and skills of individual design groups and shaped by the designers' own cultural and societal values (Razzaghi et al., 2009). Descriptive studies involving design teams demonstrate that design is not only a complex technical process but also a complex social process. Modern multidisciplinary design demands that engineers work in teams, and to be successful the team has to reach some shared or commonly held understanding. Thus, the design process needs to be addressed as an integration of technical, cognitive, and social processes (Macmillan et al., 2001).

1.4.4.5 Design is not merely a technical process (it cannot be pinned down)

Several papers consider design as more than a technical process and thus unpredictable. Design is not straightforward (Tilley & Burrows, 1995) and cannot be pinned down like butterflies (Margolin, 1992). Design is not a set of technical skills, but a process comprising unique ways of knowing and thinking (Friedman, 2007). Neither is it a rule-based or symbol-processing activity (Coyne & Snodgrass, 1993), nor can it be reduced entirely to a set sequence (Dorst & Dijkhuis, 1995). A design is not an end in itself, but an input for the next step, which may be further detailing of the design (van Aken, 2005).

Design is also not considered a single process, but a mix of activities that range from individual work to meetings and reviews in groups. Instead of the idea that designing is one "thing," design consists of many ways of thinking (Lloyd et al., 1995). Generally, many iterations and cross-linkages take place throughout the process and certainly at the front end (Turner, 1991). Furthermore, several design tasks can be carried out in different ways, using alternative technologies. The various disciplines engaged in design differ in terms of the types of problems they address and the products they design, and the design process feeds on a continuing interplay between experimentation and what the design process discloses to our understanding (Akin, 1997). Design operates not on any single episode of design, but on the dynamics of how the problem responds to changes in its articulation by the designer (Newton, 2004).

Additionally, design is not just a process or a profession. It is experienced as the situation in which designers find themselves. As a designer, one is continually faced with the very concrete challenge of a perceived design problem and must decide on the kind and content of the action to take in specific situations. In most cases, considerations linked to the content of the design situation (the perceived design problem, the designer's goals, and the perceived possibilities for the next step) will determine the "kind of action" (Dorst & Dikhuis, 1995).

1.4.5 Products of design

1.4.5.1 The goal of design is not to produce knowledge (although knowledge is produced in some form)

There have been rigorous debates on the ultimate goal of design. For example, much different from the goal of science, which is to create knowledge, some consider that the goal of design is to change the environment. Knowledge is cumulative, and although it tends to produce higher levels of refinement and sophistication in design, the principal function of design is not to produce knowledge but to have an effect. It is through design that science exceeds being pure knowledge and participates in creating an effect (Willem, 1990). Similarly, design decisions are largely influenced by experience, creativity, innovation, and other factors, and it is not always possible to utilize these factors in a formalized knowledge base (Beheshti, 1993).

While knowledge on "the way the world is" enters design decision-making at many crucial points and many different guises, in the early stages of the design process, when many critical decisions are made, designers are in no position to collect much relevant information anyway. Often all they have to work with are environments, paper designs, mock-ups and prototypes. While these yield useful information, that usefulness is directly proportional to the designers' ability to intuit potential scenarios by bringing their experience and cultural knowledge to bear upon the problem. In this case, knowledge is bound to social processes much like law and medicine (Sharrock & Anderson, 1994).

While the goal of design is not to produce knowledge, designers have to accumulate the knowledge to make effective design. While the unselfconscious designer of the past could rely on tradition and afford not to have an explicit knowledge base, the contemporary designer has to have adequate design qualifications and be able to distinguish a design from the work of all other designers (Louridas, 1999).

1.4.5.2 The product of design is not the artifact itself

Whereas debates have been presented on what might be the ultimate product of design, several authors suggest that the true product of design is not the designed artifact itself, but rather the change the artifact brings about in the world and the experience it eventually provides to its users (Jackson, 2010). The philosophical argument is that the subject area of design cannot be the actual artifacts themselves because they do not exist yet. Instead of a whole artifact, it is the artifact structure and their parts from which design is constructed. Design artifacts acquire their meaning only from the structures they describe. Design is not about artifacts at all but is merely play with symbols and patterns (Galle & Kroes, 2014).

Additionally, not all used objects are considered artifacts. If an item has been designed but does not have a function yet, it cannot be called an artifact. Artifacts are action-theoretical rather than metaphysical. They have intentional history and role. This role may be different for different agents, and one object may even play different roles at the same time. This distinguishes them from non-designed objects, such as water and air, because the artifacts' use and function are intelligible to users (Houkes et al., 2002).

Moreover, the primary concern in design is not about perfection of form and aesthetics, but rather how design opens up for use. While the object itself is still quite central, notions of form and the "finished-ness" of the object also play an important role. Roughness and improvisation must be accepted as part of the design process (Redström, 2008). In other words, the object of design is not a real object; it does not exist yet in process. What does exist are charts, acronyms, sketches, diagrams, models, mock-ups, product lines, specifications, and the like. The emergent meaning is both referential and inferential: Objects (and concepts) acquire meaning by association with other objects (and/or concepts). The mock-up itself may be minimalist in its rendering in material form and yet can be a vehicle for wide-ranging, energetic, and meaningful exchange (Bucciarelli, 2002).

Similarly, design products such as drawings are not as critical as the idea that the drawings represent (Medway & Clark, 2003). Additionally, design is not just a copy of existing designs. These still need to be modified to satisfy aesthetic and functional requirements (Lim et al., 2004). Designers use the medium of representing massing elements, and in particular, elements of regulation to manage the design process itself. This includes the management of part-whole relationships, hierarchy, topology-geometry relationships, scaffolding the design process, and structuring/restructuring of problem parameters. Each of these is a separate facet of the same phenomenon of managing the design process through massing representations (Akin & Moustapha, 2004).

1.4.6 Design activities

1.4.6.1 Non-design backgrounds

Differing views have been presented on whether a distinction needs to be made between design and non-design professions. Some consider even proximate professions such as arts as hailing from a non-design background (Chung & Whitfield, 1999). Others consider professions with technical skills in design part of the design discipline (e.g., arts, crafts, advertising, interior design) in relation to disciplines with non-design backgrounds (e.g., toolmaking, electronics, mechanical engineering; Yang et al., 2005). Some question whether there is an empirical basis for such a distinction. In one experimental study, senior design students produced more creative forms during a mental synthesis task than non-designers (law discipline) who focused on verbal concepts (Kokotovich & Purcell, 2000).

The concept of design thinking has been adopted to infuse the iterative and ambiguous nature of professions such as business management and entrepreneurial business innovation. There has also been growing interest in design thinking from sectors such as education, health and information technology, environmental sustainability, welfare, and social justice. These developments have caused concerns among design professionals because they suspect an uncritical deployment of the term *design thinking* in contexts other than design as a fad, instead of being a real opportunity to explore and extend the possibilities of design in other sectors (Stewart, 2011).

1.4.6.2 Non-design contexts

Design is not purposeful without the environment and context within which it operates. Hence, some authors suggest a more wide-ranging view of design problem-solving that attempts to foster recognition of the importance of contextualization in design cognition. They claim that in contrast to software design, design behavior cannot be isolated from its context and the need to remain aware of the social, educational, and organizational factors that mediate design behavior (Davies & Castell, 1992). Another study makes a distinction between non-design problems such as computer-simulated scenarios (e.g., operating an engineering plant and firefighting exercise) to a design problem (e.g., modifying height and inclination of a writing table). In contrast to the design problem, they claim that the computer-simulated complex problems do not require specialized knowledge (Eisentraut, 1999). Although designers do not always consciously use environmental factors in the process of designing, they have a very good mental model of the operating environment and use it to develop a workable proposition. After the design is completed, environmental context is used as a measurable factor for evaluation and optimization (Hybs & Gero, 1992).

1.4.6.3 Non-design, pre-design, and post-design tasks

A number of authors have distinguished between tasks that are aligned to design and those that are not. Non-design activities are considered to not involve design. Design activities include problem definition, gathering information, developing alternative solutions stage, generating ideas, modeling, feasibility analysis, evaluation, project realization, decision, and communication (Atman et al., 2005). This is opposed to a non-design task that might include a linguistic interpretation and explanation of noun phrases (Nagai et al., 2009). Others view that research conducted in the social sciences are more aligned with non-design problems (Yang, 2010).

There are those who also make a further distinction between design and activities that precede or follow design proper, so-called pre-design and post-design activities. The pre-design phase generally includes reasoning or organizing activities, whereas the post-design phase includes evaluation tasks (Eilouti, 2009). The pre-design reasoning activity includes precedent case retrieval, organization of information, the representation of the extracted information into more accessible forms, and the prototyping of its representations for future design solutions. Post-design includes the comparison of a generated solution to other precedents that can be considered successful solutions and the performance simulation of proposed scenarios of the generated design (Eilouti, 2009). The design proper includes mapping of a precedent to a design problem, modification of the solutions, adaptation of the revealed sub-solutions, their synthesis and development, and the documentation of the selected solutions.

Some papers refer to pre-design activities as *front-end activities*. Subsequently, design activities are classified as pre-phase (opportunity identification and idea generation), phase zero (product concept and definition, including market and technology assessment), and phase one (product definition, justification, and planning; Tzortzopoulos et al., 2006). Others consider requirement management as a part of the pre-design services (Duerk, 1993). Requirement-driven design is an information-based approach facilitating consistent design rationale tracking and evaluation, verification, and validation of design (Ozkaya & Akin, 2006).

The American Institute of Architects divides the design process into pre-design services, schematic design, design development, construction documents, construction administration, and post-design services (Ozkaya & Akin, 2006). During the pre-design phase, architects and other professionals on the planning team generally employ several interactive methods and techniques to collect information on the internal and external forces that affect a given project. Generally speaking, the planning team may conduct environmental evaluations and assessments, collect data, research best practices, interview different user groups, establish goals, and generate a list of technology and equipment needs. During this inquiry process, objectives are established, the context of the design problem is identified, and the performance requirements are defined (Maier et al., 2009).

Pre-design is also sometimes used synonymously with research. Pre-design is concerned with determining the requirements for organization and planning, thus leading to a project solution, and creating design guidelines (Johannes, 1992). One study claims that the success rate of design based on clear and well-defined specifications emerging from pre-design research is 3.3 times greater than a design without it (Chang, 2003).

In software design, some utilize pre-designed solutions of design elements such as libraries of borders. As the software matures through the release of progressive versions, changes are often made that are direct results of feedback from professionals who have used the products. These progressive changes are often developed from wish lists compiled by users forming the genetic building blocks, or memes, that will carry the genetics of the software program (Cleveland, 2004).

1.4.7 Conclusion

To understand "what design is" or "what design ought to be," this chapter examined the negative question "what design is not?" Researchers in specific disciplines have internally questioned what design means in their profession, although a multidisciplinary meta-analysis has seldom been undertaken. The summary of semantic analysis of negative connotations referring to design demonstrates that design has been traditionally debated in dialectical terms (Section 1.4.3): scientific exercise versus artistic exercise; general versus unique problems; mechanistic versus social; technically certain versus unpredictable; knowledge production versus effect; production of artifacts versus change artifacts brings; and design versus non-design activities. Such opposing views seem to align with the validity of different and seemingly contradictory views of what counts as design (Eckert et al., 2022). This constant state of flux seems to define the nature of design discipline in which the semantic core of design describes different ways of designerly knowing and the periphery of design presents several dilemmas (Aurenhammer & Ford, 2022).

Historically, the debates on the nature of design have focused on why design is not merely science, artistic creativity, or problem solving. The more recent discussions have hovered around the products of design and knowledge generation. Considering design an instrumental profession, there seem to be mixed debates on whether the design needs to result in a knowledge base or aim only for change, whether and which artifacts design should produce, and to what extent it should be a service-based profession. Most importantly, these debates reflect on the ethical role of the profession—a reflection on more empathic design (Metcalfe, 2022) and a professional duty to care (Guersenzvaig & Ventura, 2022).

As much as there are intellectual debates about the true nature of design and what is meant by a design task or activities, the recent push toward participatory and collaborative designing has further blurred design activities. With the advent of co-designing, research, and pre-/post-design activities, the profession is looking to further define who can actually function as a designer and whether or not professionals and public as a whole from non-design backgrounds can be productive partners (Ou et al., 2022).

These debates also extend to machine and technology and what role they play in pushing the boundaries of design (Sciannamе, 2022; Ostrowski et al., 2022). These debates including the role of design in social justice is a point for due consideration (Das et al., 2022; Dos Santos et al., 2022). More robust discussions on inclusivity and exclusivity of the profession will take the design profession into a more philosophical and moral reckoning in the near future.

References

Akin, Ö. (1997). Editorial. *Design Studies, 18*(4), 323–328.

Akin, Ö. (2001). *Simon says: Design is representation* [Unpublished manuscript, pdf]. School of Architecture, Carnegie Mellon University, Pittsburgh, PA. Retrieved from http://www.andrew.cmu.edu/user/oa04/Papers/AradSimon.

Akin, Ö., & Moustapha, H. (2004). Strategic use of representation in architectural massing. *Design Studies, 25*(1), 31–50.

Alexiou, K., Zamenopoulos, T., Johnson, J. H., & Gilbert, S. J. (2009). Exploring the neurological basis of design cognition using brain imaging: Some preliminary results. *Design Studies, 30*(6), 623–647.

Archea, J. (1987). Puzzle-making: What architects do when no one is looking. In Y. Kalay (Ed.), *Computability of design, principles of computer-aided design* (pp. 37–52).

Atman, C. J., Cardella, M. E., Turns, J., & Adams, R. (2005). Comparing freshman and senior engineering design processes: An in-depth follow-up study. *Design Studies, 26*(4), 325–357.

Aurenhammer, J. & Ford, C. (2022). What is design? The semantic core and periphery of design definition. In G. Goldschmidt & E. Tarazi (Eds.), *Proceedings of 13th Design Thinking Research Symposium (DTRS 13), Expanding the frontiers of design: A blessing or a curse* (pp. 24–37). Design Research Society.

Baird, F., Moore, C. J., & Jagodzinski, A. P. (2000). An ethnographic study of engineering design teams at Rolls-Royce Aerospace. *Design Studies, 21*(4), 333–355.

Ball, L. J., & Ormerod, T. C. (2000). Applying ethnography in the analysis and support of expertise in engineering design. *Design Studies, 21*(4), 403–421.

Bamford, G. (2002). From analysis/synthesis to conjecture/analysis: A review of Karl Popper's influence on design methodology in architecture. *Design Studies, 23*(3), 245–261.

Barrow, L. R. (2000). *Cybernetic architecture: Process and form. The impact of information technology*. Harvard University.

Beheshti, R. (1993). Design decisions and uncertainty. *Design Studies, 14*(1), 85–95.

Bijl, A. (1985). An approach to design theory. In *W.G.5.2 Working Conference on Design Theory for CAD* (pp. 1–23).

Bucciarelli, L. L. (2002). Between thought and object in engineering design. *Design Studies, 23*(3), 219–231.

Chan, C. S. (2015). *Style and creativity in design: Studies in applied philosophy, epistemology and rational ethics*. Springer International Publishing.

Chang, W. C. (2003). Researching design trends for the redesign of product form. *Design Studies, 24*(2), 173–180.

Chen, S. C., & Liu, Y. T. (2002). A comparative study of protocol analysis in traditional media, computer media and network environment. In K. Koszewski & S. Wrona (Eds.), *Connecting the real and the virtual: Proceedings of eCAADe20* (pp. 186–193).

Chung, S., & Whitfield, A. (1999). A comparison of the social standing of the design professions in Korea and Australia. *Design Studies, 20*(4), 381–396.

Cleveland, P. (2004). Bound to technology—The telltale signs in print. *Design Studies, 25*(2), 113–153.

Coyne, R., & Snodgrass, A. (1991). Is designing mysterious? Challenging the dual knowledge thesis. *Design Studies, 12*(3), 124–131.

Coyne, R., & Snodgrass, A. (1993). Rescuing CAD from rationalism. *Design Studies, 14*(2), 100–123.

Cross, N. (1982). Designerly ways of knowing. *Design Studies, 3*(4), 221–227.

Cross, N. (1990). The nature and nurture of design ability. *Design Studies, 11*(3), 127–140.

Cross, N. (1999). Natural intelligence in design. *Design Studies*, *20*(1), 25–39.
Cross, N. (2001). Design cognition: Results from protocol and other empirical studies of design activity. In C. M. Eastman, W. M. McCracken, & W. C. Newstetter (Eds.), *Design knowing and learning: Cognition in design education* (pp. 79–103). Elsevier.
Cross, N., Naughton, J., & Walker, D. (1981). Design method and scientific method. *Design Studies*, *2*(4), 195–201.
Das, M., Ostrowski, A. K., Ben-David, S., Roeder, G. J., Kimura, K., D'Ignazio, C., & Verma, A. (2022). Auditing design justice: The impact of social movements on design pedagogy at a technology institution. *Expanding the frontiers of design: A blessing or a curse?*, 121.
Davies, S. P., & Castell, A. M. (1992). Contextualizing design: Narratives and rationalization in empirical studies of software design. *Design Studies*, *13*(4), 379–392.
Dorst, K. (2004). On the problem of design problems-problem solving and design expertise. *Journal of Design Research*, *4*(2), 185–196.
Dorst, K., & Cross, N. (2001). Creativity in the design process: Co-evolution of problem–solution. *Design Studies*, *22*(5), 425–437.
Dorst, K., & Dijkhuis, J. (1995). Comparing paradigms for describing design activity. *Design Studies*, *16*(2), 261–274.
Dorst, K., & Royakkers, L. (2006). The design analogy: A model for moral problem solving. *Design Studies*, *27*(6), 633–656.
Dos Santos, J. A., Moreira, J., & Tschimmel, K. (2022). Design thinking workshops for social workers. In G. Goldschmidt & E. Tarazi (Eds.), *Proceedings of 13th Design Thinking Research Symposium (DTRS 13), Expanding the frontiers of design: A blessing or a curse* (pp. 175–192). Design Research Society.
D'souza, N. (2007). Describing differences between manual and digital modeling: A holographic approach to architectural design. *Proceedings of the Design Communication Association, DCA 2007* (pp. 325–338). Design Communication Association.
Duerk, D. P. (1993). *Architectural programming: Information management for design*. Van Nostrand Reinhold Company.
Eckert, C., McMahon, C. & Stacey, M. (2022). What counts as design? In G. Goldschmidt & E. Tarazi (Eds.), *Proceedings of 13th Design Thinking Research Symposium (DTRS 13), Expanding the frontiers of design: A blessing or a curse* (pp. 38–55). Design Research Society.
Eilouti, B. H. (2009). Design knowledge recycling using precedent-based analysis and synthesis models. *Design Studies*, *30*(4), 340–368.
Eisentraut, R. (1999). Styles of problem solving and their influence on the design process. *Design Studies*, *20*(5), 431–437.
Eder, W. E. (1991). Editorial. *Design Studies*, *12*(4), 194–196.
Farbstein, J., & Kantrowitz, M. (1991). Design research in the swamp: Toward a new paradigm. In E. H. Zube & G. T. Moore (Eds.), *Advances in environment, behavior and design* (pp. 297–318). Plenum.
Friedman, K. (2007). Book review: "Designerly ways of knowing," by Nigel Cross. *Design Studies*, *28*(1), 103–104.
Galle, P., & Kroes, P. (2014). Science and design: Identical twins? *Design Studies*, *35*(3), 201–231.
Guersenzvaig A., & Ventura, J. (2022).Design practice, professional responsibility, and self-enactmentIn (Eds.), (pp.104-120) ,..G. Goldschmidt, Tarazi Proceedings of 13th Design Thinking Research Symposium (DTRS 13), Expanding the frontiers of design: A blessing or a curse, Design Research Society.
Hanna, R., & Barber, T. (2001). An inquiry into computers in design: attitudes before–attitudes after. *Design Studies*, *22*(3), 255–281.
Heylighen, A., Cavalin, H., & Bianchin, M. (2009). Design in mind. *Design Issues*, *25*(1), 94–105.
Houkes, W., Vermaas, P. E., Dorst, K., & de Vries, M. J. (2002). Design and use as plans: An action-theoretical account. *Design Studies*, 23(3), 303–320.

Hybs, I., & Gero, J. S. (1992). An evolutionary process model of design. *Design Studies*, *13*(3), 273–290.
Jackson, M. (2010). Representing structure in a software system design. *Design Studies*, *31*(6), 545–566.
Johannes, R. (1992). Architectural design: A systematic approach, part 2. *Design Studies*, *13*(2), 157–199.
Jonas, W. (1993). Design as problem-solving? Or: Here is the solution—What was the problem? *Design Studies*, *14*(2), 157–170.
Jones, J. C. (1970). *Design methods—Seeds of the human futures*. Van Nostrand Reinhold.
Kokotovich, V., & Purcell, T. (2000). Mental synthesis and creativity in design: An experimental examination. *Design Studies*, *21*(5), 437–449.
Lauche, K. (2005). Job design for good design practice. *Design Studies*, *26*(2), 191–213.
Lawrence, R. J. (1993). Architectural design tools: Simulation, communication and negotiation. *Design Studies*, *14*(3), 299–313.
Lim, S., Qin, S. F., Prieto, P., Wright, D., & Shackleton, J. (2004). A study of sketching behaviour to support free-form surface modelling from on-line sketching. *Design Studies*, *25*(4), 393–413.
Liu, Y. T. (2000). Creativity or novelty? Cognitive-computational versus social-cultural. *Design Studies*, *21*(3), 261–276.
Lloyd, P., Lawson, B., & Scott, P. (1995). Can concurrent verbalization reveal design cognition? *Design Studies*, *16*(2), 237–259.
Loukissas, Y. (2012). *Co-designers: Cultures of computer simulation in architecture*. MIT Press.
Louridas, P. (1999). Design as bricolage: Anthropology meets design thinking. *Design Studies*, *20*(6), 517–535.
Macmillan, S., Steele, J., Austin, S., Kirby, P., & Spence, R. (2001). Development and verification of a generic framework for conceptual design. *Design Studies*, *22*(2), 169–191.
Maier, J. R., Fadel, G. M., & Battisto, D. G. (2009). An affordance-based approach to architectural theory, design, and practice. *Design Studies*, *30*(4), 393–414.
March, L. (1984). The logic of design. In N. Cross (ed.), *Developments in design methodology*. Wiley.
Margolin, V. (1992). Design history or design studies: subject matter and methods. *Design Studies*, *13*(2), 104–116.
Medway, P., & Clark, B. (2003). Imagining the building: Architectural design as semiotic construction. *Design Studies*, *24*(3), 255–273.
Meng, J. C. S. (2009). Donald Schön, Herbert Simon and the sciences of the artificial. *Design Studies*, *30*(1), 60–68.
Metcalfe, D. (2022). Empathic multispecies design—Using empathy to design with animal perspectives in mind. In G. Goldschmidt & E. Tarazi (Eds.), *Proceedings of 13th Design Thinking Research Symposium (DTRS 13), Expanding the frontiers of design: A blessing or a curse* (pp. 386–395). Design Research Society.
Nagai, Y., Taura, T., & Mukai, F. (2009). Concept blending and dissimilarity: Factors for creative concept generation process. *Design Studies*, *30*(6), 648–675.
Newton, S. (2004). Designing as disclosure. *Design Studies*, *25*(1), 93–109.
Ostrowski, A., Breazeal C., & Park, H. W. (2022). How do older adults engage as robot co-designers? Rapid prototyping supported by lived experiences with technology. In G. Goldschmidt & E. Tarazi (Eds.), *Proceedings of 13th Design Thinking Research Symposium (DTRS 13), Expanding the frontiers of design: A blessing or a curse* (pp. 193–209). Design Research Society.
Ou, X., Goldschmidt, G., & Erez, M. (2022). Exploring the effect of team composition on creativity in design idea generation. In G. Goldschmidt & E. Tarazi (Eds.), *Proceedings of 13th Design Thinking Research Symposium (DTRS 13), Expanding the frontiers of design: A blessing or a curse* (pp. 211–224). Design Research Society.
Owen, C. L. (1991). Design education in the information age. *Design Issues*, *7*(2), 25–33.
Owen, C. L. (1998). Design research: Building the knowledge base. *Design Studies*, *19*(1), 9–20.

Ozkaya, I., & Akin, Ö. (2006). Requirement-driven design: Assistance for information traceability in design computing. *Design Studies*, *27*(3), 381–398.

Passini, R. (1996). Wayfinding design: Logic, application and some thoughts on universality. *Design Studies*, *17*(3), 319–331.

Razzaghi, M., Ramirez Jr, M., & Zehner, R. (2009). Cultural patterns in product design ideas: Comparisons between Australian and Iranian student concepts. *Design Studies*, *30*(4), 438–461.

Redström, J. (2008). RE: Definitions of use. *Design Studies*, *29*(4), 410–423.

Restrepo, J., & Christiaans, H. (2004). Problem structuring and information access in design. *Journal of Design Research*, *4*(2), 218–236.

Rittel, H. W. J., & Webber, M. M. (1973). Dilemmas in a general theory of planning. *Policy Sciences*, *4*(2), 155–169.

Sanders, E. B. N., & Stappers, P. J. (2008). Co-creation and the new landscapes of design. *CoDesign*, *4*(1), 5–18.

Schermer, B. (2001). The art of the long view: Teaching architectural programming from a simultaneous loose-tight fit perspective. *Proceedings of EDRA, 2001*, Scotland.

Schon, D. A. (1991). *The reflective practitioner*. Ashgate Publishing.

Sciannamè, M. (2022). The designer as cross-boundaries mediator: Merging machine learning, ethics, and design for the flourishing of humans. *Expanding the frontiers of design: A blessing or a curse?*, 139.

Sharrock, W., & Anderson, B. (1994). The user as a scenic feature of the design space. *Design Studies*, *15*(1), 5–18.

Smith, G. J., & Gero, J. S. (2005). What does an artificial design agent mean by being "situated"? *Design Studies*, *26*(5), 535–561.

Stewart, S. C. (2011). Interpreting design thinking, editorial. *Design Studies*, *32*(6), 515–520.

Tilley, D. G., & Burrows, C. R. (1995). Development of computer-based techniques for fluid power systems design. *Design Studies*, *16*(4), 397–413.

Turner, B. (1991). Book review of: Successful product design, by Bill Hollins and Stuart Pugh. *Design Studies*, *12*(2), 118.

Tzortzopoulos, P., Cooper, R., Chan, P., & Kagioglou, M. (2006). Clients' activities at the design front-end. *Design Studies*, *27*(6), 657–683.

Uluoğlu, B. (2000). Design knowledge communicated in studio critiques. *Design Studies*, *21*(1), 33–58.

van Aken, J. E. (2005). Valid knowledge for the professional design of large and complex design processes. *Design Studies*, *26*(4), 379–404.

Willem, R. A. (1990). Design and science. *Design Studies*, *11*(1), 43–47.

Yang, M. C. (2010). Consensus and single leader decision-making in teams using structured design methods. *Design Studies*, *31*(4), 345–362.

Yang, M. Y., You, M., & Chen, F. C. (2005). Competencies and qualifications for industrial design jobs: Implications for design practice, education, and student career guidance. *Design Studies*, *26*(2), 155–189.

1.5 Analysing storytelling in design talk using Linguistic Inquiry and Word Count

Almila Akdag Salah[a], *Senthil Chandrasegaran*[b], *and Peter Lloyd*[b]

[a]*Department of Information and Computing Sciences, Utrecht University;*
[b]*Faculty of Industrial Design Engineering, TU Delft*

1.5.1 Introduction

Over a period of nearly 30 years, the Design Thinking Research Symposium (DTRS) series (Cross, 2018) has conducted four shared data workshops, generating data from design activity in a number of different contexts including think-aloud protocols (Cross, Christiaans & Dorst, 1996), designer-client discussion (McDonnell & Lloyd, 2009), design education (Adams & Siddiqui, 2016), and co-creation (Christensen, Ball & Halskov, 2017). In bringing different research groups and traditions together, the key achievement of these workshops has been to develop a common language with which to describe design activity. A central research thread of these workshops has been to consider the function of different words and the structure of different word forms in discourses of designing and evidencing specific ideas about what constitutes design activity with textual data has formed the basis of this inquiry. The concept of 'framing', for example, as it pertains to both Schön (1984) and Goffman (1974), has been explored by Dorst and Dijkhuis (1995), Glock (2009), Mabogunje et al. (2009), Secules et al. (2016), and Dong and MacDonald (2017). Such extensive exploration and development of a common language to describe designing through analysing common data can equally be applied to other now familiar concepts such as 'co-evolution', 'problem-finding', and 'evaluation'.

One particular concept relating to design activity that has recently found traction is that of 'storytelling'. This is the idea that significant parts of the design process, particularly in naturally occurring datasets, are taken up with narrative structures that place actors, objects, and relations in specific contexts. Arguably the early DTRS analyses on episodic memory (Visser, 1995) and precedents (Oxman, 1994) established a foundation for later work on storytelling to emerge (Lloyd, 2000) and develop (Lloyd, 2009). Most recently—in terms of the DTRS—Lloyd & Oak (2018) identified 'value tension' as an important function that stories have in design conversations:

> A story [...] provides a dynamic structure through which values can be framed and categorised, and through which design-oriented arguments can be presented.
>
> (p.109)

Concepts like that of storytelling, and the other concepts listed above, have given us a rich language with which to describe and present design activity, both within the DTRS datasets and more widely in the design research field. However, they are (necessarily) derived from the selective use of examples from design protocols. In this respect, they represent the first phase of a research process highlighting general features of design

DOI: 10.1201/b22630-7

activity. A general method of the DTRS workshops when looking at language use has been to 'fit' concepts with textual excerpts, showing how a particular linguistic mechanism might be in play. This is a manual process of close analysis and description, selecting self-defined points of interest in the discourse and describing them in abstract terms.

We might, then legitimately ask: are these concepts robust? Are they features of design discourse that can be readily and reliably identified? Can we show that they are necessary conditions of design activity?

More automated methods of textual analysis have begun to be applied to DTRS data to explore more general features of the design protocols (Menning et al., 2017). This chapter extends this recent trend by treating the DTRS data as a single corpus, amenable to computational linguistic analysis. Rather than identifying specific discursive practices within a particular dataset, we seek both to characterise these DTRS data more broadly and begin to validate previously established concepts about designing. We do this particularly in comparison to other types of textual corpuses. With more and more training datasets becoming available and with analytical sophistication increasing rapidly (Archer & Jockers, 2016), it is now possible to use comparative methods to test both specific concepts related to the DTRS data—in the case of this paper that of 'storytelling'—as well as reveal more general regularities.

In this chapter, we first describe the constitution of the DTRS corpus that we have created. We then go on to describe the Linguistic Inquiry and Word Count (LIWC) methodology, illustrating how it can be applied to the DTRS corpus and compared to other large datasets. The main body of the chapter considers the concept of 'storytelling', discussing how it can be constituted and operationalised to reveal narrative and emotional arcs in textual data. We conclude by saying that although we cannot replicate the significant results of others, we do show new methods for automated data analysis of design activity to facilitate cross-corpora comparison.

1.5.2 Building a DTRS corpus

The datasets from the four different shared-data workshops of the Design Thinking Research Symposium feature several different disciplines related to design: DTRS2—industrial design engineering, DTRS7—architecture and engineering design, DTRS10—design education and DTRS11—product design. These datasets encompass a variety of design activities from think aloud protocols (DTRS2), designer-client discussion (DTRS7), design education (DTRS10), and co-creation (DTRS11). Apart from DTRS10, which followed the design process of students and the feedback they received from their lecturers, all other datasets were collected from the meetings of design professionals.

The datasets differ in length and in the number of participants: DTRS2 is the smallest with one 2-hour 'think-aloud' design session with a single designer and another 2-hour session with three designers. The design conversation focuses on a bicycle accessory. DTRS7 has four 2-hour meetings, where two of these feature conversations between an architect and his clients. The other two are between a multidisciplinary design team working on the design of a digital pen. DTRS10 consists of 38 sessions of various lengths, and varying number of participants. As the sessions feature teacher-student discussions in five disciplines (industrial design, mechanical design, service-learning design, entrepreneurial design, and choreography) a different conversation hierarchy can be observed. DTRS11 contains 20 sessions of varying length with a focus on preparing and discussing the results of co-creation meetings. In total, it has 17 participants. Together, these datasets contain a rich reflection of design activities that occur in diverse stages of design including how designers think, discuss, reflect, evaluate, and talk to their clients.

To build the DTRS corpus from these datasets, we first got the appropriate permissions and anonymised the data by rendering all datasets into a tabulated format with dataset name, session name, speaker, and speech turn. We kept filler words such as 'ehm, uhm' etc. as well as descriptions of actions, such as [laughs], to maintain the conversational flow. The resulting corpus has 373,983 words spread across 64 sessions from the four datasets. A more detailed description can be found in Lloyd et al. (2021) and in our sister paper for DTRS13 (Chandrasegaran et al., 2022). This corpus, though not big enough to be of use as a training dataset for an artificial intelligence (AI) application, is nonetheless a valuable resource to explore hypotheses about the features of design discourse and dialogue.

1.5.3 The Linguistic Inquiry and Word Count dictionary

Linguistic Inquiry and Word Count (LIWC) is a dictionary prepared by Pennebaker et al. (2015) with the purpose of studying the grammatical, linguistic, and socio-psychological dimensions of textual data. LIWC provides a 'gold standard' of categories reflecting these dimensions through individual words (Fast, Chen & Bernstein, 2016) and has been used for more than twenty years to analyse a huge range of textual datasets. Over the course of these 20 years, the dictionary has been enhanced to its current format, which now has about 6,400 dictionary words and word stems in 91 categories and sub-categories. These categories cover linguistic (e.g., pronouns, prepositions), other grammar (e.g., quantifiers, common adjectives), and psychological processes (e.g., positive emotion, insight). Table 1.5.1 lists some of these categories with word examples.

In comparison to coding frameworks introduced in the design literature, using LIWC as a coding tool allows us to explore a wider range of linguistic phenomena. Existing design process coding frameworks tend to focus on a particular aspect of design such as different types of reasoning, how the design object is described, or the design stages themselves. In contrast, LIWC identifies more general cognitive, emotional, perceptual, and social processes. Since LIWC is a general dictionary used in various disciplines for many different purposes, a LIWC analysis of design discourse can be compared to other corpora, with the consequent potential to create new insights into what constitutes designing talk.

In the following sections, we explain how LIWC can be used to explore and analyse the concepts of 'story' and 'storytelling' as they appear in textual data. Storytelling is an activity that is, of course, not reserved for designing but is very much a part of everyday life. We tell stories for many reasons during conversations; however, the extent and instances in which we do this has not been quantitatively studied. To look at the idea of storytelling as it applies to the DTRS corpus, we need also to look at other datasets with conversations and/or stories for purposes of comparison.

1.5.4 Analysing storytelling in conversations

1.5.4.1 The structure of stories

What is it that forms the structure of a story? The search to define the basic elements of what constitutes a story stretches back to Aristotle's Poetics and the narrative arc of 'beginning, middle, and end'. In the 19th century, Freytag & MacEwan (1960) argued that narratives consisted of five key components unfolding over time: exposition, rising action, climax, falling action, and resolution. The literature studies as a field offer more

Table 1.5.1 An overview of some of the total 91 LIWC categories (Pennebaker et al., 2015) with examples

Main category	*Sub-category*	*Examples*
Linguistic Dimension	Function Words	
Pronouns	1st per. singular	I, me, mine
	3rd per. plural	they, their, they'd
Impersonal pronouns		it, it's, those
Articles		a, an, the
Prepositions		to, with, above
Negations		no, not, never
Other Grammar		
Common verbs		eat, come, carry
Common adjectives		free, happy, long
Interrogatives		how, when, went
Quantifiers		few, many, much
Psychological Processes		
Affective processes	Positive emotion	love, nice, sweet
	Negative emotion	hurt, ugly, sad
Social processes		mate, talk
	Friends	buddy, neighbour
Cognitive processes	Insight, Tentative, Differentiation	know, think, realise, maybe, perhaps, could, hasn't, but, else
Perceptual processes	See, Hear, Feel	view, saw, seen, listen, hearing, feels, touch
Drives	Achievement, Power	win, success, better, superior, bully, danger, doubt, chance
Time orientation	Past focus, Present focus, Future focus	ago, did, talked, today, is, now may, will, soon
Relativity	Motion, Space, Time	arrive, bike, go, down, in, thin, end, until, season

theories, but for the purposes of this paper, we focus on the types of narrative distinctions which are applied to quantitative textual data analysis. Boyd et al. (2020) identify three narrative markers—staging, plot progression, and cognitive tension—in several large datasets. van Laer et al. (2019) identify the degree of what they term 'narrativity' in a dataset of TripAdvisor reviews by tagging narrative content (i.e., affective consciousness, cognitive consciousness, and spatial and temporal embeddings) and narrative discourse (genres such as drama, tragedy, comedy, etc.). Finally, Reagan et al. (2016) classify computationally six types of 'emotional arcs' present in a dataset of 1327 works of fiction to reveal the affective complexity of narratives.

Of the three studies mentioned above, the study by Boyd et al. (2020) draws explicitly on the LIWC dictionary and is the analytical framework we adopt to look at the DTRS dataset. Following Freytag's theory, they divide a story arc into five segments and map these to three story elements: staging, plot progression, and cognitive tension which can be identified in terms of LIWC categories.

Staging: the beginning of a story (i.e., the first segment) is usually reserved for introducing unknowns to the audience. To do so, the writer or speaker needs to use prepositions and articles more than other word categories. Once the staging is done, the number of words from these categories is expected to reduce in the following segments.

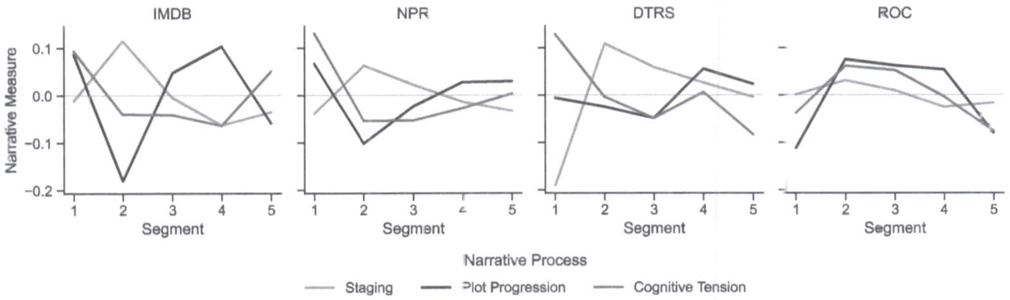

Figure 1.5.1 Staging, plot progression, and cognitive tension for stories in the DTRS, IMDB, ROC, and NPR datasets. The *x*-axis shows the segments, the *Y*-axis shows the mean standard deviation for each segment across all dataset stories. Narrative markers for each segment for a single story are calculated as the standard deviation from the mean percentage value of one narrative marker across all segments.

Plot progression: after the introduction of characters and the setting, the story moves along by using words from the function category. As an indicator of this, pronouns and auxiliary verbs are expected to rise from the second segment onwards.

Cognitive tension: each story sets up certain challenges and problems that are resolved through the unfolding of events and actions taken by the characters. These points in the story result in the usage of words from the cognitive processes category, with the peak coming in the middle of the story, after which the story moves towards the resolution.

Boyd et al. (2020) established these patterns for three different corpora in their paper resulting in the LIWC dictionary file being updated for this 'story identification' task. As shown in the LIWC network maps of Figure 1.5.1, words belong to more than one category. Boyd et al. found that this overlap between categories generates some noise when looking at a narrative, so they published two additional dictionaries to control for the overlaps. These dictionaries are called *Arc of Narrative (AON) with some overlap*, and *AON without overlap*. We used both AON dictionaries as well as the original LIWC dictionary for our analysis. The results between the three were similar, hence, we report the results for the *AON without overlap* as the results were more pronounced for this dictionary.

Additionally, Malin et al. (2014) analysed the emotional tone of stories with LIWC, again by segmenting stories first into five equal length and then calculating the positive and negative affect of each segment with LIWC's corresponding categories. They differentiate between redemptive and contaminative stories. Redemptive stories have happy endings, hence a positive emotional rise towards the end, whereas contaminative stories have a sad ending with a negative emotional rise at the end. Since we use the same segmentation in analysing the narrative arcs, we will also follow Malin et al.'s approach to examine which emotional arcs are prevalent in the DTRS and comparison datasets.

1.5.4.2 A LIWC analysis of DTRS stories

For the purposes of this chapter, we made the assumption that we would be more likely to find stories in DTRS if we looked at long speech turns. We therefore filtered the DTRS dataset first by the length of speech turns. As the conversations are between multiple speakers, the majority of the speech turns were short. We extracted all speech turns with more than 100 words, which returned 530 turns. Manual inspection of these 530 turns showed many instances where a description is given about objects, or users, but few full stories with all the story elements of staging, plot progression, and cognitive tension identified by Boyd et al. (2020). With this in mind, the 530 speech turns were then filtered by one of the authors using manual annotation to check if the turn contained story elements. This resulted in 267 turns. Sixteen of these were tagged as ‘proper’ stories, roughly corresponding to the elements outlined above, whereas the majority were deemed as story snippets, i.e., textual pieces that contain one or more story elements, but do not follow the basic principles that are found in every story. The following two excerpts illustrate the difference:

Excerpt 1: Example DTRS story
But at some point I think you feel like ‘I’m a human, I’m made to use my hands, I’m made to walk on my feet’, and when machines start to do everything I think, at some point we reach a level where you say to yourself, ’I want to use my hands, because that’s what they’re made for, I want to use my feet because that’s what they’re made for’, and I think they will reach that stage at some point, soon maybe. That’s what we feel- at least we feel in Scandinavia, that we want to return to the original human, we want to go out and use our hands, work the old-fashioned way, eat like we did in- in- in one hundred years ago, so.

(Kenny, DTRS11, 127 words)

Excerpt 2: Example DTRS snippet
This is the funeral that they had for this gentleman. And they may- these are the service sheets that they made up for him obviously showing in a sense his lifetime in pictures really and they had this on the wall itself and it just framed up and down and all the pictures just kept appearing on and off the screen and obviously in our chapel it’s so light it’s quite difficult to see some of them but this is, I mean this is this is sort of thing now this is started to move forward more now people wanted to do this and I mean in the back it says if you would like to receive a DVD please contact the family. Quite mad in a sense. But that’s what you know this is where we are now with this. They’re starting to ask for things like this and the concern I’ve got was that although this man came in to see me he had this erm screen that he had we’d have to bring in a little screen like this, a stand up screen and put it at the front which is again in our chapel quite light, difficult to see and he was then saying about the equipment that you would need. You’d need to have it really in the centre of the chapel sort of looking at the front or doing something and it would be quite difficult to do it in that. We’d have to put it at the side erm and he was then talking about all sorts of expensive equipment that we’d have to be leasing from him and you know all sorts of costs. But I mean what the, the idea is this is starting to become more the norm now. And like I say in Australia they- he saw it in Australia, and he wants to be the main agent in this country. And so obviously it’s something that he’s seen, and thought was a money spinner for himself I’m not saying that it wouldn’t be for us.

(Female A, DTRS07, Crematorium Meeting 02, 347 words)

In the first story excerpt, an idea, or rather a question is introduced as the opening of the story: what makes one a human? The staging is done via the use of the storyteller’s emotional reaction to the question, as life has brought them to the point of asking this question 'at some point I think you feel like'. Then the story unfolds, making the comparison between man and machine, and the resolution comes right after that: in order to keep your humanity, you should use your body, and not leave everything to the machine.

The second story snippet excerpt, although much longer, does not have the five segments and story elements outlined above. It looks like two stories merged into each other. In the first one, a funeral is depicted where technology is used to tell the life story of the deceased person with photographs. There is not a progression of the plot however: we do not learn the details of the deceased person's life. Instead, we hear about the technical problems encountered in the chapel: 'in our chapel it's so light it's quite difficult to see some of them'. The next part jumps back to the funeral, this time depicting the deceased's friends and family asking for those photographs. Then a new story is introduced: the company that installs the technology, and how much they ask for this service is described. For a human reader, these snippets form the moral of the text: there is demand for such technology, but it is not clear if the storyteller will adopt it. At the end, a resolution does not flow out of these story elements, even for a human observer. The computational challenge is to try and capture these story elements automatically outside of the expected narrative arcs.

1.5.4.3 An example analysis

To demonstrate the Arc of Narrative LIWC dictionary in action, the following DTRS story is -coded with *italic*, **bold**, and with underlined text to show the words related to staging coded with *italic*, the words that are expected to be used for cognitive tension coded with **bold**, and the words for plot progression coded with underlined text:

> Things will be very different I **think**, **cause** like *the* current generation that we are talking to, are people who have experienced like *the*- still have a little bit, right? I had *an* ex-colleague who was twenty-eight maybe and his dad is *a* lawyer, and he told me that when he grew up he still lived *in the* alleyways *with* no bathroom, so that everybody had *a* common bathroom, and now, you **know**, completely different, right? and twenty eight is not that old, it's like our generation, so once you hit ten years *from* now, *the* people who already grew *up with the* bathroom *in the* house, they will have *a* completely different perception *of* this.
>
> (Rose, DTRS 11, Recap with consultants)

Looking at the words highlighted in orange, we see that there is no extra staging in the first sentence, unlike the expectations. The use of article and prepositions is not higher than the remainder of the excerpt. On the contrary, towards the end, we see that articles and prepositions are used more.

Turning to the words highlighted in blue, we see that there are only three words related to 'cognitive process'. Rose's story does not ask the listener to engage in cognitive processes. She does not pose a problem or challenge that needs to be resolved throughout the story. Rose rather asks her audience to remember a time when people did not have toilets in their homes.

Lastly, if we look at words related to plot progression highlighted in green, we see that there are many of them distributed throughout the story. Indeed, there are so many that it would be better to use a quantitative approach to see the changing word use throughout the story. To do that, LIWC offers diverse options such as giving the percentage of words appearing from a given category (like function-words) in a sentence, or in a paragraph, or the whole document.

1.5.4.4 Comparison datasets

To generate a baseline for datasets that contain short stories, we made use of three existing comparison datasets. The reason for this was twofold: First, we would like to test if Boyd et al.'s (2020) hypothesis that certain LIWC categories capture certain parts of a narrative arc would work in conversational datasets, not just in single person monologues (Boyd et al. used TED Talks and Court-room orders as data). So, our first question is: how does the conversation dimension affect how stories are told?

Second, since LIWC counts words given in a specific category, words used specifically during a conversation, such as filler words, does not constitute a problem. But since we extract only one speech turn from a conversational flow, the different narrative markers of a story (staging, plot progression, cognitive tension) may occur over multiple turns and hence be missed. So, our second question is: What is the effect of the length of a story?

Boyd et al. (2020) cap the minimum length of a story at 250 words, but conversations are dynamic, with turn-taking between participants, and speech turns with more than 250 words rather rare. Still, a short story can be easily sketched with less than 250 words. If a story is told in less than 100 words, can LIWC still discern a change throughout the five story segments in relation to the three narrative elements?

Our first comparison dataset, known as the ROC stories, is a written corpus containing 50,000 stories, each five sentences long, and generated using Amazon Mechanical Turk as part of a study looking at automatic story closure by Mostafazadeh et al. (2016) at ROChester university (hence ROC). The sentences in these stories are brief to the point that a story has usually word counts of around 60–80. The ROC corpus has two important features: it imitates the causal and temporal common-sense relations that are found in daily events, coming close to mini stories that are told in everyday life. Even though these stories contain no conversational phrasing, due to their briefness they are expected to be similar to stories told within the flow of a conversation.

Our second dataset is the IMDB corpus, mentioned earlier in the chapter, which contains 1,068 movie scripts (Danescu-Niculescu-Mizil & Lee, 2011). This corpus gives insight to everyday—albeit fictional—conversations, and the comparison to DTRS dataset can highlight if there are specifics to designer conversations versus everyday language. The scripts are formatted to show the movie title, speaker, and the speech turn. For the purposes of this research no further formatting is needed.

Our third dataset contains 20 years of radio interviews on the US National Public Radio (NPR) network and contains the transcriptions of over 10,000 hours of audio interviews (Zhu et al., 2021). The transcripts are formatted with the programme name, session, and the speaker along with the speech turn. We assume that, even though both the NPR and IMDB datasets are scripted in some form, they are a close approximation to everyday conversation.

We follow the same pre-processing steps that we applied to the combined DTRS dataset. For both the IMDB and the NPR datasets, we filter for turns or utterances which were above a certain word count (WC). As these datasets are considerably larger than the DTRS datasets, we set the threshold to turns or utterances with more than 140 words. After this filtering, the NPR dataset yields 8048 utterances, and IMDB 145 utterances.

The results of filtering reflect the differences in the average turn length of these datasets, and of course the differences between the different types of conversation flows. Before filtering NPR consists of 105,258 utterances, so the dataset contains lengthy utterances or speech turns. In comparison, IMDB dataset has almost three times more turns with 304,713, but most of the turns are short, and the 140 WC threshold yields only 145 results. The utterances in the ROC dataset—each a story in five sentences—are more-or-less identical in length so we did not do any further filtering and randomly picked 10,000 stories for our analysis.

To show the similarities and differences between the stories in our three comparison datasets, we show some an example from each one of them. The five sentence-length ROC stories have the expected story-arc elements, i.e., a beginning, a middle, and an end:

An example ROC story:
Jennifer has a big exam tomorrow. She got so stressed, she pulled an all-nighter. She went into class the next day, weary as can be. Her teacher stated that the test is postponed for next week. Jennifer felt bittersweet about it.

In comparison, the IMDB dataset contains stories that are more diffuse: sometimes the beginning or end is not so clear as the conversation takes a different turn, or the story is packed with filler words asking for participation from other speakers:

An example IMDB story:
There was this kid that I grew up with; he was a couple years younger than me, and sort of looked up to me, you know. We did our first work together, worked our way out of the street. Things were good and we made the most of it. During prohibition, we ran molasses up to Canada and made a fortune; your father too. I guess as much as anyone, I loved him and trusted him. Later on, he had an idea to make a city out of a desert stop-over for G.I.'s on the way to the West Coast. That kid's name was Moe Greene, and the city he invented was Las Vegas. This was a great man; a man with vision and guts; and there isn't even a plaque or a signpost or a statue of him in that town. Someone put a bullet through his eye; no one knows who gave the order. When I heard about it, I wasn't angry. I knew Moe; I knew he was headstrong, and talking loud, and saying stupid things. So, when he turned up dead, I let it go, and said to myself: this is the business we've chosen. I never asked, who gave the go ahead because it had nothing to do with business.

NPR stories are similar to those of IMDB, especially when the speaker that tells the story is not a radio host, but they are more natural in terms of everyday conversation, with false starts, casual reflections, and repetition:

An example NPR story:
I walked in and then I looked around. And I realized I was really out of place. These people were at an audition. I mean, the girls had makeup. The guys had these sick outfits on. And I look like, pretty much, an athlete that just left the gym, un-showered - oh well. The music comes on and let me tell you guys. I was freaking living in these dance moves. I was like, I don't care what happens. They're going to remember me, even though I don't even live here. So, there was like 200 people at this audition. At the

> end, there ended up being seven guys and five girls. And they put us on video, and they were like, you know, if we want you, we'll give you a call in a couple weeks. And I literally left there like, whatever, this doesn't matter. I don't even really care. I went back home to New Jersey. Two weeks later, I was at the laundromat with my pocket full of quarters, when I got the phone call that the agency wanted me to move to LA to pursue a career in dance. So, I get there, I was dancing. I did some TV shows. I did some musicals. I was teaching a dance class and my dance class at the gym was really popular. So, my big break came when a friend called me and said, hey, there's this company named Beach Body that wants you to see if you can develop a project with them. So, I go into the Beach Body office. I ended up being there for two hours and that day I left with my contract for Hip Hop Abs.

As with the DTRS dataset, we used the LIWC AON without overlap dictionary to analyse the three comparison datasets. For the ROC dataset, each story is already clearly defined so can be used as a baseline for the other datasets. The ROC story dataset does not have any conversational noise and, even though short, contains the basic elements of staging, plot progression, and cognitive tension as well as having emotional valency. For all datasets, AON automatically calculates articles and prepositions separately and reports them as 'staging'. These words are expected to be used more in the first segment of a story. The rest of the function words (i.e., pronouns, auxiliary verbs, negations, conjunctions, and nonreferential adverbs) are calculated separately as well. These relate to plot progression and are expected to be higher in the middle of a story. Cognitive words are expected to follow a similar curve to plot progression and are the last category that is extracted using AON.

We remarked earlier that the stories told during conversations in the DTRS dataset do not necessarily neatly contain all elements that are used in defining what makes a story; these we define as 'snippets'. One simple difference is in the type of words used: during conversation words that acknowledge the audience, or the interruptions that come from the other participants of the conversation, are common. The stories are often short with sometimes just a short story snippet enough to make the other participants take on the remainder of the story. A simple description of a character, a place or an object can be enough. This raises the question: Will AON categories capture all or some part of the conversational stories in the DTRS dataset?

1.5.5 Results

1.5.5.1 The narrative arc of conversational stories

For all four datasets, the LIWC tool splits each story/utterance into five equal-sized segments. For each segment, the sum of all words relating to the narrative markers of 'staging', 'plot progression', and 'cognitive tension' are calculated and divided by the number of total words for that segment, giving a normalised percentage figure for each narrative marker. A mean percentage is then calculated across all segments for each narrative marker. The results for each segment are then expressed as the standard deviation from this mean. This means that even if stories do have differences in length or have a higher percentage of word use in a certain category, they still remain comparable. The mean of all standard deviations for a particular segment, across all stories in the dataset, forms the final number on the y-axis shown in Figure 1.5.1. Table 1.5.2 shows an example of how this works for one story only.

Table 1.5.2 An example calculation of LIWC results for one story/utterance

	Segment 1	*Segment 2*	*Segment 3*	*Segment 4*	*Segment 5*	*Mean %*
Staging (%)	12.50	12.50	20.83	4.17	38.10	17.62
Normalized Staging (Standard Deviation from mean percentage figure)	−0.40	−0.40	0.25	−1.04	1.59	–

Each segment is normalized by calculating the standard deviation from the mean of all segments for each narrative marker.

As noted previously, the DTRS, NPR, and IMDB all contains various conversational filler words. The DTRS stories have the highest number, with speech turns almost always starting with a turn-taking event acknowledging the previous utterance. As we have not cleaned the data for such conversational effects, we therefore expect to see certain differences from the 'five segment' hypotheses relating to Boyd et al. (2020). Since, even though such filler words are not counted, they do take up space. If the first two sentences in an utterance are only conversational such as: 'Well, yeah. I don't know, maybe … But if you think … Let me put it this way …', the first segment of the turn will, then, have no staging words, which will appear later. Looking at the results for DTRS (third chart from left in Figure 1.5.1) for example, we see that staging starts in the second segment, rather than the first. Cognitive tension is highest in the first and lowest in the last segment, as the start and end of a turn relates more to a change in speaker than to the story narrated within the turn. We should also emphasize the confounding factor that the DTRS data shown in Figure 1.5.1 is for both story snippets and stories.

IMDB stories do also contain filler words, but not necessarily in the same way as the DTRS data. In contrast to DTRS, IMDB stories are often presented through monologue, and hence the narrative processes can appear similar to classical stories. Lastly, NPR stories, especially if coming from programme hosts, follow a written narrative arc the best, though the 8,000 utterances do contain a wide variety. ROC stories should in theory all have the same style—all being created from the same rule of being five sentences long—and should mimic a classical narrative arc. So, we expect to see the most fluctuations from the norm in the DTRS dataset, followed by IMDB, NPR, and ROC.

Figure 1.5.1 shows that among all datasets NPR (second chart from left) comes closest to the expected narrative arc. After the first part of the story, words related to staging are used less and less, and the words for plot progression and cognitive tension are used more and more. The cognitive tension words are not used as expected, i.e., in the middle, and in general their usage is quite low.

ROC stories do follow the narrative arc for the plot progression and cognitive tension narrative processes, but not so much for staging which stays relatively constant. For both the IMDB and DTRS datasets, we can say that although the 'staging' narrative process does take place, the 'plot progression' and 'cognitive tension' narrative processes do not follow any meaningful arc.

Here we need to note that narrative arc processes for each dataset are mapped within an exceedingly small margin, and do not reflect a real consensus in the data. For example, for DTRS (third chart from left in Figure 1.5.1), the staging starts low, picks up in the first segment and drops steadily till the last segment. However, this latter movement is

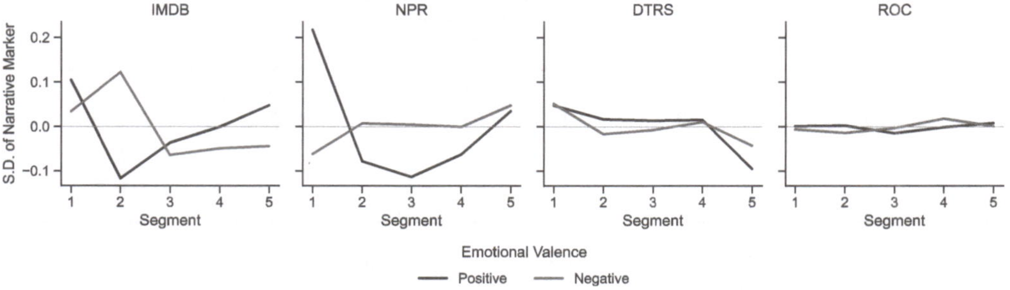

Figure 1.5.2 Positive and negative emotional arcs for the DTRS, IMDB, ROC, and NPR datasets. The x-axis shows the segments, the y-axis shows the mean standard deviation for each segment across all dataset stories. Narrative markers for each segment for a single story are calculated as the standard deviation from the mean percentage value of one narrative marker across all segments.

only due to a change of 0.08 standard deviations, much less than the changes reported in Boyd et al. (2020). If we plot each story in the DTRS dataset separately, we see many different shaped arcs, without a clear overlap. In this respect, Figure 1.5.2 shows a clearer picture than the reality.

1.5.5.2 The emotional arc of conversational stories

Even though DTRS, IMDB, and NPR are conversational datasets, their emotional tone is likely to be quite different. Of the three datasets, DTRS contains conversations recorded at various stages of the design process, and the stories (or story snippets) are told largely to further those design processes. As observed by Lloyd and Oak (2018) and described in the beginning sections of our paper, stories can be used to frame or negotiate 'value tension' within the design process. By looking at the emotional arc of conversational stories, we explore the question of whether this putative 'value tension' brings with it positive or negative emotional content in the terms discussed earlier of Malin et al. (2014).

In contrast to DTRS stories, IMDB and NPR stories are told with a variety of purposes and to a far wider audience. IMDB, mimicking everyday life, has stories that enhance character, motive, and the overall plot of the film. NPR on the other hand is a radio programme reporting newsworthy events along with everyday happenings. The appeal to a wide and diverse audience means that the use of emotional toning may be more pronounced in these datasets. The ROC stories are the shortest stories so their emotional arcs might be difficult to capture using this technique.

Figure 1.5.2 shows the results of looking at emotional valence through the five segments of the stories identified in each dataset. For the DTRS stories (third chart from the left), there is a regressive emotional arc, i.e., both for positive and negative emotions there is a fall throughout the stories and snippets. Conforming more to expectations, the IMDB stories (first chart from the left) have a rise-fall shape emotional arc for negative emotions, with the opposite fall-rise shape for the positive emotional arc. This might indicate that in general the IMDB stories have happy endings.

With the NPR stories (bottom, right), the negative emotional arc is steadily rising but more or less stable, whereas the positive emotional arc demonstrates a fall-rise U-shape which can define a more dramatic affective trajectory, but again, largely ending happily.

ROC stories do not show a clearly defined trajectory, though it is notable that both positive and negative emotional tone rises significantly during the fourth segment (sentence) of these stories.

For Figures 1.5.1 and 1.5.2, although we can discern narrative shapes for the four datasets, they are much less significant than those reported by Boyd et. al. (2020). This may be due to their conversational nature (DTRS, IMDB, NPR) or length (ROC) making the variation of results very high and leading to generally low mean standard deviation rates. Boyd et. al. (2020) report differences of 2 standard deviation, while the biggest difference we found was 0.35 for narrative processes (DTRS) and as low as 0.03 for the emotional arcs (ROC).

1.5.6 Discussion and conclusion

Our overall aim in this chapter was to create a single corpus of the DTRS shared datasets and explore this in relation to well-established concepts relating to design activity. Previous DTRS studies have largely been within a set of data, but building a corpus allows us to compare design activity with other types of activity to reveal more general regularities.

We explored (and attempted to validate) one aspect of design discourse (and DTRS data) identified in previous studies, that of 'storytelling', to see what results a corpus level analysis might yield. We drew on very recent work using LIWC to look at stories, and again used comparisons with different corpora to explore the shape of narrative and emotional arcs in conversational datasets (as well as in DTRS). Even though previous studies indicate that systematic/automatic ways of tagging stories can yield significant results, our study did not manage to replicate these results in conversational data. We discussed why the need to maintain turn-taking in conversation may interfere with clearly defined stories. Story length also proved to be problematic, with very short stories not having enough linguistic markers to reflect the same scale of the narrative and emotional arcs captured for longer stories.

Before concluding that it is not possible to capture the stories that previous studies of design activity have identified, we must look at the methodology we used to recognise and calculate narrative markers. Though consistent with the literature, fixing utterance length and having equal segmentation may not be suitable for conversational datasets, particularly smaller ones, where stories are likely to be told over multiple speech turns. As we noted, this tends to result in story snippets, rather than entire stories. In future studies, we will explore the possibility of more dynamically identifying and extracting story shapes of narrative markers independent of fixed utterance and segment size.

The advantages of using a dictionary-based approach such as LIWC has clear. LIWC itself as a tool is very easy to use, requiring no technical knowledge. The categories of the LIWC dictionary need to be understood at the conceptual level: for example, what does it mean if someone uses more first-person pronouns? Or what happens linguistically when someone is describing an object? By studying example datasets, one can get acquainted with how these categories are distributed in different type of speech, writing, or even by individual speakers. Combining the use of categories as done by van Laer et al. (2019) enables one to generate complex models to analyse text by simply counting words.

What we hope to have shown in this paper is a new way to think about, and analyse, design protocol data on a larger scale than previously. As more and more datasets become available, the techniques we have used in this chapter will become more common in determining how different kinds of design activities are constituted in text, and how they compare with other types of textual data. Our longer-term goal is to create a design dictionary as part of LIWC, to identify designerly ways of talking in large datasets.

In conclusion, although using a LIWC analysis to explore the DTRS data shows considerable promise, we found it difficult to validate the existing concept of storytelling (Lloyd & Oak, 2018). Whether this would equally apply to concepts like 'framing' and 'co-evolution' is an open question, but the conversational nature of the data is likely to be a confounding factor, so methodology needs development. The question remains open as to what type of stories designers typically tell, and how they go about telling them. Despite this, we have shown how an automated method for identifying stories could work. Constructing a corpus of the DTRS shared data and comparing it with other large datasets outside of the design discipline is a further step on the way to understanding design thinking in conversational contexts.

References

Adams, R., Siddiqui, J. (eds.) (2016). *Analyzing design review conversations*, Purdue University Press.

Archer, J., & Jockers, M. (2016). *The bestseller code: Anatomy of the blockbuster novel*, St. Martin's Press.

Boyd, R. L., Blackburn, K. G., & Pennebaker, J. W. (2020). The narrative arc: Revealing core narrative structures through text analysis, *Science Advances*, 6(32), eaba2196. doi: https://doi.org/0.1126/sciadv.aba2196.

Chandrasegaran, S., Akdag-Salah, A., & Lloyd, P. (2022). A computational analysis of tentativeness and causation in design talk. In *Proceedings of the 13th Design Thinking Research Symposium (DTRS13)*, 87–102.

Christensen, B., Ball, L., & Halskov, K. (eds.) (2017). *Analysing design thinking: Studies of cross-cultural co-creation*, Taylor & Francis.

Cross, N. (2018). A brief history of the Design Thinking Research Symposium series, *Design Studies*, 57, 160–164. doi: 10.1016/j.destud.2018.03.007.

Cross, N., Dorst, K., & Christiaans, H. (eds.) (1996). *Analysing Design Activity*, Wiley.

Danescu-Niculescu-Mizil, C., & Lee, L. (2011). Chameleons in imagined conversations: A new approach to understanding coordination of linguistic style in dialogs, *Proceedings of the Workshop on Cognitive Modeling and Computational Linguistics*, ACL 2011.

Dong, A. & Macdonald, E. (2017). From observations to insights: The hilly road to value creation, in Christensen, B., Ball, L. & Halskov, K. (eds.), *Analysing design thinking: Studies of cross-cultural co-creation*, Taylor & Francis, 465–482.

Dorst, K., & Dijkhuis, J. (1995). Comparing paradigms for describing design activity, *Design Studies*, 16, 261–274, doi: 10.1016/0142-694X(94)00012-3.

Fast, E., Chen, B., & Bernstein, M. S. (2016). Empath: Understanding topic signals in large-scale text, *Proceedings of the ACM CHI Conference on Human Factors in Computing Systems*, 4647–4657. 10.1145/2858036.2858535.

Freytag, G., & MacEwan E. J. (1960). Technique of the drama: An exposition of dramatic composition and art; *An Authorized Translation from the 6th German Edition*, Scholarly Press.

Goffman, E. (1974). *Frame analysis: An essay on the organization of experience*, Harvard University Press.

Goldschmidt, G. (2014). *Linkography: Unfolding the design process*. MIT Press.

Glock, F. (2009). Aspects of language use in design conversation, *CoDesign*, 5(1), 5–19. 10.1080/15710880802492870.

Lloyd, P. (2000). Storytelling and the development of discourse in the engineering design process, *Design Studies*, 21, 357–373. doi: 10.1016/S0142-694X(00)00007-7.

Lloyd, P. (2009). Ethical imagination and design, *Design Studies*, 30, 154–168. doi: 10.1016/j.destud.2008.12.004

Lloyd, P., & Oak, A. (2018). Cracking open co-creation: Categories, stories, and value tension in a collaborative design process, *Design Studies*, 57, 93–111. doi: 10.1016/j.destud.2018.02.003

Lloyd, P., Akdag-Salah, A., & Chandrasegaran, S. (2021). How designers talk: Constructing and analysing a design thinking data corpus, *Proceedings of the ASME 2021 International Design Engineering Technical Conferences and Computers and Information in Engineering Conference. Volume 6: 33rd International Conference on Design Theory and Methodology (DTM), 2021. V006T06A034*. doi: 10.1115/detc2021-71200

Mabogunje, A., Eris, O., Sonalkar, N., Jung, M., & Leifer, L. (2009). Spider webbing: A paradigm for engineering design conversations during concept generation. In McDonnell, J. & Lloyd, P. (eds.) *About: Designing – Analysing Design Meetings*, Taylor & Francis, 49–65.

Malin, J. J., Vine, V. J., Stanton, A., Cannava, K., Bodie, G., & Pennebaker, J. W. (2014). The arc of narrative: Using language markers to identify stories. In Poster presented at the annual meeting of the *Society for Personality and Social Psychology*, Austin, TX.

McDonnell, J., & Lloyd, P. (2009). *About: Designing – Analysing design meetings*, Taylor & Francis.

Menning, A., Grasnick, B., Ewald, B., Dobrigkeit, F., & Nicolai, C. (2017). Combining computational and human analysis to study low coherence. In Christensen, B., Ball, L., Halskov, K. (eds.) *Design conversation in analysing design thinking - Studies of cross-cultural co-creation*, CRC Press, 291–309.

Mostafazadeh, N., Chambers, N., He, X., Parikh, D., Batra, D., Vanderwende, L., Kohli, P., & Allen, J. (2016). A corpus and cloze evaluation for deeper understanding of commonsense stories, *Proceedings of the 2016 Conference of the North American Chapter of the Association for Computational Linguistics: Human Language Technologies*, 839–849.

Oxman, R. E. (1994). Precedents in design: A computational model for the organization of precedent knowledge, *Design Studies*, 15(2), 141–157. doi: 10.1016/0142-694X(94)90021-3

Pennebaker, J. W., Booth, R. J., Boyd, R. L., and Francis, M. E. (2015). Linguistic inquiry and word count: LIWC 2015. Pennebaker Conglomerates.

Reagan, A. J., Mitchell, L., Kiley, D., Danforth, C. M., & Dodds, P. S. (2016). The emotional arcs of stories are dominated by six basic shapes, *EPJ Data Science*, *5*(1), 1–12. doi: 10.1140/epjds/s13688-016-0093-1

Schön, D. A. (1984). Problems, frames and perspectives on designing, *Design Studies*, 5(3), 132–136.

Secules, S., Gupta, A., & Elby, A. (2016). Piecemeal versus integrated framing of design activities in Adams, R., Siddiqui, J. (eds.) *Analyzing Design Review Conversations*, Purdue University Press, 155–176. doi: 10.1016/0142-694X(84)90002-4

Van Laer, T., Edson Escalas, J., Ludwig, S., & Van Den Hende, E. (2019). What happens in Vegas stays on TripAdvisor? A theory and technique to understand narrativity in consumer reviews. *Journal of Consumer Research*, 46, 267–285. doi: 10.1093/jcr/ucy067

Visser, W. (1995). Use of episodic knowledge and information in design problem solving. *Design Studies*, *16*(2), 171–187. doi: 10.1016/0142-694X(94)00008-2

Zhu, C., Liu, Y., Mei, J., & Zeng, M. (2021). MediaSum: A large-scale media interview dataset for dialogue summarization, *ArXiv:2103.06410 [Cs]*. http://arxiv.org/abs/2103.06410

1.6 Designing as self-enactment

Care and responsibility in design practice

Ariel Guersenzvaig[a] *and Jonathan Ventura*[b]
[a]*Elisava Barcelona School of Design and Engineering UVic-UCC, Spain;*
[b]*Shenkar – Engineering. Design. Art., Israel*

1.6.1 Introduction

Consider this, construction workers around the world die every day from the lack of adequate protective gear, the negligence of irresponsible managers, or general institutional recklessness. Naturally, the death of a worker is woeful, and the private and public sectors must tackle the issue and implement a plan to reduce this ever-growing death toll fast. However, from a moral perspective, who is responsible for these deaths? The answer is far from self-evident. Because of the number of actors involved, it is very difficult to hold one specific party reasonably responsible. Yet, going over an intuitive culprit list, a case could be made that architects should be at the top or even the middle of the list. As planners and designers of the built environment, don't they possess a degree of responsibility for the way their vision is materialized? And on the same note, aren't graphic designers at least partially responsible for the side effects of the consumerism they help promote: health-related problems, poverty, and the overall addiction to the rush of purchasing something new? How are they responsible for the staggering environmental impact of purchasing an endless parade of useless objects just for the sake of doing so for the right price?

One possible way for a designer to quickly deal with these quandaries would be to disregard them altogether and pass over the matter. Conversely, a different approach would be to try to find a less harmful way to earn a living by exiting the profession altogether and avoiding thus the blemish of its moral entanglements. Yet, an exit could hardly count as a solution at all as it solves none of the underlying problems it purports to address. Quitting the design profession—whether to move to a spiritual retreat or to work at a 'green' company outside the design field—might be individually and socially beneficial, yet effect not so much a deeper change at the professional level as the situation than needs change is fundamentally systemic, not individual.

In the following pages, we shall take a different path toward responsibility—facing the fact that professional design in its current state inexorably involves, at some point or another, working in, with, and for the industry. We bite the bullet and don't argue for an exit of our pledged profession, but for a reconsideration of what it can mean to be a responsible designer, i.e., one that *cares* for the discipline and seeks change from within rather than exiting it to harbor a stance of a grudging outsider.

As the COVID-19 pandemic has shaken our fundamental views on human nature and on social and communal responsibility, many of us—as design scholars and practitioners—believe the very foundations of our discipline need to be re-examined and reformulated. In a way, our field finds itself at a historic crossroads, similar to the one presented to the founders of the

DOI: 10.1201/b22630-8

Ulm School of Design during the 1960s in post-WWII Germany: should designers be agents of the industry, facilitating the production of goods, or should they become a social agent of change? (Segal & Ventura, 2019a).

Since the dynamics of consumption—perhaps even its very essence—has changed in the pandemic, the latter alternative seems to be not only the most attractive but also the most relevant in this critical juncture. However, for us as practitioners, changing things is easier said than done as we still find ourselves working with and for the industry. Yet, this juncture is a chance, if not to rewrite history, at least to challenge and reframe the perspective from which we look at our profession. Once again, we strive to redefine our discipline echoing the Sisyphean struggle to offer answers for the vexed questions of design—not 'how to design' (which is a *technical* 'how') or even 'what to design', but rather deciding 'why to design' and 'when not to design'. At the same time, we need to figure out how can we keep connected to one's ethical bedrock while still earning a living as a practitioner (this is a *moral* 'how'). Facing the current reality, and from a normative perspective these questions can be posed as: what kind of responsible designers should we imagine?

In this chapter, we propose a novel approach for discussing issues of responsibility in design, centering our focus on the *professional* dimension of the practice of design—which we regard as the moral kernel of the discipline. We will offer a *normative* ethical perspective to the issue of professional responsibility considering it from the standpoint of virtue, not deontology. Through an interpretive approach, we shall make the case that a designer's personal investment in the design profession can be seen as a way of moral development and self-enactment. This investment, in turn, furnishes the designer with strong motivating reasons for responsible action, i.e., reasons 'for which someone does something, [reasons] that, in the agent's eyes, count in favour of her acting in a certain way' (Álvarez, 2017).

To make our case more persuasive, we will provide throughout the text different quotes from various graphic designers elaborating on their intricate relationship with clients, the practice, and the industry. These snippets are included to strengthen the analytical validity of our argument and its general cogency—they are not presented here as *proof*.

Finally, we are hopefully confident that the conceptual apparatus we introduce here will be of use to the DTRS community and, more generally, to all design researchers seeking to theoretically ground empirically based investigations on responsibility and, more generally, on design ethics, i.e., the ethics of *designing*. For some recent work on the topic, see e.g. Gray et al., 2022.

1.6.2 Premises for analysis

Before we dive deeper into the intricate possibilities of integrating responsibility and professional practice, let us introduce and clarify several key concepts, some of which, due to the length and nature of this chapter, will not be argued in detail, yet are concisely laid out here as premises for the analysis that will follow.

The thrust of our argument rests on seeing professions as essentially moral projects. What's more, a profession needs thus to be understood as implying more than just getting paid for a job or doing a job skillfully in a technical sense. Naturally, not all occupations are professions. In philosophical terms, professions could be described as those occupations that typically require expert knowledge and are primarily concerned with helping others attain certain strategic goods that play a crucial role in their lives. Along this line, some philosophers have argued that 'a good profession is one which involves a commitment to a key human good, a good which plays a crucial role in enabling us to live a humanly flourishing life' (Oakley & Cocking, 2001, p. 74). Though there is no specific and agreed-upon definition of what exactly is a profession, the notion is commonly associated with occupations 'that both require advanced study and mastery of a specialized body of knowledge, and undertake to promote, ensure, or safeguard some aspect of others' well-being' (Whitbeck, 2011, p. 77).

That professions are not primarily about getting rich can be inferred from the cases of teaching and nursing, which both epitomize commitments to key aspects of human existence while highlighting a negative correlation between social and moral importance and paid salaries. Interestingly, the moment professions transform their primary venue from the good of society to the good of private players, their ethical structure comes under strain and societal calls emerge for stronger regulations and oversight mechanisms. For example, when medical research is conducted under the umbrella of big pharmaceutical companies for mere financial gains, or an applied anthropologist conducts marketing research for a big beverage firm, the very ethos of these professions comes under deep questioning.[1]

As hinted above, an obvious assumption we make, which is nonetheless worth declaring, is that design *is* a profession. We posit that it falls well within the margins stipulated above for what counts as a profession. Design rests upon a considerable learned body of practical and theoretical knowledge, and its influence on others' well-being is enormous and undeniable (for a comprehensive discussion, including arguments in favor and against considering design a profession see Guersenzvaig, 2021, pp. 61–119). Design, however, differs from nursing, teaching, and other established professions in two crucial manners. First, it is a rather young *professional* discipline. Second, it was historically developed to help and contribute to the industry and economic profit. Famously, Franklin Delano Roosevelt harnessed the evocative power of design to jumpstart the American economy out of its 1929 slump. Later, the Eisenhower administration took a similar path and presented mass consumption as a civic duty. Design was mobilized during the Cold War, for instance when top designers of those days such as Buckminster Fuller, the Eameses, and George Nelson took active part in the American National Exhibition in Moscow in 1959 (Castillo, 2010, 148–149).

For this chapter, we are rather lenient in our understanding of design. On the one hand, it can be understood in a narrow sense associated with form-giving, planning, and creating new material and immaterial artifacts. On the other hand, design can also be construed more broadly, i.e., as the prefiguration and reconfiguration of new ways of living, or, in other words, as the envisioning of new possibilities, new ways of being and doing (Dong, 2008; Manzini, 2015).

Indeed, contemporary design deals not only with problem solving, but rather with reshaping, reframing and resignifying daily situations and activities. It is a truism that design is everywhere: from the coffee mug that kickstarts our day, through the vehicle transporting us to our daily chores, to the digital platform on which we read the news, find entertainment, and communicate with others. The purview of design is located, to paraphrase Ernesto Rogers' famous dictum, from the spoon to the global supply chain. The upshot is that the designer transcends the simpler task of coupling shapes, materials, and colors to acting as an interpreter, gatherer, and co-shaper of (social) meanings. For example, French designer Philippe Starck in a somewhat humorous yet truthful manner accepts the role of the designer as a central player in the global consumer culture. Relating to his famous *Juicy Salif*, designed in 1991, Starck commented that its purpose is not to help squeeze lemon juice, but something totally different:[2]

> Sometimes you must choose why you design—in this case not to squeeze lemons, even though as a lemon squeezer it works. Sometimes you need some more humble service: on a certain night, the young couple, just married, invite the parents of the groom to dinner, and the groom and his father go to watch football on the TV. And for the first time the mother of the groom and the young bride are in the kitchen and there is a sort of malaise—the squeezer is made to start the conversation.
>
> (quoted in Lidwell & Manacsa, 2011, p. 100)

The Salif lemon squeezer exemplifies how from a classic Marxist lens design can't be neatly situated in either the superstructure (i.e., culture, norms, identities, and so on), or in the economic base (i.e., the means and relations of production), but rather as bi-directionally connecting the two as an active shaper and maintainer of both. Along these lines, design is not a mere byproduct of the superstructure. It goes beyond economic causes alone and can rightly be seen as the framing, planning, and conceiving of 'the complex wholes that provide a framework for human culture [...] to support human fulfilment' (Buchanan, 2001, p.38). Seeing design as a strategic activity at the intersection of economics, social life, culture, and politics underscores its moral import, which is eloquently illustrated by Deyan Sudjic (2008, p. 48):

> Design is the language that a society uses to create objects that reflect its purposes and its values. It can be used in ways that are manipulative and cynical, or creative and purposeful. Design is the language that helps to define, or perhaps to signal, value.

While lenient in our understanding of design, we are not referring here to design as a basic human capability or functional trait, but to design as a *professional practice*. Our locus of concern here is professional designers, i.e., design practitioners deeply rooted, albeit not exclusively, in an economic-market-oriented professional system. Naturally, we must surmise that designers are just one amidst many other players that participate in a design project. Yet, as mediators, professional designers orchestrate a complex motion between generating need, desire, and socio-cultural norms, to technology, modes of production, market value, revenue, and more. In a previous publication (Segal and Ventura, 2019b), the call for 'value-oriented design' has been emphasized as a key notion to navigate between the landmines which are design practice, that is, working inside our trade yet cultivating a clear moral and ethical core. Here we integrate the concepts of responsibility and professionalism to navigate between the challenges of the practice.

In his landmark study of professions, Schön (1983) emphasizes the substantive field of knowledge commanded by a professional and the mastery that is required to apply that knowledge. Among the various classic professions, Schön famously highlights the importance of design as a 'special' profession, in which the crucial link of the imagined correlates with the real, where the theoretical is interwoven with the applied. Along with design's unique temporal trait—designers design for a near or distant future while being situated in the present—designers operate as *reflective* practitioners. Their reflective skills are much needed to navigate the various *practical* difficulties inherent in their profession. Needless to say, reflexivity is one of many significant features embedded in design thinking.[3]

In line with all this, another crucial premise is that, due to the reflective and situated nature of design practice, the ethical considerations and the decisions that are made during design activity are mostly inseparable from the context in which they arise and can hardly be explored and resolved beforehand by applying declarative ethical principles or guidelines.

As in other professions, design consists of various and sometimes conflicting elements, intertwining theory and practice. According to the philosopher Alasdair MacIntyre, a *practice* can be understood as a specific type of social cooperative activity with specific intrinsic rewards, its own standards of excellence, methods, tradition, vocabulary, and values (MacIntyre, 2007, pp. 187–203). This will be the primary meaning of the notion of practice in this text.

Practices are always embedded in larger social contexts and require at least a minimally shared conception of the goods that are pursued by their practitioners. At the same time, practices are intrinsically integrated into the practitioner's own quest to lead a good life as an individual. In this sense, the practices to which we belong mirror our worldview, belief systems, and core values. To put it differently, partaking in a profession—in as far as it is a practice—means not only earning a living, but being a part of something that exceeds us as individual and shapes our current reality and daily life. What's more, practices are spaces of dialogical encounter, where the intersubjective commitments to key shared purposes are constantly being discussed, developed, and revised, determining thus the practice's ongoing traditions and standards (Guersenzvaig, 2021, pp. 159–169; Guersenzvaig, 2022).

1.6.3 Design and responsibility

It is almost customary to state that every professional must develop a sense of responsibility because the impact they have on our daily existence. However, how this translates into practice is not obvious to all. Professionals such as Zaha Hadid reject having anything to do with the death toll of workers in her sites, deflecting the charge and redirecting it toward regulators, among others (Riach, 2014). Due to its elusive and somewhat broad meaning, the concept of responsibility is worthy of a few brief clarifications.

Responsibility is often understood as having to do with blame or praise, as when a person is retrospectively *held responsible* for having done something or for failing to act as expected—e.g., having failed to deliver a project on schedule. Conversely, responsibility can also be understood prospectively as a relation between a person and a state of affairs or act, in the sense of being *responsible for* doing something—i.e., a person having an *obligation to* an individual or collective entity such as a client, a company, or society at large. Examples of obligations would be not overcharging clients, avoiding conflicts of interest, not denigrating the reputation of a fellow professional, considering implications of professional work and minimizing the adverse impacts, and so on. For a detailed treatment, see Melé (2020, pp. 60–65) and Van de Poel (2015).

Responsibility as an obligation is frequently encountered in discussions about responsibility in design, yet it is 'problematically understood and defined, mostly not going beyond the terms of "professional due diligence"' (Fry, 2004, p. 146). Two main concerns have dominated the understanding of responsibility as an obligation: on the one hand, those linked to proper professional conduct toward clients or the public; and, on the other hand, concerns related to a wider social responsibility of designers (Fry, 2004, p. 146). In short, practice-oriented responsibility in design usually has to do with the professional inclination to present a good product, i.e., one that is suitable to the requirement presented by the client. Responsibility constitutes thus almost a *technical* notion. Furthermore, abiding by local or national legislation is indeed a part of the designer's legal responsibility—health and safety, accessibility, inclusivity, and sustainability procedures are all a part of a designer's professional responsibility.

Along these lines, the concept of responsibility typically suggests tragic vignettes like safety issues in construction or designing weapons. However, while responsibility might occasionally have to do with life and death quandaries, it's more often related to more prosaic issues such as whether to manufacture more chairs into a world cluttered with plastic garbage or designing ads for online gambling platforms.

Next, we want to propose a complementary, alternative way to look at professional responsibility, which is seeing it as consisting of three distinctive, yet integrated dimensions.

The first dimension integrates the notion of duty, praise, and blame that we covered above. It consists of an almost objective set of norms and conventions stemming from the technical characters of the profession. Arguably, sub-disciplines of design will entail a partially differing set of responsibilities.

The second consists of the subjective layer of professional responsibility stemming from a person's integral set of identity-oriented world views, principles, and values—what some would colloquially call their 'personal ethics' or 'ethos'—which can also be linked to how the profession is individually exercised. One designer could abide by a strict vegan code for their designs, another by fair trade, while someone else will focus on universal accessibility.

The third dimension is the most crucial one in our account. Here responsibility must be understood as a *virtue* connected to the professional practice—especially to its larger purposes, excellences, and intrinsic rewards. The central aspect to grasp from responsibility as a virtue is that being responsible is a character trait of the practitioner that's not determined by a set of declarative obligations the practitioner is bound to, but rather by the entailments of belonging to a professional practice. The layer of subjective values acts in combination with the shared purposes of professional activity and accumulates to a more holistic understanding of design responsibility: responsibility as a virtue.

In the remaining sections, we shall leave aside the first two dimensions to focus our attention on the third: responsibility as a virtue intimately linked to the professional practice and its shared purposes. Our analysis concentrates thus on how shared professional purposes manifest in will, thought, and action at the individual level. Despite this focus on the individual designer, it's critical to keep in mind that this account maintains that *the practitioner's ethical reasoning and willing are relational*, i.e., to a significant degree determined in relation to others. In the professional sphere, when a professional designer wills an end, they do so guided by the profession's regulative ideals and overarching purposes in interdependence with other designers, who mutually and dynamically shape these (Guersenzvaig, 2022). In the last section of the chapter, we will briefly revisit the importance of the communal in responsibility, but this clarification suffices for now.

1.6.3.1 Professional responsibility and care

To elaborate further, we need to introduce the concept of 'care'. Care must be understood as an ongoing, relational practice that 'shows us how to respond to needs and why we should. It builds trust and mutual concern and connectedness between persons'

(Held, 2005, p. 42). Care is something in itself, but it also is a key component of responsibility as a virtue. For the ethicist Jessica Nihlén Fahlquist (2015, p. 194), 'to be a responsible person is to feel responsible and to feel responsible is to care about others' wellbeing and how one's actions affect our fellow human beings'.

Thus, for example, inclusive design cares for marginalized social groups, not by labeling them as such, but rather by following the dictum that 'by understanding the extreme you innovate for the mainstream'. Similarly, they do not act out of pity or compassion, but rather out of a concern for social justice, following the principle that every person deserves the same services. Rather than fighting stigma by creating other stigmas, inclusive designers might reframe marginalized groups to a part of the mainstream as a strategy to product and service innovation (Coleman et al., 2016). Moreover, when designing for global development a responsible designer neither aims at 'empowering' nor 'assisting' the local populations from a position of superiority, but at 'working with' design partners toward establishing caring relations based on mutual concern, harnessing their unique knowledge to create together a suitable outcome.

Be that as it may, the essence of imbuing care among designers as practitioners may need a process of reframing and restructuring the very role of designers in society. A short detour through history might illuminate the predicament in which design as a discipline found itself not so long ago.

Concretely, three globally influential circumstances offered a change for designers: the economic collapse of 1929, the war effort of WWII (specifically in the production of an array of vehicles and weaponry), and the post-war world trying to redefine itself after the mid-1940s. For the purposes of this article, suffice it to say that these events triggered a complex shift linking design and capitalist values, as an inherent and almost automatic correlation (Ventura & Shahar, 2021). In other words, design has since then become a marketing instrument meant to motivate consumers to purchase more goods and replace their old models every few months with a newer model. Thus, raising the very essence of care and responsibility vis-a-vis design is not a simple nor a trivial matter.

Within the third, a key moment lies in Germany's efforts to return to the global center stage as a humanistic cradle of design. Heralding a new and improved German stance and heritage, this effort brought to life the New Bauhaus or the Ulm School, three decades after the Nazis shut down the Bauhaus—the first socially sensitive school, still influencing designers 100 years after its inauguration. Interestingly to this context, the school's two headmasters symbolized the two paths of design as a discipline. Contrary to the Bauhaus' approach and mired by its near history, Max Bill, Ulm's first headmaster, opted to focus on functional, affordable objects, rather than politics, values, and ideology. However, following internal and local strains, at the beginning of the 1960s the school's second headmaster, Tomas Maldonado, raised various issues presented in this chapter, more than 60 years ago (Maldonado, 1965; 2018)—what is the role of the designer in society? Should design as a discipline trigger political debate and social change? Why should designers design and how?

The story of responsibility then follows an unfortunate short path through the second half of the 20th century, offering few consolidated answers. Yet at no time as the current reality do these questions resonate more clearly—how could designers care for and about, and what must they be responsible for?

Surely, adjacent paths, triggered by these puzzling questions, from the Scandinavian social approach in the 1960s, through the UK strive for healthcare design, designing the other 99%, culminating with social design and the appraisal of care vis-a-vis design, all stem from these influential historical events (Clarkson et al., 2013; Sparke, 2013). As designers moved away from being 'mere' problem solvers to also focus on the 'fuzzy front end' of open-ended processes of social innovation (Manzini, 2015, pp. 40–43), so did the very scope of the profession change.

As a consequence, roughly two main approaches came to dominate the discipline, albeit not remotely equal in influence—the designer as a marketing agent versus the designer as a social agent. From a historical point of view, however, dealing with care has not been the purview of designers, nor was it a point of interest till the later decades of the 20th century when the aforementioned shift took place.

Contrary to theoretical disciplines—and many other professions—because of their connection with production, designers can transform caring *about* into actual caring *for*. For instance, caring about the environment can be transformed into actual caring for, that is, into *actual* care. In this manner, designers hold a unique stance to influence the daily lives of millions. For instance, an environmental agenda from IKEA's designers to stop manufacturing chairs could have a global impact, just as developing a global standard for laptops' charging cables. These examples do not even begin to exhaust the field of potential courses of action. Furthermore, caring about the environment can be actualized differently: some scholars argue that better than chasing innovative, sustainable solutions it is to just stop designing new things altogether to focus on maintenance (Vinsel & Russell, 2020).

1.6.3.2 Designing as self-enactment

Our argument goes in a different direction than priming maintenance or social and environmental justice, which are not in the least incompatible with our claims. So, while the focus on others and the environment is surely all-important (see Ventura & Bichard, 2017 for further treatment), here we want to make and defend a different assertion from arguing for the necessity to consider others as active subjects of care in design. *We posit that design as a profession is itself a reasonable object of care for a responsible designer to have. Responsible designers care about and for others also because they care about and for their profession.* We'll explain.

Considering the ability of design to influence many aspects of others' lives, the designer's role must be examined along a continuum ranging from the designer as a technician—mainly bounded by legal and general moral duties—to the designer as a *responsible* professional in the virtue sense we explored above, in which duties are still important but do not fully envelop what responsibility is about. Because *being a responsible professional* plays a key part in the designer's own quest to lead a good life, the profession itself—design—can be an object of care. This attachment provides sufficient motivation for acting responsibly. In other words, responsibility is linked not only to an external focus (others, the planet as a whole, etc.) but *also* to an internal one related to the role one's profession plays in one's life plans for becoming the person one seeks to become. We call this process of becoming *self-enactment*. Naturally, this too requires further elaboration.

A person that is 'responsible' is personally invested in what they care about. Philosopher Harry Frankfurt posits that '*we identify* with what we care about' (Frankfurt, 1999, p. 111). When our object of care flourishes we see that as a personal gain. Conversely, when our object of care is harmed, we experience that as a personal loss. This can be illustrated with the example of a design instructor. An invested instructor sees it as their personal gain when a student advances, learns new skills, and acquires abilities they didn't have. Contrarily, a teacher who cares will often experience a sense of personal loss when their students fail to connect and apply their learned subject matter. The upshot is that a responsible teacher is personally *invested* in their students' learning process, and it is because *they care* that a sense of loss or gain is experienced.

Something similar happens in professional design activity. As we advanced above, it is because the practice of design personally matters to us as designers that we are driven to act responsibly. This could be initially articulated through the 'external-internal motivation' concept in pedagogic philosophy (Harpaz, 2013). While monetary incentives are welcome, it is an intrinsic, personal set of values or aims that will equip a learner with the necessary motivation to thrive as a learner. While this may seem the case in other professions as well, design is once more different in the designer's special position as mediator and facilitator. Being a key agent mediating between policymakers, industry and marketing leaders, and the 'ordinary people', to use De Certeau's (1998) description, only raises the importance of caring for design on the designer's part.

For the responsible designer, hiding behind empty excuses such as 'I just work here' does not cut it when influencing the visual and material reality of thousands or even millions of people. Typically, however, designers 'are better able to discuss the principles of the various methods that are employed in design thinking than the first principles of design' (Buchanan, 2001, pp. 36–37). It is perhaps because of this that some often eschew deep discussions of 'ideology' and 'ethics' and prefer to draw an all too easy red line over designing weapons, ads for tobacco products, or pornographic websites. On the other hand, working mindlessly without impacting our field or social sphere, or keeping a safe distance from the theoretical outskirts of the discipline does not cut it either. A responsible designer seeks to walk a different path altogether, bringing together responsibility and care.

Responsible professional designers are personally affected when their object of care—their profession or, even, design as a whole—is advanced or hindered. Hence seeking to realize specific purposes and design solutions matters to them. Designer Michael Bierut, who worked with leading companies and agencies, expresses the concern designers feel when their own beliefs conflict with the clients' or a shifting reality and how this affects them personally:

> In terms of design, I really admire and envy designers who always must do it their way and can walk away from a job if it's not done on their own terms. I remember early on in my career, I worked with a guy who was absolutely secure in his convictions; though he liked it when people agreed with him, it wasn't necessary in order for him to feel that he was right. Whereas if I go into a client meeting, and I can't sell something, I feel like I've failed and my convictions get shaken.
>
> (quoted in Millman, 2007, p. 6)

In practice, responsibility as a virtue starts by asking oneself questions about what choices and decisions are compatible with desirable aspects of one's identity, and about the pursuit of what purposes contribute to one's development as a person. A question like 'could I stand to be the sort of person who can do that?' is a question we, albeit tacitly, often ask ourselves—yet perhaps not often enough. Moreover, due to its broad and overreaching nature, design can and often does lead to thorny uncertainties around equality, gender, identity, legal boundaries, and more. As such, dealing with responsibility is not a hypothetical or theoretical conundrum, but a significant matter steering one's identity and the ongoing process of *becoming oneself*, in both the professional and the personal spheres.

Therefore, the classic interrogation of 'what would you not design?' might trigger an answer of 'what is not necessary', rather than the abovementioned redundant clichés (e.g., 'weapons'). Moreover, setting aside the designer's perception of an ego-driven creator in favor of a more complex and relational conception of designing might help in strengthening one's character, securing one's set of values and kernel of responsibility. Indeed, while professionally participating in the care for others with special needs or those who are marginalized may fall to the hands of dedicated professional roles (e g., inclusive designers), the endeavor of defining one's relationship with and influence on the industry and the market becomes an inevitable process for all responsible design practitioners.

1.6.3.3 From caring about to caring for

As we explained, the external and internal sides of responsibility are basic necessary traits of design as a practice. Yet, when moving from this starting point to a deeper and more significant stance, we must reconsider the essence of care. The notable addition to this factor is the consideration of an even more complex and richer notion of empathy, identification, social virtue, beliefs, and emotions. While the status of emotions as a valid source of moral knowledge is a hotly and widely contested issue in philosophy, the character-based perspective we present here leaves plenty of room for considering them as a relevant input for moral reasoning. Again, transferring the designer's creativity from connecting emotions and creating evocative objects, to the deeper and relevant link between care, empathy, and phenomenology. In other words, the designer's understanding of the design partners' lifeworld, viewed through a research-based understanding of feelings, memories, and experiences helps in securing their subjective set of values as well as their sense of responsibility, which is firmly anchored in the very profession they are committed to. This shift could be articulated through a simple choice of color in the design process—am I choosing orange because it sells more, or because it will enable my design partners confront shame and stigma?

In short, what we care *about* and, more importantly, what we care *for* helps define our professional expertise—social design or marketing-oriented design, for example— but it is paramount in our ethical life. Ethics is not primarily about following rules, but about pondering on the effects our everyday existence has on the world as well as about our reasons for actions and purposes. This is why design as a daily, disposition-led action is so relevant. The graphic designer Milton Glaser devised a 12-question test he called 'The Road to Hell'. The test accentuates the importance of reflecting on 'what not to do' when designing and it becomes more difficulty the deeper you go. For example, *would you design*:

1. a package to look bigger on the shelf?
7. a package aimed at children for a cereal whose contents you know are low in nutritional value and high in sugar?
8. a line of T-shirts for a manufacturer that employs child labor?

(Glaser, 2004)

While coming to an answer to the latter question might be fairly uncomplicated,[4] the first two also tackle the topic of harms but from a more elusive perspective. Should designers design a technically well-executed product, for which they are handsomely paid, to deceitfully yet legally influence or even harm customers? Glaser's words are ethically relevant in that they remind us of what to care *about*, or at least what is worth preserving and protecting. He is not providing a set of dos and don'ts to follow or adhere to blindly, yet all of his questions have the principle of 'do no harm' as a backdrop. With his test, Glaser makes us question what we are aware of: 'In regard to professional ethics, acknowledging what it is we do is a beginning' (Glaser, 2004).

To elaborate on this theme, imagine the following scenario: an interface designer, suppose their name is Chris, feels uneasy and troubled regarding a client asking for the inclusion of so-called 'dark patterns' in the design for an app they are working on.[5] Imagine also that Chris seeks to negotiate with their manager or client and argues for a different design, one that is not manipulative and does not trick (potential) customers of a service into engaging in determinate behaviors that are detrimental for them.

In line with our argument, it is the *way* Chris approaches and deals with this situation that is central to ethics. The result they might obtain in their negotiation is naturally relevant but is also highly contingent on many factors outside Chris' sway—first and foremost, power dynamics, influenced by race, gender, or age, as well as an institutional culture and so on. The upshot is that Chris shows moral and emotional maturity by recognizing and facing a difficult, ethically fraught situation. In this way, Chris' character is further developed and strengthened as they seek to balance as a whole person the multiple demands they face: not doing something as a professional designer—a manipulative design—they reject as a person—manipulation. Chris is willing and able to engage. They avoid being mindless and shutting off the uneasiness not necessarily because they have an obligation toward the end-user, but because in their view a responsible designer would not design to manipulate others. What's more, Chris would not bear to become that type of designer who does that.

This scenario is mirrored in the assertion offered by the designer James Victore that 'the unspoken part of what we do is compromise. Clients don't just come to me and say, [...] "we'll let him do what he wants"' (quoted in Millman, 2007, p. 107). Victore thus truthfully summarizes the daily essence of design, one which is typically not told to design students: design practice is not a magical and instantaneous process, but a rather arduous, conflicting, and time-consuming process of negotiation and conversation. The designer materializes or visualizes an intricate compromise, mediated by the value systems of a plethora of design partners with blurred argumentation positions, which serve as the basis for the 'conversational moves' that move the project forward (McDonnell, 2009, pp. 35–36).

Another vignette could be articulated through industrial design. Imagine Lucy, another hypothetical designer, who receives a commercial brief to design a plastic chair. Several possibilities arise. First, they can design the chair restricting themself to the consideration of the technical aspects of the brief without appraising further implications. Second, they can outright reject the brief *Quixotically*, stressing that the project is beyond rescue because the world is already filled with chairs and there is no need for another. Third, they can suggest alternative and more plausible approaches. For instance, subtracting an amount of plastic from the mold (i.e., 'dematerializing' the chair, which makes it lighter), designing for longevity and repairability, placing the chair into a circular business model (such as 'Product Service Systems Models'), and so on. Evidently, there are many other possible courses of action, some of which might end up in the decision of not designing the chair at all or in firing the client or in Lucy quitting their job. All of this, however, would be outside the point we try to make here, which is agnostic about advocating sacrifices to uphold one's values whatever the cost. Our point is that Lucy's consideration of the possibility of different alternatives is a rejection of inevitability that initiates responsible behavior and instantiates responsibility. Joseph Weizenbaum (1976, p. 241) affirmed that 'the myth of inevitability is a powerful tranquilizer of the conscience. Its service is to remove responsibility from the shoulders of everyone who truly believes in it'.

Through the conception of alternatives, Lucy embraces responsibility as they resist thoughtlessness and fatalism, not in the harsh Marxist binary perception, but rather as a protest through everyday choices and actions (De Certeau, 1998, p. 34) that allow her to face the difficult issues and come to terms with them. This scenario focuses on environmental aspects, but Lucy, being the responsible designer they are, is not naive about design and knows that its effects are often even broader.

1.6.4 The perils of compartmentalization

In general, personal identity is a composite of the different roles we adopt—professional, friend, parent, sibling, daughter, employee or employer, lover, citizen, and so on. This compounded nature entails that the ethical deliberations one performs end up generating irresolvable tensions and conflicts between one's different social roles. However, truly acting in a way one can live with means making it personal and owning the decisions one makes.

This means reflecting on the power dynamics of the situation and on which important things are at stake and must be balanced, as well as on what gets lost in the process and for the preservation of what greater good. This reflection requires moral and emotional maturity. Especially in the form of emotional regulation skills, which help us avoid being overwhelmed and alienating ourselves from what is important to us in order to cope with the complexity of work life. Victore (quoted in Millman, 2007) articulates the difficulties of navigating internal conflicts between one's personal and professional roles:

> Professionally, I don't really have any worries. Any. I like what I do. But I am worried about what the state of the profession will be in the future. I'm worried about the state of the world. My concern now is to make a little bit of money. And for the first time in my life, I feel guilty about it.
>
> (p. 108)

> I like what I do, and I seem to have a reputation for altruism and telling the truth; but at the same time, all that work I do for free. Or I pay for it with my own money. And now I'm worried about making a living for my family. And this bothers me because I don't know how to do both. And I want a hot rod!
>
> (p. 109)

With candor, Victore highlights the classic dilemma of designers rooted in the industry—trying to maintain a sense of self by navigating between one's values or moral compass and one's red lines, all while upholding the professional commitment and a sense of shared purpose that is related to a practice. To cope with this difficult position, people often—willingly or not—turn to *compartmentalization*; that is, dividing one's life into 'segments, each with its own norms and modes of behavior. So work is divided from leisure, private life from public, the corporate from the personal' (MacIntyre, 2007,

p. 204). This means doing something as a professional that one finds ethically inadmissible as a friend, citizen, or parent. When designers compartmentalize, they typically tell themselves that 'design is just a job', 'I am not my job', 'I'm a designer, my job is to help companies with whatever they ask me, whether I like it or not', and so on and so forth. While allowing one to function and—often only fleetingly—resolve one's dilemmas, a fragmented self is seriously diminished as a moral agent, and it is incapable of bridging the limitations foisted by its own different roles.

Again, as we have seen in the Ulm School's internally conflicting worldviews, design as a discipline needs to debate whether it is in exclusive service of the industry, or perhaps something else altogether. This second option does not necessarily entail becoming a pro-bono designer or a secluded hermit. To start, we need to—as Glaser proposed—acknowledge what it is we do. We also need to learn to recognize and avoid the harmful influence of a fragmented self, only then we can try to determine a path that would be acceptable in both a professional and personal manner.

Every designer constantly negotiates and compromises in their daily work about technical, formal, or aesthetic issues. Through 'conversational moves' between stakeholders in which information is exchanged, a course of action is collaboratively negotiated to make a project move forward (McDonnell, 2009). This stems from the fluid and temporal nature of the discipline, in which designers find themselves working in various fields of production and facing conflicting demands from different actors. But it is our contention that this negotiation normatively involves the responsible designer as a whole person, implicating their different roles when relevant. The importance of negotiation is illustrated by the words of the designer Paula Scher (quoted in Millman, 2007, pp. 50–51):

> There are all kinds of problems and compromises that [one] must negotiate. Things that have to be held on to, things that have to be protected to make something move forward. And it's very, very, very hard work.

These words are important because they can be understood as highlighting that Scher is not merely *wishing* to hold on to these elusive things—these things truly matter to her and are worth working hard for. She makes them especially her own. And they seem to be truly important because she fully *identifies* with them. It would be a personal loss for her if what she cares about was diminished.[6] Designing as self-enactment lies also in mundane tasks like perfecting a billboard with poor legibility, as it does in developing a virtual solution for an inclusive clinic or a banking app that allows those who suffer from gambling addiction to block transactions relating to gambling (some banks already offer a tool like this, see Monzo, n/a). A responsible designer—because they care—would realize that even if this blocking feature won't resolve the underlying gambling addiction, it will add a layer of 'friction' that can contribute to reducing the impulsivity that is afforded by the immediacy of digital tools.

As we see, even among the Olympus of the designers, there are relevant and important issues larger than answering a client's brief. Therefore, we can surmise that becoming a responsible designer means learning to recognize, appreciate and care about and for those things *that need to be held on to and protected to make a project move forward.*

But what are these elusive things? At its most general, the virtue of responsibility is expressed in truly caring for certain things because they are seen to be directly connected to the attainment of the *internal goods of design*—the exclusive rewards that design as a practice offers its practitioners, which can *only* be obtained by designing (Guersenzvaig,

2021). Perfecting a billboard's legibility and designing an elegant interaction are just two examples of things that can only be achieved through design. Internal goods have more to do with those things that make a project move forward than with a high income or prestige. Money or fame, however important, don't make the cut to count as exclusive to design, as they can be attained through many other means. Goods like these are *external* goods (see MacIntyre, 2007, Ch. 14 for a canonical discussion of internal vs. external goods). The contrast between internal and external goods is clearly articulated by Scher (quoted in Millman, 2007, p. 44): 'To me, success is not about money, it's about what I design'.

Clarity about goods and a sturdy belief system is all well and good, yet when working in the industry, designers find themselves in complex situations leading to difficult choices and oftentimes requiring *rotten* compromises. Even a star designer like Bierut acknowledges a certain—yet not irresolvable—fuzziness in the conflicts between internal and external goods. Should we focus on being paid, on sustaining our reputation, or perhaps on something bigger that is more intrinsically related to design? In the following quote, he also expresses his concern with the struggle between professional responsibility and the need to earn a living or simply keep the lights on in your studio:

> Most of us would say that our ideals, whether newfound or long held, give way at the end of the day to the pressures of running our businesses; that the sanest course of action is to push environmental activism or social consciousness as far as you can and then back off to fight another day; and that a client's a client and an invoice is an invoice. In the end it's all about money, isn't it? Well, maybe not. Maybe it's about something else, something that hasn't changed, something to do not with money, but with the very structure of the relationship between designers and their clients.
>
> (Bierut, 2007, p. 19)

Milton Glaser stresses the external nature of money in relation to his work and highlights how money, while important to him as to most designers, is not the primary driver behind his work. At the same time his words make explicit the relation between designing and his sense of self:

> Money has never been a motivating force in my work. I am very happy to have made enough money to live as well as I do, but I never thought of money as a reason to work. For me, work was about survival. I had to work in order to have any sense of being human. If I wasn't working or making something, I was very nervous and unstable.
>
> (quoted in Millman, 2007, p. 32)

While designers differ from artists in their need to correlate with their client's brief, this should not serve as an excuse to annul all other evaluative systems in favor of maintaining the relationship with clients. Designers must acknowledge and articulate a system of responsibility and values, for both themselves and the discipline. Responsibility as a virtue revolves around a readiness to care and a willingness to respond to a plurality of often conflicting demands. Of course, this readiness to care is bound by the *effective ability* a designer has to responsibly deal with those demands. In other words, internal capacities such as imagination, perseverance, or judgment, as well as the external institutional dimension of one's circumstances (Williams, 2008, p. 459).

What *can* be done in actuality, then? We sketch two boundary conditions for action. First, leaving designers to decide between two unproductive courses of action—either leaving the industry or embracing carelessness is—to reiterate—simply not a reasonable option. Designers work *in, with,* and *for* the industry. Any viable professional ethics must recognize this fact and preserve the ability to make a living through the profession while staying connected to the industry. Second, as many authors have vigorously argued from different political and theoretical perspectives, *it is necessary to act in order to change our relation to the discipline and change the discipline itself* (e.g., Boehnert, 2018; Bruinsma, 2000; Costanza-Chock, 2020; Dreyfuss 2001 [1955]; Garland, 1964; Irwin, 2015; Maldonado, 2018; Margolin, 2002; Monteiro, 2019; Papanek, 1984). Moreover, the understanding and triggering of a holistic professional change must come from within design practice because crucial insights into the ethics of designing can be obtained by practitioners examining the practice from within to recognize what designers *already* do that amounts to what design can be at its best (Guersenzvaig, 2022). Change from within also requires reflecting on and taking responsibility for design's role as a marketing-oriented and consumption generator agent to the industry. It is from this awareness about the practice that we can change our relation to the discipline and transform the *caring about* into the actual *caring for* that design is about.

In sum, avoiding compartmentalization demands from us to 'stay with the trouble', to state it using Donna Haraway's formulation. Attempting to change the practice while maintaining integrity does not require, even in urgent times like these, to 'clear away the present and the past in order to make futures for coming generations' (Haraway, 2016, p. 1). A more modest yet promising individual and collective commitment to change is to engage in practices of responsibility—and responsibly looking at the past and the present is included in these—without futilely expecting as a precondition that our practice is pure and free from everything we repudiate because it will never be. Design is too complex and fraught to be cleansed and purged of its ethical entanglements.

1.6.5 Conclusion

Our approach avoids the well-trodden yet unfruitful, paths of succumbing to the industry and coping with it from a disengaged and fractured self, on the one hand. And exiting the profession altogether, on the other. We suggested a different course of action in which a responsible designer is aware of what bounds and constrains them morally and furthermore invests a lot of effort in changing the constraining conditions. All this out of a readiness to care and a sense of integrity. The words of Milton Glaser serve, again, to highlight this avenue for professional responsibility:

> I think I stand for an attempt to think responsibly about the practice of design in terms of its effects on society. I don't say that self-righteously. I believe that you have to think about the consequences of what you do. I try to be conscious about this and raise the issue wherever and whenever I can. I also think I stand for an openness about what the design profession can be, and how things beyond the practice affect and modify both society and the way we think about design.
>
> (Milton Glaser, in Millman, 2007, p. 39)

Glaser's words signal the need to embrace introspection and reflection on the nature and effects of design, but they also underscore the need to implicate ourselves in the effort to change the constraining conditions under which we operate. This effort consists primarily of engaging in difficult negotiations and persuading others—the client, project managers, marketing execs, etc.—into a consensus about what counts as possible. The narrower the conception of what counts as a valid expected outcome—that is, one that is accepted by the client as a valid project *result*—the more constrained the space of true possibilities and, in turn, the designer's true options for moving from caring *about* to caring *for*.

The effort to change the constraining conditions resonates with what Paul Dourish (2001, p. 172) calls the 'designer's stance'; i.e., the designer's own conception of what they are doing and the particular perspective they adopt regarding the role they play in a given design situation. A responsible designer has a stance that enables them to design as to aim to ensure that the possibility of care becomes an actuality. It has to do with a sense of integrity and identification with what they care about—in this case, it is design, their profession, its internal goods, and its larger purposes.

One of the reasons *why* a responsible designer strives to broaden the solution space becomes clear: it's because it personally affects them whether care is realized or not. The responsible designer finds motivating reasons for ethical action because they care, and they care because designing is a constituent part of who they are and who they aspire to become. What they do as a practitioner is an important part of their narrative quest—their very own developing story as a person. Designing responsibly brings them closer to becoming the person they seek to be. All of which will help nudge the scales from a short-term focus on external goods such as money or prestige to the purposeful and long-term investment in the internal goods of design. The virtues of responsibility and practical wisdom allow the designer to know what to do by developing attachments to what matters and by enabling them to rank goals, purposes, and rewards that are to be pursued and held to. This is what self-enactment is about.

This account enables us to retain the notion of responsibility without implying that a designer is wholly *responsible for an outcome*. Arguing that designers have full control over both process and results would be absurd. Instead, responsibility is about actualizing care when one has the effective capacity to respond and about not eschewing to seek out ways to be less constrained by the circumstances when one doesn't have those effective capacities. When responsibility is understood as the readiness and willingness to convert care into design under complex and dynamic circumstances, acting responsibly does not seem farfetched, but rather plausible *even* in a market-led environment.

1.6.5.1 Clarifying remarks

Before concluding we want to make three clarifying remarks even though we don't have the space left at this point that would be necessary to consider these threads with the attention they deserve.

First, caring in the sense outlined in this chapter is not *necessarily* selfless or altruistic, neither does it emerge out of a sole aptitude for empathic concern for others—though it can and in some cases *should* be. Caring is, however, unquestionably personal. A caring stance is characteristic of how a responsible designer engages in the pursuit of the internal goods of design, which is—to reiterate—part of their narrative quest as a person. Being responsible and being invested in one's profession is fundamental for enacting one's life plans as a professional, as well as for advancing design as a profession. In this sense, our approach shows a plausible way in which responsibility can go beyond being a remote aspirational ideal for the design profession, but a purpose that is both reasonable and worth pursuing. This understanding, however, entails constantly reevaluating and adjusting the broader, abstract goals we give ourselves—such as belonging, happiness, authenticity. transcendence, success, and many others—and how they fit together with our own narrative quest and the other practices we participate in.

Second, we wish to zoom in to and acknowledge once more the dynamics of current professional design, which is embedded in for-profit organizations. There are huge political, organizational, and economical constraining conditions that a professional designer faces in the context of contemporary modern capitalism. Everyone who has a job knows that market forces affect and tend to reduce all spheres of human activity to that which can be exploited for profit. We believe, however, that all this does not invalidate our proposal, which is more about how to navigate these tensions than about resolving the problems of designing in times of capitalism.

Moreover, influencing the field through its hardcore industrial sphere creates another sort of impact than choosing the safer approach and influencing through design venues that are already deeply rooted in moralistic-humanistic roots (i.e., inclusive, empathic, human-centered design, etc.). Designers—and especially our students as designers-to-be—would do well to realize that, beyond market forces, our deeper reality is anchored in interdependency and vulnerability, which inescapably require care (MacIntyre, 2001; Held, 2005). What's more, capitalism itself is sustained by all sorts of care activities, which take place against a backdrop of unequal, exploitative class and gender relations, without which capitalism can't survive. Think of the unpaid or low-pay care work, which is mainly and disproportionately carried out by women: all that cooking, cleaning, and caring for children, people with special needs, and older adults. For many scholars, this is no coincidence but a basic condition of capitalism (Müller, 2019). This instrumentalization, however, doesn't detract any value or goodness from care itself—whether realized through design or child-rearing—but it makes us recognize, question, and reject the parasitic nature of the relationship.

In 1964, Dutch graphic designer Jan van Toorn lambasted: 'It is not our job to please business' (Crouwel & Van Toorn, 2015, p. 55). Can he be right? It certainly *is* our job as designers to deliver work that meets the requirements stipulated in the design brief and for which we were hired in the first place. This might seem to contradict Van Toorn's defiant words. Yet, despite this apparent contradiction, he has a point: meeting business goals can't possibly be the designer's *only* job. Professional designers have many other goals and purposes, among which contributing to others' flourishing is paramount—of course, how this task ought to be carried out *in practice* is a perennial matter of discussion among practitioners.

To clarify, we are not arguing against increasing a company's revenue or market share, neither against seeking fame for oneself, for that matter. We posit that this turns problematic when achieving external rewards like these become the designer's main concern that systematically overrides more strategic professional purposes. When this happens the practice itself is debased and derailed. The pernicious consequences of pleasing business *no matter what* are evident: not only does the practice of design suffer but its claim to professional status becomes moot.

Third, despite self-enactment being our unit of analysis, it would be mistaken to view this account as putting forward a view of designers as self-absorbed selfish co-operators. Responsibility, as we presented it here, is a disposition, rather than a matter of convenience. Furthermore, an individualist interpretation can't possibly be a correct reading, as our account relies on an at least minimally shared conception of the goods and purposes that are pursued by the profession *in practice*. While responsibility as a virtue surely arises from discovering what it is to be ourselves, many of the issues calling for responsibility are not resolvable individually.

Sharing a purpose doesn't tell the whole story because actual and genuine participatory engagement with the practice is indispensable to realizing the ends we pursue as designers, which are connected to a communal narrative. Yet, acting upon shared purposes is not, in the words of Charles Taylor (1991, p. 8), 'just a matter of changing the outlook of individuals, it is not a battle of "hearts and minds."' The sheer size and influence of the market forces we alluded to in the previous paragraphs nudge us in the direction of atomism and hopelessness, where we often abide by the rules of market efficiency, even when they conflict with our own views (Taylor, 1991, p. 97). It would be naive or even cynical to deny this. Nevertheless, there is no use in fatalism, thoughtlessness, and passivity other than further reinforcing the vicious circle of inevitability and the abdication of responsibility.

Coming to instantiate responsibility and genuinely sharing a practice together with others are ways of resisting those forces while at the same time *caring for* it from within. Active engagement in a common pursuit strengthens the practitioners' identification with the practice as well as the mutual attachment to fellow designers, all of which weaken the atomism and fatalism that make us feel hopeless. At the same time, our very selves are enacted and widened. Thus, a shared purpose transcends each of us as individual practitioners and substantively connects us to one another, making self-enactment a collective quest as well as an individual one: as professional designers, we enact ourselves with and through others.

Notes

1 In many professions, unlike design, a strong ethical tradition has been established over time, often resulting in formal oaths and codes of ethics. Yet, for an internal questioning to take place more is required than simply formulating a code, which remains inconsequential when not inscribed in the matrix of a robust professional ethics (Guersenzvaig, 2021, p. 57).
2 Interestingly, Starck's client in this case—Alessi—attested a different approach toward the product, claiming it works wonderfully (the actual case is that as a functional juicer, the Salif is a terrible product, but as a discursive object or social status symbol it is exemplary).
3 Here, the term 'design thinking' must unequivocally be understood in a narrow sense, as a distinct form of intelligence and cognitive processes related to the discipline of design— not as a set of techniques. There is a vast body of scholarly research in this area including especially the DTRS series (See Cross, 2010).
4 We call it 'uncomplicated' because defending child labor is a morally untenable position. Despite being a curse for every child and the world as a whole, child labor is still one the most critical unresolved issues humanity faces. In 2020, according to UNICEF, 160 million children were forced to work.
5 Dark patterns are design tropes that seek to promote business objectives such as increasing revenue by exploiting cognitive biases and psychological mechanisms.
6 A disclaimer: we want to illustrate a point and we might be wrong in our interpretation of Scher's words, the point we make about self-enactment stands regardless.

References

Álvarez, M. (2017). Reasons for action: Justification, motivation, explanation. In Zalta, E. N. (ed.), *The Stanford Encyclopedia of Philosophy* (Winter 2017 Edition). Available at https://plato.stanford.edu/archives/win2017/entries/reasons-just-vs-expl/, Retrieved on 6/11/2021.

Bierut, M. (2007). *Seventy-nine short essays on design*. Princeton Architectural Press.
Boehnert, J. (2018). *Design, ecology, politics: Towards the ecocene*. Bloomsbury.
Bruisnma, M. (2000). *First things tirst 2000 a design manifesto*. Available at https://maxbruinsma.nl/index1.html?ftf2000.htm, Retrieved on 12/05/2021.
Buchanan, R. (2001). Human dignity and human rights: Thoughts on the principles of human-centered design. *Design Issues*, *17*(3), 35–39.
Castillo, G. (2010). *Cold war and the home front: The soft power of midcentury design*. The University of Minnesota Press.
Clarkson, P. J., Coleman, R., Keates, S. & Lebbon, C. (2013). *Inclusive design: Design for the whole population*. Springer.
Coleman, R., Clarkson, J. O. H. N. & Cassim, J. (2016). *Design for inclusivity: A practical guide to accessible, innovative and user-centred design*. Routledge.
Costanza-Chock, S. (2020) *Design justice: Community-led practices to build the worlds we need*. MIT Press.
Crouwel, W. & Van Toorn, J. (2015). *The debate: The legendary contest of two giants of graphic design*. The Monacelli Press.
Cross, N., (2010). Design thinking as a form of intelligence. In Dorst, K., Stewart, S., Staudinger, I., Paton, B. & Dong, A. (eds.), *Proceedings of the 8th design thinking research symposium (DTRS8) interpreting design thinking, Sydney*, 19-20 October 2010, pp. 99–105.
De Certeau, M. (1998). *The practice of everyday life: Living and cooking* (Vol. 2). Minnesota University Press.
Dong, A. (2008). The policy of design: A capabilities approach. *Design Issues*, *24*(4), 76–87.
Dourish, P. (2001). *Where the action is: The foundations of embodied interaction*. MIT Press.
Dreyfuss, H. (2001 [1955]). *Designing for people*. Skyhorse Publication.
Frankfurt, H. G. (1999). *Necessity, volition, and love*. Cambridge University Press.
Fry, T. (2004). The voice of sustainment: Design ethics as futuring. *Design Philosophy Papers*, *2*(2), 145–156.
Garland, K. (1964). *First things first*. Available at http://www.kengarland.co.uk/KG-published-writing/first-things-first/, Retrieved on 12/05/2021.
Glaser, M. (2004). *Road to hell in 12 steps*. Available at https://www.miltonglaser.com/files/Essays-Ambiguity-8192.pdf, Retrieved on 22/06/2021.
Gray, C. M., Hasib, A., Li, Z. & Chivukula, S. S. (2022). Using decisive constraints to create design methods that guide ethical impact. *Design Studies*, *79*, 101097.
Guersenzvaig, A. (2021). *The goods of design: Professional ethics for designers*. Rowman & Littlefield.
Guersenzvaig, A. (2022). An inquiry into the practice of Design. In Maletta, S., Mazzola, D. & Simoncelli, D. (eds.), *Practical rationality and human difference*. Mimesis International.
Haraway, D. (2016). *Staying with the trouble: Making kin in the Chthulucene*. Duke University Press.
Harpaz, Y. (2013). *Teaching and learning in a community of thinking: The third model*. Springer Science & Business Media.
Held, V. (2005). *The ethics of care: Personal, political, and global*. Oxford University Press.
Irwin, T. (2015). Transition design: A proposal for a new area of design practice, study, and research. *Design and Culture*, *7*(2), 229–246.
Lidwell, W. & Manacsa, G. (2011). *Deconstructing product design: Exploring the form, function, usability, sustainability, and commercial success of 100 amazing products*. Rockport Publishers.
Macintyre, A. (2001). *Dependent rational animals: Why human beings need the virtues*. Open Court.
Macintyre, A. (2007). *After virtue: A study in moral theory*. University of Notre Dame Press.
Maldonado, T. (1965). Die ausbildung des architekten und produktgestalters in einer welt im werden/The emergent world: A challenge to architectural and industrial design training. *Ulm Zeitschrift der Hochschule für Gestaltung*, (12/13), 2–10.
Maldonado, T. (2018). *Design, nature, revolution: Toward a critical ecology*. Minnesota University Press.

Manzini, E. (2015). *Design, when everybody designs: An introduction to design for social innovation*. MIT Press.
Margolin, V. (2002). *The politics of the artificial: Essays on design and design studies*. The University of Chicago Press.
McDonnell, J. (2009). Collaborative negotiation in design: A study of design conversations between architect and building users. *CoDesign*, *5*(1), 35–50.
Melé, D. (2020). *Business ethics in action*, 2nd. Red Globe Press.
Millman, D. (2007). *How to think like a great graphic designer*. Allworth Press.
Monteiro, M. (2019). *Ruined by design: How designers destroyed the world, and what we can do to fix it*. Mule Design.
Monzo. (n/a). *Gambling block*. Available at https://monzo.com/i/gambling-block, Retrieved on 10/11/2021
Müller, B. (2019). The careless society—Dependency and care work in capitalist societies. *Frontiers in Sociology*, *3*(44).
Nihlén Fahlquist, J. (2015). Responsibility as a virtue and the problem of many hands. In van de Poel, I., Royakkers, L. & Zwart, S. D. (eds.), *Moral responsibility and the problem of many hands*, 187–208. Routledge.
Oakley, J. & Cocking, D. (2001). *Virtue ethics and professional roles*. Cambridge University Press.
Papanek, V. (1984). *Design for the real world*, 2nd Revised. Academy Chicago Publishers.
Riach, J. (2014). *Zaha Hadid defends Qatar World Cup role following migrant worker deaths*. Available at https://www.theguardian.com/world/2014/feb/25/zaha-hadid-qatar-world-cup-migrant-worker-deaths, Retrieved on 12/05/2021.
Schön, D . (1983). The reflective practitioner: How professionals think in action.Basic Books.
Segal, K. & Ventura, J. (2019a). *The short(er) history of design*. Books in the Attic (in Hebrew].
Segal, K. & Ventura, J. (2019b). From material to abstract: Contemporary Israeli design manifested through the culture of wood. *Israel Affairs*, *25*(5), 826–853.
Sparke, P. (2013). *An introduction to design and culture: 1900 to the present*. Routledge.
Sudjic, D. (2008). *The language of things*. Penguin.
Taylor, C. (1991). *The malaise of modernity*. House of Anansi Press.
Van de Poel, I. (2015). Moral responsibility. In van de Poel, I., Royakkers, L. & Zwart, S. D. (eds.), *Moral responsibility and the problem of many hands*, pp. 12–49. Routledge.
Ventura, J. & Bichard, J. A. (2017). Design anthropology or anthropological design? Towards 'Social Design'. *International Journal of Design Creativity and Innovation*, *5*(3-4), 222–234.
Ventura, J. & Shahar, D. (2021). The social designer: Ideology, the crucial link between education and practice. In P. Rodgers & C. Bremmer (eds.), *Design education in the anthropocene*, Routledge.
Vinsel, L. & Russell, A. L. (2020). *The innovation delusion: How our obsession with the new has disrupted the work that matters most*. Currency.
Weizenbaum, J. (1976). *Computer power and human reason: From judgement to calculation*. W. H. Freeman & Company.
Whitbeck, C. (2011). *Ethics in engineering practice and research*, 2nd edition. Cambridge University Press.
Williams, G. (2008). Responsibility as a virtue. *Ethical Theory and Moral Practice*, *11*(4), 455–470.

Part 2

Design thinking in the studio

DOI: 10.1201/b22630-9

2.1 Contributions toward design as a methodology for material-reflective reconfiguration

Ruth M. Neubauer and Christoph H. Wecht
New Design University, St. Pölten, Austria

2.1.1 Introduction

Design has an immense potential as a method for seeing the world, understanding relationships, and for shaping those relationships toward more desirable ways of working and living (Buchanan, 2015). Design experts aspire to be orchestrators of social change as "they collaborate actively in creating conditions that facilitate it" (Manzini, 2015, p. 3). Organizations readily adopt design thinking as a method of problem solving, for navigating production processes, for promoting new mindsets, and as a radical reinvention of the human experience (Dell'Era, Magistretti, Cautela, Verganti, & Zurlo, 2020). Courses in design thinking have popped up around the globe; prominent among those are the Stanford d.school's and IDEO's (Melles, 2010). Many transdisciplinary design projects have been implemented with these design thinking approaches, and many curricula use these processes as a guide to teaching design thinking. However, teaching and practicing design as a general method for change is a new challenge. Design has a strong practical tradition as a discipline that is done by the designer, and it is theorized as a form of individual thinking and doing (Margolin, 2013). This individual ability is in the academic design literature often understood as a product of mental capabilities (Kimbell 2011, Badke-Schaub Dorst 2010, 2012; Tonkinwise, 2018). It is uncertain how this conceptualization scales up to teams practicing design thinking (Badke-Schaub, Roozenburg, & Cardoso, 2010). Furthermore, design as a material practice – as recognized in fine art practices (McDonnell, 2011) – is underrepresented in design research. Although it has been established that materials and toolkits play an important role in the design process (Ball & Christensen, 2020; Sanders & Stappers, 2012; Dorothé Smit et al., 2016), there are insufficient answers to how ideas 'materialize', in the full meaning of the word. There is more work to be done to fully capture the capacity and agency of design in order to teach and practice design thinking as a method for change. How may designing be theorized so it can be taught and practiced as a general method of making change, in the wider context, outside of the designer studio, done by 'non-designers'? How can we prepare practitioners in a way that captures the material essence of designing? With the help of theories of materiality, we attempt to illuminate these questions and contribute toward a theoretical substantiation that explains how designs materialize and tangibly impact the world.

This work focuses on the 'material-reflective' quality of designing. The research draws on concepts that illustrate design as an activity where practitioners and materials act together through the design artefacts that are made. These design artefacts, being simultaneously reflective and material artefacts, actively progress an ephemeral thought to a tangible design (Neubauer, 2022). A key idea we use in this chapter is three-dimensionality, which on the

DOI: 10.1201/b22630-10

one hand embodies the impression of a three-dimensional object that helps imagine design as a tool. On the other hand, it amplifies the idea that design works in all dimensions and has the capacity to facilitate change in the world.

2.1.2 A design thinking curriculum

During a research project, we utilized the curriculum of a bachelor's degree program in design thinking at the New Design University in Austria to research and develop the idea of design as a three-dimensional method. We, the authors, are teachers in this program, and we are design scholars of different backgrounds – one of us from a traditional art and design background, the other from a technology management background. We are thus faced with theorizing design in a way it works for both perspectives and so it can be taught to our students who are not traditional design students but management students. The research question, therefore, was appropriate for this setting: The theorizing of design as a generally accessible method for change, harnessing the core capacity of design as a material practice.

2.1.2.1 Theoretical framing and research methodology

Design is a material practice, and as an activity, it is as much material as it is practical. This circumstance makes the theorizing of its own practices not something that comes easy, but all the more it is necessary for understanding design. Knowledge is often ascribed to individual mental capacities, but for the purpose of understanding what is known in material practice, it is more useful to adopt the 'sociomaterial' understanding of practice-based theories, which describe knowledge as distributed in networks of human bodies, tools and materials, along with the objectives, rules and norms that guide the activities (Gherardi, 2017). Knowledge is here understood as embodied – it emerges in situated action and is anchored in the material world (Gherardi, 2012, p. 25). When Schön (1983) explored the knowledge that professionals employ when they work, he realized that they are in "a reflective conversation with the situation" (p. 93). In this conversation, the practitioner makes a move and the situation "talks back" (p. 79). The practitioner employs a reflection in action to assess the consequences of a move for its desirability, suitability, and any "new problems or potentials" it may create (p. 101). A design move "transforms the design situation relative to the state in which it was prior to that move" (Goldschmidt, 1996, p. 72). Design moves interlink with previous moves through "evaluative, sometimes repetitious acts that define, constrain and clarify the problem space of the design task" (p. 89). In, what Schön calls, the "language of designing", parallel ways of talking and drawing (p. 80), the design moves become audible and visible and are anchored in artefacts, such as drawings, maps, models, or stories. These accounts describe design as a material-reflective activity, which is materially anchored in bodies and materials, and reflectively builds on current situations in order to change them.

According to concepts of materiality in science and technology studies (STS), the world is made up of sociomaterial relations, which have material and discursive qualities (Barad, 2007; Law, 2004; Mol, 2002; O'Riordan, 2017; Suchman, 2007). These relations are temporarily stabilized products of interactions. As humans interact, material and discursive relations are made and broken; they move and are negotiated, and, most importantly, they stabilize (always temporarily) as facts, objects and environments. They stabilize until they are destabilized again through an intervention that gets them to form new relationships; new configurations of relations. Design can be read as an activity that cultivates the work of intervention in existing sociomaterial relations (Neubauer, 2022). Design artefacts are proposals and dedicated speculations on future configurations and as such they intervene in the current configuration of relations (p. 87). They are not only reflections on possible futures but also active players in bringing this future into being, by helping designers knot the necessary material relations (p. 88). Artefacts are thus cultivated as design imaginaries that have the agency to bring ideas into being.

In the introduction, we have outlined the aim to research and develop design as a methodology for change. Here, we describe how we also intend to use design as a methodology to research. Therefore, design is our object of research as well as the instrument of investigation. We anticipate that through this, we further develop the conceptualization of design as a methodology for change.

Practically, we go about this researching/developing through teaching design to students. In doing this, we draw on Schön's approach to teaching as a "reciprocal inquiry in which each party shapes a problematic situation and is then shaped by it" (Schön, 1991, p. 23). As the students study the learning materials prepared by the teachers (shaping a "problematic situation"), both the teachers and the students are "shaped" by the students' subsequent learning response. A new situation arises, which elicits new learnings and thus, new knowledge is created. Teaching, therefore, can be used for design-based research (Ejsing-Duun & Skovbjerg, 2018). In the tradition of Research through Design (RtD), we work practically on the tasks posed by the learning/research materials, and we use the emerging interactions and artefacts to reflect on the new situations arising (Stappers & Giaccardi, 2017).

The learning materials and reflection methods differ according to each of the three tracks, that have been formed according to the three years of the study program. Track 1 focuses on design as a practice of reflective action, track 2 studies methods in design, and track 3 teaches design as a set of methods that itself can be and will be changed during the design project. The three tracks jointly teach/research design as a three-dimensional tool where each dimension charts a particular aspect of design.

The practical activities through which we teach/research 'design as a methodology' will be described in the following three tracks, that are organized according to each year in the three-year bachelor program. Inspired by Frayling's three approaches to design research (1993), we set the scene differently in each track of research.

Track 1: In the 1st-year design studio, we teach the basic principle of reflective action. Hereby, we conduct research *into* design practice (Frayling, 1993, p. 5), gaining knowledge about design activity by enlisting ethnography (Gobo, 2008; Pink, 2009; Pink, Horst, et al., 2016; Pink, Sinanan, Hjorth, & Horst, 2016).

Track 2: In the 2nd-year design studio, we teach different methods of designing. We research *for* design practice (Frayling, 1993, p. 5) by gathering materials and embodying the knowledge in the artefact (Volonté, Rampino, & Colombo, 2015).

Track 3: In the 3rd-year design studio, we teach the designing of methods of designing. We research *through* design (Frayling, 1993, p. 5) and design practice being its own method of doing and reflecting (Vaughan, 2017). While reflecting on our spatial and material conditions of design practice, we intervene in our practice to create new conditions for our design practice, thus innovating our design practice itself.

In the following illustrations of the three studios, we will describe in more detail each of the dimensions in this attempt of developing design as a three-dimensional methodology, as well as the studios' substantiation as knowledge-producing research tracks. According to the RtD methodology, the following text is simultaneously a documentation and an embodiment of our design. Therefore, the following text does not only represent the knowledge of the research but is also one of the design artefacts embodying ‘design as a methodology’.

2.1.2.2 Track 1: Reflective action

In the 1st-year design studio, we research and teach design as a reflective form of action. We draw on a practice-based design methodology (Stappers & Giaccardi, 2017) and on ethnographic observation (Gobo, 2008), based on Frayling’s approach to research *into* design (Frayling, 1993, p. 5). During the design project, the students are provided with the ‘reflective template’ that guides them in developing their idea, as well as their way of working (Neubauer & Wecht, 2020). The template is based on design as a “reflection-in-action” (Schön, 1983), It invites them to inquire about their idea, make design moves and reflect on the new situations arising (p. 79). The students are encouraged to keep a dialog with their idea through drawings and words. The template prompts the students to formulate a question (e.g. what would it look like from that angle?) to which they can respond with a drawing or a word description. The question is a theoretical reflection, while the answer is a practical attempt to act on the reflection. The template encourages students to continue with question-answer pairs until they have explored and developed the idea far enough to become tangible. It encourages the students to inquire about their own perceptions and to engage with their concepts deeply. Once the student has made a design move, the situation responds and shapes their next move, and so on. We observe how the idea changes shape with each move, and how the template encourages a designerly relationship between the students and their work (see Figure 2.1.1).

As researchers in this studio, we observe how well the students are able to adopt reflective action, using the templates we have given them. The reflective template is developed and changed as part of the research project, in reflection on the usability of the template (Neubauer & Wecht, 2020). We test the template’s ability to guide the students in making their design artefacts, and we iterate it based on the insights.

The theoretical background given to the students is the understanding that objects and environments are not solid by nature, but that they are continuously in a state of emergence. Objects can be many different things, depending on how they are embedded in practice. A bottle can hold wine, or it can be a candleholder. A bottle can also become a murder weapon, a thank-you gift, or an addiction. Objects and their functions materialize when they are enacted as part of practices (Barad, 2003; Gherardi, 2010; Orlikowski, 2007). We emphasize in our teaching that a design is not simply implemented, but that it emerges in action. In our abstraction of the emergence of designs, we follow physicist Barad, who

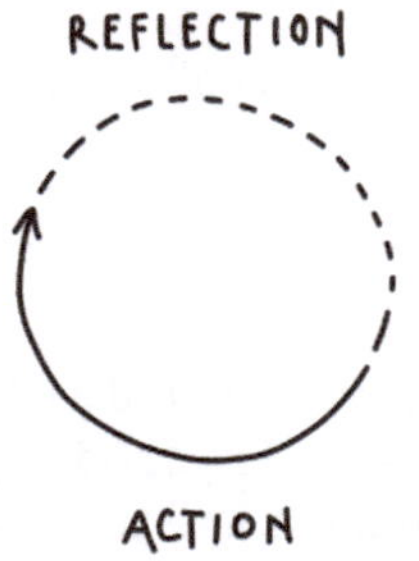

Figure 2.1.1 Designing as a reflective action.

postulates that objects or subjects do not exist by themselves but exist in the relationships that make them so (Kleinman, 2012). Objects can be understood as relations that bundle through being enacted. 'Intra-actions', making material and discursive relations, create the objects (Barad, 1998, 2003, 2007). 'Agential realism' is a term established by Barad (1998, p. 106). It describes how matter comes into being. The more that particular relations are enacted, the more 'matter' and the more 'real' an object becomes. Applying agential realism to design means that the object of design is an emerging product of the interactions around it. We may speak about an 'object-in-design' as the design of objects is never finished but continues to emerge (Telier et al., 2011, p. 15).

Accordingly, we teach the students that reflective action critically evaluates the relations of a design. Design is demonstrated as an activity that does not take the qualities of objects as naturally given, but makes each quality a careful consideration. The reflective template guides the students to use their design artefacts (the drawings and descriptions) to inquire and consider the makeup of their design.

2.1.2.3 Track 2: Reflecting on method

With 2nd-year students, we teach and research design methods. In this studio class, we go beyond tangible objects and mainly focus on the design of an experience. And, as we will outline, our research has opened insights in an unexpected way.

The brief is to design a sustainable clothing practice. During the project, we explore the method of making a story map. Design practitioners from industry practice join us to tell us about their experience with mapping and other methods in practice. With these reflective activities going on in parallel, we get to work and use story mapping to build a map. Groups of students are responsible for different parts of the map. The map embodies the collective knowledge we have developed through gathering information on the production, the retail, and the use practices around clothes.

Mapping is a method to externalize relationships (Kalbach, 2016). We teach mapping as a method which enables many stakeholders to contribute their experiences, including future users of the design, as well as other experts that are critical in the realization of a design. In story mapping, knowledge is collated from many different groups of stakeholders (Patton, 2014). An overview is created of the diverse experiences (preface). Mapping serves as the negotiation process of integrating experiences that may differ or even conflict. The mapping activity is guided by rules of how to represent actions, materials, and people in their practices (pp. 1–35). The rules are considered an important aspect of the effectiveness of mapping as a method. As outlined throughout the description of the method, working with simple cards with words on them is supposed to give equal access to the design activity to a range of stakeholders (see, for example, p. 117).

As researchers, in this track, we explore how different methods influence designing (see Figure 2.1.2). We draw on a practice-based design methodology (Stappers & Giaccardi, 2017) and on ethnographic observation (Gobo, 2008), based on Frayling's approach of gathering materials *for* design (Frayling, 1993, p. 5). Any new knowledge is supposed to be embodied in the design artefact. The map serves as the design artefact embodying the knowledge that the students collect. In part, we are practically exploring the mapping, and as teachers, we reflect ethnographically on the students' experience of using the method.

But we are disrupted by something unexpected. The students have begun to critically question the usefulness of the mapping. They report that it seems laborious and that they are not certain of its value in helping them create good ideas for a clothing solution. The

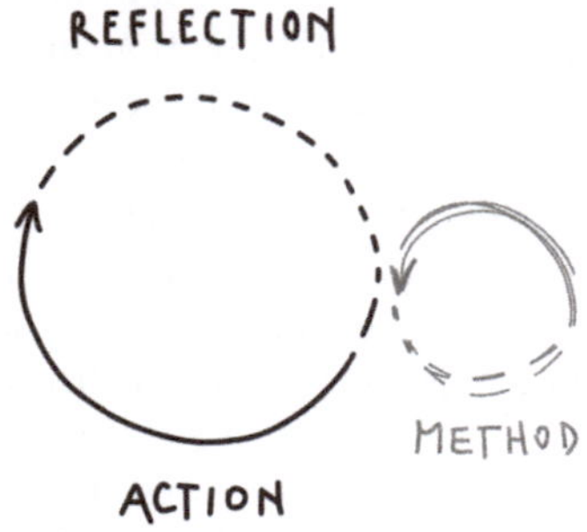

Figure 2.1.2 Designing as reflecting on the method of acting.

mapping has become a problem for them. Even though some critical engagement with the method has been expected by us, the teachers, and is a planned part of the learning, we have not expected the students to challenge our teaching. It emerges that the students struggle to meet deadlines and to complete tasks. They are unclear of the time plan for the semester: they are not clear how much studio time is remaining, on what days meetings are scheduled, how long each meeting will last, and what tasks need to be completed when. Due to the outbreak of the coronavirus, our studio classes have taken place online, and the digital environment has added uncertainty and unfamiliarity. Dates, times and virtual spaces have turned out to be invisible constraints that create discomfort for the students. We realize that temporal and digital structures are an instrumental part of the virtual design studio (Neubauer & Wecht, 2021). These constraints have created particular working conditions. For example, we use Miro, an online whiteboard, for mapping, instead of the walls of our studio at the university as initially planned. Due to the online situation, we have also pulled together two studio classes and rescheduled the timetable to suit the new situation. But this has been organized in an ad hoc fashion by the teachers, and the students were not consulted. It has taken some time to uncover this source of frustration. The constraints of time and digital space have been implicit – noticeable but invisible.

Taking this real-life problem as an opportunity, we begin to use mapping to solve the problem. The mapping method allows us to respond to the local conditions that have emerged as problematic for some (Schiffer, 2020). With the new constraints appearing (the time and space constraints), we extend the story map with another map – a map of our studio schedule. We map an overview of our ways of working, including the time and place conditions that the students have made us aware of. This has made visible the parts of the project that were not visible before.

This experience demonstrates to us the agency of mapping as a method – the method has allowed us to adapt our way of working; it has provided a space for negotiation. The learning experience has been augmented through experiencing first-hand how a method can fail to work and how it can be adjusted. Upon reflection, it has been a good learning outcome for teaching mapping as a design method, and it has demonstrated to us that method – how designing is done – is a form-giving factor in design activity.

2.1.2.4 Track 3: Reflecting on method and action

In this track with 3rd-year students, we research *through* design, as Frayling defined it (1993, p. 5). This practice-based approach is reflective of its activity (Stappers & Giaccardi, 2017).

In this studio class, we work on the research and development of 'design as a methodology', through the vehicle of the design thinking curriculum.

The project kicks off with a heated debate about what design is, and whether it can be a method or a set of practices, an approach, a strategy or a posture or mindset. The debate broaches the big question of in what way 'design thinking' can be systematized and applied. We ask the crucial question: how, if not in a systematized way, are we able to learn and apply design?

We explore conceptualizations of design, introducing the theoretical concept of design ability as particular 'designerly ways of knowing' (Cross, 1982). We also explore practice-oriented concepts of knowledge (Gherardi, 2012, 2017; Gherardi & Nicolini, 2006; Orlikowski, 2007, 2010; Suchman, 2002, 2007). We consider how the creation of knowledge in practice-based design research takes place through a performative process of 'make-it-to-see' (Vaughan, 2017, p. 35). The aim of the project is the (re)design of our design methods (e.g., the reflective template or design kits), and for this to examine our own practices, using the categories: 'physical materials', 'digital materials', 'physical space and time' and 'virtual space and time'.

Utilizing the concept of reflectivity in design, following Schön (1983), we design the reflective template, which is then given to 1st-year students to work with. We also stage interventions in our own practices, through inquiring and changing material relations. One intervention explores the tension between distributed and local space by transporting us back to the same lecture room at the university during a distributed studio class, giving us all the same background image when talking in Zoom, our video meeting software (see Figure 2.1.3).

The remote studio classes take place on Zoom, with everyone sitting in their own workspaces in a different location of the world. When we come together in Zoom for the

Figure 2.1.3 Intervention, bringing us all into the same room during distributed working.

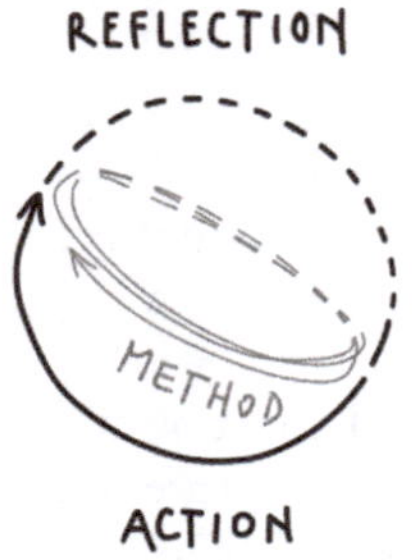

Figure 2.1.4 Designing as a reflecting on action and on method.

design studio class, the interface produces a different background for everyone. As we experience the distributed locations to impact on our ability to collaborate, the students design an intervention to attune us better for collaboration, and to bring us closer together. The students' attention turns to these different backgrounds, placing us in different locations. The meeting on Zoom being the current work method, the students look for ways to improve it. The different backgrounds are part of the material relations of the method. Intervening in these material relations (and thus in the method) the students bring us into the same workspace by giving everyone the same video background. The intervention does of course not bring us physically into the same space, but it creates an experience of seeing everyone in the same space. We explore how this experience impacts us and our method of designing. This is an example of how we reflect and act on the methods of design, inquiring and reconfiguring our design practice through it.

Other interventions supply us with design kits for prototyping. Yet another intervention invites us to 'reset' our local work environment in order to restart work in a refreshed way; to take a break, open a window, fetch a glass of water and taste it. This intervention inspires a reference to artist Marina Abramović and her instruction of water drinking in an embodied way (Louisiana Channel, 2013). On an ongoing basis, we reflect on these interventions; in conversation during the studio class, on our research website, and with conference contributions (Neubauer & Wecht, 2020, 2021; Dorothé Smit, Neubauer, & Fuchsberger, 2021).

We bring together designing as a 'reflecting on action', as well as a 'reflecting on method' (see Figure 2.1.4). In this studio class, in this track of research, all tracks come together and make up the three-dimensional body of design thinking. With this work, we respond to a major challenge that we see for design thinking scholarship: How can designing be conceptualized, something that is traditionally done by a designer in the studio – and make it accessible as a general method for change?

2.1.3 Design as a reconfiguration of material conditions

This work conceptualizes design thinking as three-dimensionally active; a process that does not only act toward the world in reflective action but also toward itself, renewing and adapting itself through the reflections and actions of the practitioners and their own practice. In the third track, the design practice itself has become the object of design – the thing we are designing – through reflection and reconfiguration. We discover that design practice is capable of adapting and reconfiguring both – the object of design as well as its method of working. Design turns into a three-dimensional tool that simultaneously does reflect on action while it does reflect on its methods of action (see Figure 2.1.5).

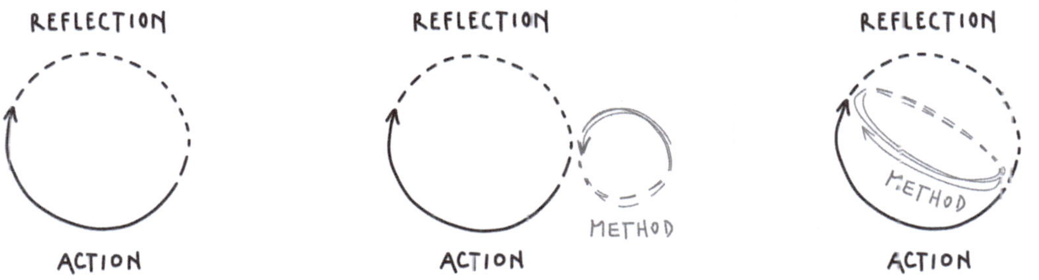

Figure 2.1.5 The three-dimensional approach to design thinking.

In track 1, we research design as the reflective action that brings into being the objects of design. We use the theoretical framing of sociomaterial relations that are open to reconfiguration (Suchman, 2007). The students learn to have a material-reflective relationship with the qualities of their idea, and through a practice of inquiry and action to re-knot the material and imaginative relations of their objects, and so bringing them into being. Objects emerge through the enacting of their material and imaginative relations (Barad, 1998). Designs materialize in the active reconfiguring of existing sociomaterial relations (Neubauer, 2022).

In track 2, we research method and its significance on the implementation of designs. Methods shape objects of investigation (Law, 2004). As we experienced first-hand with the students challenging our method, method impacts on design situations and its agency needs to be made visible. The principle of reflecting on method creates a situation that remains open to the effects of ways of working in design practices

In track 3, we turn the principles of reflective action onto design practice itself, which turns design into a tool that is capable of acting toward the world, as well as acting on its own local parameters of practice. Design practitioners enact a reflective approach through which they are aware of their methods of producing design artefacts, which they are able to adapt. For example, the students open the inquiries of how they will we use mapping, or how they will collaborate remotely. As the practitioners reflect on the conditions of their design work, their practices can be reconfigured in turn.

With these three tracks of teaching and researching design, we summon three-dimensionality as an affirmation that design is a tool for change. In highlighting method as a form-giving capacity of design, this research works toward a concept of design as a methodology for reconfiguration. It contributes to the research of design by providing a conceptualization of its agency and practice.

2.1.4 Conclusion

The design concept we have outlined is reflective of material conditions and enables acting on these conditions in order to change them. We argue that design has the capacity to change material conditions, in particular, because it also enables a reflection on its methods of action and the adaptation of these.

The educational design studio is a site of developing knowledge, of learning, and of practicing design (Tovey, 2015). Through the three tracks of teaching and researching, we innovate our curriculum of design thinking, and we also develop conceptualizations of design. The outcome is the concept of design as a tool that provides awareness of its methods, and that is, while changing the conditions of the world, capable of redesigning itself as a method; being in a reflective conversation with itself. This self-awareness of design feeds into the canon of design scholarship that calls for more responsibility and accountability in design practice, in regards to social, environmental and ethical concerns (Escobar, 2018; Manzini, 2015; Terry, Tonkinwise, & Kossoff, 2015; Tonkinwise, 2018). We conclude that design is a material-reflective tool that can change the material and imaginative conditions of the world, but that it needs to remain open to its own configurations of design practice, in order to be able to (re)configure its impact.

There are many design processes, some of them very successful. They guide practitioners to reflect on the world and on the experiences of people with empathy. However, there is also a need to reflect on its own conditions and methods. It is a scientific paradox that in research the object of investigation is changed through the methods of investigating (Law, 2004). Design, which is the tool for change, can utilize this nature of investigative action, and turn this 'aberration' of science into a feature. However, as conditions change, new, uncertain situations will be created, and people might change and their objectives. This special feature of design might take a little while to gain traction among the change-making organizations of the world. This kind of design might only be embraced if a feeling prevails that the common approaches have reached their limits and that there is not much left to lose.

Acknowledgments

We thank the students of Management by Design (MxD) for taking part in this research and sharing with us their experiences, thoughts, and insights. This research was in part funded by the New Design University, Sankt Pölten, Austria.

References

Badke-Schaub, P., Roozenburg, N., & Cardoso, C. (2010). Design Thinking: A paradigm on its way from dilution to meaninglessness? In K. Dorst, S. C. Stewart, I. Staudinger, B. Paton, & A. Dong (Eds.), *DTRS 8: Interpreting design thinking* (pp. 39–49). DAB Documents.

Ball, L. J., & Christensen, B. T. (2020). How sticky notes support cognitive and socio-cognitive processes in the generation and exploration of creative ideas. In *Sticky Creativity* (pp. 19–51). London: Academic Press.

Barad, K. (1998). Getting real: Technoscientific practice and the materialization of reality. *Differences: A Journal of Feminist Cultural Studies*, *10*(2), 87–126.

Barad, K. (2003). Posthumanist performativity: Toward an understanding of how matter comes to matter. *Journal of Women in Culture and Society*, *28*(3), 801–831.

Barad, K. (2007). *Meeting the universe halfway: Quantum physics and the entanglement of matter and meaning*. Duke University Press.

Buchanan, R. (2015). Worlds in the making: Design, management, and the reform of organizational culture. *she ji The Journal of Design, Economics, and Innovation*, *1*(1), 5–21.

Cross, N. (1982). Designerly ways of knowing. *Design Studies*, *3*(4), 221–227.

Dell'Era, C., Magistretti, S., Cautela, C., Verganti, R., & Zurlo, F. (2020). Four kinds of design thinking: From ideating to making, engaging, and criticizing. *Creativity and Innovation Management*, *29*(2), 324–344. doi:10.1111/caim.12353

Ejsing-Duun, S., & Skovbjerg, H. M. (2018). Design as a mode of inquiry in design pedagogy and design Thinking. *International Journal of Art & Design Education*, *38*(2), 445–460. doi:10.1111/jade.12214

Escobar, A. (2018). *Designs for the pluriverse: Radical interdependence, autonomy, and the making of worlds*: Duke University Press.

Frayling, C. (1993). Research in art and design. *Royal College of Art Research Papers*, 1(1), 1–5.

Gherardi, S. (2010). Telemedicine: A practice-based approach to technology. *Human Relations*, *63*(4), 501–524.

Gherardi, S. (2012). *How to conduct a practice-based study: Problems and Methods*. Northampton, MA: Edward Elgar Publishing Limited.

Gherardi, S. (2017). Sociomateriality in posthuman practice theory. In A. Hui, T. Schatzki, & E. Shove (eds.), *The nexus of practices: Connections, constellations, practitioners* (pp. 38–51). New York: Routledge.

Gherardi, S., & Nicolini, D. (2006). *Organizational knowledge: The texture of workplace learning*. Oxford: Blackwell Publishing.

Gobo, G. (2008). *Doing Ethnography*. London: SAGE.

Goldschmidt, G. (1996). The designer as a team of one. In N. Cross, H. Christiaans, & K. Dorst (eds.), *Analysing design activity* (pp. 65–91). England: John Wiley & Sons Ltd.

Kalbach, J. (2016). *Mapping experiences*. Sebastopol, CA: O'Reilly.

Kimbell, L. (2011). Rethinking design thinking: Part I. *Design and Culture*, *3*(3), 285–306.

Kimbell, L. (2012). Rethinking design thinking: Part II. *Design and Culture*, *4*(2), 129–148.

Kleinman, A. (2012). Karen Barad, Intra-actions. *Mousse*, *34*, 76–81.

Law, J. (2004). *After method: Mess in social science research*. Oxon: Routledge.

Louisiana Channel (Producer). (2013). Marina Abramović: How to drink a glass of water.

Manzini, E. (2015). *Design, when everybody designs an introduction to design for social innovation*. The MIT Press.

Margolin, V. (2013). Design studies: Tasks and challenges. *The Design Journal*, *16*(4), 400–407.

McDonnell, J. (2011). Impositions of order: Strategies for establishing coherence—a comparison between design and fine art practices. *Design Studies*, *32*(6), 557–572.

Melles, G. (2010). Curriculum Design Thinking: A New Name for Old Ways of Thinking and Practice? In K. Dorst, S. C. Stewart, I. Staudinger, B. Paton, & A. Dong (Eds.), *DTRS 8: Interpreting design thinking* (pp. 299–308). DAB Documents.

Mol, A. (2002). *The body multiple: Ontology in medical practice*. Durham, NC: Duke University Press.

Neubauer, R. M. (2022). Materializing the agency of design in innovation practices. *Design Issues*, *38*(1), 81–91.

Neubauer, R. M., & Wecht, C. H. (2020). *Learning by design: How practice-based design studio education might reconcile teaching, learning and research*. Paper presented at the AMPS Proceedings Series 22.1. Teaching – Learning – Research, Manchester School of Architecture, UK.

Neubauer, R. M., & Wecht, C. H. (2021). Materiality of space and time in the virtual design studio. In E. Bohemia, L. M. Nielsen, L. Pan, N. A. G. Z. Börekçi, & Y. Zhang (eds.), *Proceedings of the DRS LEARNxDESIGN 2021* (pp. 756–765). Jinan, China: Design Research Society.

O'Riordan, K. (2017). *Unreal objects: Digital materialities, technoscientific projects and political realities*. London: PlutoPress.

Orlikowski, W. J. (2007). Sociomaterial practices: Exploring technology at work. *Organization Studies*, *28*(9), 1435–1448.

Orlikowski, W. J. (2010). The sociomateriality of organisational life: Considering technology in management research. *Cambridge Journal of Economics*, *34*, 125–141.

Patton, J. (2014). *User story mapping: discover the whole story, build the right product*. Cambridge: O'Reilly.

Pink, S. (2009). *Doing sensory ethnography*. London: SAGE.

Pink, S., Horst, H., Postill, J., Hjorth, L., Lewis, T., & Tacchi, J. (2016). *Digital ethnography: Principles and practice*. London: Sage Publications.

Pink, S., Sinanan, J., Hjorth, L., & Horst, H. (2016). Tactile digital ethnography: Researching mobile media through the hand. *Mobile Media & Communication*, *4*(2), 237–251.

Sanders, E. B. N., & Stappers, P. J. (2012). *Convivial design toolbox*. Amsterdam: BIS Publishers.
Schiffer, A. (2020). Issues of power and representation: Adapting positionality and reflexivity in community-based Design. *International Journal of Art & Design Education*. doi:10.1111/jade.12291
Schön, D. A. (1983). *The reflective practitioner: How professionals think in action*. New York: Basic Books.
Schön, D. A. (1991). Teaching and Learning as a Design Transaction. In N. Cross, K. Dorst, & N. Roozenburg (Eds.), *Research in design thinking* (pp. 21–36). Delft University Press.
Smit, D., Neubauer, R., & Fuchsberger, V. (2021). Distributed collaborative sensemaking: Tracing a gradual process. Paper presented at the *ACM TEI International Conference on Tangible, Embedded and Embodied Interaction*, Salzburg.
Smit, D., Oogjes, D., de Rocha, B. G., Trotto, A., Hur, Y., & Hummels, C. (2016). Ideating in skills: Developing tools for embodied Co-Design. Paper presented at the *Proceedings of the TEI '16: Tenth International Conference on Tangible, Embedded, and Embodied Interaction - TEI '16*.
Stappers, P. J., & Giaccardi, E. (2017). Research through design. In M. Soegaard & R. Friis-Dam (eds.), *The encyclopedia of human-computer interaction* (pp. 1–94). The Interaction Design Foundation.
Suchman, L. (2002). Practice-based design of information systems: Notes from the hyperdeveloped world. *The Information Society*, *18*(2), 139–144. doi:10.1080/01972240290075066
Suchman, L. (2007). *Human-machine reconfigurations: Plans and situated actions*. New York: Cambridge University Press.
Telier, A., Binder, T., De Michelis, G., Ehn, P., Jacucci, G., Linde, P., & Wagner, I. (2011). *Design things*. Cambridge, MA: The MIT Press.
Terry, I., Tonkinwise, C., & Kossoff, G. (2015). Transition design: An educational framework for advancing the study and design of sustainable transitions. Paper presented at the *6th International Sustainability Transitions Conference*, University of Sussex, Brighton.
Tonkinwise, C. (2018). The magic that is design. In T. Fisher & L. Gamman (eds.), *Tricky design: The ethics of things*. London: Bloomsbury Visual Arts.
Tovey, M. (2015). Design education as the passport to practice. In M. Tovey (ed.), *Design pedagogy: developments in art and design education* (pp. 37–49). Farnham: Gower Publishing Ltd.
Vaughan, L. (ed.) (2017). *Practice-based design research*. London:Bloomsbury Publishing.
Volonté, P., Rampino, L., & Colombo, S. (2015). Basics of a design research epistemology. Paper presented at the *European Academy of Design Conference*, Boulogne Billancourt, France.

2.2 Machine learning and ethics to expand 21st-century design boundaries

Martina Sciannamè
Politecnico di Milano, Design Department

2.2.1 Introduction

2.2.1.1 The specific and pervasive character of design

Defining design is a very challenging task and it becomes even trickier if we want to draw some boundaries. DTRS13 (Goldschmidt & Tarazi, 2022) timely addressed this issue questioning the legitimacy of expanding design beyond its disciplinary borders in relation to the necessity to mark its distinctiveness.

On the one hand, some peculiar characteristics of design can be outlined. With their semantic analysis of design definitions, Auernhammer and Ford (2022) proved that there are some recurrent, thus shared, themes when it comes to understand what design is. They state four core elements. First, design is (i) an activity, a way, or a process, which can be identified as (ii) creating artifacts, meanings, and solutions to problems. These generative activities are aimed at (iii) new and purposeful changes and configured as (iv) means to an end.

On the other, if we consider design in its broadest sense, as the act of meaningfully conceiving and/or creating something for a purpose – diffuse design in the words of Ezio Manzini (Manzini & Coad, 2015) – we easily recognize how this capability has characterized humankind since the beginning of time. Then, its spreading across different disciplines is just a natural and intuitive outcome, which might benefit any aspect of human life. What is important, though, is to maintain and emphasize the undoubted specificities of designers while their roles can expand across different fields.

Throughout history, designers have been identified as the professionals who care about the beauty and the appearance of things, or as those focusing on their function, yet, going to the very roots of the practice of design in its multiple forms and embedding the viewpoint of activity theory (Kaptelinin & Nardi, 2009), we may say that what primarily define designers is their capability to mediate between humans and the world. Based on what they have available, they can envision things that have never existed before and create something valuable both in a concrete and an abstract sense, spacing between different scopes, which makes design a pervasive activity (Louridas, 1999). Especially from this point, the current figure of the designer needs to be reshaped in the eyes of the lay public as well as within the disciplinary educational institutions.

Designers' main value lies in their ability to transform the world and shape new meanings according to a predefined purpose (may this be for pleasure, better usability, or survival). Hence, they can easily adapt to different contexts, and it gives them the possibility to play critical roles in today's pressing issues: either to solve problems, no matter how ill-defined they are, but especially to frame them correctly thanks to their systemic reasoning.

DOI: 10.1201/b22630-11

In fact, more than ever before, we are becoming aware of the challenges we are facing, and that may increase in the near future. Friedman (2019) even listed a set of specific challenges for design that span throughout all levels of human condition: from performative issues to systemic, contextual, and even global challenges (Meyer & Norman, 2020). What emerges from the authors' perspective is the necessity for designers to work across disciplinary boundaries to actually pervade and contribute to any sort of issue we are called to respond to, not only by being incrementally involved in multidisciplinary teams but also by taking on organizational and managerial positions.

2.2.1.2 ML: A promising context for a cross-disciplinary mediation

In this historical moment, we are assisting to the revival of artificial intelligence (AI) and – more prominently – of machine learning (ML) systems, which are wildly spreading into everyday life, due to the favorable technological development (i.e., improved computational power and availability of large amounts of data) and are expected to have significant impacts on human life as electricity did (Kelly, 2016). Dating back to the 1950s, AI is the field concerned with building machines "that can compute how to act effectively and safely in a wide variety of novel situations" (Russell & Norvig, 2020), and ML – its most popular subfield – is getting a lot of attention as it allows to create systems that address uncertain issues (i.e., for which writing step-by-step instructions would require a great effort or may not be possible) only by giving them a specific goal and huge amounts of examples to learn from, to finally let them derive how to reach the goal and improve their performance over time. In practical terms, it means that, among several other ML applications, it is now possible to easily automate burdensome and repetitive human tasks, personalize contents according to people's preferences and behaviors, interact with devices by using natural language, and infer information from massive quantities of data. Indeed, ML could benefit people and their ecosystems in a revolutionary way as it unveils whole new possibilities to interact with the world. The main issue, though, is that the technology-driven diffusion of ML systems often is not in line with people's needs, desires, or expectations. It does not comprehend the "wholeness of humanity" (Antonelli, 2018) nor the ethical implications it arises, while design has the potential to address this matter in a human-centered way. Therefore, this field may be a perfect playground for designers, who have the skills to anticipate the impacts of a technology and to provide it new meanings, but, unfortunately, they seem unprepared to effectively leverage ML capabilities and foresee opportunities as they lack understanding, experience, tools, and methodologies to deal with it (Dove et al., 2017; Yang et al., 2020).

Thus, besides a revised definition of their role (as suggested in 1.1), designers need to be given suitable means to amplify their field of action and have a role in this current hot topic. Specifically, as Meyer and Norman (2020) report, design education should "draw more extensively on knowledge developed in other established fields, translating that understanding into a form useful to practicing designers."

2.2.1.3 Design education as a place to initiate change

To support the argumentation, the chapter proposes an example of expanding the borders of design across disciplines by merging ML knowledge, ethical perspective, and traditional design skills (like empathy, system-level thinking, and a transformative influence on the world) in the educational context, to enable designers to impact human life. In fact, synthesizing Auernhammer and Ford's (2022) results, designers have the potential to steer the spreading of this technology by making sense of it through their creations toward beneficial outcomes and trigger unexpected innovation. For this, new designerly processes and ways of knowing about this subject matter are needed and should be introduced in design education.

The experimentation stems from a research project aimed at providing future designers the essential skills to navigate across disciplines in order to design responsible and meaningful systems integrating ML. Specifically, it focuses on design education and aims at developing a methodological contribution to enhance cross-fertilization and interdisciplinary communication between design and ML. As it deals with a quite uncharted territory, it has an exploratory and constructivist character, implementing an action research approach.

Two are the main purposes of the proposed activity: (i) providing design students the tools to understand and exploit ML by translating technical knowledge in a designerly way, and (ii) framing their education in a value-sensitive design perspective (van den Hoven, 2013).

As a matter of fact, it is in the light of awareness and responsibility – about technological potential and limitations as well as about ethical and sociological implications – that designers can play their mediating role to address the criticalities of human life.

2.2.2 An experimental method for expanding the frontiers of design

2.2.2.1 Setting up the didactic activity: Aim, target, and modalities

Even though some attempts to bring AI/ML (Futurice, 2017; Piet, 2019) and ethics (Artefact, 2017; Gispen, 2017) to designers are emerging, to the best of the author's knowledge, this represents the first empirical experiment on merging together the three disciplinary perspectives on ML systems development. Specifically, the study aims to assess the response of designers to this multifaceted topic. Configured as an introductory didactic activity to present the core ingredients of ML as design material, it targets doctoral students with some or no prior knowledge of the subject matter. They are relevant testers because they are close to the intended audience for the research, and they have more analytical skills, didactic experience, and consciousness about theories and methods to build an insightful discussion and peer evaluation of the activity. Further iterations and applications may also involve MSc design and engineering or computer science students.

The expected impact of this activity is to understand and open a discussion about the possibilities to build awareness on the ML topic and pave the way toward a value-sensitive approach to designing ML-infused solutions that could be implemented in design education. For this, the expected learning outcomes for the participants include that: (i) in terms of knowledge, students should gain familiarity with the basic capabilities, limitations, and implications of ML systems, as well as the main values and possibilities for a responsible and trustworthy design. (ii) In relation to skills, they should learn to pursue a value-driven design process, activate fruitful discussions in team collaboration, and identify ML as a design material. While (iii) concerning values, they should understand how to be responsible designers in the contemporary situation and that ML can be an asset to face big challenges.

Due to its introductory character, the activity has been imagined as a brief (one to two hours) and engaging experience, possibly anticipating a more extensive didactic path. Thus, a cooperative board game format seemed the perfect fit to gently introduce the main topics, give procedural information, and stimulate reflection and discussion in a practical context. In fact, games create a parallel world, a safe place in between the physical and the game dimension itself (Huizinga, 1938), where players can freely express themselves, make mistakes, and learn with no concrete repercussions. The cooperative nature of the game, in this case, is essential as collaboration amplifies the team dimension that characterizes a design environment (to which the participants are used) and naturally encourages discussion and exchange.

As the pandemic period required, the activity has been developed and tested with both physical and digital supports in two different sessions, each with two Ph.D. design students, for a total of four participants. For the qualitative purposes of the research, the limited number of testers allowed the researcher to participate as an observer, it left enough space for the involved students to actively interact with the game and freely express their reflections, and it allowed for an in-depth look at the experience of the activity itself, but of course, this presents some limitations in terms of representativeness.

The first in-presence session (Figure 2.2.1) involved two students with no prior knowledge of the subject matter. For the second one, instead, the selected Ph.D. students' research areas are somehow related to ML as a technology. For this session, the Miro platform (Figure 2.2.2) has been chosen as digital support because it effectively enables most of the actions that they would have performed in live collaboration.

Figure 2.2.1 In-presence playtest (photo by the author).

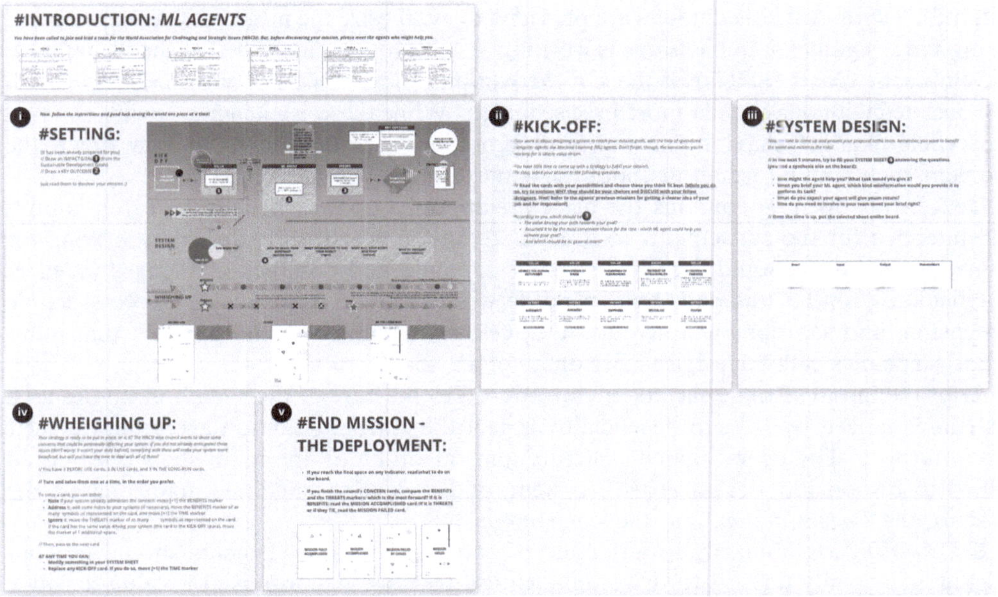

Figure 2.2.2 Overview of the digital board supporting the online activity (created by the author).

2.2.2.2 An introductory board game: Theory, tools, and procedures to structure the activity

As the focus of the activity is to push designers to the edge of disciplinary boundaries to embrace and mediate between designerly approach, ML contents, and ethical perspectives, the design process represents the perfect space for the combination of the three ingredients. In a preliminary analysis (Figure 2.2.3), the author explored how the three disciplines could contribute to each of the five stages of the design thinking process, as illustrated by (Dam & Siang, 2021). The ideation phase resulted in being the one in which the mergence is most functional and significant. Therefore, the whole game explicitly refers to it.

SET-UP. The initial situation presents the players with their role as designers at the service of a fictional World Association for Challenging and Strategic Issues (WACSI). They are called to join and lead a team with the aim of designing a system to accomplish a

	EMPATHISE	DEFINE	IDEATE	PROTOTYPE	TEST
DESIGN	researching space for intervention	framing the problem	envisioning possible solutions	setting up tests and anticipating outcomes	monitoring and assessing impacts
ETHICS	influencing the research	driving towards a meaningful purpose	fostering acceptability, sustainability and desirability	orienting choices	defining parameters and trade-offs
ML	/	/	opening up new possibilities	setting technical constraints	supporting the evaluation system

Figure 2.2.3 Author's analysis of disciplinary contributions at each stage of the design thinking process.

given mission. Before starting, an *impact goal* and a *key outcome* – as intended in (Design Kit, n.d.) – with available datasets are provided to synthesize the mission. The *impact goals* – long-term, significant impacts one is working to achieve – are identified as the Sustainable Development Goals (SDGs) of the United Nations (UN - Open Working Group, 2015) and declined with one of their more precise targets. While the *key outcomes* – near-term and observable change/behavior one wants to promote – are defined to narrow down the problem, including information about the context and target audience.

Hence, this opening contains the basic assumptions upon which the research is built and intended for the participants to internalize: in the near future, designers striving for human flourishing should face increasingly dynamic and complex challenges (Weil & Mayfield, 2020). To this end, human needs will extend to a societal and eco-systemic dimension, and solutions will involve cutting-edge technology – an approach that influential academics called DesignX (Friedman et al., 2019).

After the initial setting, players are introduced to the core of their mission: to design a system, enhanced by ML, to beneficially impact the current situation according to their given purpose. The overall game structure aims at outlining the main steps one should follow to responsibly design an ML system, and it develops in three main phases: the *kick-off*, the *system design,* and the *weighing up*.

KICK-OFF. Preliminarily to any concept construction, few elements should be considered and openly defined to drive the ideation process. According to a value-sensitive design approach (van den Hoven, 2013), shared values should be expressed from the very early stages to be embedded in the technology and affect affordances and constraints of the system. Additionally, thinking explicitly about the values is morally significant and may lead to meaningful results. Hence, the players are invited to select one (i) *value* that will drive their system. Of course, it should not be the only one embedded but the most prominent one. The values proposed (Figure 2.2.4) are an adaptation of the ethical principles behind the European Guidelines for Trustworthy AI (High-Level Expert Group on Artificial Intelligence, 2019b), which are highly comprehensive and founded on fundamental human rights. They are as follows: respect for human autonomy, attention to fairness, increase of intelligibility, prevention of harm, and promotion of flourishing. The latter has been added in antithesis to the previous one to suggest a more proactive attitude toward life, well-being, growth, progress, and prosperity; while intelligibility replaces explicability because it implies that clarity can be achieved either by explanation or intuitively, expanding its sense to include UX issues as well.

Other essential components relate to the material at hand: ML systems. Their most common capabilities are introduced and synthesized in the form of (ii) *ML agents*. They represent the keystone of knowledge transfer. As in (Meyer, 2010) argumentation, they are conceived to be functional to move pieces of knowledge from one field to

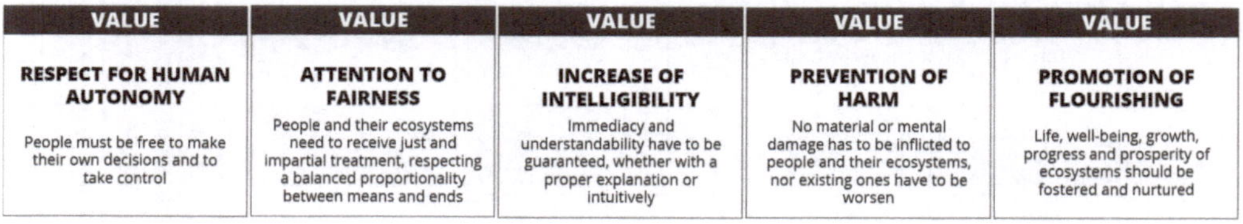

Figure 2.2.4 Value cards provided in the game (created by the author).

another by transforming it into a more familiar language for the recipients. Indeed, the characterization of ML systems as agents is actually borrowed by (Russell & Norvig, 2020), a worldwide reference textbook for AI – and, while maintaining technical correctness and clarity on their computer nature, an emphasis on the parallelism with human agents helps designers understand how to approach them as part of larger systems. Each *ML agent* embodies an ML task – or problem in the words of (Russell & Norvig, 2020) – and summarizes the question: what is the basic functioning principle of the system, generally speaking? The key tasks presented are classification, regression, sequence prediction, generation, clustering, and action selection – the explanation of which is beyond the scope of this argumentation. A previous workshop, organized in collaboration with a MSc student in Digital and Interaction Design at Politecnico di Milano (Arnone, 2020), proved this level of information, in-between the technical construction and the contextual capabilities of ML systems, to be essential and sufficient to enable design students to exploit this technology in concept elaborations effectively. The same experience revealed that a formal yet simplified definition of an ML system (summarizing its internal process in terms of input, procedure, and output) combined with case studies provide enough information for operational comprehension. Therefore, the *ML agents* sheets (Figure 2.2.5) include a synthetic question expressing the core of the task, a brief definition, some system's contextual capabilities to exemplify current applications, and two case studies (a positive one that complies with SDGs-level challenges and a questionable one, threatening ethical principles).

Finally, the players are supplied with (iii) five general *intents* to beneficially impact human life (Figure 2.2.6) – automate, augment, empower, inspire, and specialize – among which they should choose the one responding to the question: "how would your system improve the current situation?"

Then, *value*, *ML agent*, and *intent* constitute the proposed founding material to orientate the design process and outline a meaningful and responsible solution. Their initial definition

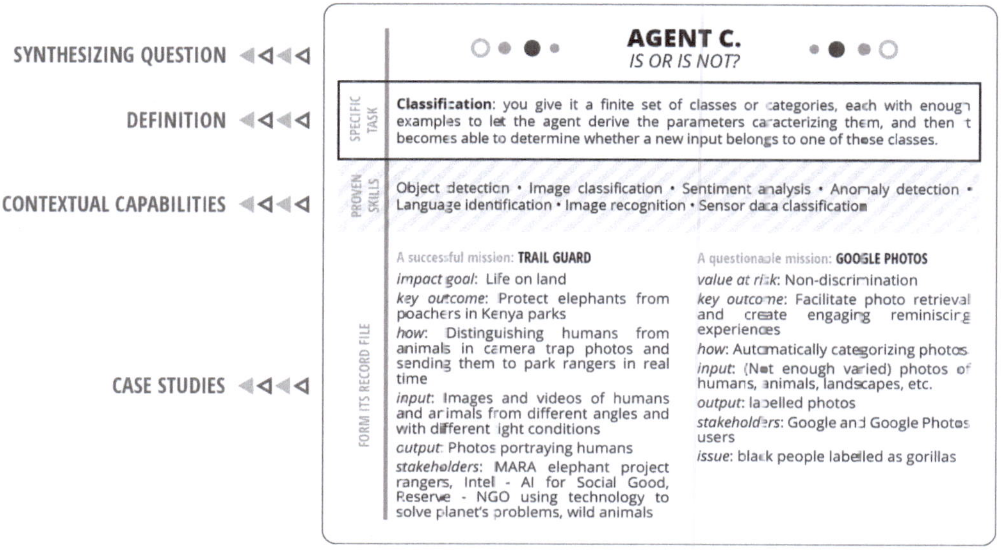

Figure 2.2.5 Example of an ML agents sheet (created by the author).

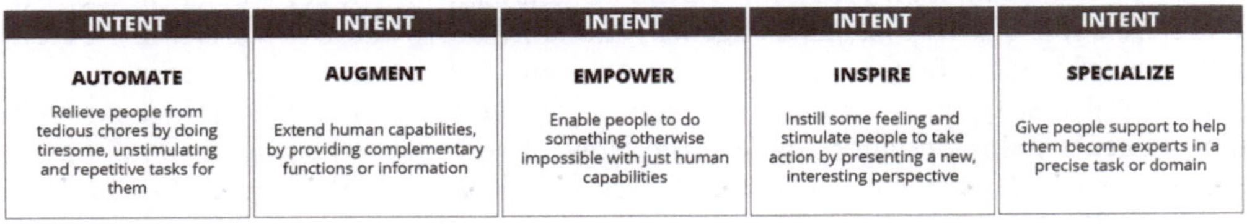

Figure 2.2.6 Intent cards provided in the game (Created by the author).

is aimed at stimulating reflections and at keeping the team aligned on the same principles. It does not need to be permanent; instead, variations are part of an iterative process.

A limitation of this approach is that players are forced to include one *ML agent* in their design for the didactic purpose of understanding them through a direct application in a concrete example. This does not imply that ML is the preferable solution to any problem and neither that a single *ML agent* is enough to tackle complex tasks nor cannot work in cooperation with others. Unfortunately, as the activity pursues immediacy, a short time frame, and a playful format, these issues are just briefly mentioned, and a proper explanation does not find place in the game. Nevertheless, it should complement the educational path.

SYSTEM DESIGN. Once the *kick-off* phase is concluded, and the consequent foundational elements have been selected – always with the mission goals in mind – the backbone and the boundaries of the system to design should be clear. Indeed, the discussion for the selection should embed the envisioning of possibilities. That being the case, the elaboration of a more defined concept should be relatively immediate by following the established guidelines. Thus, for this activity, a time limit of five minutes is provided to avoid redundant overthinking, and the completion of a *system sheet* is required. The latter should support the participants in formalizing the system's structure by pointing out its characterizing features. Specifically, it follows the configuration of the case studies attached to all the *ML agents* (as well as their definitions) and asks designers to describe: (i) the system task – how the *ML agent* should help to reach the goal; (ii) input – what information should the agent be supplied with to perform the task; (iii) output – what the agent is expected to give in return; (iv) stakeholders – who need to be involved in the team to set the right premises. Actually, (i), (ii), and (iii) are the basic constituents of any AI system as outlined in the definition of the European Commission (High-Level Expert Group on Artificial Intelligence, 2019a). Instead, the identification and potential involvement of stakeholders in the design process is not only a widely suggested ethical practice but also a consolidated approach in the field of design (participatory design, or co-design). Ultimately, to anticipate possible difficulties in the systematization of the idea, examples have been prepared for the players to compare or adopt once the time is up. After the comparison, final adjustments can be made before moving on to the closing phase.

WEIGHING UP. Among morals, meaningfulness, and usability, the last part of the game encapsulates the human-centered soul of the approach. It introduces the criticalities that may emerge in an ML system, presenting them – in the form of cards – as the *concerns* raised by the fictional WACSI wise council. Each card (Figure 2.2.7) includes a possible limitation of the system and the related implications, values, and options that designers have to avoid such hindrance. The *concern cards* are distinguished by

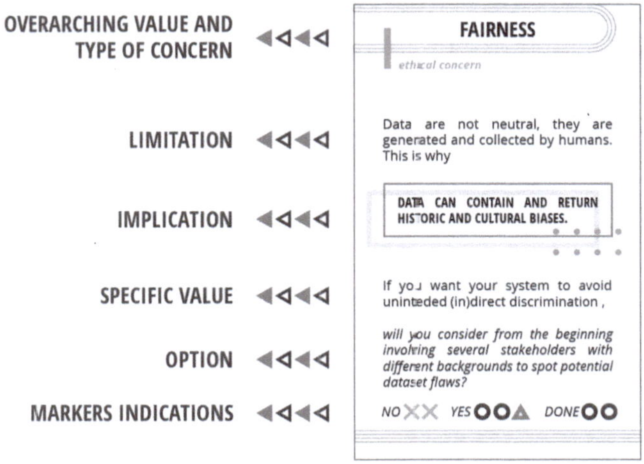

Figure 2.2.7 Example of a concern card (created by the author).

disciplinary competence: some issues may be primarily identified or handled by ML developers, some by designers, and others by ethicists or social scientists. Still, they share the common goal of benefiting human beings and the world around them.

Concerns rooted in ethical discussions, which constitute the largest part of the deck, are the result of the analysis of the existing ethics guidelines comprehensively collected in the AI Ethics Guidelines Global Inventory (Algorithmic Watch, 2020). Currently, it includes 173 guidelines, among which only those in English, belonging to certain sectors, addressing AI in general (not specific applications such as AI in healthcare), and listing some principles AI systems should embody were considered to derive relevant values, implications, limitations, and options for the cards. As a result, 51 guidelines were selected: 7 from academia, 20 from civil society, 1 from intergovernmental organizations, 1 from international organizations, and 22 from the private sector. Further references include some consistent tools (Artefact, 2017; Calderon et al., 2019; Futurice, n.d.; IDEO, 2019) and the research work developed in the project Meet-AI for UX concerns (Spallazzo et al., 2021).

Moreover, time is another significant variable for facing concerns. Whether they occur during the design, development, or use of an ML system, a responsible approach requires that they be addressed early in the process, both to prevent risks and to avoid late and costly interventions. To highlight this point, the concern cards have been divided into three categories according to when the cause of concern originated. In the *before use* deck, the cause of concern precedes the model creation (then the implications are patent also before the development of the system is complete); in the *in-use* deck, the cause of concern affects the use of the system; and in the *long run* deck, it develops over time.

With the purpose of stimulating discussions about possible unforeseen outcomes and reflections on responsible practices to be accounted for already in the ideation phase, the ultimate part of the game makes the players address up to nine different *concern cards* (three from each time category) randomly drawn. They can be read and solved in any order to underline that there is no preferable procedure as long as all kinds of issues are well-pondered.

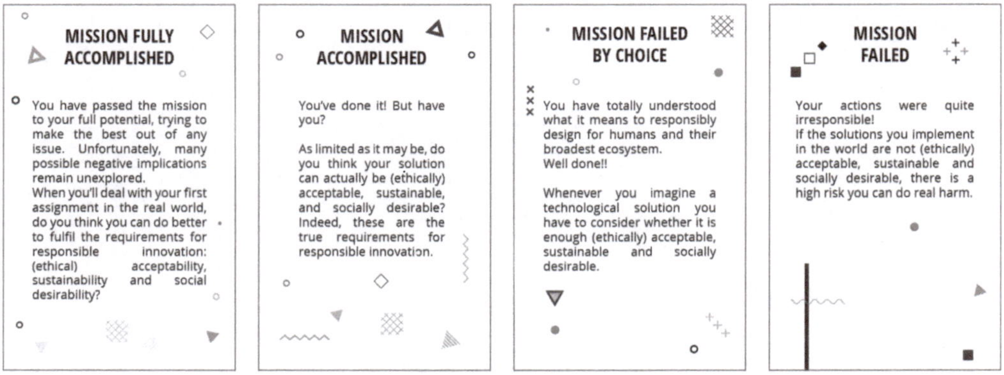

Figure 2.2.8 Available mission end cards (created by the author).

As anticipated, all the cards present the readers with a possible action that they can handle to increase the beneficial impacts of their system or that they can disregard, exposing the system to be a potential threat. Of course, not all the concerns apply to the imagined system or are preferable to comply with. The focus here is on balancing the possible good or harm a system can cause, keeping into account the time factor (time will not be sufficient if all options are fulfilled, as tackling all concerns in a project would be an endless work), and acting accordingly. To do so, players can add some notes to their system sheet (if necessary), replace any of the *value*, *ML agent,* or *intent* cards (optional, this action that would cost time), and advance a marker on the *benefits* and *time* indicators if they decide to address the concern, just move the *benefits* marker if they considered the issue before reading the card, or advance on the *threat* indicator when they decide to ignore it. Instructions about the movements of the markers are provided on the cards and vary according to their potential impact.

MISSION END. The end of the game is declared when players solve all the concern cards or reach the final space of any indicator (getting to use all the time at their disposal or attaining the maximum amount of benefits or threats). The level of positive or harmful impact achieved by the designed system determines the completion of the mission. Four different epilogues are available (Figure 2.2.8), though they all converge in presenting the fundamental requirements for responsible innovation that students can guard for future reference: being (ethically) acceptable, sustainable, and socially desirable (von Schomberg, 2013). Ultimately, the game results in leaving open space for further considerations. In fact, whether a system satisfies these requisites – regardless of the outcome of the mission – is up to the players to understand and, hopefully, discuss.

2.2.3 From theory to practice: Discussing the development of the game and the playtests

In the following, the results emerging from the observation of the playtests and the subsequent semi-structured interviews on the translation of ML and ethics concepts for designers are discussed. The aim is to highlight the issues that affect the didactic purpose of the activity, while insights about the game materials and mechanics are beyond the scope of this argumentation.

2.2.3.1 Limitations: Where the translation needs refinement

Since it was the first tentative experimentation of merging three disciplinary perspectives to pursue a common goal, the playtests have been a very informative occasion to spot some weaknesses in the instructional framework.

Above all, conveying ML's value as a design material that can help facing important societal issues seemed feeble. In the first session, the *impact goal* was not made explicit as one of the SDGs, although clarifying that in the second playtest did not change the participants' perception. In fact, being Ph.D. students in the Department of Design at Politecnico di Milano, they are quite used to dealing with such high goals. Certainly, *ML agents* were embraced in both sessions as manageable tools to address the mission, regardless of the scale of the problem.

Their communication, however, highlighted some difficulties. Even if the combination of a definition and case studies effectively delivered the core principles and the level was appropriate, the first impact of a technical presentation was a little disorienting for the players with no prior knowledge. At the same time, a more visual language could enhance assimilation and memory. The most problematic point, though, is the length of *ML agents* sheets. Although containing just essential information, they require some effort from players. If introduced in the *kick-off* phase, they break the rhythm of the activity, but also making them anticipate the game (as tested in the second version), it remains a burdensome task. A collective introduction, aided by visual supports, may facilitate the knowledge transfer, leaving *ML agents* sheets the role of reminders.

Similarly, the tone of the *concern cards* needs simplification for greater accessibility by master students (the ultimate target of the activity). Additionally, marking the differentiation of the cards' contents (limitations, implications values, options) may increase the intelligibility of the translation.

Overall, the game experience appeared balanced with respect to the capabilities of the testers and the provided knowledge. Some uncertainty emerged only in the *system design* phase of the second playtest when prior (non-expert) familiarity with the topic brought up issues that went beyond the intended task. This was a further indicator that expert assistance is essential at this stage of the game development. Of course, some minor graphic adjustments might smooth out the fruition. Indeed, the order of the contents on the board has been modified for the second digital playtest to limit external intervention, increase the focus on each phase of the game, and the visibility of important definitions that the participants should acquire for a precise understanding of the elements in play. Though some confirmations by the facilitator were always sought, and the two-page rules booklet predominantly guided the players, who quite ignored the additional contents on the board.

2.2.3.2 Challenges: Among unexpected outcomes and space for improvement

The premises of this experimentation reveal its ambitious nature. Merging ML, ethics, and design in a single brief activity to start expanding the borders of current design education necessarily required simplification.

The problem is exemplified by the fact that ML systems are rarely linear and often necessitate integrating multiple systems to achieve seemingly simple goals. More *ML agents* may be combined to get to a solution, yet the players are asked to select only one and describe the system accordingly. Although this helps limit the time and complexity of the activity by encouraging novices to focus on one agent at a time, it lacks completeness

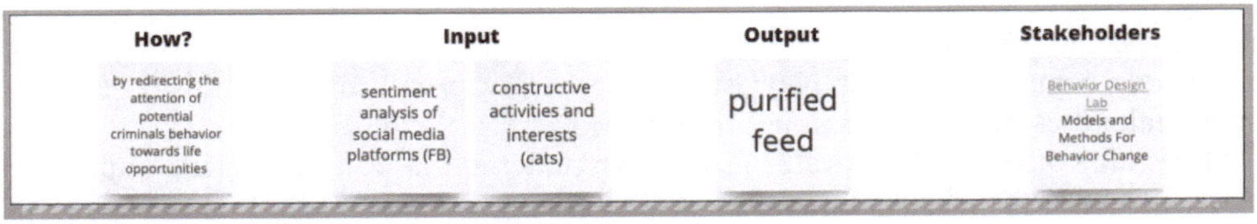

Figure 2.2.9 System design sheet developed in the second playtest.

In fact, some confusion arose during the second session *system design* phase. In this case, the participants aimed at building a recommendation system to orient potential criminals toward positive activities instead of perpetrating in noxious environments. To do so, they identified the *Sequence Prediction (SP) Agent* – able to predict values or outcomes in a sequence based on historical information (data in which it is possible to detect patterns of activities or behaviors over time) – as means to reach their goal. As visible in Figure 2.2.9, they wanted to feed the system with twofold sequential data: sentiment analysis of posts on social media (like Facebook) and videos and ads of constructive activities capable of capturing users in a virtuous circle. Though, to produce both kinds of information, other *ML agents* are necessary: specifically, a classification system able to determine whether hate or violence emerge from written posts (to detect which people to target with positive contents) and another one to identify contents that instill positive addiction toward constructive subjects (to decide which suggestions the *SP Agent* might propose). At this point, the recognized overlapping of sequence prediction and classification tasks induced a short circuit and required the facilitator's intervention. Yet, to understand the basic principles of ML, is it necessary for the ideated systems to be totally correct, or are some mistakes allowed in favor of immediacy?

The example highlights how the gradual expansion of the frontiers of design has to be reasoned thoughtfully: the completion of the system design sheet leverages consolidated design skills, but the modalities of knowledge transfer may challenge the process.

Another central issue is enabling the participants to elaborate on a system concept by integrating the multidisciplinary perspectives acquired. It affects both the framing of the ML problem for the *set-up* and the *system design* phase. In the *set-up*, defining the *key outcomes* that players need to reach with their systems means finding a balance between specificity and freedom. In the first case, the *key outcome* should be very detailed and imply only one ML solution; in the latter, it would be broader, leaving space for several possibilities (maybe not straightforward or not involving ML). Another concern that can be raised is that forcing an ML-infused solution could suggest a technology-driven approach. For the *system design* phase, the focus is on the guidelines for an actionable *system design* sheet: they need to be easy to understand and follow, and, ultimately, they should empower players to depict a system in both a sufficient but not too detailed way.

Since the activity entails the total inexperience of the students in designing with ML, it was challenging to understand which level of guidance they might need with the few tools provided. In the end, to measure the capabilities of the testers, the *key outcome* cards contained only a quite general brief (i.e., prevent deadly attacks orchestrated by organized crime in public spaces) and suggested datasets as possible inputs for the system to design (e.g., video and wiretapping recordings, police reports, expert profiling models, propaganda on social networks, direct testimonies). Additionally, sample sheets were prepared and included in the game as a backup plan. The *system design* sheets, instead, were supported by the structure of *ML agents* sheets. Surprisingly, during both experiences, the participants had no trouble in quickly delivering a system idea. The suggested datasets in the settings and the examples after the *system design* activity were not even necessary (in fact, the latter were not included in the second playtest, which caused no problem). Instead, the case studies from the *ML agents* sheets were important references to comprehend the requirements properly. However, what needs to be clarified is whether the systemic vision of Ph.D. students may have affected these results or if it also applies to master-level students.

Eventually, the playtests revealed some challenges also in the *weighing-up* phase. Although they had the chance to modify the elements defined in the *kick-off* (*value, ML agent, intent*), this was never considered. On the contrary, deciding to address a concern was so easy to comply with, while ignoring it caused negative feelings. Especially in the first session, the cards were perceived as prescriptive and triggered reactions like: "How to say no?" In this light, the purpose of the *weighing-up* phase could be reinforced toward a more conscious and thorough questioning of the idea. Then, instead of having students acknowledge some possible concerns about ML systems, the *concern cards* could be re-framed to go deeper into project matters, stimulating a more realistic iterative process.

Overall, the main challenges for enabling designers to work across disciplinary boundaries lie in the interstitial points between them, more precisely in knowledge transfer and operationalization. Especially in the introductory phases of a disciplinary expansion, trade-offs are required: it is necessary to understand what the most appropriate levels of completeness, depth, guidance, and active involvement with the contents for designers are to deal with.

2.2.3.3 Strengths: Rudiments for a hybrid designer

The enthusiast exclamation "I learned something today!", pronounced by one of the participants at the end of the game, perfectly synthesizes the most valuable outcome of the didactic experience. Indeed, testers in both sessions proved and confirmed that they had finely assimilated the pieces of knowledge that the activity aimed to instill, despite the substantial cognitive load to which they have been exposed. Moreover, the connection of the three disciplinary perspectives was discerned, and the participants demonstrated no problems in handling them, proving their openness toward expanding the frontiers of design.

In fact, despite the aforementioned issues with the dense contents of *ML agents* sheets, all the Ph.D. students involved in the experiment appeared at ease with the given materials and tasks, and these were effective in providing the basic equipment to face the system design. For instance, in this sense, the custom additions to the values and intents usually encountered in the literature received very positive feedback and have been incorporated into the system's foundations (Figure 2.2.10). In fact, toward the *key outcome* of preventing deadly attacks orchestrated by organized crime in public spaces, both groups embarked on a proactive approach. The first one elected *promotion of flourishing* as driving value because it was considered more comprehensive than *prevention of harm:* they stated it could help people appropriate their territory and avoid crimes. The second chose *inspire* as system intent to encourage potential criminals to pursue constructive interests.

Already during the *kick-off* phase, several ideas were brainstormed, triggered by an effective combination of ML agents' capabilities, intents, and values, so that the design of the system flowed smoothly. Also, to define the trickier elements (i.e., input and stakeholders), the players needed no assistance and identified coherent and original ones.

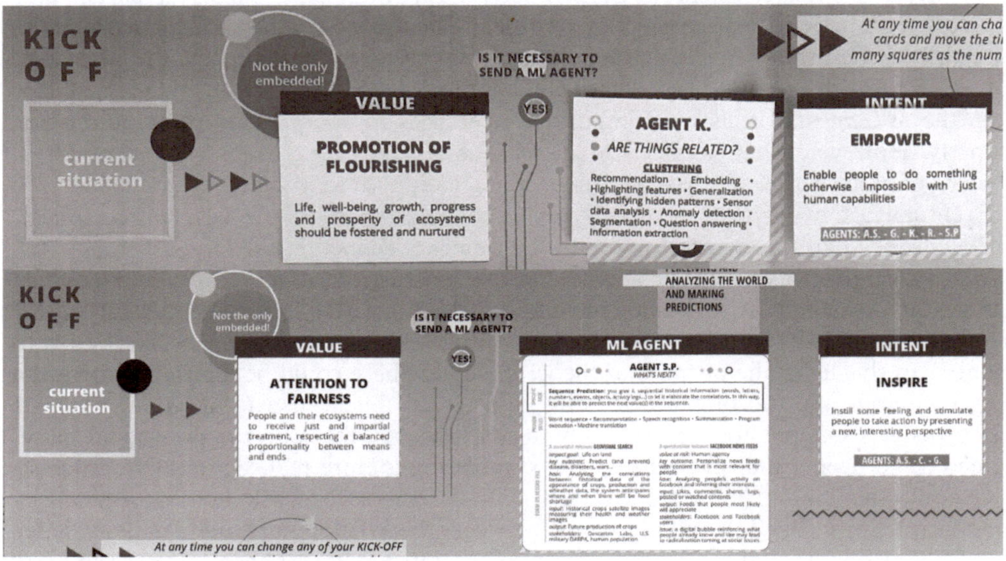

Figure 2.2.10 Selected elements for the kick-off phase in the in-presence (above) and online (below) playtests.

Additionally, from the very beginning, ethical concerns were explicitly discussed, as a value-driven approach to design may already be embedded in the mindset of the involved designers. Indeed, this was the case for both the Ph.D. students with no prior knowledge of ML and for those who might have been exposed to this subject. The former imagined a detector of the conditions that nurture crime (given context and municipality investments), insisting on including different stakeholders for a plural perspective. The latter proposed a recommendation system to deter criminals from malicious activities and drive them toward more constructive ones. In this case, the players were perfectly aware of their system's insidious and manipulative nature, though they considered this solution more ethically acceptable compared to capillary and undifferentiated surveillance. Therefore, even without a formal ethics education, the Ph.D. students manifested a responsible approach to design. Being early-stage researchers, their sensitivity may be increased; still, these kinds of considerations are at the basis of human-centered reasoning.

Albeit the expected outcome of enabling students to activate fruitful discussion should be verified in less guided contexts, the cooperative game turned out to be a fertile ground for debate and represented a successful example for opening the frontiers of design and enriching its tools. Throughout the entire activity, in both the physical and digital environments, communication is effectively fostered by the materials and the assignments players are faced with. The selection of foundational elements, the outlining of the system, and the final weighing are all collaborative activities and entail making decisions together by exchanging personal perspectives. As intended, also the *mission end* cards successfully accomplished this result, encouraging a retrospective discussion on the designed system and the whole experience.

2.2.4 Conclusion and future work

As Manzini stated in Frascara (2020), designers can operate in a variety of fields, and they can contribute to conceive and realize varied artifacts. A reason for this may be that "designers bring multiple talents to the solutions of complex issues, but first and foremost [...] empathy" (Friedman et al., 2014).

Empathy allows designers to understand the needs of all the people involved in a project, which implies absorbing their perspective. This is why, if properly equipped, they can comfortably navigate among disciplines and build upon their knowledge and methods, as corroborated by the empirical experiment above depicted. If the translation of technical knowledge (the primary objective of the activity) can be a delicate practice as it requires finding the correct language to efficaciously engage and empower designers to comprehend and make use of the transferred contents, instilling a value-sensitive perspective (secondary goal) appeared a far more natural outcome.

Indeed, going back to Alain Findeli's study "Ethics, Aesthetics, and Design" (Findeli, 1994), he already stated that ethics is an expansion of the currently accepted design definition. He sustained that "ethical deliberation is not very different from any other decision-making process," suggesting that designers not only have the means and the capabilities to handle ethical questions, but they also inevitably act in the field of ethics when they design an artifact. In fact, like any human activity, design is not neutral: choosing a technological mediation (in the broadest sense of creating instruments for people to interact with the world) is itself a matter of ethics, not of technology. Further assonances can be reported in that designers' criteria of choice do not arise from the truth (as in science), but from more qualitative evaluations of appropriateness, acceptability, correctness or desirability, and the now common practice of involving different stakeholders in the design process, that the author defines a "multicriteria approach," is already a moral action. What distinguishes a design decision from an ethical one is, in the words of Findeli (1994), a "total moral engagement on the actor's part," which, in line with the opening of this paragraph, can be reframed as the empathy representing a major strength of designers. To support once again the innate capability of designers to include ethical principles in their reasoning, Findeli (1994) underlines that "the systemic apprehension of a complex reality comes from intuition," a skill that designers have to train with appropriate methods but that is rooted in their previous education and practice.

After all, just as design pervades every aspect of human life, any field can be embraced and shaped by design, reinforcing the idea of a transdisciplinary design (Blevis et al., 2015) that transcends disciplinary boundaries. Expanding the frontiers of design, then, could contribute to the creation of a new kind of skill: that of mediating among approaches and value systems of a variety of disciplines (Kirschener & Norman, 2021). As self-conscious bricoleurs (Louridas, 1999), designers will not just merge the materials at hand; they could also identify and bring together different kinds of professionals to foster collaboration across multiple disciplines.

In the end, extending the breadth of design borders will not threaten a loss of identity. As demonstrated by the portrayed didactic experience, what characterizes designers is their attitude, their approach, and the processes they are used to apply to frame and solve problems. From its origins, design was outlined through the distinctive traits of other disciplines combined – namely, art, science, and technology. Today, this disciplinary distinction should be blurred and enclosed within a more general model to define a designer, whose essential skills would be perception (including visual intelligence) and (moral) action (Findeli, 2001). With such foundational equipment, the designer could easily merge into different disciplinary realms to deal with the most pressing problems and act as a bonding agent.

To support this vision, it would be interesting to expand the proposed introductory activity. It may be the building block for a more comprehensive educational project or include more vertical topics and become modular and flexible. Likewise, its finality could also be broadened, and further relevant insights could be gained from the submission of the game to students with different but inherent backgrounds. In the current state, if ethicists may not find cues to operationalize their knowledge, ML students could be interesting testers. Could a different perspective on *ML agents* enrich them? Can they get to design a system with the provided materials? How would they react to ethical concerns? Responding to this sort of questions may convey a deeper understanding of computer scientists or engineers' approach to the same problem, eventually suggesting how to facilitate communication with designers and make this game a basic tool for interdisciplinary problem-based teams (Friedman et al., 2019).

Acknowledgment(s)

I would like to thank the participants of the study for their commitment and feedback and the blind reviewers and network of DTRS13 for the valuable and constructive suggestions.

References

Algorithmic Watch. (2020, April). *AI ethics guidelines global inventory*. https://inventory.algorithmwatch.org/

Antonelli, P. (2018, August 2). *AI is design's latest material*. Google Design Library. https://design.google/library/ai-designs-latest-material/

Arnone, O. (2020). *Machine learning for designers. Enable designers to merge users' needs with machine learning capabilities in the idea generation and problem framing phases* [Master Thesis in Digital and Interaction Design, Politecnico di Milano - School of Design]. https://www.politesi.polimi.it/handle/10589/173637

Artefact. (2017). *The tarot cards of tech*. http://tarotcardsoftech.artefactgroup.com/

Auernhammer, J., & Ford, C. (2022). What is design? The semantic core and periphery of design definition. In G. Goldschmidt & E. Tarazi (Eds.), *Expanding the frontiers of design: A blessing or a curse?* (pp. 24–37). DesignTech, Faculty of Architecture & Town Planning Technion Israel Institute of Technology.

Blevis, E., Koskinen, I. K., Lee, K.-P., Bødker, S., Chen, L.-L., Lim, Y., Wei, H., & Wakkary, R. (2015). Transdisciplinary interaction design in design education. *Proceedings of the 33rd Annual ACM Conference Extended Abstracts on Human Factors in Computing Systems* (pp. 833–838). 10.1145/2702613.2724726

Calderon, A., Taber, D., Qu, H., & Wen, J. (2019). *AI blindspot: A discovery process for preventing, detecting, and mitigating bias in AI systems*. MIT.edu. http://aiblindspot.media.mit.edu/

Dam, R. F. & Siang, T. Y. (2021). 5 Stages in the Design Thinking Process. https://www.interaction-design.org/literature/article/5-stages-in-the-design-thinking-process

Dove, G., Halskov, K., Forlizzi, J., & Zimmerman, J. (2017). UX design innovation: Challenges for working with machine learning as a design material. *Proceedings of the 2017 CHI Conference on Human Factors in Computing Systems* (pp. 278–288).

Findeli, A. (1994). Ethics, aesthetics, and design. *Design Issues*, *10*(2), 49–68. 10.2307/1511628

Findeli, A. (2001). Rethinking design education for the 21st century: Theoretical, methodological, and ethical discussion. *Design Issues*, *17*(1), 5–17. 10.1162/07479360152103796

Frascara, J. (2020). Design education, training, and the broad picture: Eight experts respond to a few questions. *She Ji: The Journal of Design, Economics, and Innovation*, *6*(1), 106–117. 10.1016/j.sheji.2019.12.003

Friedman, K. (2019). Design education today—Challenges, opportunities, failures. *Chatterjee Global Lecture Series*. https://www.academia.edu/40519668/Friedman._2019._Design_Education_Today_-_Challenges_Opportunities_Failures

Friedman, K., Lou, Y., Norman, D., Stappers, P. J., Voûte, E., & Whitney, P. (2014, December 6). Why designX? Designers and complex systems. *Core77*. https://www.core77.com/posts/27936/Why-DesignX-Designers-and-Complex-Systems

Friedman, K., Lou, Y., Norman, D., Stappers, P. J., Voûte, E., & Whitney, P. (2019, August 1). *DesignX: A future path for design*. Jnd.Org. https://jnd.org/designx_a_future_path_for_design/

Futurice (2017). *The intelligence augmentation design toolkit*. https://futurice.com/ia-design-kit

Gispen, J. (2017). *Ethics for designers*. Ethics for Designers. https://www.ethicsfordesigners.com

Goldschmidt, G., & Tarazi, E. (2022). *Proceedings of DTRS13 Conference: Expanding the frontiers of design: A blessing or a curse?* DesignTech, Faculty of Architecture & Town Planning TechnionIsrael Institute of Technology.

High-Level Expert Group on Artificial Intelligence. (2019a). *A definition of AI: Main capabilities and scientific disciplines*. European Commission. https://ec.europa.eu/digital-single-market/en/news/definition-artificial-intelligence-main-capabilities-and-scientific-disciplines

High-Level Expert Group on Artificial Intelligence. (2019b). *Ethics guidelines for trustworthy AI*. European Commission. https://ec.europa.eu/futurium/en/ai-alliance-consultation/guidelines

Huizinga, J. (1938). *Homo ludens*. Routledge & Kegan Paul.

IDEO (2019). *AI & ethics: Collaborative activities for designers*. https://www.ideo.com/post/ai-ethics-collaborative-activities-for-designers

Kaptelinin, V., & Nardi, B. A. (2009). *Acting with technology. Activity theory and interaction design*. MIT Press.

Kelly, K. (2016, June). *How AI can bring on a second Industrial Revolution*. https://www.ted.com/talks/kevin_kelly_how_ai_can_bring_on_a_second_industrial_revolution

Kirschener, P., & Norman, D. A. (2021). *The teaching and learning of design. Future of Design Education*. https://assets.website-files.com/6010c2ec764764887cfefee7/60515ff38234e90fac4c64cb_Kirschner%20and%20Norman.%20Teaching%20and%20Learning%20of%20Design%2021.01.21.pdf

Louridas, P. (1999). Design as bricolage: Anthropology meets design thinking. *Design Studies*, *20*(6), 517–535. 10.1016/S0142-694X(98)00044-1

Manzini, E., & Coad, R. (2015). *Design, when everybody designs: An introduction to design for social innovation*. The MIT Press. https://www.jstor.org/stable/j.ctt17kk7sv

Meyer, M. (2010). The rise of the knowledge broker. *Science Communication*, *32*(1), 118–127. 10.1177/1075547009359797

Meyer, M. W., & Norman, D. (2020). Changing design education for the 21st Century. *She Ji: The Journal of Design, Economics, and Innovation*, *6*(1), 13–49. 10.1016/j.sheji.2019.12.002

Piet, N. (2019). *AI meets design toolkit*. https://www.aixdesign.co/toolkit

Russell, S., & Norvig, P. (2020). *Artificial intelligence: A modern approach* (4th ed.). Pearson.

Spallazzo, D., Sciannamè, M., Ajovalasit, M., Ceconello, M., & Arquilla, V. (2021, to be published). Exploring AI-infused products qualities to unleash. AmI. 12th International Symposium on Ambient Intelligence, Salamanca.

UN - Open Working Group. (2015). *THE 17 GOALS | Sustainable Development*. https://sdgs.un.org/goals

van den Hoven, J. (2013). Value sensitive design and responsible innovation. In R. Owen, J. Bessant, & M. Heintz (eds.), *Responsible innovation* (pp. 75–83). John Wiley & Sons, Ltd. 10.1002/9781118551424.ch4 von Schomberg, R. (2013). A vision of responsible research and innovation. In R. Owen, J. Bessant, & M. Heintz (eds.), *Responsible innovation* (pp. 51–74). John Wiley & Sons, Ltd. 10.1002/9781118551424.ch3

Weil, D., & Mayfield, M. (2020). Tomorrow's critical design competencies: Building a course system for 21st Century designers. *She Ji: The Journal of Design, Economics, and Innovation*, *6*(2), 157–169. 10.1016/j.sheji.2020.03.001

Yang, Q., Steinfeld, A., Rosé, C., & Zimmerman, J. (2020). Re-examining whether, why, and how human-AI interaction is uniquely difficult to design. *Proceedings of the 2020 CHI Conference on Human Factors in Computing Systems* (pp. 1–13). 10.1145/3313831.3376301

2.3 Factors, strategies, and criteria in design convergence

Naz A.G.Z. Börekçi
Middle East Technical University, Department of Industrial Design

2.3.1 Introduction

In the double-diamond design process model, based on the linguist Banathy's (1996) dynamics of divergence–convergence model, and later on popularised by the Design Council, UK (2019), a series of *divergence followed by convergence* acts take place, the first set for framing the design problem, and the second set for naming the design solution. Divergence in design has been an objective in the pursuit of situated solutions to a design problem, as this provides designers with a widened solution space containing numerous and diverse alternatives (Jones, 1980). Convergence, on the other hand, is the narrowing down of these alternatives, requiring an assessment, selection and transformation of those that they find best matching the problem (Jones, 1980; Lawson, 2005), and this may prove to be more difficult. During design exploration, we expect our students to generate many ideas and suggest methods for them to support divergence. On the other hand, we seldom offer them methods of convergence for proceeding from many ideas to few, as this seems to be inherent in their design activities carried out during the course of the design process, and the related activities depend on the nature and requirements of the design problems dealt with. Based on the interest that this issue has raised for me, with the study described in this chapter, I inquire into how design convergence takes place for students, as they prepare for the delivery of their design output.

2.3.1.1 Aim of the study and research questions

Referring to convergence as pinned in the double-diamond design process model, I address the following questions:

- What are the acts of design convergence?
- How does design convergence take place in reference to these acts?

The study first reviews the conduct of a short-term design project on 2-in-1 hand-held stick vacuum cleaners (2-in-1 SVCs from hereon) carried out three years in a row in a graduate design methods course. In the first year when this project was applied, the decision-making activities of the students took my attention, and therefore, in the following two years, I followed the same design process for the same project topic, and additionally, I gave the students of those years, a survey as the follow-up of the project outcomes, questioning their acts of design convergence. The process of the project was planned based on the double-diamond design process model, with a series of activities

DOI: 10.1201/b22630-12

employing tools and methods supporting divergence and convergence. Towards the end of the five-week project, as they were preparing for their project delivery, students were given the survey forms to reply in written. The survey in both years contained open-ended questions, and responses were collected for a total of 30 students. The questions were related to the *design convergence* activities of the students, following design divergence through which they had generated numerous alternative sub-solutions for 2-in-1 SVCs. The aim was to identify the influences that contributed to and thus explain their process of design convergence, first for developing six alternative design solutions, then for sifting them into one final design solution.

This chapter begins with a literature review on design convergence in order to provide a theoretical framework on the design convergence activities to set basis for the analysis of the survey responses. Then, the design process for the 2-in-1 SVC project is described in reference to the activities of divergence and convergence, illustrating with examples from the class activities and project deliverables made for each stage. Finally, the survey responses are analysed in reference to the project, and the findings are presented explaining how design convergence has taken place for the participants.

2.3.1.2 Design convergence

Convergence is the cognitive effort of *narrowing down* by eliminating alternatives and refining those selected. Design convergence involves assessment, decision-making, and design synthesis. The main thinking type involved in this process is convergent thinking.

2.3.1.2.1 Convergence as a stage in the design process

Jones (1980) names the three stages of the design process as divergence, transformation, and convergence, which are methodologically separated according to the requirements of a well-working process at the systems level. In divergence, the design problem is identified, objectives are defined, and the primary uncertainties are resolved. In transformation, design exploration is carried out, yielding alternatives to evaluate in terms of the secondary uncertainties related to the problem area. In convergence, the alternatives are reduced until a final design solution is obtained.

The stage of convergence has certain features, as Jones (1980) explains. As the problem is dealt with more and more closely, those alternatives that are not worth investing in, need to be identified quickly and put aside, making *persistence and rigidity of mind and method* critical (Jones, 1980). Design exploration in the transformation stage takes place more at a general level, therefore some sub-problems may have been avoided during this stage. During convergence, these sub-problems may prove to be critical and have to be dealt with, calling for *negotiation*, which may result at the cost of some of the design decisions made earlier. The representations of the design alternatives will have to be *more in detail*, for an effective assessment. The design iterations will require approaching the design solution both with an *in-out* and an *out-in* strategy, meaning the external form will need to be considered simultaneously with the internal arrangement of components, towards a match (Jones, 1980).

2.3.1.2.2 Convergent thinking

Convergence involves design acts which integrate and further synthesise ideas through deductive generation towards developed and finalised designs. In divergence, alternative ideas are generated, and in convergence, synthesised ideas are explored (Lubart, 2016). Basadur (1994) states that it is convergent ability that transforms divergent thinking into creative output; accordingly in their study, Zhu, Shang, Jiang, Pei, & Su (2019) provide evidence that divergent thinking can play a role in creativity only when convergent thinking capacity reaches a certain level.

Studies have shown that convergent thinking is more impactful on problem-solving in creative contexts; it is also acknowledged that convergent thinking provides control in chaotic activations caused by divergent thinking (Webb, Little, Cropper, & Roze, 2017). Convergent thinking may benefit from less impulsiveness and risk-avoidance when creative problem-solving (Shen, Hommel, Yuan, Chang, & Zhang, 2018). Creativity requires divergent thinking for lateral transformations, as well as convergent thinking for vertical transformations; divergent and convergent thinking are seen in the creative phases of the design process as happening in cycles, linking forward movements with past movements, and in each cycle, ideas are generated through divergence then explored and evaluated through convergence, for transformations to be complete (Goldschmidt, 2016). Hence, it is possible to refine and finalise design solutions. Ability to deliberately control the activation and usage of divergent and convergent thinking correlates with the success in creative ill-defined problem-solving and of its result (Wolf & Mieg, 2010).

2.3.1.2.3 Role of knowledge and information in convergence

Cropley (2006) explains that knowledge is of particular importance in convergent thinking, as it is the source of ideas, pathways to solutions, and criteria of assessment. Knowledge prepares the mind to check and test alternatives, compile further information, and make strategic decisions; this preparation is important so that during convergence, the existing can be refined based on new insights, and intuition from the decision maker's part can also play its part (Cropley, 2006). In this process, speed, accuracy, unambiguity, logic, and recognising the familiar are among keywords that play role in decision-making, and are all closely linked to knowledge (Cropley, 2006). Successful design convergence takes place when designers are able to make use of other knowledge types besides design knowledge and are able to avoid "conservative thinking styles" (Jung & Chang, 2014, p. 1), meaning those that obligate designers to keep to known rules and procedures (Zhang, 2009), prevent them from taking risks and restrict them from moving away from previously recognised solutions. In the achievement of design breakthroughs, a correlation is suggested to exist between the triple activity mode of "examining-drawing-thinking" (Akin & Lin, 1996, p. 59) and novel design decisions.

During design convergence, information (knowledge, designs, variables, etc.) will need to be organised, and hierarchical classification schemes are a way of doing this (Müller-Wienbergen, Müller, Seidel, & Becker, 2011). Such schemes allow the organisation, categorisation, and ordering of information, and tools such as the interaction matrix, classification trees, and weighting trees (Cross, 2000; Roozenburg & Eekels, 1995; Wright, 1998) can be given as examples. Lawson (2005) explains that measurement in design requires a balance of quantifiable and qualifiable judgement of alternatives against criteria. For example, ordering design solutions according to preference will require assessing each design against each criterion, then combining these assessments. Three issues will surface as difficulties: 1) not all criteria will be of equal value, requiring a weighting among them, 2) some criteria will require objective judgement, whereas others will require subjective judgement, and 3) these judgements will have to be combined for an overall assessment. Quality and quantity in design evaluation is intertwined, and inevitably, this will also include value judgements. Values may be subjective, such as goodness and usefulness, as well as objective, such as those defined in standards and regulations. Different reasoning strategies will come into action in leading design ideation, depending on the evaluation based on different types of values (Christensen & Ball, 2016). As values will be relative, it is important to develop a common scale of measurement considering the advantages and disadvantages of a decision for all parties involved, and with which the decision makers and the stakeholders agree to an optimal degree (Lawson, 2005). A group (e.g., design team) determines the criteria against which to assess the alternatives based on information, and the quality of information given prior to the process affects convergence. Particularly when the process is not rushed and time is allowed, team coherence takes place, leading to better design performance, and better formulation of the assessment criteria (Fu, Cagan, & Kotovsky, 2010). Convergence will be considered successful when the decision makers are convinced that they have chosen the best possible solution to an ill-defined problem.

2.3.1.3 Framework of convergence for this study

Convergence, reducing the alternatives into one, is a function of the acts of *critical assessment* made in terms of many factors, *design synthesis* made through the integration of relevant information with detailed design representations, and *decision-making* made by matching solutions with the design problem, all requiring a decisive attitude that is based on well-informed judgements. *Decision-making*, as a mechanism of convergence, may be subjective, e.g., based on personal attitudes, as well as objective, e.g., based on a match between the project objectives and design decisions. Convergent thinking lies in the core of convergence and is best supported with methods and approaches that facilitate, but also control, both the generation (through divergence) and the exploration (through convergence) of alternatives, and the compilation and processing of knowledge towards design synthesis. Having thus answered the first research question, "what are the acts of design convergence", this study further investigates the involvement of these acts in the process of a design project carried out with graduate design students, with an attempt to explain how design convergence takes place.

2.3.2 The design project: 2-in-1 hand-held stick vacuum cleaners

Three academic years in a row (2017–18, 2018–19, and 2019–20), a short-term design project was conducted on the renewal of the 2-in-1 hand-held stick vacuum cleaner (2-in-1 SVC), for a graduate course on design methods. This course is given for 14 weeks in a semester, and the weekly teaching hours is 4. The course aims to teach various design methods and provide a medium in which these can be applied on short projects. Following 4 weeks of lectures, two projects are conducted in the course; one follows a problem-oriented process, and the other follows a solution-oriented process. The process described in this chapter was followed for the first project of the semester, integrating various design methods into a structured, problem-oriented design process, from idea generation to design development and evaluation. In the above-mentioned academic years, the same design process has been used for the conduct of the 2-in-1 SVC project, and the project took 5 weeks in total (20 hours).

Table 2.3.1 Participants of the project

Participants	*Female*	*Male*	*MS*	*PhD*	*Design background*	*Non-design background*	*Turkish*	*Other nationality*	*Total*
2017–18	7	7	11	3	11	3	12	2	14
2018–19	10	4	13	1	13	1	14	0	14
2019–20	9	10	17	2	19	–	17	2	19
Total	26	21	41	6	43	4	43	4	**47**

2.3.2.1 Participants

Throughout the three academic years, a total of 47 students participated in the project (Table 2.3.1). The courses were conducted entirely in English. The survey respondents were students from the 2018–19 and 2019–20 academic years. A total of 33 students attended the course in these years. Thirty were master of science students, and three were PhD students. Thirty-two were from a design background (30 industrial design, 1 architecture, 1 graphic design) and one was from a non-design background (mechanical engineering). Thirty-one students were Turkish, and two were from other nationalities (1 Egypt, 1 Pakistan). Thirty students responded to the survey (three Turkish students from a design background did not respond).

2.3.2.2 Design process and project conduct

The design process was composed of three stages. The students were expected to make an individual final submission. Each step of the stages required output submitted individually, except for the morphological chart, which was carried out in teams. The students were asked to work in groups (interactively, or simply in parallel), even when a task required individual submission, in order to help each other, keep up with the pace of the class, enhance the cognitive effort, and enrich the content of the output. For each academic year, the project process was well-documented with the anticipation of such a study. The submissions made for each step of the process were digitally collected, forming a large set of visual material, for future reference. The following sections provide a general description of the process stages, covering all three academic years.

2.3.2.2.1 Problem analysis

The design process begins with *problem analysis*. The project briefing is made, the goal is set for product modifications, project dimensions are discussed, and students are asked to consider various issues such as the user, use environment, usage, maintenance, and storage. Following, students carry out in-class product trials with numerous 2-in-1 SVCs brought to class for *function analysis* (Cross, 2000; Pahl & Beitz, 1996; Roozenburg & Eekels, 1995) and *operational sequence analysis* (Kirwan & Ainsworth, 2005) (Figure 2.3.1). This is followed by a *quantified structure analysis* (Tjalve, 1979), where the interaction between the components is studied first in 2D as diagrams and then in 3D with simple mock-ups, showing the configurations of the components in their correct proportions and rough forms (Figure 2.3.2).

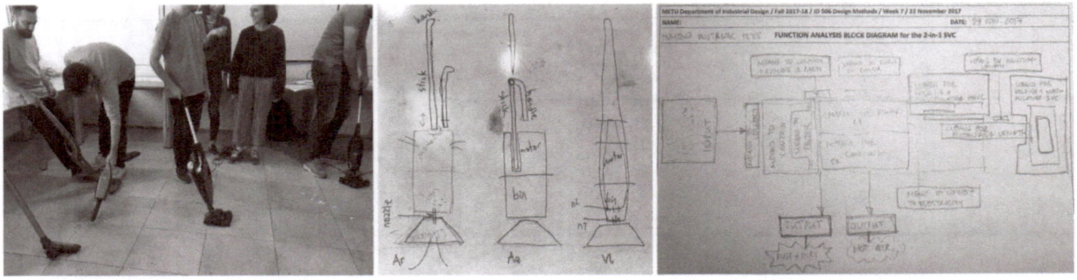

Figure 2.3.1 Left: Product trials in class, 2019–20. Centre: Study of component configurations during product trials, on whiteboard, 2018–19. Right: A function analysis study, 2017–18.

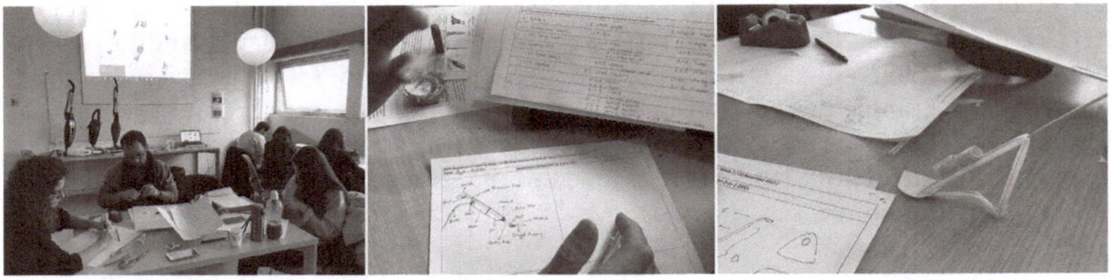

Figure 2.3.2 Left: Classwork for quantified structures, 2017–18. Centre: 2D quantified structure study, 2019–20. Right: 3D quantified structure study, 2017–18.

2.3.2.2.2 Idea generation

The process continues with *idea generation*, relying heavily on the *morphological chart* method (Cross, 2000; Pahl & Beitz, 1996; Roozenburg & Eekels, 1995; Wright, 1998) for design divergence and convergence. This method involves idea generation for the components of a decomposed design problem, not looking for novelty in particular, but aiming for quantity and variety. The method asks for the design problem (in this case, the 2-in1 SVC) to be decomposed into its sub-functions (i.e., product components each fulfilling an individual function). Identification of 8 to 12 sub-functions is recommended for this task (Cross, 2000; Roozenburg & Eekels, 1995; Wright, 1998). These sub-functions are placed in the rows of the morphological chart. Then for each sub-function, six ideas (sub-solutions) are generated in the cells of the row, each different in terms of the means that can fulfil the sub-function. Offering alternative sub-solutions for all the sub-functions yields a chart with numerous sub-solutions (e.g., 12 sub-functions × 6 sub-solutions = 72 sub-solutions) to pick from and combine into an overall design solution, with the combination possibilities being almost endless.

This functional decomposition calls for a thorough understanding of the essential function of the product and how it operates (Srinivasan, Chakrabarti, & Lindemann, 2012; Wright, 1998), and is done in a *brainstorming* session with the class. The sub-functions are open to design interventions that can reflect as improvements on the overall final design solution. Once the chart is conjointly completed, students may now individually select sub-solutions for the sub-functions and combine them into overall design

Figure 2.3.3 Left: Sub-functions identified for the morphological chart, 2017–18. Centre: Team working on their morphological chart, 2019–20. Right: Completed morphological chart, 2018–19.

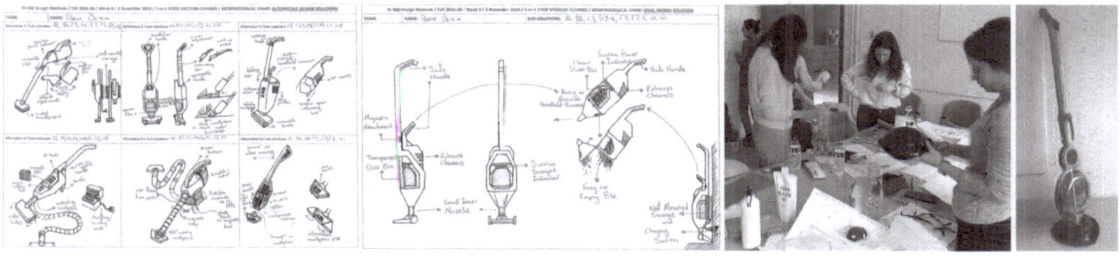

Figure 2.3.4 Left: Convergence step 1, six alternative design solutions from the morphological chart; Second: Convergence step 2, one final design solution, Doruk Öktem, 2019–20. Third: Model making in class, 2018–19. Right: Full-scale model for 2-in-1 SVC, Yağmur Merve Arık, 2017–18.

solutions (Figure 2.3.3). This calls for a two-step convergence, the first step resulting in six alternative design solutions represented in 2D (Figure 2.3.4, left).

With a break from design activities, another brainstorming session is carried out with the class, where project objectives are determined altogether, and then ranked in order of importance, using an *interaction matrix* (Jones, 1980). In the second step of convergence, students reflect on their six alternative design solutions and select one to develop into a final design solution (Figure 2.3.4, second). The students are expected to finalise their design individually and present as a full-scale paper-based model (Figure 2.3.4, right). Model making is done in class, with shared tools and materials, and an opportunity for interactive discussions among class members (Figure 2.3.4, third)

2.3.2.2.3 Evaluation

The process concludes with two rounds of *evaluation*. The list of rank-ordered objectives is used as reference to evaluate the design ideas represented as full-scale models, during an *expert opinion* jury session (Figure 2.3.5, left, centre). Each idea is presented to the jury and class. Jury members give feedback to each student on their design solution. In the background, peers listen to the discussions and occasionally participate, filling in individually, in real time, a shared online document, applying the *searching for visual inconsistencies* method (Jones, 1980) for the visual and functional contradictions in the design and for suggestions on how to improve these, in reference to the project objectives,

Figure 2.3.5 Left: Jury for expert opinion, Ayşe Kaplan 2018–19. Centre: Jury for expert opinion, Hakan Dinçer, 2019–20. Right: Peer evaluation during jury; students filling the online visual inconsistencies form, 2017–18.

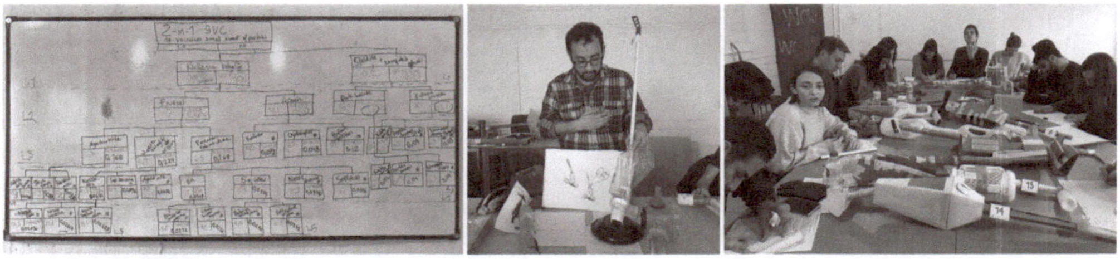

Figure 2.3.6 Left: Objectives weighting tree for RIRs, on class whiteboard, 2018–19. Centre: Presentation to class, Yashar Kardar, 2017–18. Right: Class listening to peer presentation, with exhibited models for RIR evaluation, 2019–20.

personal opinions, and jury feedback (Figure 2.3.5, right). This online form is accessible to the students, who are then expected to improve their design according to the qualitative feedback, given by the peers and the experts, for the following week.

The models are revised and submitted for the following week's final evaluation. This is a quantitative peer evaluation, based on the *relative importance ratings* (RIR) for the project objectives, determined altogether through brainstorming, using the *objectives weighting tree* (Cross, 2000; Wright, 1998) (Figure 2.3.6, left). The objectives are revised and explicitly defined during the activity, which leads to an increase in their number (e.g., 14 objectives determined earlier, may result in 23 elaborated objectives for the final evaluation) and each is given an RIR. The final design models are exhibited, and the students act as panel members, evaluating each design solution on their charts, in reference to the rated objectives (Figure 2.3.6, right). Each chart produces quantitative scores for the designs, which are then combined, to yield the accumulative scores. These scores provide a rating of the appreciation given to the final design solutions. Rank ordering the final design solutions according to their scores opens up a final interactive discussion on the qualities of the designs before the project is closed.

2.3.2.3 Convergence in the design process

For this study, the design process followed was reviewed for the acts of divergence and convergence, and a series of diamonds were identified on which three design iterations were based (Figure 2.3.7).

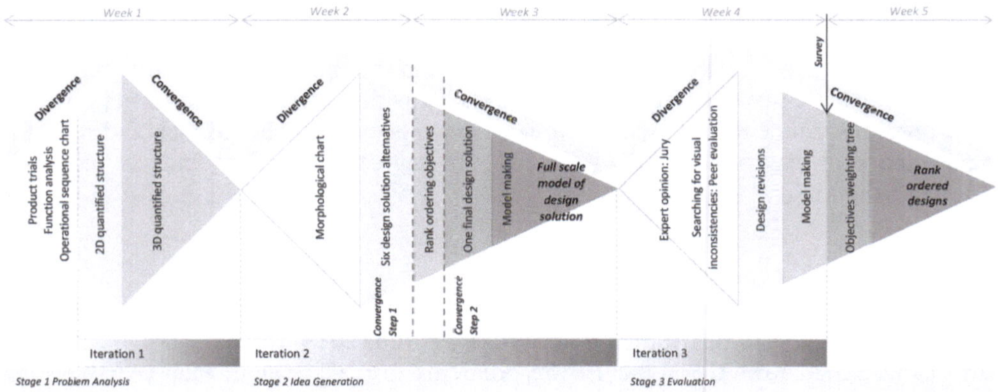

Figure 2.3.7 The triple-diamond design process model for the 2-in-1 SVC project.

2.3.2.3.1 Divergence–convergence in Stage 1: Problem analysis

The first takes place in the *problem analysis* stage, where product trials and the accompanying function and operation sequence analyses contribute to defining the problem space, followed by an exploration of 2D quantified structures to open up the solution space. The stage concludes with an act of convergence, where students carry out reflection on the numerous 2D quantified structures and transfer a number of these into 3D mock-ups.

2.3.2.3.2 Divergence–convergence in Stage 2: Idea generation

This stage is characterised by a divergence effort made for the decomposition of the product into its sub-functions and design exploration for sub-solutions using the morphological chart. Having completed the chart interactively in teams, students then individually carry out acts of convergence in two steps, where they first combine the sub-solutions into six overall alternative design solutions, and then transform these alternatives into one final design solution (Convergence Step 1 and Convergence Step 2, indicated in red in Figure 2.3.7, and with arrows in Figure 2.3.8). Between these steps, the brainstorming session for project objectives provides an opportunity for students to reflect on the criteria on which they can ground their decisions for idea exploration, assessment, and selection.

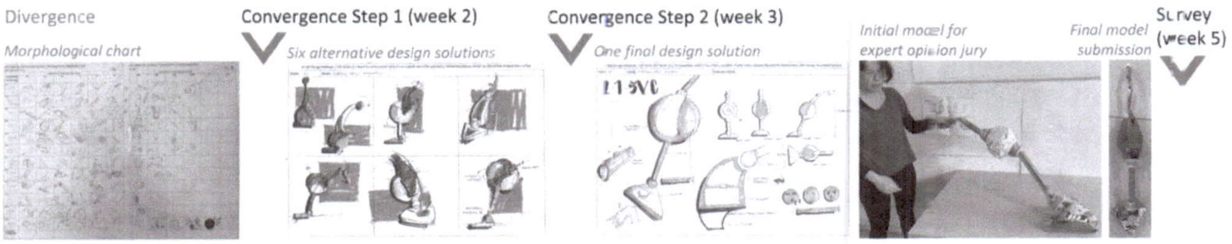

Figure 2.3.8 Convergence step 1, convergence step 2 and survey timing displayed with student outputs (Kardelen Aysel, 2018–19).

2.3.2.3.3 Divergence–convergence in Stage 3: Evaluation

This stage is characterised by the expert and peer reviews offered to the students on their design ideas. In terms of divergence, feedback given in the expert opinion jury, parallel to the peer feedback given on visual inconsistencies, provides reference to students for their design revisions. Students are expected to consider this qualitative feedback and reflect it on one final full-scale model submission, to the degree which they find fit, requiring convergent thinking.

2.3.3 The survey

With the aim of investigating the acts of design convergence within the design process, a survey was prepared and distributed to the students of the 2018–19 and 2019–20 academic years. Table 2.3.2 shows the number of respondents to the survey in each year.

The focus of the survey was the two steps of convergence taking place in the idea generation stage, where students narrowed down their design ideas first into six alternative design solutions, and then into a final design solution (Figure 2.3.7). The survey was distributed to the students at the end of their model-making activities for their final submission; this was found to be a suitable time for reflection as students had concluded their designs at this stage. The same six open-ended questions were posed in the survey in both years.

The first four questions inquired:

1 The factors that affected the selection of sub-solutions from the morphological chart to bring together into six alternative design solutions.
2 How the sub-solutions were processed into overall alternative design solutions.
3 The factors that affected the development of the final design solution.
4 How the alternatives were processed into one final design solution.

The final two questions were on the priorities of the students. The questions inquired:

1 Which project dimensions were considered for the alternative design solutions?
2 Which project objectives were considered for the final design solution?

2.3.3.1 Analysis of the data

The survey was distributed to the class as a two-page print form (one double-sided A4), and the students were given time to respond and submit (Figure 2.3.9, left). The collected data was written verbal statements of qualitative nature, and this data was analysed using thematic analysis (Braun & Clarke, 2015) and content analysis (Krippendorf, 2018). The responses for each question were read through across all forms and were summarised and

Table 2.3.2 Number of respondents to the surveys

	Questions 1–4 (No. of respondents)	*Questions 5–6 (No. of respondents)*
2018–19	11	11
2019–20	13	19
Total	**24**	**30**

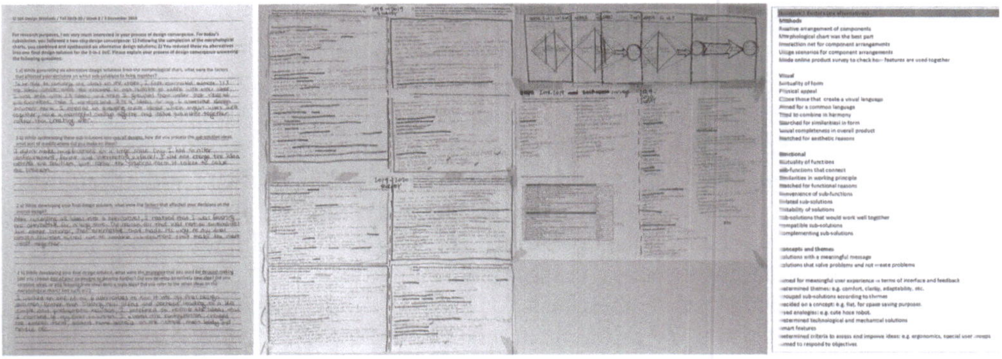

Figure 2.3.9 Left: Example survey form filled by respondent, 2019–20. Centre: Annotations from survey forms on sheet showing codes (themes) in colour. Right: Example page from digital spreadsheet grouping topics (content).

transferred by hand as annotations onto a sheet (Figure 2.3.9, centre). The annotations were colour coded according to thematic content and then transferred onto a digital spreadsheet, where they were grouped into topics according to the questions (Figure 2.3.9, right). These topics were revised for content related to the students' acts of convergence. The aim was to derive the approaches of students to design convergence, and all hints were considered of equal value; the analysis did not include frequency count for mention, and no hierarchy between the themes was sought.

2.3.3.2 Findings

The findings reveal numerous themes collected under three main titles, namely *factors* ("what") that the students considered, *strategies* ("how") that they followed, and *criteria* ("why") they used for judgement during design convergence.

2.3.3.2.1 Factors

Factors were the issues that students considered for convergence: they were related to decision-making in particular. Factors were identified from the responses given to the survey questions 1 and 3, but responses to all questions were also cross-checked for related content. Five factors were identified from the survey responses.

1 *User:* According to the survey responses, the user came forth as the most prioritised project dimension and was mentioned in the answers of all 24 respondents (of questions 1–4). The user was in many instances described as a single person living alone in a small flat. 1) "Ease of usage", "effortlessness", and "healthier usage conditions" for the users and their "well-being" were considered for the 2-in-1 SVCs, supported with weight reduction for ease of portability from room to room and ergonomic features a) preventing fatigue from repetitive arm movements and long-term handling, and b) allowing easy gripping of the handle. Accordingly, adjustability of stick height, adjustability of handle for hand-held component, balance, and a "flexible centre of gravity" that could be changed according to the component being used (i.e., hand-held, or stick vacuum), were suggested for the design solutions. 2) "Practicality" was suggested

with an optimum level of performance in terms of motor capacity, and foldability for storage. 3) Easy and "speedy" physical interaction with the 2-in-1 SVCs was expected, in terms of a) converting from stick to hand-held vacuum cleaner; b) removing the hand-held component from the stick; c) assembling and disassembling components and accessories (e.g., filter, bin, nozzles, and accessories); d) charging batteries; e) accessing and putting away accessories; and f) cleaning, maintaining, and repairing the device.

2 *User experience:* Positive user experience was a second major consideration, with design solutions for 1) easy-to-access controls on body, handle, and stick; 2) effective feedback mechanisms such as displays, touch screens, sound and light for warning, to prevent interruptions during usage (e.g., remaining battery level) and provide process information; 3) understandable interfaces for product mode, speed, and suction power selection and adjustments for different accessories used for different surfaces or dirt types; 4) making the device compact, flat or keep upright for space saving purposes; and 5) not having "different accessories lying around" but instead findings means to keep them together neatly during storage.

3 *Use environment:* The use environment was considered the home, with usage in different parts of the home that have 1) different surface types, such as carpets, wood flooring, or ceramic tiles; 2) different levels of difficulty, such as stairs, corners, or underneath cupboards; 3) different heights, such as floor, furniture upholstery, top of door frames, and curtains; and 4) sources that generate various dirt types. These were mainly resolved with 1) angular or curved stick structures that allow reaching to "difficult areas", "narrow corners", and "small areas", and 2) nozzle accessories and features that a) were either multiple and different, or few and multifunctional, and that b) could detect the changes of surface types, dirt types and amount of dirt, and adapt adjustments (e.g., suction power), settings (e.g., wet mode), and features (e.g., retracting brush) accordingly.

4 *Design features:* Design features were described as critical for an improved device performance and positive user experience. The considerations were 1) manoeuvrability of the 2-in-1 SVCs as both hand-held and stick; 2) manipulability of the components during usage and maintenance (e.g., removing, emptying, and replacing bin); 3) having a self-standing structure for the 2-in-1 SVC during usage as well as during storage; 4) having an overall "consistent visual identity" for the components such as nozzle, filter compartment, dust bin, and handle in terms of "design", "geometry", and "aesthetics", with an emphasis on visual appeal and solving use-related problems (e.g., nozzle design allowing to reach narrow corners); and 5) differentiation from, and as alternative, resemblance to the products available in the market. Differentiation was sought through features such as transparent bin, exhaust panels, multifunctional and adaptable nozzle accessories, direct access to filter via a separate compartment, and solutions for positioning, winding, and storing the cable.

5 *Design directions:* Students described concerns for developing solutions "with a meaningful message", and "that solved problems rather than create ones". To achieve this, they reported 1) determining goals for themselves (e.g., ergonomics, special user groups); 2) deciding on a concept and following it through (e.g., flat, for space-saving purposes); and 3) developing themes (e.g., comfort, adaptability, and cute house robot). These goals, concepts, and themes were used to generate ideas for sub-solutions, assess and improve them, group them (e.g., technological sub-solutions and mechanical sub-solutions), then bring them together into overall alternatives. 4) All students mentioned that they aimed to respond to the collectively determined project

objectives, in the second step of convergence (developing one final design solution). The most frequently mentioned project objectives were user comfort, ease of use, enhanced user experience, user friendliness, and usefulness. One student mentioned that combinations of these objectives provided "added value" to the final design solution.

2.3.3.2.2 Strategies

Strategies were the ways in which the students carried out convergence; they were related to design synthesis in particular. Strategies were identified from the responses given to the survey questions 2 and 4, but responses to all questions were also cross-checked for related content. Five strategies were determined from the survey responses.

1 *Strategies for selection:* Students reported 1) selecting the sub-solutions to bring together based on "problem hierarchy", in terms of the most problems resolved with a component. These problems were those identified during product trials, as well as through past personal experiences. Students also mentioned 2) selecting according to "function priorities", and "most important and relevant features". One student mentioned having a main functional focus (e.g., form that can be both hand-held and stick without any component removal) and bringing together sub-solutions that fitted this idea. 3) Supra-functional qualities (e.g., user comfort), personally set criteria (e.g., realisability, cost-efficiency, compatibility), and project objectives were also used in selection. A number of students indicated 5) "liking an idea", "having a favoured idea", and "making a personal preference" as reasons for choosing sub-solutions. Finally, 6) the technical or visual qualities of the sub-solutions were also given as a reason for selection. Some students indicated that they selected according to visual appeal and "aesthetic match", whereas others indicated that they selected according to types, such as "technological sub-solutions", and "mechanical sub-solutions".

2 *Strategies for elimination:* Students also eliminated sub-solutions to limit the number of alternatives to consider. As strategies, some students reported 1) assessing whether sub-solutions were "unrealistic", "non-practical", or "considered extra" (e.g., additional feedback mechanism, nozzle extensions for different surface types), and others reported 2) assessing whether a sub-solution would fit an overall form as a design feature, or not, during design synthesis.

3 *Strategies for building configurations:* While transforming the designs, students reported that they 1) grouped sub-solutions first, according to a) determined concepts or themes, b) specifications or selected functions, c) working principles for the overall design, d) visual appearance for the form, and e) those that they "would like to work on". Once the sub-solutions were grouped, some students; 2) studied the interactions between them and configured the ensemble accordingly. Others 3) determined a main component and built other sub-solutions around it, or 4) combined each sub-solution with its neighbouring component first, making changes to the sub-solutions and building the form on the way. Some students indicated that they 5) decided on the overall form first (indicating that they "already had rough ideas for the forms" to begin with), then decided on the "location" (configuration) of the components on the body. Others stated that they 6) grouped the sub-solutions according to the main functions to which they related (i.e., hand-held vacuum and stick vacuum), brought them into two combined larger components, and then studied alternative configurations for the two ensembles towards overall design solutions. One student indicated that she 7)

determined one or two features that she would like to have common on all six alternatives, and chose the remaining sub-solutions in order to variate the alternatives. All respondents, except for two, indicated that they 8) checked and modified components for compatibility with each other and also with the overall resulting configuration. Two students indicated that they 9) made no changes in the sub-solution designs and combined them directly.

4 *Usage of methods:* It was seen that many students used methods in assessing sub-solutions, building the alternatives, and developing the final design solution. A quotation from one of the respondents illustrates her methodological approach to the process: "I reviewed the morphological chart, chose some sub-solutions to combine, so I had rough ideas for overall form to begin with. Then in my mind, I modified the underlying ideas in each into different configurations, eliminated sub-solutions accordingly, and adapted the sub-solutions I wanted to use. I used my project criteria to assess the alternatives and further develop their designs". Some students mentioned 1) the methods used in class for the project in the decision-making and design synthesis processes: a) relative arrangements of components, tried out during quantified structure analysis, were reported to form the basis for the structural configurations onto which the sub-solutions were built, b) interaction nets were mentioned to be used in determining how components could come together, and c) the morphological chart method was commented upon as being rather useful in generating numerous and diverse ideas, opening up possibilities for variety in the alternatives rather than repetitions of close design ideas. 2) One student mentioned having used scenarios for empathy building and design assessment. 3) One student mentioned having made an online product survey "to check how features are used together" during convergence for six alternative design solutions.

5 *Usage of tactics:* It was seen that all students used certain design tactics in both steps of design convergence, while making design moves during the generation of their six overall design solution alternatives, as well as the development of their final design solution.

Tactics for transformation

a *Combination:* Students mentioned that combination was an effort in bringing solutions that "made sense together", and bringing "the best features" from all alternatives into a new design solution. They reported combining 1) features of two main alternatives, 2) features from multiple alternatives, and 3) features from all six alternatives.

b *Addition:* In some cases, the combination of features into the final design solution required that some sub-solutions were eliminated, and new ones were added instead. Students reported 1) using new sub-solutions from their morphological chart, as well as 2) devising entirely new ideas according to the requirements of the transformation.

c *Adaptation:* Students indicated that for the final design solution, form had to be adapted according to the "component operation", "aesthetics" of the overall design, and "assembly style" of the components. One student reported 1) transferring the structural assembly of one of the six alternative ideas onto the alternative selected as the final design solution. One student reported 2) transferring minor but useful details from other alternatives onto the alternative selected as the final design solution.

d *Modification:* Modifications on features placed on the overall design solution were made for "design quality", 1) in order to ensure match between the design solution and project objectives, 2) based on experience from product trials, and 3) according to personally set criteria (e.g., easy separation of hand-held vacuum component from stick structure).

e *Negotiation:* Transforming sub-solutions into an overall design required that design features were negotiated for aesthetic as well as functional compatibility. For this, students mentioned developing details 1) for neighbouring component connections, and 2) for components to nest, host, or accommodate each other. In some cases, this design negotiation resulted in 3) the elimination of some sub-solutions, and addition of new ones, or 4) changes of form or features of a component and/or of the overall structure of the design.

Tactics for evaluation

a *Diversification:* Students indicated that they aimed for variety and diversity among their six alternative design solutions. For this they 1) combined different sub-solutions for each alternative, 2) considered different specifications for each alternative, and also 3) made use of the various goals, themes, and concepts they determined for themselves.

b *Differentiation:* Students indicated that they strived for differentiation in their final design solution from products available in the market, products used in the class for trials, and alternative design solutions that their peers generated. For this, they reported 1) aiming for "developing an entirely new idea", 2) changing features of their solutions in terms of design, and 3) responding to the usage problems they identified during product trials in order to "add value" to their designs.

c *Revision:* While making design decisions, students reported reviewing 1) the objectives list, 2) the morphological chart for overlooked ideas, and 3) their alternative design solutions. While making these revisions, students indicated that they 4) cross-checked their designs with their personally set criteria, 5) identified unattended features, or 6) emphasised features that would make their final designs distinct.

d *Refinement:* Some students indicated that before representing the final design solution, they refined their design ideas by carrying out 1) 2D design explorations on the side, as well as 2) off-scale 3D design explorations for component configurations, and then 3) full-scale 3D design explorations to act out the vacuuming-related tasks in real size and review the design decisions accordingly.

2.3.3.2.3 Criteria

Criteria were the values that the students constructed for convergence; they were related to critical assessment in particular. Criteria were identified from the responses given to the survey questions 5 and 6, but responses to all questions were also cross-checked for related content. Eight criteria were revealed from the survey responses.

1 *Aesthetics:* An important criterion reported by the students was 1) visual appeal, both in selecting the sub-solutions and in representing them in an overall form. Many students a) chose sub-solutions and adapted them into an overall form, whereas many others b) first considered the overall form and then chose sub-solutions accordingly. Another criterion was 2) achieving a "coherent design" through the configuration of

the components for the 2-in-1 SVC as a whole, as well as when the components are separated (i.e., components do not look incomplete when hand-held is removed). For achieving these two criteria, students used sub-solutions that they found having a "common visual language" and "matching in aesthetics" in terms of form and style, describing their efforts as accomplishing "form harmony in the combined components", "visual completeness", and "unity in form".

2 *Functionality:* This group of criteria was related to the suitability of purpose of the components and practicality of the overall device. 1) While selecting the sub-solutions to bring together, most students reported aiming for the a) "mutuality", "compatibility", "suitability", "matching", and "convenience" of the components, and that they would b) "complement each other", have "similarities in working principles", be "connected" or "related", and "work well together". 2) While bringing the sub-solutions together, students reported that they made adjustments to the "working principles", "underlying ideas", and "means in order to achieve the essential function".

3 *Product performance:* This group of criteria was related to the effectiveness of components and efficiency of overall device. As the device was composed of two main component assemblies with specific functions (stick structure and hand-held component), a main expectation was 1) good performance from both, in terms of usefulness and covering functional requirements. Other expectations were 2) high capacity in a) vacuuming dirt, b) motor power, c) dust bin volume, and d) battery life and 3) low environmental cost of usage.

4 *Features and specifications:* This group of criteria was related to addressing must-have features and their specifications. Many students 1) took the "essential function of the device as the basis" in selecting sub-solutions and transforming them into design alternatives. Students indicated that they 2) decided on the critical features first, or made a list of specifications first, and then determined the sub-solutions to combine accordingly. A fewer number of students reported 3) considering the two functions of the device, namely hand-held vacuum and stick vacuum, in parallel and as separate pursuits.

5 *Variety:* Most students reported to aim for variety in the overall features in each design idea, to be able to generate distinct alternatives. One approach for this was 1) "incorporating as many components as possible", by selecting different sub-solutions for each alternative design solution. Another approach was 2) addressing different specifications in each alternative. One student indicated her approach as 3) combining "more-than-one sub-solutions for a same sub-function" in order to enrich the design details.

6 *Simplicity:* This group of criteria was about avoiding visual and functional complexity. Two students reported to aim for simplicity, one 1) making variations for each alternative in reference to a common list of specifications, and the other 2) repetitively using some features that were common in all alternatives.

7 *Realisability:* This group of criteria was about the transformability of ideas into overall design solutions. Students reported 1) examining the morphological chart to select sub-solutions that could be modified for realisability, 2) considering "extreme" or "not suitable" ideas to learn from them, but not using them directly, and 3) ignoring ideas not in line with their personally set criteria.

8 *Feasibility:* This group of criteria was about the development of the ideas for making them "realisable in terms of production". The criteria included 1) manufacturability, 2) low production cost, and 3) possibility of "immediate launch to market".

2.3.4 Discussion and conclusion

This chapter presented a study with graduate design students that attempted to explain design convergence. As per the first research question of the study, acts of design convergence involve decision-making, design synthesis, and critical assessment. As per the second research question of the study, design convergence takes place in reference to these acts through the involvement of various factors, strategies, and criteria. Factors are the considerations behind the design decisions made and provide directives for informed design moves. Strategies are the ways in which design synthesis is carried out and following them leads to design transformations as combinations of design moves and knowledge integration. Criteria are the values strived to attain or make judgement against during critical assessment and adopting them allows to situate and contextualise the design transformations. Based on its theoretical framework, this chapter argues that these factors, strategies, and criteria formed the mechanisms for design convergence that took place for the 2-in-1 SVC project carried out in the graduate design methods course.

These mechanisms have been identified from responses to a survey carried out among the graduate design students who had approached the end of a rigorous design process for a familiar and non-specialist product (2-in-1 SVC). It can be argued that these mechanisms came into action during design convergence because the problem area was familiar to the students to begin with and the problem frame was established with relevant information and well-defined limits. Also, the design process was methodically planned step-by-step, making it easier for the students to apply acts of convergence and recognise and describe them upon reflection.

It is believed that these mechanisms of design convergence were facilitated with certain provisions.

2.3.4.1 Preparedness with methods, materials, and mindset

It was seen that following a methodological approach in the project conduct made it easier for students to make design moves in each step. Students knew that any outcome for a particular step would be related to the design task to be carried out in the following step, explained in the literature as task interdependence (Sutton & Hargadon, 1996). Working in groups also supported the students in making timely decisions and set the bars for an optimum level of quality in the submissions (Börekçi, 2017).

Being prepared for convergence with the necessary skills and experiences, quality of information, and number of ideas were also found influential. Product trials were useful in building empathy in students. They were able to try out the products, as well as see their peers use them, and compare the various features in terms of design and performance. Idea generation through divergence, using the morphological chart, which aimed for quantity, variety, and diversity in means, formed an expanded solution space with options for the students to choose from. Studies for quantified structures, carried out before the morphological chart method, seem to have influenced the students' assessments in structural configurations and therefore overall form possibilities, allowing the students to simultaneously explore external form and internal component arrangements. It was seen that, during design divergence, many of the students had already made assessments as to the workability of the ideas, their possible combinations, and the design directions they wanted to take. Another beneficial experience was working with mock-ups that supported 3D exploration, particularly full-scale models that allowed acting out with gestures.

2.3.4.2 Experience in evaluation and decision-making

Due to their undergraduate educational experiences with backgrounds in design and engineering, the students were familiar with evaluation in design processes, and they were also experienced in giving and receiving criticisms, including self-criticisms. It can be said that students had judgemental awareness, which was helpful in decision-making, design synthesis, and critical assessment. Even when students mentioned "personal preferences" or "liking" as a reason for their decisions, when selecting design features, there was an explainable and informed judgement involved. This could be related to a feature's aesthetic or functional appeal, an advantage that was seen in its use, its workability towards improvement, its match with a higher value, and its qualities that constituted an attractive challenge for the students to decide to undertake.

It was seen that the students were selective from among the numerous sub-solutions on a morphological chart, meaning, rather than eliminating, they tended to review all (less complex) ideas and pick up those that they found fit. As for the six alternative design solutions, students were seen to be more eliminative, meaning they tended to put aside an overall alternative more easily. This may have been in order to limit the cognitive effort by leaving a lesser number of (more complex) alternatives to assess. This may also be due to having already assessed the qualities of these alternatives while bringing the components together into an overall design solution and having made a judgement on whether the outcome could be utilised or matched with the project objectives.

Design convergence is a critical stage in the design thinking process and requires a significant cognitive effort that evaluates the outcomes of design divergence by determining the most appropriate idea, transforming it, and situating it within the problem context, thus concluding the design process. It is hoped that this chapter may be of interest to design educators, practitioners, and researchers who wish to further understand design convergence and find ways of supporting the related acts. I would like to thank the students who took part in this study.

References

Akin, Ö., & Lin, C. (1996). Design protocol data and novel design decisions. In N. Cross, H. Christiaans & K. Dorst (eds.), *Analysing design activity* (pp. 35–63). John Wiley & Sons.

Banathy, B. H. (1996). *Designing social system in a changing world.* Plenum Press.

Basadur, M. (1994). Managing the creative process in organizations. In M. A. Runco (ed.), *Problem finding, problem solving, and creativity*, pp. 237–268. Ablex.

Börekçi, N. A. G. Z. (2017). Visual thinking styles and idea generation strategies employed in visual brainstorming sessions. *Design and Technology Education: An International Journal, 22*(2), 1–19.

Braun, V., & Clarke, V. (2015). How to use thematic analysis with interview data. In A. Vossler & N. Moller (eds.), *The counselling and psychotherapy research handbook*, pp. 183–197. Sage Publications.

Christensen, B. T., & Ball, L. J. (2016). Dimensions of creative evaluation: Distinct design and reasoning strategies for aesthetic, functional and originality judgments. *Design Studies, 45*(Part A), 116–136.

Cropley, A. (2006). In praise of convergent thinking. *Creativity Research Journal, 18*(3), 391–404.

Cross, N. (2000). *Engineering design methods: Strategies for product design* (3rd ed.). John Wiley & Sons.

Design Council, UK (2019). *The double diamond: 15 years on.* Retrieved from https://www.designcouncil.org.uk/news-opinion/double-diamond-15-years

Fu, K., Cagan, J., & Kotovsky, K. (2010) Design team convergence: The influence of example solution quality. *Journal of Mechanical Engineering*, *132*, 111005-1.

Goldschmidt, G. (2016). Linkographic evidence for concurrent divergent and convergent thinking in creative design. *Creativity Research Journal*, *28*(2), 115–122.

Jones, J. C. (1980). *Design methods: Seeds of human futures*. John Wiley & Sons.

Jung, J. H., & Chang, D. R. (2014). The thinking styles of creative and talented persons in design convergence. *Archives of Design Research*, *27*(2), 215–229.

Kirwan, B., & Ainsworth, L. K. (2005). *A guide to task analysis* (ebook). Taylor & Francis eLibrary.

Krippendorf, K. (2018). *Content analysis: An introduction to its methodology* (4th ed.). Sage Publications.

Lawson, B. (2005). *How designers think: The design process demystified* (4th ed.). Routledge.

Lubart, T. (2016). Creativity and convergent thinking: Reflections, connections and practical considerations. *RUDN Journal of Psychology and Pedagogics*, *4*, 7–15.

Müller-Wienbergen, F., Müller, O., Seidel, S., & Becker, J. (2011). Leaving the beaten tracks in creative work - A design theory for systems that support convergent and divergent thinking. *Journal of the Association for Information Systems*, *12*(11), 714–740.

Pahl, G., & Beitz, W. (1996). In K. Wallace (ed.) *Engineering design: A systematic approach*. (K. Wallace, L. Blessing, & F. Bauert, trans. from German). Springer-Verlag London Ltd.

Roozenburg, N. F. M., & Eekels, J. (1995). *Product design: Fundamentals and methods*. John Wiley & Sons.

Shen, W., Hommel, B., Yuan, Y., Chang, L., & Zhang, W. (2018). Risk-taking and creativity: Convergent, but not divergent thinking is better in low-risk takers. *Creativity Research Journal*, *30*(2), 224–231.

Srinivasan, V., Chakrabarti, A., & Lindemann, U. (2012). A framework for describing functions in design. In *Proceedings of International Design Conference DESIGN 2012*, 1111–1121. Dubrovnik, Croatia, May 21-24, 2012.

Sutton, R. I., & Hargadon, A. (1996). Brainstorming groups in context: Effectiveness in a product design firm. *Administrative Quarterly*, *41*(4), 685–718.

Tjalve, E. (1979). *A short course in industrial design*. Butterworth & Co.

Webb, M. E., Little, D. R., Cropper, S. J., & Roze, K. (2017). The contributions of convergent thinking, divergent thinking, and schizotypy to solving insight and non-insight problems. *Thinking & Reasoning*, *23*(3), 235–258.

Wolf, K., & Mieg, H. A. (2010). Cognitive determinants of the success of inventors: Complex problem solving and deliberate use of divergent and convergent thinking. *European Journal of Cognitive Psychology*, *22*(3), 443–462.

Wright, I. (1998). *Design methods in engineering and product design*. McGraw-Hill.

Zhu, W., Shang, S., Jiang, W., Pei, M., & Su, Y. (2019). Convergent thinking moderates the relationship between divergent thinking and scientific creativity. *Creativity Research Journal*, *31*(3), 320–328.

Zhang, L.-F. (2009). Anxiety and thinking styles. *Personality and Individual Differences*, *47*(2009), 347–351.

2.4 Innovative teaching methodology

A learning process on how to translate design thinking knowledge and tools to non-designers

Ana Thudichum Vasconcelos, Carla Paoliello, and Ana Lia Santos

FBAUL - Lisbon University Fine Arts Faculty

2.4.1 Introduction

The academic project analyzed here is based on the educators' experience supported by design thinking's (DT) potential for searching innovative problem-solving solutions. DT tools enable people to promote transformation processes, leading to positive changes and improvement of society and community's quality of life.

The research questions were: How could DT be disseminated among non-designers (N-D) and, what are the design skills that could be interesting to propagate? It was important to identify how DT tools can empower N-D and to find out its transformation scale. The design exercise was to create DT workshops for N-D, with the title Design Workshop for Non-Designers (DWS4ND). The main objective was to create DT awareness and to positively and actively share its goals, principles, and tools with the community involved. The students became real actors in the process. There was an invitation to discuss DT practice and to bring personal insight into the process of transmitting DT knowledge.

2.4.2 What interest might DT have for N-D?

Design is a human activity. It is an active process of learning and critical problem-solving. Brown (2010) stated, "design is a powerful tool" capable of providing working methodologies to intervene in many different areas.

In our perspective, design promotes skills that can generate social transformations (Vasconcelos, 2017a, 2017b). Through the recovery of our design capabilities, we can instigate democratic actions that, collectively, aim to a better, more just, and egalitarian world (Porto, 2019). Design through DT methodologies can trigger innovative processes wherein critical thinking combines with critical making. DT is an approach capable of generating innovation in today's world (Vasconcelos & Porto, 2019; Cardoso, 2012) due to the great potential that DT has to solve complex problems, which can be a formative contribution to both designers and N-D.

Jones (1992) was the first to use the term *design thinking* with an outside-in perspective. The original version of his book dates back to 1970. He stated that DT contributes to generating information and insights (p. 45). He also reflected on the advantages of adopting collaborative systems. Later, Buchanan (1992), in "Wicked Problems in Design Thinking" focused on the human being. He favored multidisciplinary collaboration, and the perception of thoughts and processes to build paths that lead to innovative and creative solutions. As stated by Brown (2010), there is a set of principles that different people can apply to a wide variety of problems.

DOI: 10.1201/b22630-13

We find DT skills in the development of creativity (Vasconcelos, 2014), empathy (Paoliello, 2018), entrepreneurship (Paoliello et al., 2013), collaboration, and communication, suggesting that DT develops soft skills through innovative processes. These features are useful for all, increasing self-satisfaction and quality of life.

2.4.3 How can DT – knowledge and tools – be transmitted to N-D?

We recognize two distinct strands in DT knowledge transmission. One is promoted by non-designers, and the other is from within the design culture. In the first group, we observed two trends. One related to the teaching of DT in the scope of management and business courses. For example, there are the Stanford Design Thinking Research Program Management; the Executive Education, MIT, Mastering Design Thinking; and the Business School and Aalto University Executive Education that promote "skills to bring new value-driven creative solutions into your company" (https://www.youtube.com/watch?v=HHl0zw4P3tQ).

From the design point of view, Emily Pilloton's "Project H" stands out. As described by Rawsthorn (2013):

> It was intended to nurture skills that could help them (students) with problem-solving, communication, leadership, and innovation in the hope that their experience of studying design and the design process would give them practical and intellectual tools to draw on for the rest of their lives.

There is also the work carried out in the context of the educational activity promoted by the Smithsonian, Cooper-Hewitt, National Design Museum, with the seal of the designer and director of the museum, Bill Moggridge. And also, children's workshops developed by Bruno Munari (2004), documented in his book *The Tactile Workshops*.

Another perspective is in "Bright Minds, Beautiful Ideas" (Schwartz, 2003), which documented how design can go beyond problem-solving in an inspiring experimentalist and creative proposal.

In this chapter, DT vision fits into a humanist perspective. The purpose is to raise awareness and disseminate DT, in the same sense of fostering and disseminating knowledge about design (Bonsiepe, 2011; Papanek, 1992; Norman, 2000), through the creation and communication of several WS for N-D. The proposal presented here emerges as a way of creating a dynamic knowledge transmission platform, based on intellectual curiosity and unconventional attitudes toward the presentation and resolution of problems through DT. In short, it is a rich field for experimentation.

It is the opposite of what is described in Fleischmann et al. (2012, p. 107), where they stated that "undergraduate design programs typically focus on creating employable 'problem solvers' rather than 'solution finders'". In our approach, design students have an active role as creators of these innovative pedagogical processes. They become facilitators of DT methodologies and more than that, they become active agents of fertilization. The learning process is also positive for the students because DWS4ND is a living and critical laboratory in the classroom. When implemented (even if only in a restricted circle), it hints at a bridge between academy and the community, as well as being a link between designers and N-D.

We adopted the Education by Design model by Oliveira (2017) and Fontoura (2002), aiming at a contribution in expanding the existing competencies of design students and N-D to create thinking focused on innovation and collaborative work.

2.4.4 Plan and methods

The project began in the academic year 2019/2020. The teaching experience was then repeated in the following years 2020/2021 and 2021/2022. The educational project was implemented in Project IV, a curricular unit in the 5th semester of a Product Design program in the Fine Arts Faculty of Lisbon University.

The first perception of this academic activity's unusual success, because of its positive impact on students, motivated us to look for local institutions in which to install the proposal. The local institutions were the Center for Modern Art of the Calouste Gulbenkian Foundation, the National Museum of Natural History and Science, the Museum of Art, Architecture and Technology – MAAT, and the Portuguese Institute of Sport and Youth – IPDJ. They have all welcomed the proposal and showed interest in integrating the WS in their cultural programs and activities. Unfortunately, this was put on hold due to the pandemic restrictions.

The WS development method presented to the students is divided into four phases: (1) Creation process and development, (2) Testing and Feedback incorporation, (3) Communication, and (4) Academic WS evaluation. The same process was repeated, with different students, each first semester of the two following years. A selected set of WS became our representative models. To help students better understand what was expected to be delivered at each stage, the deliverables of the first three phases were named similar to software development. That is because they correspond to the same kind of process development: phase (1) delivers an Alfa version/not functional version of the WS; phase (2) is a Beta version/ready for testing and incorporation of feedback functional version of the WS; phase (3) is a Release Candidate (RC)/the best Beta version to be presented, but not yet launched, of the public WS version. Phase (4) corresponds to Academic evaluation and delivers the seriation of the WS proposals by the following criteria: overall quality, fulfillment of the proposed objectives, coherence and control of the design, applicability, and potential for replication. The evaluations not only the final one but also all that accompanied the different stages of the process also yielded data for investigation.

2.4.4.1 Breakdown of phases

1 Creation process and development. The questions for this part of the study were: how could we motivate students to deepen the design culture, broaden their references, and debate knowledge in the design process? And also, how to pass DT tools to an N-D public? The answers established the basis for WS creation. This process includes several steps. First, the promotion of an open debate about students' ideas and concerns related to design. Then the students were given a list of 100 relevant designers from the 20th and 21st centuries. From that list, each student chose a designer to carry out an in-depth study to identify his/her work philosophy. Design is inputted at a multi-level perspective as the students present each professional to the class.

   Afterward, the students form groups of three, compare the chosen designers and, with the help of the Uffe Elbaek model, choose the one with whom they most identify and that they consider appropriate to serve as a base reference for the WS development. This phase is completed with the presentation of each group's concept for the WS. This approach requires that students deepen their design process knowledge and methods since each group had to set the WS goals through a participatory process and answer four questions about their WS proposal: what, why, to whom, and when. After that, they mapped their concept through a divergent and convergent thinking approach. They were also invited to build User Profiles to define the target participants and Value Proposition Canvas to streamline the WS practical aspects: the materials used, the WS duration, the cost inherent to the activity, and to evaluate the space where the WS would happen. As referred above, the deliverable was an Alfa version of the WS, an understandable but not yet functional version.

2 Internal and external testing and feedback incorporation. The questions for this second part of the study were: How to map the WS process? Does it make sense? Does it achieve the initial goals? how to improve the process? Are the methods, means, and tools chosen correctly? The answers establish the WS process, testing the feedback incorporation and streamlining it until it is functional. The testing phase follows the WS creation in the classroom. The importance of testing is related to the feasibility of the WS proposal and the measurement of its adequacy to N-D. Students were encouraged to use online Surveys, Feasibility Testing Scenarios, Storyboarding, House of Quality (HOQ), and Service Blueprints to prepare, document, and evaluate the WS testing. The first series of assessments happened in the classroom context where creation and testing coexist. The same students who created the WS were the ones who tested it. Then, other internal tests were carried out, now between pairs of student groups. The pairs exchanged experiences with each other, allowing the improvement of their proposals. This phase allows the selection of the most relevant WS activities. The more feedback data they received, the easier they identified pain points to solve and better understand the WS potential. We recognize that it would be important to have external testing with stakeholders from the institutions involved; however, the circumstances have postponed this phase. The solution was to carry out these external test sessions with the students' acquaintances and relatives: 12 groups of students tested at least two or three times with groups of 10–15 people, a total of around 500 people per semester. The deliverables of the WS Beta version were functional but not yet complete. They were fully operational and, more importantly, tested with a reasonable number of target people.

3 Communication. The initial idea was that students would be the ones to implement the DWS4ND in the partner institutions. They would be responsible for launching their WS in a real situation with the partner institutions' audiences. As this was not possible, they were asked to prepare a final communication (a Presentation Deck) in the classroom exactly as they would do for one of these institutions. It is not the same, but it gives students the time and space to check and reassess all stages of the WS development, while preparing the final communication. The Beta version of the WS was again adapted and refined until it became de RC (release candidate) version. It is the best Beta version, which is fully functional and ready to be published, but it is not the final version, because it was not released to the public.

4 Academic Evaluation. The WS dynamic is not about transposing formal design educational structure to a non-formal one. It is exactly the opposite: the objective is to design activities that allow N-D to perform beyond their expectations, leveraged by DT tools, namely those associated with soft skills, creativity, and innovative processes. In this way, the continuing investigation intends to deepen the way DT could interact with N-D. How could it intensify collaborative work? And how could we investigate more relationships "through design"? Our great challenge lies in the exploration of a new DT pedagogical processes, one that relies on innovative teaching processes. At the end of the process, and in each semester, we observed the existence of high quality,

variety, and diversity of WS proposals. The WS creation process allowed the idea of appropriation which is fundamental to each student's growth. That is translated into several original proposals that build potential in terms of the impact on N-D.

Phase 4 evaluation resulted in a synthesis of the other phases' assessments, as the overall result incorporates reference research, concept idealization, development, and feedback from testing with an N-D audience and streamlining for communication. The deliverables of this phase are a) the assessment of the process expressed in reviews and papers, b) the work of the students presented, and c) the N-D feedback. We trust in the potential of WS because this exercise has been developed in different semesters, produced 26 WS proposals, involving around 1,000 people, and the majority of stakeholders showed positive reactions throughout the process. Those positive reactions are evident in the WS results, especially because the success of the DWS4ND is mensurable by the ability of the N-D to complete the task successfully. This indicates DT transmission. In the showcase of students' work, it is visible that people with no previous design skills or inclination can, through DT, perform tasks that in other situations they would have thought to be difficult or even beyond their capabilities.

It is also important to consider, with due reservation, that the final semester grades can also be linked to the success of this exercise in the specific curricular unit because in the last two years, there was a 10% improvement in the average of students' grades. The students were also more engaged and produced more and better work in class. But to prove cause-effect, different assessment protocols will be used as this study will continue.

2.4.5 Results: The most representative WS showcase

In this part, we present eight of the most representative WS carried out by the students during the last years to showcase the DT concepts studied. The following examples illustrate the process and the result of the WS. Under the DWS4ND umbrella concept, everything in the WS was designed. Once each group of students sets their scope of actions (design, in general, is too large a subject to be addressed in a WS) and locks their focus (design-related specific set of objectives to achieve) and goals (set of information or skills to embed in the WS to pass to N-D), then they proceed to design a Process Tree and the Activity Flow. Later they evaluate the means, develop the methodology, and create tools to be used by N-D. The user interface is carefully polished until the overall WS experience produces consistent results when tested with different subjects. The level of performance of the N-D involved in the WS testing is important, because achieving it is a very difficult part of the WS development, and students invest a lot of time and resources in it. After all, it is there that the fulfillment of objectives is effectively measured.

One WS was elected to illustrate the overall process in detail, and the other seven WS are presented with a summary framework and visual contents that were developed by students for phase 3 – Communication.

2.4.5.1 A WS that illustrates the overall process

One of the top eight WS stands out for having the development process clearly illustrated and a more systematic testing-incorporation feedback loop than in other WS.

B.I VISUAL 2020/21 WS by Bruno Vilar, Inês Leite, and Renata Cabaça is about the representation and visualization of an abstract concept, in this case, identity. It shows the importance of designing a visual language in cultural communication. They started by

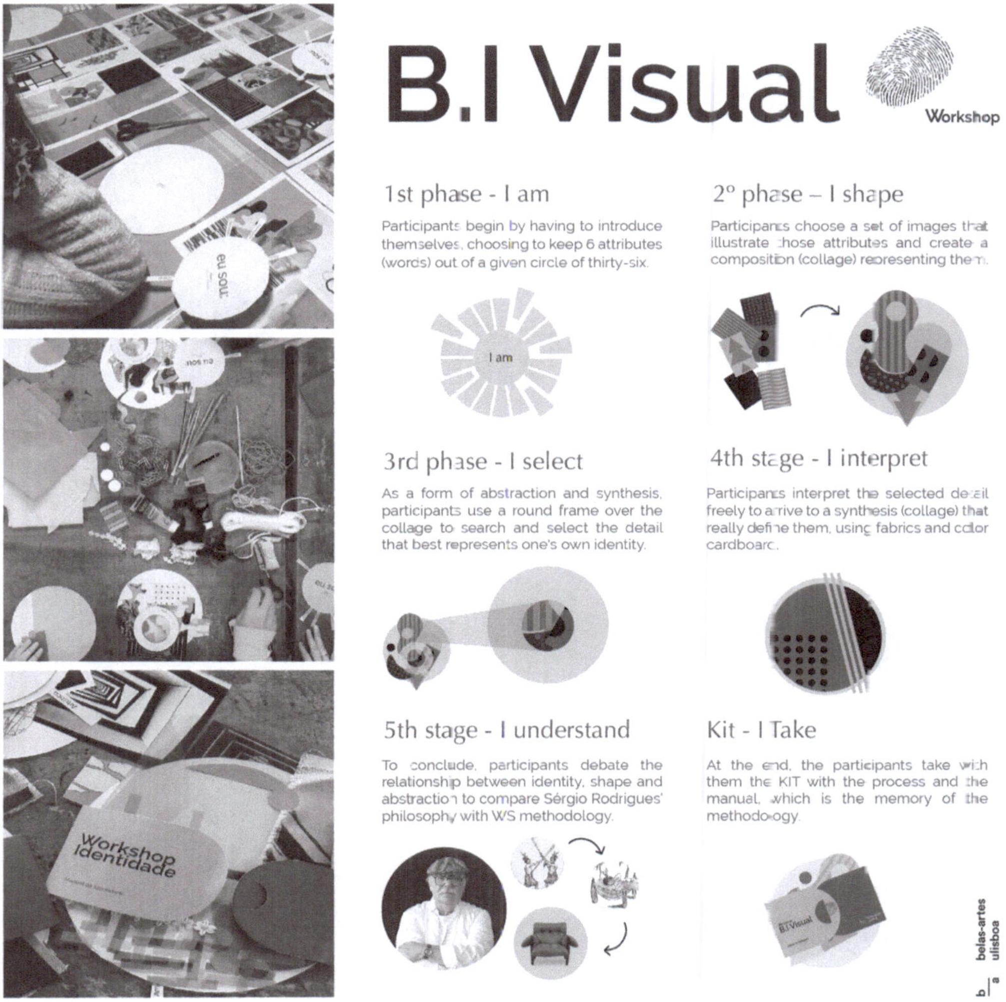

Figure 2.4.1 B.i visual WS by Bruno Vilar, Inês Leite, and Renata Cabaça (2020/21).

choosing the designer Sergio Rodrigues and the way he always pursued the incorporation of Brazilian identity into his objects. Then a system was developed that allows N-D to create a visual presentation of their own identity, and then take the process with them and apply it in everyday life situations (Figure 2.4.1).

2.4.5.2 The other seven most representative WS

The following seven WS also obtained the highest classification, i.e., grade A (18–20 out of 20) in overall quality, fulfillment of the proposed objectives, coherence and control of design, applicability, and potential for replication.

Technology for humans not robots 2020/21 WS: by Elisabeth Duran, Mariana Rosa, and Miguel Teles started with Jonathan Ive's input to draw a WS that would allow N-D to better understand the interface design that is present in objects and that allows humans to

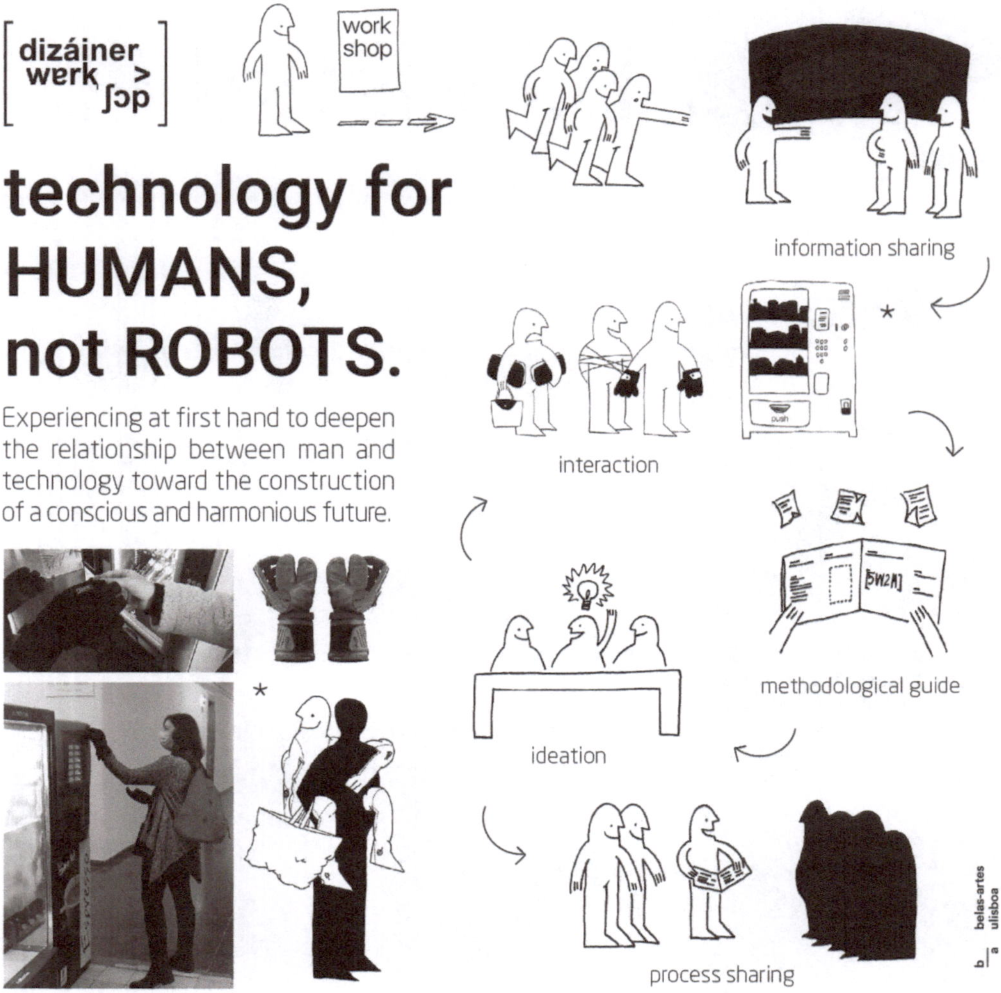

Figure 2.4.2 Technology for humans not robots WS by Elisabeth Durin, Mariana Rosa, and Miguel Teles (2020/21).

use them. The participative dynamics and tools of this WS motivated a discussion about intuitiveness and human/machine design empathy tools (Figure 2.4.2).

"Everythinker": is a word invented to combine "Everything" and "Thinker", which Ângela Pinto, Sara Jesus, and Shihan Wang proposed for the 2020/21 WS title. They departed from the work of Kenya Hara to explain the concepts of "void" and "white" in design and their role in creativity. The work was developed to the extent of becoming a product, i.e., a set of cards that allows the process to be replicated autonomously by N-D (https://wangshihan1996.wixsite.com/everythinker and Figure 2.4.3).

Modular 2020/21 WS: by Catarina Martins, Inês Silva, and Lara Beato is an approach to readability and a reading workshop for N-D and it aimed at enabling them to

Figure 2.4.3 Everythinker WS by Ângela Pinto, Sara Jesus, and Shihan Wang (2020/21).

understand the basic rules of these two fundamental aspects of Typography. Using Jonathan Hoefler as a reference, they created a very open and free-of-fear process for experimental typeface composition (Figure 2.4.4).

Photostorming 2020/21WS: by Ana Pedro, Diogo Mindouro, and Eduardo Rebelo had Naoto Fukasawa as a reference. They understood that observation and changing perspectives were important to the development of the idea. Their WS invited N-D to behave as designers, providing N-D with new ways of thinking through creative ways of looking at the world. They also promoted the use of everyday objects and spaces in a different context (way) and encouraged emerging narratives about those activities (Figure 2.4.5).

Ar-chi-tec-ture 2019/20 WS: by Stijn Haegen and Marta Bat. They elected Peter Zumthor as reference to challenge N-D to work with the tools common to architecture and

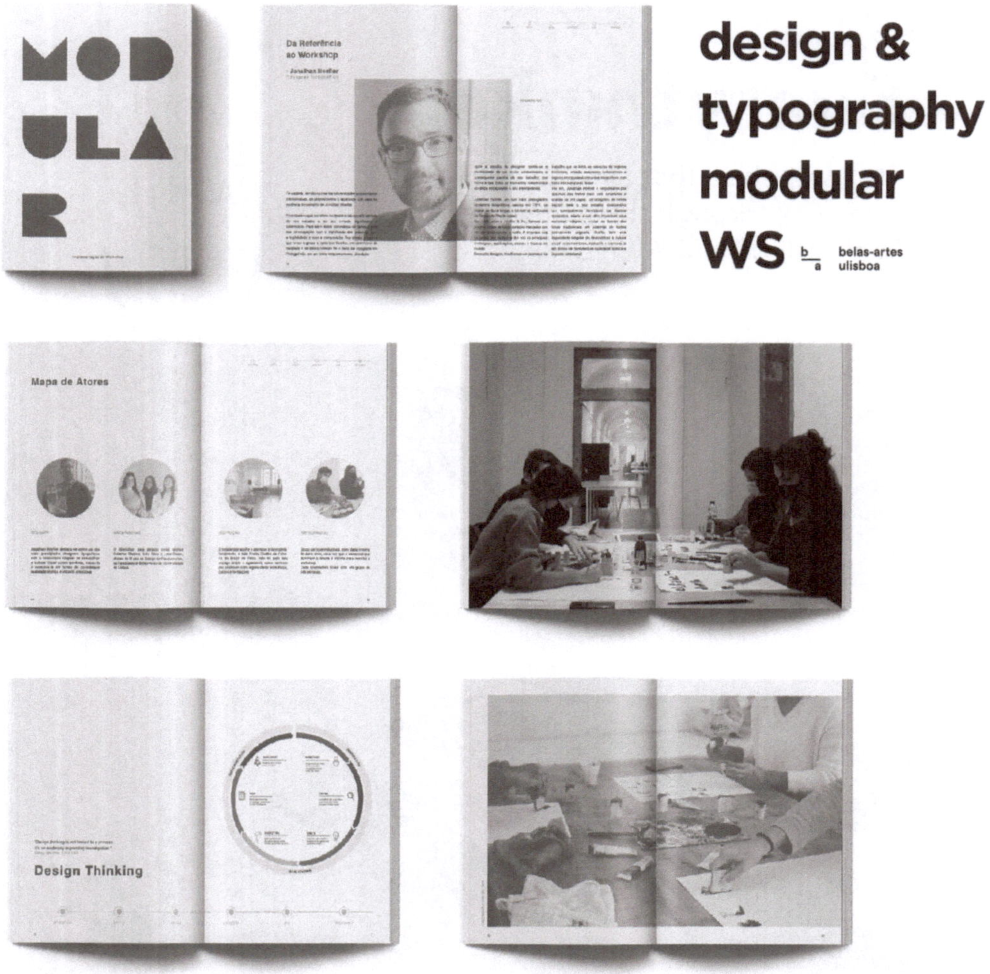

Figure 2.4.4 Modular WS by Catarina Martins, Inês Silva, and Lara Beato (2020/21).

spatial design, namely: line, plane, scale, color, texture, light, and perspective, to build conceptual habitats as a way of encouraging new relationships with our surroundings: from a specific to a general situation, from the solved to the act of solving (Figure 2.4.6).

Object (id)entity 2019/20 WS: by Maria Amélia Silva, Gustavo Calé, and Mariana Martins. The activity started with Andrea Branzi's dichotomous premise "object-culture" and moved toward the elaboration of a cultural scenario, and after that it explored the integration of an artifact in the projected context. The general objective was to reflect with N-D about up-to-date design-related society, through brainstorms, sketching, and theoretical discussions (Figure 2.4.7).

Alternative realities 2019/20 WS: by Inês Guimar, João Calado, and Xavier Guerra developed Dunne & Raby´s reference concept of the speculative. They aimed to introduce this concept to a larger and more varied audience. They proposed that the WS should

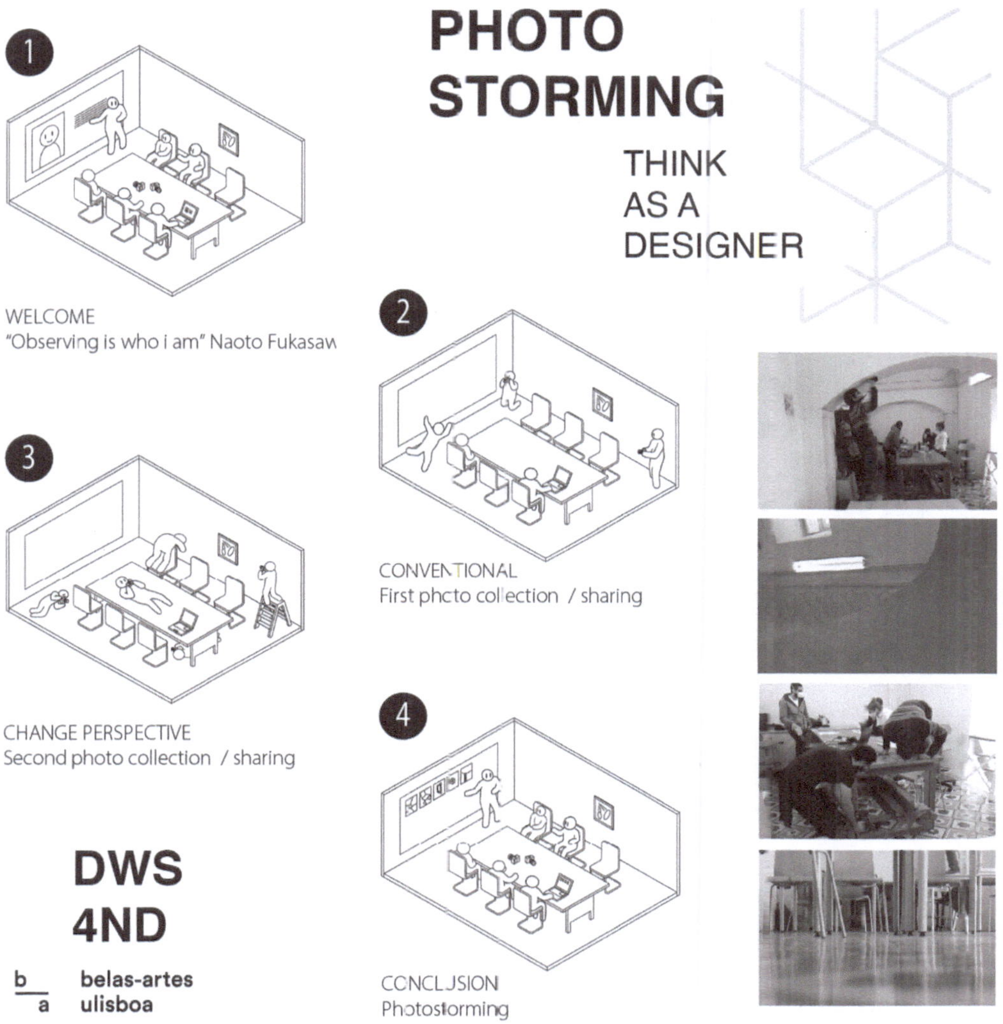

Figure 2.4.5 Photostorming WS by Ana Pedro, Diogo Mindouro, and Eduardo Rebelo (2020/21).

take place in a circular (safe) space where N-D can experiment with the free use of imagination by using a toolkit designed specifically for the WS: diagrams, future scans, brainstorms, and prototypes to help them draw alternative realities (Figure 2.4.8).

2.4.6 Discussion and conclusion

Students' creation of WS for N-D involves a process that takes place at two distinct levels. At the first level, it is based on the input from a personal identification with the work philosophy of a reference in the design world. At the second level, the focus is on interpretation and appropriation of what DT enables and the definition of what a DWS4ND should be.

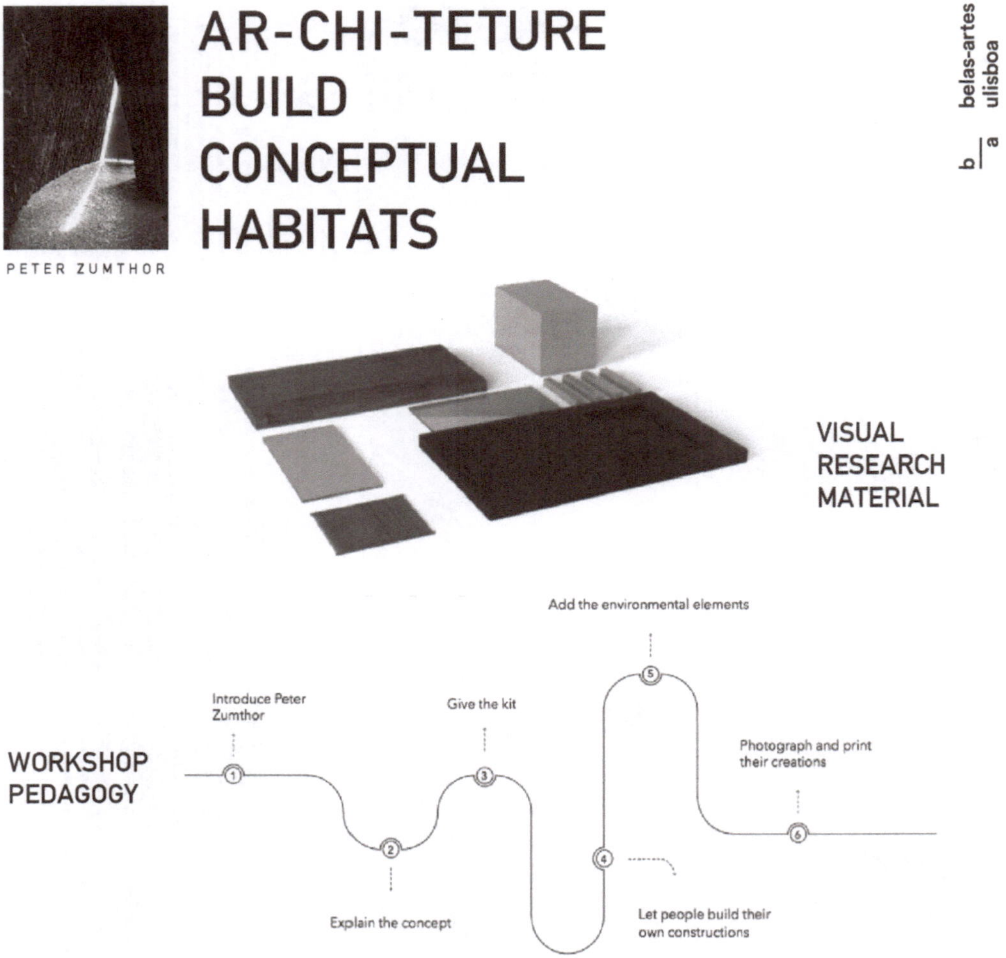

Figure 2.4.6 Ar-chi-tec-ture ws by Stijn Haegen and Marta Bat (2019/20).

This means that the students had the task of finding their voice and choosing the message they wished to pass and the posture to adopt with design and design dissemination. Once they defined what the WS should be, the reference designer became less important. From that moment, they felt free and autonomous to work on the concepts learned and even formulate new ones.

Throughout the workshop development process, the students chose what is important to them, as they grew more aware of the importance of including others (the team, the N-D, the community) in the chain of decisions. They understood that design concepts and tools will be used by others in an explicitly design-related context. Their relationship with design became deeper and more personal. In the end, the majority of students showed a sense of ownership of their ideas, and a willingness to make them heard. Maybe it is because they became more prepared to be designers and ambassadors of the design

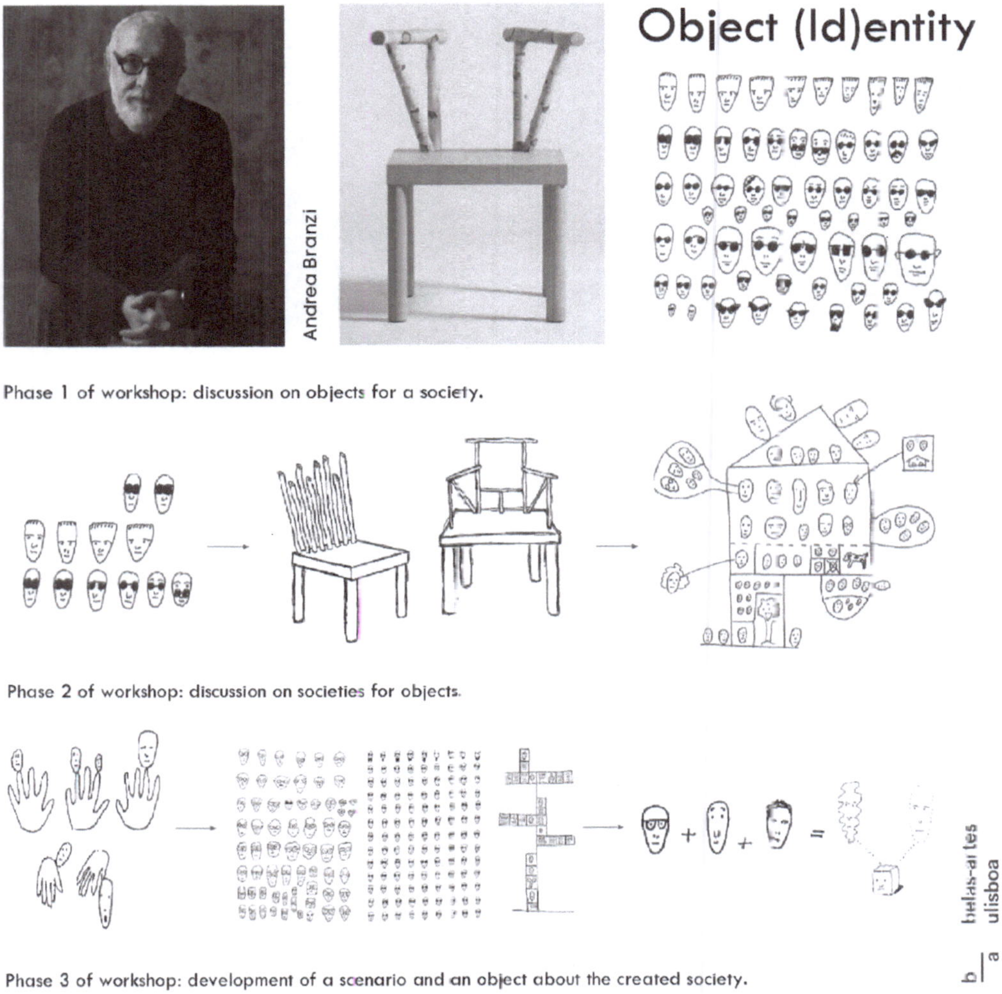

Figure 2.4.7 Object (id) entity WS by Maria Amélia Silva, Gustavo Cale, and Mariana Martins (2019/20).

culture as they switched from "creative ways of knowing, thinking and acting" as expressed by Alain Findeli (2010, p. 289) to "creative ways of sharing".

Although it was not (yet) possible to implement the project in its entirety, there has been a positive impact in nearby social circles as well as in the class. This teaching pedagogy places students at the center of the educational action. Their behavioral transformation was so tangible that this study will be continued. The feedback, although it was still informally quantified, is positive enough to see that there is a real opportunity for cultural and social intervention through a program of DWS4ND implemented by students. Additionally, it is a form of cooperation between the academy and other structures of knowledge and cultural dissemination, such as museums and public institutions dedicated to promoting development and design assets. This is a first step to introduce designers in contexts where they are absent, and to transfer DT to everyday public life.

Figure 2.4.8 Alternative realities WS by Inês Guimar, João Calado, and Xavier Guerra (2019/20).

It is a way to start the dissemination of design, as we so strongly defended, and the best part is doing it by creating the educational strategy to empower the next generations of designers to carry it on.

References

Bonsiepe, G. (2011). *Design, cultura e sociedade*. São Paulo: Blucher.

Brown, T. (2010). *Design thinking, uma metodologia poderosa para decretar o fim das velhas ideias*. Rio de Janeiro: Elsevier.

Buchanan, R. (1992). Wicked problems in design thinking. *Design Issues*, 2(8), 5–21.

Cardoso, R. (2012). *Design para um mundo complexo*. São Paulo: Cosac Naify.

Findeli, A. (2010). Searching for design research questions: Some conceptual clarifications. In Findeli, A., & Russel, K. (eds.), *Questions, hypotheses & conjectures - discussions on projects by early stage and senior design researchers*. iUniverse. New York: Bloomington.

Fleischmann, K., Visini, G., Daniel, R. (2012) We want to add to their lives, not take away …' In Rodgers, P. (ed.) (2012). *Articulating design thinking*. Farington, Oxfordshire: Libri Publishers.

Fontoura, A. M. (2002). *EdaDe - A educação de crianças e jovens através do design*. Tese. Universidade Federal de Santa Catarina.

Jones, J. C. (1992). *Design methods*. New York: John Wiley & Sons.

Munari, B. (2004). *The tactile workshops*. Verona: Corraini.

Norman, D. A. (2000). *The design of everyday things*. London: The MIT Press.

Oliveira, I., Trigueiros, P., Providencia, B., Duarte, M., Martins, C., & Fernandes, I. (2017). Artefacto #0000: Study of traditional ceramic jug "Infusa", for innovation by design, in Northern Portugal. In *Senses & sensibility'17: Design beyond borders and rhizomes* (pp. 90–97). Universidade da Madeira, Funchal: UNIDCOM| IADE. Retrieved from http://hdl.handle.net/1822/47354

Paoliello, C. (2018). Sobre uma cidade, um passeio, um grupo de avós, e uma residência artística. In A. G. Pinto, P. Reaes Pinto, & T. V. Furtado (eds.), *Cross media arts. Artes sociais e transdisciplinaridade (Social arts and transdisciplinarity)* (pp. 526–543). Casal de Cambra: Caleidoscópio.

Paoliello, C. *et al.* (2013). *Artesanato Identidade para O Bagre*. Belo Horizonte: Instituto Cidades Criativas, v.1.

Papanek, V. (1992). *Design for the real word: Human ecology and social change*. New York, London: Thames and Hudson.

Porto, R. G. (2019). *Processos de design e inovação social: Expectativas, êxitos e desvios em projetos financiados*. Ph.D. thesis, Faculdade de Arquitectura, Universidade de Lisboa Publishing.

Rawsthorn, A. (2013). *Hello world, where design meets life*. London: Hamish Hamilton.

Schwartz, A. I. (2003). *Bright minds, beautiful ideas, parallel thoughts in different time, Bruno Munari, Charles and Ray Eames, Martix Guixé, Jurgen Bey*. Amsterdam: BIS.

Vasconcelos, A. T. (2014) Estimular a criatividade em design. A cultura material como meio de exploração poética. In Pereira, J. D. L., Lopes, J., Marcelino S., & Maltez M. A. (eds.). *As artes na educação*. Chaves, Edição Intervenção – Associação para a Promoção e Divulgação Cultural. ISBN 978-989-97571-6-5.

Vasconcelos, A. T. (2017a). Best practices in design education leads to a new curriculum. *The Design Journal* 20(1), special issue (online): Di Lucchio, L., Imbesi, L , & Atkinson P. (eds.), Proceedings of the 12th European Academy of Design Conference, Sapienza University Rome, 12-14 April 2017.

Vasconcelos, A. T. (2017b). Soluções (sedutoras) de design para uma alimentação sustentável. In Oliveira, R., Amâncio, S., & Fadigas, L. (eds.), *Alfaces na avenida. Estratégias para (bem) alimentar a cidade*. Universidade de Lisboa, Colégio Food, Farming and Forestry. http://colegiof3.ulisboa.pt/docs/Alfaces_na_avenida-Estrategias_para_bem_alimentar_a_cidade-ColegioF3_2017.pdf

Vasconcelos, A. T., & Porto, R. G. (2019). O design para a inovação sSocial e a criatividade, na promoção dos objetivos de desenvolvimento sustentável. In Lopes, J., Marcelino., Oliveira, G., & Pereira (eds.), *As artes na educação especial. Teorias, metodologias e práticas sociais, culturais e educativas para a inclusão cultural* (pp. 165–174). Chaves: Cultural, Intervenção - Associação para a Promoção e Divulgação.

2.5 Empathy goggles

The advantages of a physical empathy exercise when teaching design process and group work

Petra Ahde-Deal and Mette Laier Henriksen
Copenhagen School of Design and Technology

2.5.1 Introduction

> Social and emotional skills, such as empathy and respect for others, are becoming essential as classrooms and workplaces become more diverse.
>
> (OECD, 2019)

This chapter discusses empathy in teamwork and understanding the users in a design process. We address the impact of empathy when designing with blind and partially sighted people and discuss our experiences on how to engage students when starting a new project in an eye-opening physical empathy exercise. We also explore the impact of using an empathy exercise in group work and in the design process.

We discuss this topic in the light of the different skill sets needed by future design professionals and base our discussion on the future skill sets defined by UNESCO and Organisation for Economic Co-operation and Development (OECD) to give strength to our argumentation (Care et al., 2019; OECD, 2019). We can see that the design profession is in transition. We cannot see a strong future for the traditional solo designer as a profession. On the other hand, the future of the design professional seems quite hybrid and the designer of the future will have many transversal skills and will be able to navigate in the complexity of creating new products, systems, and services. We strongly believe in design education and therefore do not suggest that ordinary people become designers but rather to utilise them as experts in the design process. In our opinion, design work is true teamwork that requires many approaches, methods, and tools. As Ricardo Sosa, we also believe "that advancing knowledge and transforming practice in the teaching of empathy and creativity are key to equip people to ask powerful questions, generate innovative ideas, and make more responsible and sophisticated decisions in everyday life" (Sosa, 2019, p. 1). Furthermore, Hess and Fila (2016) suggest that embedding empathy in the design curriculum is very much aligned with OECD's and UNESCO's approach to transversal skills for future work forces.

2.5.2 Wearables class

We will illustrate our argument with teaching examples from two classes called Wearables at the BA Jewellery, Technology and Business programme at KEA – Copenhagen School of Design and Technology (later KEA) held in 2020. In these classes, our jewellery students not only studied the vulnerable user group, partially sighted and blind people, but also designed together with them and produced products following universal design principles, which would improve anyone's everyday life (Story, 2011).

DOI: 10.1201/b22630-14

The classes were conducted in collaboration with the Danish Institute for the Blind and Partially Sighted (IBOS) as they had approached us with a wish for improving the design of assistive aids and tools provided for blind and partially sighted. IBOS is a knowledge and rehabilitation centre in the field of vision impairment (IBOS, 2020).

Our consultant at IBOS, Birgit Christensen, told us that many assistive aids for this target group are very expensive as it is a small market, and at the same time, the appearance, sound, and quality of the materials used are often poor, as the producers of the products believe it is only the function that is important. Many blind and partially sighted feel isolated and secluded from everyday life as they are provided with these technical aids to replace human contact. The assistive aids are supposed to help them, but many have a really loud, annoying sound, which makes it stigmatising to use them in public and they are made of materials that feel cheap, uncomfortable, and fragile. Often, when they need to use an assistive aid, they cannot find it, it doesn't work, and they cannot change the battery themselves or read the manual to fix it, so instead of helping the user, the assistive aid makes them frustrated and feel more secluded (observation notes).

We found this case very fitting for our pedagogical aim to learn to teach the importance of empathy in the design process as it adds an engaging aspect to the wearables classes when the user has a real need. We also had a focus on teaching the students to work in interdisciplinary groups to give them readiness to empathise not only with the users but also with group members from other programmes at KEA.

Our definition of users is inspired by Kenneth Balfelt's definition of user participation: "The users are not architects and vice versa. However, users are experts on the context and on their own needs" (Balfelt, 2019, p. 79). Our blind and partially sighted participants are not laypersons, but experts in their everyday life as non-seeing people and users of aids for the partially sighted. They therefore constitute an invaluable source of information and inspiration for our students.

We define wearables "as a novel multi-domain field which incorporates electronics embedded into objects worn on the body." Our "aim of the Wearables class has been to encourage students to broaden the design space in this field. In our programme, we believe (...) there are fundamental similarities between the electronics and jewellery domains (see example in Figures 2.5.4 & 2.5.7). They both deal with extreme detailing, high precision, size constraints, and precious conductive metals" (Ahde-Deal & Henriksen, 2019, p. 2).

2.5.3 Methodological discourse on empathy in design

In this chapter, we focus on the aspects of utilising empathy in the design process, and our work is mostly built on empathising with the users. Empathy is a very well-studied aspect in design research, user-inspired design, and in many defined design processes. In this chapter, we follow the design thinking process (Brown, 2009) as it is the main design process we teach at KEA. To better understand the importance of empathy in the design process, we have looked for help from prior scholars' work who build on the discourse of designing with empathic inspiration from the users (Kumar, 2013; Koskinen, 2003; Mattelmäki, 2006; Brandt, 2011; Christensen & Ball, 2017). We also wanted the students to learn to take users into the design process by learning from the users' empathic actions and behaviours (Koskinen et al., 2011; Redström, 2017; Luck, 2009). In our teaching cases, the students applied different ethnographic design research methods involving the users.

Several scholars propose that the use of empathy in design should not be merely "to be like someone," especially when it comes to designing for people with disabilities and designers simulating their disability. Instead, we should aim towards exploring how "to be with someone" (Bennett & Rosner, 2019) and support the designer's ability to design with others within an empathic sociability (Devecchi & Guerrini, 2017). Empathic sociability is defined by Richard Sosa as "a sense of empathy that connects to creativity as the capacity to change with others" (Sosa, 2019 p.2). It was important for us to involve the aspect of designing together with the users at this stage. Empathy is an essential aspect in applying co-experience in the design process and when involving the users in the actual design work (Battarbee, 2004; Sanders & Stappers, 2013, Kumar, 2013). Concerning the design of activities to another vulnerable user group, at disabled children's summer camp, Hess and Fila (2016) regard empathy as the cornerstone for making everyday experiences joyful. We have experienced the same when guiding students to utilise empathy in the process of understanding the users. It is challenging to engage with the end users without the empathic approach (Hess & File, 2016).

2.5.4 Empathy

Imagine having lunch without seeing what you are eating or with whom you are having your lunch. Having your senses cut down to sound, smell, taste, and touch. That is the every day of the research participants in these teaching cases.

In the spring 2020 class, we had additional students from the KEA Bachelor programme, Product Development and Integrative Technology PTI (later, technical students), and in the autumn 2020 class, we had the jewellery students working together with technical experts. In both classes, the task was to produce a prototype of wearables for a concept inspired by blind and partially sighted people, which would help them socialise more in their everyday life and which would also be of interest for anyone else to use.

Our ambition was to teach empathy to better understand the everyday life of the end users in this interesting, yet vulnerable user group. One way to get to know the users is to visit their everyday life and to try to experience their joys and frustrations. We let the students get as close to the users' everyday life as possible by making the students prepare a Make tools session (Sanders & Stappers, 2013) and a Wardrobe interview with the blind and partially sighted experts in their homes (Klepp & Bjerck, 2014) and by getting the students to collaborate with the users throughout the projects. The ethnographic design research methods were used to gain empathy in our teaching cases, to help the students understand the importance of learning from and with the users (Møller, 2019; Fletcher & Klepp, 2017). We succeeded in the best possible way, conforming with the COVID-19 regulations and recommendations.

Let's go back to the lunch. On the first day of their classes, the students were divided into groups to work together on the project. We wanted them to feel how it is when your sight is limited, and therefore, we designed an empathy exercise. The empathic design pathways Hess and Fila (2016) describe having been inspirational for the empathy exercise we describe in this chapter. It is important to apply an empathy approach in a complex path of learning about experiences around the issues and concerns that are important for the users (Hess & Fila, 2016). Together with our class's design expert, Miss Blindspot, a partially sighted Danish blogger (Nahnsen, 2020), we prepared 30 pairs of ski goggles to represent different levels of visual impairment with black tape, black paint marker, and glue. The goggles simulate different medically defined eye diseases and conditions, from being completely blind to being able to see partially (Figures 2.5.1, 2.5.2, 2.5.3, and 2.5.5).

Figures 2.5.1 A group of students right after receiving the goggles. The person with the pink cane is wearing goggles that prevent any seeing, the other goggles simulate conditions such as blind spots, tunnel vision, and partial sight due to brain damage.

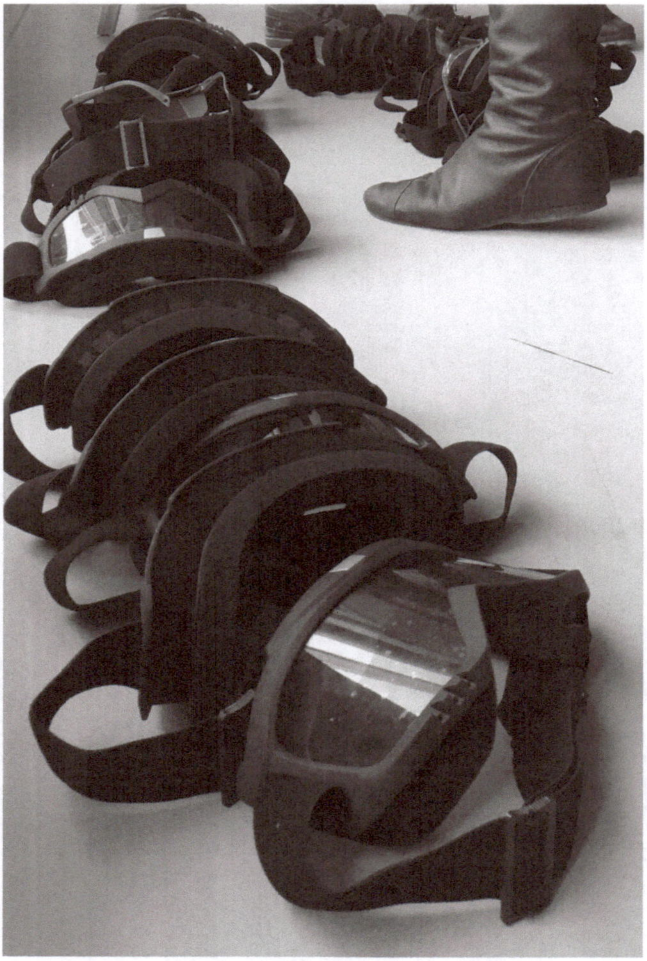

Figures 2.5.2 Goggles ready to be distributed to the students.

We distributed the goggles among the students once they had been introduced to their new group members whom they did not know prior to this class. They were put in groups of five. In each group, there was one totally blind person with a cane and one assistant who was able to see normally. The rest wore goggles with different conditions. They wore the goggles for one hour while having lunch together and navigating around campus. According to Miss Blindspot, it was important to give the students an opportunity to experience and engage in social activities involving many skills in order for the students to better empathise with the partially sighted people.

Bennet and Rosner argue that empathy built through immersion "may steer designers toward narrow and inaccurate conceptions of disability experiences." Several scholars believe that building empathy in design cannot be based on empathy alone, other competencies need to be present, such as compassion, humility, and noticing. Techniques which Bennet and Rosner list as important are "codesign exercises in which designers enlist disabled people as partners" (Bennett & Rosner, 2019 p.4). We applied this to the

Figure 2.5.3 Students out for lunch in the autumn class (Pictures from students' final presentation).

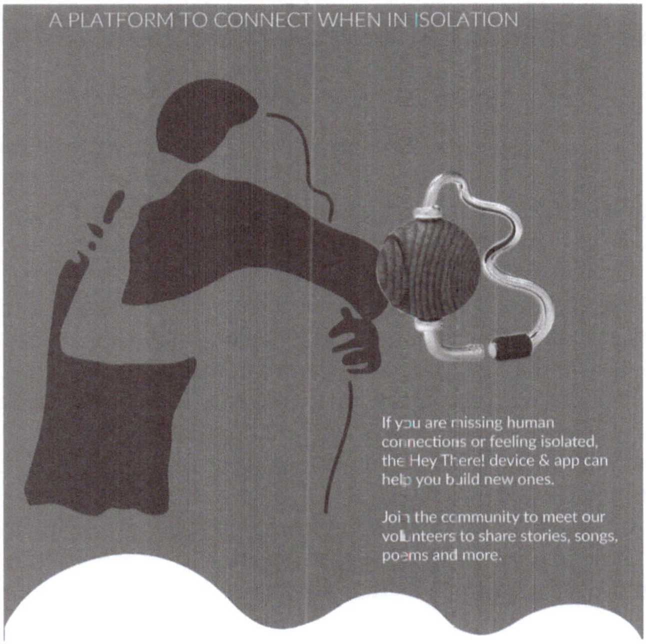

Figure 2.5.4 Student project, autumn class. Hey There! A small piece of silver jewellery with a wooden ball with a button containing the electronics. You can hang it in your necklace or your keyring or carry in your pocket.

Figure 2.5.5 Students from the spring class wearing goggles with different eye conditions.

class through the Make tools sessions (Sanders & Stappers, 2013) and Wardrobe interviews (Klepp & Bjerck, 2014).

Although it was evident that with the empathy goggles exercise, the students would not entirely understand the everyday of a partially sighted or blind person, they were able to better connect with them through the experience. As the student Pauline says, "Of course we are never going to feel like they feel, but I think it is good to just have a glimpse of how it feels. And if it was just this and we wouldn't be in any contact with people with the impairment, it for sure wouldn't be enough, but I think it is a good starting point, a good way to make us think" (Interview 1, autumn). This joint extraordinary experience also enabled the students to create a consensus in their groups from the start of their collaboration.

During the empathy exercise, Pauline had felt that her personality changed a bit, as she felt isolated and unable to speak to people because she couldn't see what was happening. Her group ended up picking "this subject that was about breaking social barriers and how you can feel isolated while having the impairment, so it (the introduction with Miss Blindspot and the empathy exercise) kind of guided us towards that direction" (Interview 1, autumn).

Each group was connected to one user. They were well prepared to first meet and interview their user and try to understand their everyday, and to afterwards plan and conduct user research following the design thinking principles and conducting the Make tools sessions (Sanders & Stappers, 2013) and Wardrobe interviews (Klepp & Bjerck, 2014).

We conducted the class twice in 2020. Each class was influenced by COVID-19 restrictions and rules on teaching and acting in physical places. During the spring 2020 class, the students were able to meet with the users and conduct their user studies physically. Teaching was normal until the lockdown, which in our case meant until the final prototypes were to be made. The consequence of this was that many of the prototypes worked only partially, and in the end, the main emphasis was on the concept. We were not able to iterate with the physical models or with the prototypes. This was a pity since physical interaction to get further communication with the users would have given so much more in this setting of working with blind and partially sighted people.

The autumn class was also influenced by restrictions. Although there wasn't a complete lockdown, we had to manage most of the teaching online. Further, contact with the users did not run as smoothly as before COVID-19 as some, both students and users, were more cautious in the current situation about meeting physically.

2.5.5 Designers' professional skills in the future

Within design, demarcation can provide new approaches to the profession. We can implement such approaches when we have enough capacity to maintain and control frontiers. We are living in a moment of time when we need to update the definition of the designer profession. To re-define it, we must look into design education and critically seek new ways of re-thinking education. We cannot educate solo designers for the future labour market. The focus needs to be on functioning teams whose members can understand each other (Bedford & Arns, 2017). We educators need to focus on the hybridity of the design profession. This means that we must educate not only team players but also design professionals who can rapidly adapt to new ways of thinking, new approaches, and new technologies and who have the will and motivation to apply those (Educators' Summit, 2020).

At KEA, we have a strong focus on practice-based learning, and we emphasise hard skills within all our programmes. This means that our students in the design area are very competent in using computer programmes to design products and produce graphic materials for their concepts. They are also competent in giving form to the products through an understanding of the material features and the production processes. However, they may graduate with a slight lack of transversal skills. UNESCO has emphasised a concern about the lack of transversal skills (Care, et al., 2019). According to UNESCO, transversal skills are as follows: Interpersonal Skills, Intrapersonal Skills, Media and Information Literacy, Global Citizenship, and Critical and Innovative Thinking. All the transversal skills, and especially the Interpersonal and Global Citizenship Skills, require empathy.

As a global citizen, one ought to understand how people in different cultures, religious settings, and countries live. What are their needs, desires, and perhaps fears? Being a global citizen does not only mean understanding people who are physically far away, but it also gives readiness to understand people nearby who are different. Being able to design across the different kinds of needs that people with different backgrounds have, is an essential skill for future designers that we educators must cherish (Halskov & Christensen, 2018). Understanding others requires interpersonal skills. By interpersonal skills, Care et al. (2019) mean skills like communication skills, teamwork and collaboration skills, organisational skills, being able to be social and collegial, and being able to work with empathy and compassion (Care et al., 2019).

OECD discusses three different skill categories. "The OECD Learning Compass 2030 distinguishes between three different types of skills: cognitive and metacognitive skills; social and emotional skills; and physical and practical skills" (OECD, 2019). In our education, we are especially interested in becoming better at teaching social and emotional skills. OECD describes those as skills needed in workplaces and classrooms, which are getting more diverse. Diverse ethnically, culturally, and linguistically. In these classrooms, and further on in the workplaces, empathy, self-awareness, and respect for others, among the ability to communicate, are essential according to OECD's Concept Note: Future of Education and Skills 2030. The Note emphasises the transformation that the skill sets are in, as computers and other technologies take over some of the labour tasks. These tasks are mainly repetitive tasks without analytical or creative aspects.

Focusing on the soft skills: social and emotional skills described by OECD and transversal skills described by UNESCO (Care et al., 2019; OECD, 2019) should include a focus on the mindset of future professionals in the design field. According to our experience with teaching wearables (Ahde-Deal & Henriksen, 2019), we have learned the power and importance of interdisciplinary teamwork. It is especially important to be able to adjust one's own mindset to fit to the team's mindset.

As Josefine from the spring 2020 class says, "We were in a group with two (PTI students) who actually was really engaged in the project and actually really good. So that gives your engagement a little boost, so you want to do it." (Interview 1, spring).

In our programmes, there is much diversity in the groups. That is a richness that we can cherish and nurture with the right mindset and a focus on the soft skills throughout education.

So, our mission of bringing empathy into the class was to summon up these inherent soft skills that many people have. We would like to emphasise the need to teach these skills in the design education, especially in our institution. However, it is problematic to teach empathy and compassion and still be social and collegial in action and also true to the user we are designing with and for. We do not have any exact solutions to that, but we are happy to present our aim to summon up empathy in the design process and teamwork.

As Bea said about their cooperation with their technical expert: "It's trying to also just have empathy with each other and be patient trying to explain and explain again what we actually want, but I think we are capable of doing that" (Interview 2, autumn).

2.5.6 The class setting

In 2020, we had two courses with students learning wearables with IBOS. The methodological and theoretical learning goals were the same both times. The execution of the classes was slightly different in the spring due to an unexpected combination of online and offline teaching.

Due to a very insightful class in spring 2020, we continued our collaboration with IBOS and Miss Blindspot in the autumn 2020 class and repeated the project in a slightly different setting. We redid the empathy exercises and the close collaboration with the blind and partially sighted users. We have reflected on and evaluated our experiences with empathy exercises within these two similar settings.

In the spring class, we had 35 jewellery students and 12 technical students working in eight groups. In the autumn, we had 29 jewellery students in seven groups and two technical specialists (one teacher and one student helper) from PTI. We have interviewed students from both classes and asked them to complete a small survey. These interviews, evaluations, and reflections form the base of our paper for DTRS13 (Table 2.5.1).

The two classes were alike in most of the content and the methods taught and applied. But there was a difference in the students' overall motivation because of all the changes at school due to COVID-19.

Table 2.5.1 An overview of the content of the two classes with similarities and differences

IBOS project– mandatory wearables project at KEA	*Spring class – Danish 6th semester; full-time study*	*Autumn class – International 5th semester; full-time study*
Specifications of the classes	35 students: 8 groups/7 weeks /10 ECTS	29 students: 7 groups/8 weeks/13 ECTS
COVID-19 alterations of projects	• The last 2 weeks online - guidance and final presentations: no feedback from the user, only from IBOS. • No access to workshops/ 3D printers: no physical prototype.	• Normally 5th semester students are in internship, but due to COVID-19, this class will have their internship in the 6th semester and school in their 5th semester. • Introduction of one day instead of two days (involving less people from outside school). • A lot of online guidance during entire project due to limitations in number of students allowed in classrooms, a fire that took place on our floor at school and changing COVID-19 guidelines.
Collaboration with PTI (technological experts)	• 12 student consultants from PTI, 1–2 consultants in each group.	• 2 technical consultants (one teacher and one student helper) responsible for guiding 3–4 groups each.
External presentations for class	• Stegeager & Hastrup – former BA students who specialised in assistive aids for the partially sighted using the Wardrobe method. • Miss Blindspot - partially sighted blogger and owner of empathy goggles representing 5 different eye diseases. • Birgit Christensen, leader of IBOS assistive aids exhibition • Visit to IBOS assistive aids exhibition with introduction and test of various aids.	• Miss Blindspot - partially sighted blogger and owner of empathy goggles representing 5 different eye diseases. • Birgit Christensen, leader of IBOS assistive aids exhibition. • Visit to IBOS assistive aids exhibition with introduction and test of various aids.
Empathy exercises	• Empathy goggles worn for an hour over lunch, follow-up discussion with Miss Blindspot and IBOS representative. • Preparing and conducting an adapted wardrobe interview/ Make tools session with user in their home.	• Empathy goggles worn for an hour over lunch, follow up discussion with Miss Blindspot and IBOS representative. • Preparing and conducting an adapted wardrobe interview/ Make tools session with user in their home.
Meetings with target group	• Group interview with all users. • Individual interviews with their specific user. • Wardrobe/Make tools session in the same user's home. • Test session on the proposed design concept.	• Group interview with all users. • Individual interviews with their specific user. • Wardrobe/Make tools session in the same user's home. • Test session on the proposed design concept.
Final presentation of project for IBOS and users	• Online presentations and demonstration of the technical prototypes. • Online feedback from IBOS representative and teachers. No blind or partially sighted users present.	• Online presentations and demonstration of the technical prototypes. • Online feedback from IBOS representative, one partially sighted user and teachers.

The students in the spring class had just come back from their internship and were motivated for new learning. However, towards the end, some of the students lost a little of their motivation due to the lockdown. There was also frustration during the last two weeks of the project, since there was no possibility to finish the prototypes properly, nor to test or have feedback from the users. However, the students kept their engagement in the project.

The autumn class had experienced many changes to their education due to COVID-19. The major deviations were: Their internship was moved to the sixth semester and their fourth semester entrepreneurial project was prolonged to the beginning of the fifth semester, as it was not possible to finish it during lockdown. When they started on the IBOS project, the students were already tired from COVID-19 and a long entrepreneurial project, which many of them hadn't completed as they still had orders to finish. This meant that there was a difference in the two classes' motivation and engagement at the start of the project. Therefore, we do not consider the autumn project as successful from an overall view, but we still believe the insights the students gained from the different methods taught and practiced have given them valuable learning to take with them.

In spring 2020, the teachers, students, and users were truly engaged throughout the whole project, even though it ended up being interrupted by COVID-19. The students took a genuine interest in their users and were in contact with them during the design process. The cooperation with students from different programmes was a success, due to the engagement built into the empathy exercise. The concepts designed by the students were all comprehensive and responded to their users' needs, while being relevant for products for the non-blind also. So, what made it such a success? What made the students so engaged? One student replied in the spring survey: "It has been the most interesting project (at KEA) for me, as it was about making a difference for some people, which affected the commitment" (Spring survey).

2.5.7 Methodological framework

We noticed that both users and design students gained inspiration from each other. In these classes, design research methods were combined, as the empathy exercise cannot stand alone. Orientating our students to expect reciprocation shifted their "empathy building toward opportunities for mutual sense-making, multiple first-person narratives, and shared accountability" (Bennett & Rosner, 2019, p.9). The applied methods involving users in these classes were Make tools sessions (Sanders & Stappers, 2013) and Wardrobe studies (Klepp & Bjerck, 2014). Both methods were adopted for the setting in which the students were going to use them.

In the generative design method Make tools, users are invited to physically build rapid prototypes presenting solutions to their problems. It is therefore a generative approach in the sense that the participants can express their ideas and dreams and share their insights through a selected set of stimulus items (Sanders & Stappers, 2013). As experts on the subject, the participants can gain influence as co-designers of the project. Many Make tools session exercises are based on the visual sense, which does not fit well with gaining insight into the experiences of the blind and partially sighted, but the students adapted the method to touch, sound and smell. They also included strong colours and light for partially sighted research participants.

In the ethnographic research method, Wardrobe studies, "materiality is at the core" and the focus is on practice theory which "emphasises bodily movements, things, practical knowledge, and routine" (Reckwitz, 2002, p. 259). The participants are interviewed in their home with their wardrobes, to gain insight into their emotional connection and interaction with the wardrobes (Klepp & Bjerck, 2014).

As the Wardrobe method often focuses on clothes and accessories, we broadened the spectrum to include favourite objects in the participant's home, objects the participant liked because of functionality, material, or emotional value. We also did this to gain insight into the everyday practicalities of a blind or partially sighted person and further, to observe how they interacted with their objects and whether they had any needs or issues in their everyday life (Figure 2.5.8).

One example of an everyday obstacle was found by a group of students who noticed how their user had to scroll through all the apps on his phone every time he wanted to show them a special app he used (observation notes). This action put the user in a situation where everyone around him would get to know all his apps and personal information related to the apps, including passwords, because he had to use the voiceover function on his phone. Later, this group started working on a concept of discrete use of apps (see Figure 2.5.6).

The beauty of combining many research methods is that you build a deeper understanding of the users' everyday life with a holistic approach, meaning that as many aspects as possible can be viewed in the user study. The students did not only look into a specific issue that blind and partially sighted people have but also started understanding their life in a broader perspective. Therefore, the students did not design new canes or patches that display the condition, they went further with their designs.

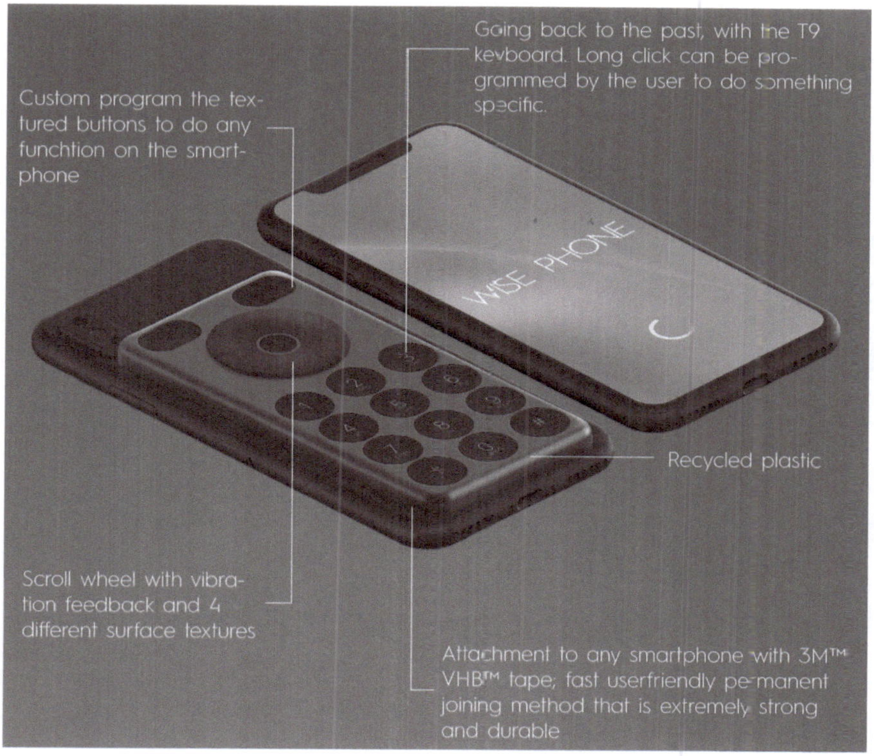

Figure 2.5.6 Student project, autumn class. Wisephone: "Make your smartphone smarter".

Figure 2.5.7 Student project, spring class. "BEAWAVE can help you concentrate in different environments. BEAWAVE helps you to improve your conversations, strengthen your attention, and most importantly improve presence in your everyday life".

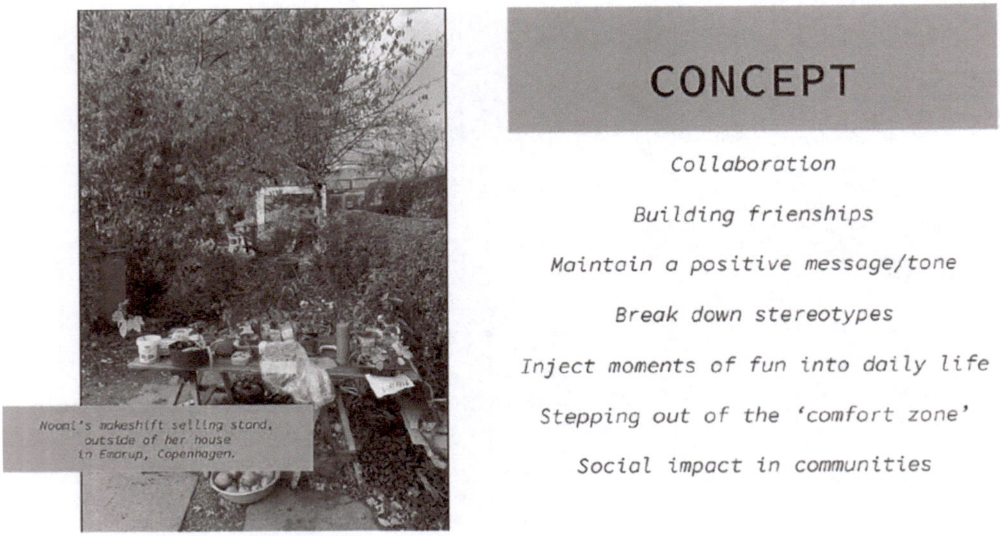

Figure 2.5.8 Make tools/Wardrobe session, autumn class (Picture from students' final presentation).

2.5.7.1 Method to evaluate the teaching

We conducted semi-structured interviews with students from each class. Two in the spring and two in the autumn. In the spring 2020 class, we carried out a short survey (11 students replied) to get an overall impression of the whole class and the empathy exercise.

Likewise, we conducted a survey of the autumn 2020 class just before the end of the project (13 students replied).

Our observations included the students' level of activity and engagement during their classes at KEA. Due to COVID-19 regulations, we could not carry out many physical observations during class in the autumn class, as the whole class was only gathered physically at KEA when we had presentations. So here, the observations were mainly gathered from what they presented to us at feedback sessions, and from online guidance.

2.5.8 Empathy in the classroom

The students created a connection with their blind and partially sighted users, and in both classes each group worked closely with one user. The users gave the design students access to understand their everyday, their needs, and struggles. None of the blind and partially sighted persons were educated designers, nor had they worked or currently work in the design field.

The students had an introduction day, where they met Miss Blindspot and heard her compelling story about how it is to become partially sighted and learn to live with it; they did the empathy goggles exercise and heard insights by IBOS in a presentation of eye diseases and assistive aids provided to the blind and partially sighted people in Denmark. The students were thus given a basis on which to start brainstorming ideas for possible wearable concepts. This was combined with a visit to IBOS, where the students saw an assistive aids exhibition and met their users for the first time. They started with a group interview with all the users, followed by individual interviews with their appointed user.

A week later, after having analysed their first interviews and prepared their Make tools kits (Figure 2.5.9) and Wardrobe interviews, the students visited the users' homes and combined the two methods in a new interview. Users are not designers, but they bring valuable and essential knowledge to the process as specialists. 'The users' analyses, knowledge, and needs are translated and processed by the project group, thereby facilitating a process where users and the project group both contribute what they are best suited to

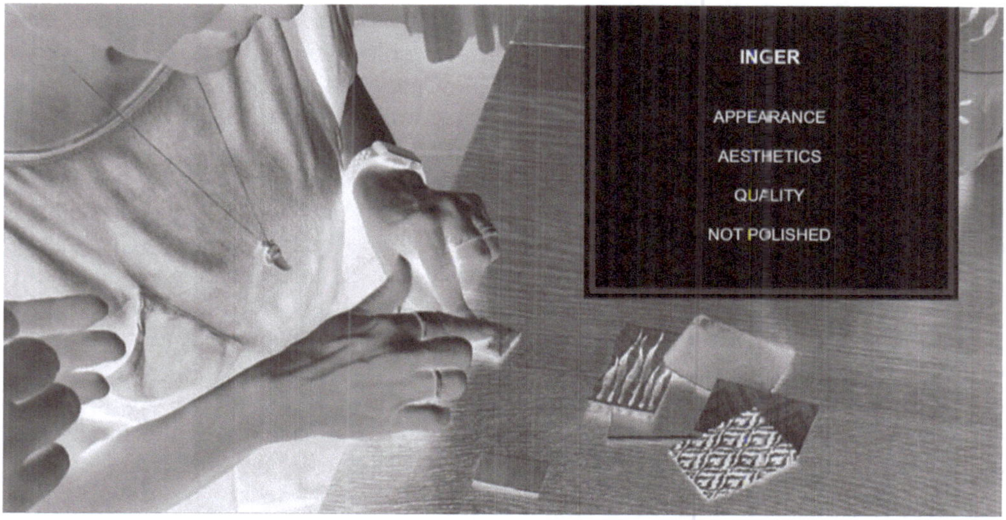

Figure 2.5.9 Make tools session, spring class (Picture from students' final presentation).

bring to the process" (Balfelt, 2019, p. 79). The participants act as co-designers in this project. They feed the ideas and the designers transfer those into products and services.

As Elisabeth explained about their user, "We found out with our user that he liked a lot of colours and he liked when he could help decide how the things should look. So, it gave us something we could use in the design process" (Interview 2, spring). The combination of the empathy exercise and the close connection to their users through ethnographic research provided the students with knowledge they could not have gained by reading or by only observing their users. Accessing the users' homes gave the students insights that would have been hard for them to study otherwise, for example, in an interview.

Bea: "it is just like with the empathy goggles I guess; it gives us more empathy that we are able to see the person more in their own context" (Interview 2, autumn). We tend not to be able to observe even our own routines, and therefore visiting the blind and partially sighted people's homes gave a lot of information that would have been impossible to ask for.

Many of the groups were surprised regarding the use of colours and patterns in the clothing and interior, just as they were surprised to discover how much the users took an interest in natural materials. Pauline:

"it was very interesting to see him do things inside his house, serving us coffee, having a friend over and seeing how the house was (…) put together, (…) we noticed that he didn't have any pictures or like any of his family or anything" (Interview 1, autumn).

Some of the groups had male participants of 50 years and older and they discovered that some of the Make tools exercises they had designed for their user did not really work as expected, as Bea says: "Bo was very engineering, (…) he was like, judging upon a functional aspect which was quite interesting because we prepared a more emotional interview with him, where he should connect colours with emotions and so on and that didn't really quite work with him, but then we got to know Tina, his girlfriend, and she was way more like talking about emotional aspects" (Interview 2, autumn).

As most of our students are female and in their twenties, it was a challenge for them to plan exercises that connected with their user if the user was older and male. Nevertheless, through a combination of exercises, the visit to their user's home and the conversation with their user, the students gained genuine respect and interest for their user.

We were able to witness excitement in the student groups that was rooted in their own bodily experience and their learning from the Make tools sessions, Wardrobe interviews and observations. Through the combination of these methods, the design students, whose life is quite different from that of their users', were able to empathise with people who cannot see and are therefore trapped in their homes or routines.

What most students observed was that blind and partially sighted people often find it difficult to change or alter their routines. Although some of the blind and partially sighted users impressed the students with being very active horse-riding, sailing, or travelling abroad alone, they also spend more time in their homes than those who can see, because it takes a lot more planning and energy for them to actually go outside. Therefore, the wearables designed for them often had features helping them move more and break their routines in a safe and social setting.

None of our insights claim that it would be easy for those who can see to change their routines or leave home. However, people who are not blind or partially sighted have more choices. Their obstacles with changing routines or leaving home might derive from something else than it being physically difficult. They may not want to go out, whereas if you are physically trapped in your home, you might still want to leave your home. Most of us have experienced some unwanted homestay during the COVID-19 lockdown.

2.5.9 Insights and takeaways

By conducting empathy exercises, we learned about the importance of empathy. We learned that our students gained a solid base and many valuable insights by the empathy exercises which they would have had difficulties achieving otherwise. They acquired these insights by a combination of many features: a thorough introduction day to the blind and partially sighted and to the empathy goggle exercise and, subsequently, by a combination of interviews with the blind and partially sighted users and the use of the ethnographic design research methods, the Make tools sessions and Wardrobe interviews. We experienced the importance of adapting the Make tools sessions and Wardrobe interviews to the specific participant that the students visited, and also that no interview ever goes as planned. The students were able to design relevant products through compassion, humbleness, observation, and relying on everyday experts' insights, and using the participants as valuable partners in the design process (Bennett & Rosner, 2019).

We also observed that inspired by their learnings from this project, the students from these classes applied the empathy methods later in their design research in different and new settings. Furthermore, we learned that the interdisciplinary focus was clearer and more fruitful for the students when they cooperated closely with students from another KEA programme (spring 20) and not only with teachers (autumn 20), although both classes learned about communication with specialists from a different field

As Josefine contemplated: "I think collaborating with some other students (from a different educational field) has been really, really interesting because it gives a very realistic picture of how you as a designer in the labour market most likely will get to work" (Interview 1, spring).

Even though, due to circumstances out of our hands (COVID-19), the engagement of the students in the autumn was not as high as in the spring class, the jewellery students still learnt valuable insights from the combination of wearables, the empathy exercises and an untraditional user group, all of which they can use further on in their studies, and hopefully later in their working career.

2.5.10 Conclusion

In this chapter, we have presented a teaching case that has a strong emphasis on empathy. We have shown how we used empathy exercises to make both the design process richer and the teamwork in the classroom more fruitful. It is our ambition to make our experiences with these empathy exercises available to other design educators. We have learned to consider the hybrid nature of future design professionals when teaching wearables to interdisciplinary teams at bachelor level in a design education programme.

We have built our empathy exercise on our learnings from earlier scholars' notions on empathy. It is common to all these notions that using, utilising, and applying empathy in the design process or in design research is fundamental (Kumar, 2013; Koskinen, 2003; Koskinen et al., 2011; Mattelmäki, 2006; Battarbee, 2004; Redström, 2017; Møller, 2019; Sanders & Stappers, 2013; Fletcher & Klepp, 2017, Hess & Fila, 2016; Halskov & Chistensen, 2018; Luck, 2009). We are contributing to empathy in design and in the design research discourse with our empathy goggles exercise. We are hoping that our experiment and experience presented in this chapter will help other design educators to experiment with empathy in their curriculums.

In this chapter, we first browsed the so-called soft skills that UNESCO and OECD have foreseen as essential in future. We then explored the empathy exercise that we developed as a consequence of our experiences in classrooms with a great diversity of students. We conducted the empathy exercise twice in 2020 under the influence of COVID-19. What we have learned from these teaching cases illustrates the need for team members to be able to see the world from someone else's perspective.

Josefine: "It is a reality picture of cooperation across professional groups, and then the empathy exercises- it was clearly a totally different way to create empathy with the user" (Interview 1, spring).

Our students gained an extra layer in their user-oriented design process. This was achieved by adding a wearables design project to the jewellery class which focused on technology and function, and which included cooperation with technical specialist students. When designing wearables, some students discover that they have widened their design perspective and can design and cooperate in a much broader field than previously. Elisabeth: "It shows a little how you also can work (…) it doesn't have to be just with jewellery, but you could get hired in a company working with technology (….) And be able to contribute with the knowledge I have gained" (Interview 2, spring).

We believe that the empathy research methods applied and the insights gained by the jewellery students throughout this project provide a valuable addition to the students' transversal skills in their future work, as students as well as design professionals. We further believe that this will not only enhance their chances in the future job market, but also improve their empathic and collaborative skills and their ability to tackle future global challenges.

References

Ahde-Deal, P. & Henriksen, M. L. (2019). Teaching wearables. In *Proceedings of DRS Learn X Design Conference 2019*, 9-12 July, Ankara, Turkey.

Balfelt, K. A. (2019). *Artistically developed methods for user involvement.* [e-book] Museet for Samtidskunst. Available at:<https://www.kennethbalfelt.org/userinvolvement> [Accessed 17June 2020].

Battarbee, K. (2004). *Co-experience: Understanding user experiences in social interaction.* Doctoral dissertation. Helsinki: University of Art and Design Helsinki.

Bedford, D. A. D. & Arns, J. W. (2017). *Unpacking a design thinking process with discourse and social network analysis.* In Christensen, B. T., Ball, L. J. & Halskov, K. (eds.), *Analysing design thinking: Studies of cross-cultural co-creation* (pp. 349–371). London: CRC Press.

Bennett, C. L. & Rosner, D. K. (2019). *The promise of empathy: Design, disability, and knowing the "other" CHI 2019*, May 4–9, 2019, Glasgow, Scotland, UK.

Brandt, E., Redström, J., Eriksen, M. A. & Binder, T. (2011). *XLAB.* The Danish Design School Press.

Brown, T. (2009). *Change by design: How design thinking transforms organizations and inspires innovation.* Harper Collins Publishers.

Care, E., Vista, A., & Kim, H. (2019). *Assessment of transversal competencies: Current tools in the Asian region.* UNESCO Bangkok; Washington DC: The Brookings Institution, 56 p. ISBN: 978-92-9223-625-0 (print version), ISBN: 978-92-9223-626-7 (electronic version).

Christensen, B. T. & Ball, L. J. (2017). Fluctuating epistemic uncertainty in a design team as a metacognitive driver for creative cognitive processes. In Christensen B. T., Ball L. J. & Halskov K. (eds.), *Analysing design thinking: Studies of cross-cultural co-creation* (pp. 249–270). London CRC Press.

Devecchi, A., & Guerrini, L. (2017). Empathy and design. A new perspective. *The Design Journal*, 20(sup1), S4357–S4364. DOI: 10.1080/14606925.2017.1352932

Educators' Summit (2020). Sustainability in education. Panel debate on *What role does the physical vs the virtual play in educating and preparing our students for the future?* KEA & IP Apparel online event on 5th November 2020.
Fletcher, K. & Kleppp, I. G. (2017). *Opening up the wardrobe: A methods book*. Novus Forlag.
Halskov, K. & Christensen, B. T. (2018). Designing across Cultures. *CoDesign, International Journal of CoCreation in Design and Arts*, 14, 2018 (2). 10.1080/15710882.2018.1459101.
Hess, J. L. & Fila, N. D. (2016). The manifestation of empathy within design: Findings from a service-learning course. *CoDesign, International Journal of CoCreation in Design and Arts*, 12(1). 10.1080/15710882.2015.1135243.
IBOS - Institute for Blind and Partially sighted (2020). IBOS.dk [online] Available at: https://www.ibos.dk/ [Accessed 17 June 2020].
Klepp, I. G. & Bjerck, M. (2014). A methodological approach to the materiality of clothing: Wardrobe studies. *International Journal of Social Research Methodology*, 17(4), 373–386, 10.1080/13645579.2012.737148.
Koskinen, I. K. (2003). Emphatic design in methodic terms. In Koskinen, I. K., Battarbee, K. & Mattelmäki, T. (eds.), *Empathic design, user experience in product design* (pp. 59–65). Helsinki: IT-Press.
Koskinen, I. K., Zimmerman, J., Binder, T., Redström J., & Wensveen, S. (2011). *Design research through practice: From the lab, field, and showroom*. Morgan Kaufmann.
Kumar, V. (2013). *101 Design methods – A structured approach for driving innovation in your organization*. John Wiley & Sons, Inc.
Luck, R. (2009). Does this compromise your design? Interactionally producing a design concept in talk. *CoDesign, International Journal of CoCreation in Design and Arts*, 5(1). 10.1080/15710880802492896.
Mattelmäki, T. (2006). *Design probes*. Doctoral dissertation. Helsinki: University of Art and Design Helsinki.
Møller, T. (2019). *Wearing health products - A wearer-led accessory approach to wearable health design*. PhD Thesis. Design School Kolding.
Nahnsen, C. (2020). *Miss Blindspot*. [online] Available at: https://missblindspot.com/ [Accessed 17 June 2020].
OECD (2019). *Future of education and skills 2030*. Concept note. Available at: https://www.oecd.org/education/2030-project/teaching-andlearning/learning/skills/Skills_for_2030_concept_note.pdf [Accessed 12 November 2020].
Rederström, J. (2017). *Making design theory*. The MIT Press.
Reckwitz, A. (2002). Toward a theory of social practices: A development in culturalist theorizing. *European Journal of Social Theory*, 5, 243–263. 10.1177/13684310222225432.
Sanders, E. B. & Stappers, P. J. (2013). *Convivial toolbox: Generative research for the front-end design*. Bispublishers.
Sosa, R. (2019). Teaching (with) empathy and creativity in design. In Börekçi, N., Koçyıldırım, D., Korkut, F. & Jones, D. (eds.), *Insider knowledge, DRS Learn X Design Conference 2019*, 9-12 July, Ankara, Turkey.
Story, M. F. (2011). The principals of universal design. In Preiser, W. F. E. & Smith, K. H (Chapter 4), *Universal design handbook*, 2nd edition. The McGraw-Hill Companies, Inc.

2.6 Are we that different?

Empathy, design pedagogy, and cultural objects

Kenny Segal and Jonathan Ventura

Department of Inclusive Design, Hadassah Academic College, Jerusalem

2.6.1 Multi-cultural design and the essence of wood

> Do not make something if it is not useful, but if it is both necessary and useful, do not hesitate to make it beautiful, as long as the decorative elements are an inherent part of the design and do not interfere with function.
>
> (Sprigg et al., 1991: 130)

For the last twelve years, we have been teaching the course "Design and Culture through Wood" at the Department of Inclusive Design at Hadassah Academic College in Jerusalem. Throughout the years, we have been aiming at four main pedagogic goals: first, to teach our students that contemporary design starts primarily at the socio-cultural elements of each material cultural in which it is created. The essence of multiculturalism, in our eyes, is the ability to truly understand varying cultures and embed their point of view in our own, which brings us to the second goal. As a department specializing in inclusive design, we wish to highlight the importance of empathy and care not only through the 'classic suspects' of healthcare or education design, but a holistic approach relevant to design practice in general. Third, as design thinking (Brown, 2009) became an overarching methodological and strategic approach for businesses and non-governmental organizations (NGOs), we wished to present our own local interpretation of the concept. Understanding local cultures, vis-a-vis industrial design's tendency to present global and standardized outcomes imbues the concept of design thinking with a different meaning. Using theory, cultural, historical, philosophical, and religious knowledge helps students reflect and re-evaluate not only the outcomes of this course, but their approach to the discipline. Our final goal was to articulate contemporary Israeli and Jewish culture vis-a-vis contemporary design. The studio focuses on three distinct cultures of wood: The Shakers, a protestant sect originating in the North-Eastern parts of the US in the late 1700s; traditional Japanese culture, focusing on the major religions in Japan, along the ensuing aesthetics and philosophies; and finally, the various cultures of Africa, and specifically those of the East and West coasts along Central and South Africa. During this studio, the students must choose a culture, an era, or a tribal culture, identify, and articulate a central philosophy, religious belief, or aesthetic trait and generate a design brief, that is, translating this abstract knowledge into material and visual configurations of a designed object. Naturally, the essence of the studio is not to copy or duplicate another culture's material artefacts, but rather to re-interpret these through a contemporary Israeli point of view.

We can summarize our overall pedagogic core values as consisting of several elements: first, we wish to present the students with an alternative to the classic industrial design

DOI: 10.1201/b22630-15

process of identifying a need/problem/opportunity, outline a brief, conduct research, map-out the field, etc. Returning to the core values of design thinking as presented by IDEO, we offer a different interpretation of the model. During the first weeks of teaching, we tell the students to stop thinking about value, function, market research and a detailed mapping of design partners and shareholders. Instead, we focus on core socio-cultural values, philosophical or religious principles and poetic yet practical vernacular design. A second principle, which also echoes the core values of IDEO, is the anthropological principle that understanding comes from a combination of empathy and a lack of judgment. Third, the pedagogic choice of focusing on a single material takes out of the equation a big part of the classic design process. We chose this path since it allows us to focus primarily on the cultural content of each project, rather than on various attributes of different materials. In other words, the designed objects server us to deliver a deeper message of empathy and care. However, as each of the taught cultures developed differing design technologies, this element is manifested through the student's understanding of the link between material and production methods. Finally, this unorthodox approach towards design, in which the classic design process is reversed, derives from our understanding of the current and future global and local realities, as well as our notion of the designer's role in society. Apart from the current pandemic, the world is rife with political shifts, refugee currents, economic inequality, rampant power struggles by tech companies, and more. In this complex reality, and as a direct result of our definition of inclusive or social design in society, we feel empathy needs a broader and deeper place in design practice. We believe, as a department, that designers should be more concerned about design's influence on society, the climate and our current material and visual reality. As such, we stress the importance of the designer as an agent of socio-cultural change, through understanding the individuals and communities with which we work. These elements became embedded in the very core of this studio.

Indeed, a central and crucial element of the studio, as it evolved, however, was not only the ability to shift from the abstract to the material, but rather to harness varying scenes of socio-cultural artefacts to develop empathy among young designers. As an inclusive design department, our core values of empathy reside between healthcare and social design, yet we found it extremely interesting to also harness classic design approaches to imbue our students with empathy towards different cultures, thus criticizing their own (Segal & Ventura, 2019a).

Our department's title switched from 'industrial design' to 'inclusive design', a switch based on shifting values from the classic approach towards design, focusing on supplying the industry with functional solutions, to a more socially- and people-oriented approach. While industrial and product design differ in the scale of manufacture, links to industrial partners and more, a common denominator would be a somewhat pragmatic and functional approach focusing on identifying needs or searching for problems in existing products, thus offering new/innovative alternatives. Questions relating to the definition of product design today still present designers with difficulties, such as what is design? Is it a separate discipline from industrial design? Does product design stem from art or is the opposite correct? Our view is that product design is a primordial activity stemming from the basic needs of humanity's survival which, with time, developed to become an encompassing cultural phenomenon (Segal & Ventura, 2019). Based on this view, speaking of the field of product design, describes only the physical outcome and not the overall cultural activity. We will refer therefore to Design – an interdisciplinary activity to be recognized and measured through the prism of culture. Furthermore, considering what we call the design situation (see Ventura, 2016), rather than design practice, we imbue the material product a holistic and broader socio-cultural scenario in which it is designed. This, in turn, includes norms, a specific vernacular context, religious and philosophic attributes, as well as the various intended design partners. Conversely, Julier (2006: 70) emphasizes correctly that design culture strives "to imply collectively-held norms of practice shared within or across contexts". Manzini (2016: 54) adds another layer to Julier's definition of design culture, surmising that "it is mainly the culture of the designers themselves and of the communities in which they operate: the culture on which design itself is based and thanks to which innovative meanings can also be proposed".

We do not wish to engage the age-old debate on the definition of culture. This has rallied anthropologists and other researchers for decades. The contemporary variant of culture defines it as a "social domain that emphasizes the practices, discourses, and material expressions, which, over time, express the continuities and discontinuities of social meaning of a life held in common" (James, 2017: 53). This definition follows the footsteps of anthropologist Clifford Geertz (2012; 1973) enrolling into this complex concept socio-cultural norms, as well as material and visual products of a specific culture. Conversely, culture is not only constructed of things, abstract notions, ideas, and stories, but a common continuous structure. While classic key elements executing this notion were either religion or national government, in the last decade, design has a foothold as well. In other words, just as literature or visual arts, material and virtual designed products mirror one's cultural values. Following design's temporal natural, we are talking not only of a social group's past cultural values, but those of the present and in some cases the future as well.

It follows then, that design is a material or virtual manifestation "of social meaning of a life held in common". This is an exciting proposition as we can study designed artefacts and through them get an insight into the "continuities and discontinuities" of local social meanings, norms, and values (see, Ventura & Ventura, 2017). The obvious tool for studying norms and values of societies is cultural anthropology but a more specific tool would be required to fulfil the proposal of studying culture through objects. The relevant tool in our case, therefore, would be design anthropology which focuses on cultural variation through the study of objects using accepted anthropological methodologies (e.g., see: Otto & Charlotte Smith, 2013; Julier, 2013). Specifically, Gunn et al. (2013: xiii) claim that

> practitioners of design anthropology follow dynamic situations and social relations and are concerned with how people perceive, create, and transform their environments through their everyday activities [...] design anthropology practices occur across different scales and timelines and involves many disciplines, each bringing their own distinct ways of knowing and doing.

Gunn and Donovan (2012: 9) stress the importance of the socio-cultural relationship between the designer and the end-users. A crucial point they consider in the fluxing relationship between design and anthropology is the difference between what they term dA, Da, and DA:

> dA – the theoretical contribution is for anthropology rather than design. Design follows the lead of anthropology in terms of adopting theoretical understandings or becoming the subject of anthropological study. Da – fieldwork is in the service of design. Framing originates from problem-oriented design approaches rather than

engagement with peoples. Anthropology is put in service of design, for example ethnographic studies are used for establishing design requirements. DA – disciplines of design and anthropology are engaged in a convergence of efforts each learning from the other. DA is a shift from informing design to re-framing the social, cultural and environmental relations in both design and anthropology.

Gunn and Donovan (ibid: 1) rightfully claim that a central resemblance between anthropology and design is the ability to interpret daily activities: "central to engaging with others is finding ways of imagining oneself into another person's world". This very idea of understanding one's perspective through imagination stands at the very heart of empathy as a human trait. Indeed, using previous definitions of design anthropology as a basis, we follow its evolution to social design (Ventura & Bichard, 2017). We propose that design anthropology does not focus solely on supplying research methodologies and theoretical venues to practitioners, but rather imbue the discipline with core values needed to tackle the wicked problems of design (see Buchanan, 1992). One of these deals with the moral essence of design, as well as the designer's role in society, to both of which cultural empathy is much needed. One of the key traits of empathy is understanding another's point of view, thus interpreting a specific situation from a different approach. While designers conduct these methods as part of the classic design process, through our course, the essence of cultural understanding is not dependent on market research, product value or functionality, but rather on understanding other cultures for the broader and deeper value of Humanism. Therefore, using wood as a common denominator allows us to isolate other variables to focus on culture, understanding and interpretation.

Our solution, then, was to level the playing field and allow for comparative research by choosing a classic, raw material and mapping its implementation in specific cultures that have a tradition of working with said material. The material we chose was wood, a magical natural material that has been used by mankind to survive for an estimated 200,000 years. Through this material 'looking glass', a whole new dimension is added to the understanding of objects and design. Alongside the object, its cultural function, and the behaviour of the users, we decipher the relevant cultural values, traditions, and rituals, local crafts, and techniques – all on one platform of raw material enabling a broader and deeper comparative study.

Incorporating this methodology within the practical side of design culminates in an approach that sees the designer as a creative visual-material interpreter. In the case of design practice, the object, understood through social/cultural meaning, allows for the designer to create new meaning or objects by changing context, or social meaning while utilizing all the different disciplines required in one complex object. So, through the looking glass of wood, using the methodology of design anthropology to understand a complex social meaning, designers can create new cultural artefacts by implementing an interdisciplinary approach in a multi-layered cultural object.

To illustrate our pedagogic values, as well as our practice-based approach, we have outlined a few of the students' projects, chosen out of over 450 projects accumulated over 12 years. These were chosen to highlight the importance of design anthropology to the practice of design, as well as the ability to harness empathy in design. We based our interpretation on a previous research methodology labelled "visual-material content analysis" (Ventura & Ventura, 2018). Naturally, deciphering these unique objects offered quite a challenge, one that brought us to a complex yet articulated solution. Using design anthropology as a basis, thus considering a specified socio-cultural layout for each object, we needed a broader outline for material interpretation. Although semiotics is a typical answer, we believe it is not enough, due to its fluid and in some cases too flexible nature (see Ventura & Shvo, 2018). Thus, adding interpretive analysis, based primarily on design phenomenology and design hermeneutics, we imagine each object as a text or a narrative holding a sensorial as well as a descriptive nature. Following this method, we interpret the chosen objects according to several attributes: the configuration, aesthetics (shapes, colours, textures), construction technologies (derogation, joinery, etc.), thus allowing for a richer interpretation of the socio-cultural values embedded in these designed objects. This approach presents researchers with various possibilities for integrating empathetic design in pedagogy and practice. These include a deeper, thorough examination of socio-cultural attributes and decipher their meanings in a material object. Furthermore, as a pedagogic tool, one can either interpret existing objects (in a curatorial or historical courses, for example), or create new objects with the students (in industrial design, product design, jewellery design, fashion, or architecture, for example). The key notion would be to harness empathy not solely through the classic spheres of inclusive design or disability studies, but through a broader plethora of objects.

2.6.2 Japanese shakers

A recurring trigger for religious tolerance in our course came from bringing together classic Japanese values with the three monotheistic religions based in Jerusalem. As all three (Islam, Judaism and Christianity) demand a single devotion to its own core value system, the animistic, flexible, and liberal approach presented through Japanese artefacts confronted Israeli students with acute questions. An interesting cultural 'mirror' presented in various projects, then, conducts a 'face-off' between two seemingly similar cultures, yet upon a deeper examination, broadens the student's empathy and understanding towards differing value systems.

In the first project, designed by Gleb Tukschok, we identify several key points echoing two distinct cultures: first, the strict relying on what would later be termed by the Bauhaus school as "truth to materials" (Pevsner, 2005), i.e., setting up materials on the front stage without interference such as patina or gloss. Second, while different in philosophy, both the Shakers and Japanese traditional aesthetics, share a principle of simplicity and austerity. A classic semiotic approach, therefore, would have focused on the aesthetic similarities, yet these stem from completely different value systems. While classic Japanese design's austerity generates from Zen Buddhism as a philosophical way of life, the Shakers' origin of approach lies in a strict Protestantism calling for hard work, asceticism, efficiency, and modesty before God. In other words, while the first resolves around the tension between the individual and their community, the second depicts a tension between the individual and their God.

While Shaker in design, we can clearly see the Japanese philosophy in this series (Figures 2.6.1 and 2.6.2). The water container (bellow in the middle) seems roughly hewn, an almost natural piece of wood, yet the bottom of the bowl shows a true craftsmanship. Following a Zen approach, this bowl looks as effortlessly un-designed as possible, leaving it as natural as nature intended. The same is true concerning the whisk (*chasen*), in which each filament is made of piece of wood polished and shaped to perfection, bolstering the same effortless design. Again, while the base of the object is made of maple, the pale wood echoing Japanese design, the filaments are made of American walnut, an echo to Shaker design. The teaspoon again resembles a natural twig, picked from the nearest forest, yet a closer look reveals the Shaker precise functional and ergonomic contours of a carefully

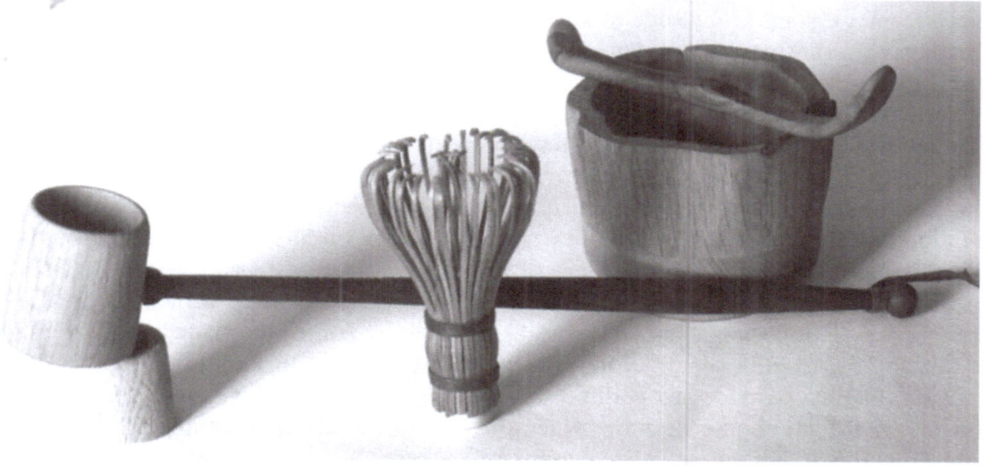

Figure 2.6.1 Japanese Shaker series, utensil for pouring hot water (right), wooden chasen (centre), and hisaku (left).

Figure 2.6.2 Japanes Shaker series (details).

designed object. Instead of the classic bamboo, this unique spoon is made of a combination of oak (classic in Shaker design), Indian rosewood and touches of wenge. The array of wood, added to the technique of stack lamination, enabled for a feel of *wabi sabi* we can clearly see in the almost natural and twig-like design of the spoon Finally, the *hishaku* seems like any other, yet a close look at the handle reveals the classic design of a Shaker ladder-back chair, the fine mark of a professional cabinetmaker. While the lower part of the vessel echoes the mastery of Shaker artisanship, the upper part resembles the un-perfect pottery of Japanese craft. The essence of empathy and cultural openness in this project derives from the very way wood is perceived in both cultures. The nuance of woodwork emphasizes the similarities and differences between the two cultures, while keeping a distance, thus allowing for a reinterpretation rather than a redesign or a simply act of copying another cultural style.

2.6.3 African–Shaker Chairs

The notion of bridging two differing cultures through an act of design is even more acutely pronounced when juxtaposing African cultures and the Shakers. When comparing two cultures, or to be more exact – three (the designers' Israeli identity as well), shifting one's stance from a distant knowledge to a deeper understanding, brings forth empathy almost effortlessly. In this project, two chairs' configuration combined with a cultural philosophy regarding the place of a person in the greater social structure, demonstrates an interesting mix between African and Shaker aesthetics. Naturally, questions regarding our place as individuals in a broader societal context have been the backbone of sociological theories and social psychology as disciplines. However, understanding these dilemmas in other cultural contexts imbues young designers with empathy and broadens their professional horizons.

Famously, the Shakers believed in a combination of hard work, asceticism, and scripture-based faith. Therefore, their furniture mirrors this central belief system, favouring functionality and frugality over lavishness or indeed what Veblen (2007 [1899]) termed 'conspicuous consumption'. Among their many contributions to modern design, we can find the reclining chair, along with the ladder-back chair and more (Figure 2.6.3). Since cleanliness was close to godliness according to Shaker belief, furniture had to be easy to clean, light enough to move when dusting, not to gather dust and of course – be highly functional (Kassay, 1980; Andrews & Andrews, 1964).

Figure 2.6.3 Shaker reclining African chair (notice the figurheads on the ladder back).

Two of the most famous Shaker chairs are the ladder-back chair and the rocking chair. Aside from the ladder-back functionality to hang the chair on a peg while cleaning the floor, this frugal design is also economic in wood use and stronger. The rocking chair, in turn, was used for knitting, as well as for a hard-earned rest from the day's hardships, preparing one to the next day's challenges. Among else, these two iconic chairs influenced various leading designers, especially so in Scandinavia, as would attest the famous works of Danish designer Hans Wenger (Becksvoort & Sheldon, 2000).

Usually, these chairs were made from strong and cheap wood, such as oak and ash, while the seating was made of either wood or cloth weaving. As each Shaker furniture was a materialization of a specific function, each was built according to specific craft knowledge and logic. As the rocker was to be moved around the living room and lifted when cleaning, it had to be light. Similarly, the Shaker-milking stool had to be small enough to fit next to the cow, as well as strong enough to support the weight of the person sitting on it. Therefore, the Shaker stool was made of a single piece of wood, moulded into a specific ergonomic shape according to the intended milking person, and fitted with three legs for easy use (Fine Woodworking, 2001).

Following the strict and functional philosophy of Shakers design yet wanting to add a twist to these timeless pieces, the designer Roy Aryeh made a detour to Western Africa and Japan. The Malawi chair is well-known for its simplicity, craftsmanship, and straightforwardness. This chair is built from two crafted pieces of wood, joined together by the angle of each facet. Usually, each African chair is adorned with a personal addition – a depiction of a bird, an animal, a person, or favoured possession of its owner (Sieber & Rubin, 1970). Another influential motif in these two chairs is the mask tradition of the Fang tribe of Gabon, usually used for tribal rituals. These masks are a testimony of classic Western African minimalism and figurative art. Lengthy in proportion, heart-shaped facial features, along narrow eyes, and lengthy nose.

The last contribution to these two chairs is the Japanese bath stool (*koshikake*). This Japanese stool is traditionally made of cypress and measures around 30 cm, upon which people soap and rinse themselves prior to entering the tub (Koizumi, 1986). This simple yet sturdy stool is made of three pieces of wood joined together by a single peg.

The result of these three cultures of wood is an African–Shaker rocker and a Japanese Shaker milking stool. The rocker is made of maple and a crisscross cloth for the seating (Figure 2.6.4). This unique configuration is based on the basic configuration of the Shaker rocker with a touch of the Malawi chair, as well as two protective figurines on top of the ladder-back. The chair's strength is a result of an elegant use of Japanese joinery. In turn, the milking stool is based on the classic Japanese bath stool yet is made from red eucalyptus by derogation carving. Contrary to classic Japanese design, the single joint is made in a somewhat exaggerated manner, to symbolize the user's importance, according to Western African tradition. Surprisingly, these three cultures are materialized in these two chairs in such a manner that seems almost natural. Indeed, this seemingly natural ease is a key pedagogic strategy attesting to the students' application of empathy towards differing cultures in their practice.

Another interesting fusion between the Shakers cabinetmaking and African woodwork can be seen in a chair designed by Chen Adler. The chair, titled "Women Rock" imbues the configuration of a classic Shaker rocking chair with cultural traits of the Ethiopian Karo women. The chair incorporates two models of powerful femininity as derived from both cultures. Austere and ascetic in nature, the Shaker religion hold women in high regard, as Ann Lee, founder of the Shakers, served as the head of the congregation, as

Figure 2.6.4 African–Shaker milking stool (left) and Adler's Women Rock (right).

well as other women who served in various high-ranking positions along the years. In fact, celibacy freed women from raising children and added to the existing equality among the Shakers. While the congregation would not be considered feminist in contemporary terms, its tendency towards gender equality far succeeded contemporary societies (Andrews, 2012; Foster, 1991).

Adler's chair is based on the Karo femininity model of scarification, the use of red ochre, and beads. Adding these cultural features to the classic Shaker rocker created a surprising result. The Afro-Shaker rocker was planned in such a manner to clearly highlight two classic chair configurations: the shaker rocker and the African chief's chair. Using a mixture of maple and American walnut allowed for an immediate recognition of each configuration. The chief's chair bolsters the classic African slant of the backseat, as well as 'scarification' marks on the seat, symbolizing the chief's power. In such a manner, the elegance and austerity of the Shaker rocker are highlighted by the presence of the innate power of the chief's chair.

The essence of empathy, and indeed the holistic nature of design anthropology, is manifested in these chairs in several manners. First, as mentioned, the notion of comparing the three cultures, as well as the designer's own Israeli identity accentuates their equal importance. Second, using these cultures and wood as a cultural 'anchor' allows the designers to tackle broader and thornier issues, such as gender, family structure, and an individual's role in society.

2.6.4 Kokeshi as a platform for tolerance

> 'Beauty' is a currency system like the gold standard. Like any economy, it is determined by politics [...] In assigning value to women in a vertical hierarchy according to a culturally imposed physical standard, it is an expression of power relations in which women must unnaturally compete for resources that men have appropriated for

themselves. 'Beauty' is not universal or changeless, though the West pretends that all ideals of female beauty stem from one Platonic Ideal Woman.

(Wolf, 2002: 12)

Indeed, as part of our broader approach to use various practice-based methodologies to approach socio-cultural subjects through design, the fluid perception of beauty and the importance of gender in various cultures is paramount. In her classic book *The Beauty Myth* (2002), Naomi Wolf highlights the culturally oriented definition of feminine beauty in Western society. The culturally embedded norm of feminine beauty, accordingly, is a result of patriarchal conventions embedded in Western society, culminating in contemporary consumer culture and popular media. Famously, Wolf stresses that beauty is viewed as an objective goal women must embody, and men must possess these model women. Griselda Pollock demonstrated the depiction of femininity and the role of women as objects created for the enjoyment of spectators (men of course) in art history (Pollock, 1988; Parker & Pollock, 2013). Indeed, a useful starting point to imbue students with empathy towards other cultures lies in the comparison between various notions of beauty (Figure 2.6.5). Is the lip-disc worn by the women of the Mursi tribe in Ethiopia any different from silicone enhancements or Botox shots in Western society?

Of course, delving into the various aspects of embodiment and theoretical attributes of power and the body are beyond the capacities of this chapter,[1] yet, we will shed light on several issues regarding the body and femininity, nonetheless. Consumer culture turned the feminine body into an unnatural and technologically modified type unreachable and separated from the 'normal' body. The image of the young, slim, muscular body consisting of 0% fat became the globally aspired norm, depicting large breasts, a slim waistline, and flat muscular stomach (Grogan, 2007).

A central materialization of these feminine bodily perceptions is seen in dolls, and especially so in the case of Barbie. Following various researchers, we can clearly see that Barbie's body is unrealistic, unnatural, and unattainable, yet, it is the best-selling doll, influencing body images of men, women, and children alike. As is the case with many other toys and especially with dolls, children report dissatisfaction with their body image vis-a-vis

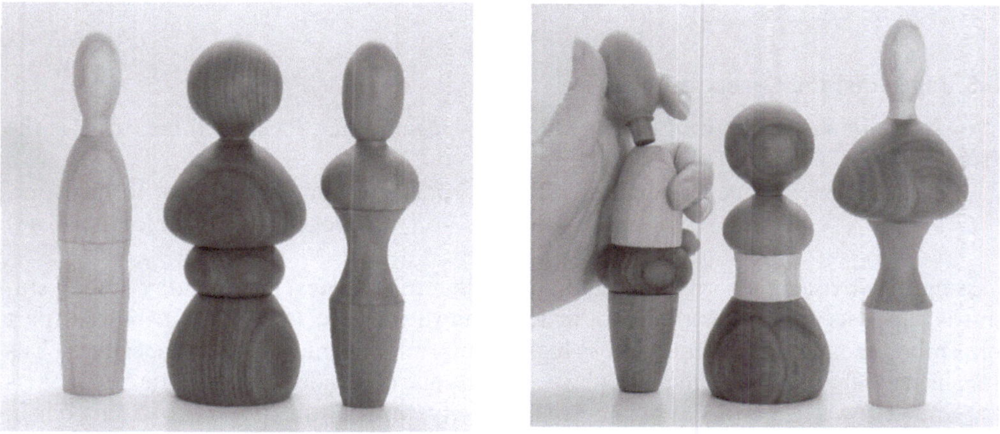

Figure 2.6.5 The three models of beauty: Japanese (left), African (centre) and Western (right); on the right - Mixing the three models of beauty as a game to eradicate the inherent beaty types.

their portrayed depictions (Worobey & Worobey, 2014). Mattel, building and manufacturing Barbie, sent several child-models across the US calling to 'get in shape' as Barbie. Furthermore, one of the most important features of Barbie are her accessories, ranging from makeup to hairdo's to bags and clothes, broadcasting a distinct and exact femininity and sexuality (O'Sickey, 1994). The result is a Western depiction of feminine beauty, specifically distributed in various media (Gimlin, 2002), spreading her plasticity and global appeal, a manifestation of an almost posthuman technological entity (Toffoletti, 2007).

In this project, designed by Hila Raam, we can see a materialization of three models of femininity, based on a more flexible definition of the term. The empathy in this project stems from the understanding of the Westernized patriarchal view of the feminine body, as well as questioning what are body image norms and what should be designers' contribution to this important question. The famous commodified Western depiction of feminine beauty, then, is the classic 60-90-60 figure, depicted, among else, in the Barbie doll. Typically, the ideal feminine beauty in Japan calls for long, shiny black hair, white skin, and slender body (Spielvogel, 2003).[2] Alternately, the classic African feminine model, as seen in various art pieces and archaeological figurines, boasts large breasts and bottom, echoing fertility and health (Lifshitz, 2009; Shakarov & Senatorova, 2015). In this project, the platform of *kokeshi* dolls is used to illustrate and question inherent cultural depictions of feminine beauty. As in other cultural platforms, this starting point allows us to 'clothe' this platform in various options, thus allowing comparison and criticism.

Following the classic craft of the *kokeshi*, each doll is made of four parts (a head, torso, waist, and legs) along the doll's traditional size. It typically depicts classic Japanese beauty: long and slender features, flat chest, and pale skin, and made of maple. The maple adds not only the paleness but also a general flawlessness to its grain, further accentuating the pursuit of flawless skin in Japanese tradition. The African doll bolsters generous breasts, bottom, and waist. This *kokeshi* is made of walnut, a darker tree rich with grain features and natural ornamentation, alluding to tribal painting and scarification. The Western doll depicts a muscular, statuesque, and somewhat unnatural physique. The slender waist, large breasts, smaller head, and muscular legs are further accentuated by the unnatural tan of the reddish eucalyptus. The playful element of mixing the three dolls alludes to the arbitrariness of the judgment of beauty and our own definition of beauty vis-a-vis other cultures.

2.6.5 Japan and the essence of tolerance

> "I can see you are a man of the way (Dao)", Dōgen said. "Please tell me, what is the true Way?"
>
> "The universe has never concealed it", the cook said and turned back to his work.
>
> (Dōgen in McDaniel, 2014: 48)

Using the *kokeshi* platform as a material configuration to develop empathy among students is highly useful since it allows for articulating various socio-cultural attributes. Apart from embodiment, religious tolerance is highly important in a multi-cultural society such as Israel in general and Jerusalem particularly. In fact, one of Japan's most unique attributes is its tolerance towards non-local religions. Conversely, three of the most central religions in Japan are not indigenous: according to US Department of State, 51% of Japanese identified with the Shinto belief, 34% are Buddhists and a meagre 2% are Christians.[3] However, contrary to other cultures, in Japan, a person can believe in two or more sets of faiths, unlike

Judaism in which the faithful must be solely Jewish. Religious tolerance and freedom of belief is rooted in the essence of the state taking precedence over all else (Isomae, 2014) In that matter, it is understandable that Christianity did not take root in Japan due to its exclusivity ethos. Indeed, while during the 17th century Christianity was banned in Japan, prior to the Meiji reforms, it enjoyed a new-found tolerance (Nosco, 1996).

The history of these four religions and their impact on Japan reflects the history of Japanese culture. Indigenous in nature, Shinto ("way of the gods") is one of the two central religions in Japan, bolstering around 100,000 shrines and 20,000 priests recognized by their unique attire. Earliest records of Shinto date back to the 6th century and survived as a socio-cultural and historical religion till today. The above-mentioned religious flexibility is seen in the fact the while Shinto is the largest religion in Japan, supported by around 100 million citizens, only a small percentage of the populace identifies themselves as Shintoists. Conversely, it would be wrong to define Shinto as a classic religion, but this unique belief-system is embedded in history and practice alike, as well as daily life and identity formation. These include the shrine rituals (since *kami* reside in shrines), as well as holidays and festivals (Breen and Teeuwen, 2010).

Unlike Shinto, Daoism and Buddhism are not indigenous to Japan. Dao was brought to Japan from China, as a philosophic and religious way to reach the Way. This Chinese belief system was developed between the 2nd and 6th centuries and revolved around the writings of Lao Tzu (especially the famous *Tao Te Ching*). While contemporary Daoism consists of an almost endless variation of approaches, the central ethos revolves around reaching an understanding through oneself (Kirkland, 2004). Daoism is hard to define since it was never united under one rule. In addition, Chinese religion is a mixture of the three teachings (Buddhism, Confucianism and Daoism). Historically, Daoism and Confucianism were related, yet in contemporary Japan few will identify themselves as Daoist. The roots of Dao go back as far as the 4th–2nd centuries BCE (Oldstone-Moore, 2003). Buddhism was introduced to Japan much later. The first official record of Korean king Song Myong sending an image of the Buddha along with several sutras date to 538, yet researchers claim Buddhism reached Japan unofficially much earlier. Prince Shotoku studied Buddhism, erected shrines and is considered the father of Japanese Buddhism (Matsuo, 2007).

The basic configuration of all the figurines, designed by Harel Oberman (Figure 2.6.6), follows the dimension and proportions of the classic *kokeshi* doll. Their head is similar

Figure 2.6.6 Religious pluralism kokeshi: Shinto (right-to-left), Christianity, Buddhism and Dao; African kokeshi - Karo (left), Surma (centre), and Maasai (right).

and made of cherry, while the bodies vary according to the principal Japanese religions. The Shinto *kokeshi* is made of maple and follows the abstract shape of the Shinto priest' robes. The maple echoes the white fabric. The Catholic *kokeshi* is identified by the distinguished white collar and black clothes, made of wenge timber. The Buddhist *kokeshi* depicts the a-symmetric orange robes, leaving the right shoulder bare. The doll is made of a combination of maple and red eucalyptus. Finally, the Dao *kokeshi* echoes the ornate and flowing robes of the Daoist monks, made of eucalyptus and cherry. While very un-Israeli in its beginning, this project led our students to reflect upon the need of religious empathy in a multi-cultural city such as Jerusalem.

The material phenomenon of the *kokeshi* doll has taken a turn from its traditional depiction into the current style described as *kawaii* ("cute"). This new style is manifested in the use of pink colouring and limbless figures. As we have seen previously, the overall length of the doll is around 20 cm and it is made of two distinct parts, a torso, and a head, joined by a process of heat and pressure. Today *kawaii* refers to an amalgam of concepts such as beautiful, loveable, cool, funny, quirky, and more. The rise of Japanese culture of cute in 1980s meant that girls did not only want adorable accessories, but to become *kawaii* themselves. Several key design features turn something into kawaii: pastel colours, big round eyes, a large head, and general roundish features (Okazaki & Johnson, 2013). Indeed, this cuteness serves as a sort of a 'poisoned apple',[4] depicting a sugary coating hiding explosive topics ranging from social and economic inequality to gender bias, religious intolerance, and territorial politics. Through these seemingly alien cultures, the students reflected and confronted their own agendas and belief systems.

Cuteness is not automatically linked to aesthetics, since, as Harris (2001) put it, it is closely linked to the grotesque and the malformed. Yet, marketing agents link more and more products with cuteness, emanating images of helplessness and happiness. In an almost sadistic manner, the cute is a result of an awkward situation in which a blundering and helpless figure finds itself. Furthermore, kawaii gives people the chance to hold on to childhood and postpone the pressures of adulthood (Avella, 2004). While either the traditional *kokeshi* is usually lightly adorned or not at all, the creative *kokeshi* is a direct result of the culture of *kawaii*, identified by brightly coloured graphics and a shiny patina. In this project, designer Roni-Lee Berri chose to turn to three major African tribes (Maasai, Karo, and Surma) to recreate the traditional *kokeshi*'s minimalism via African symbolism.

Their unique red cape (*dagang*) easily recognizes the Maasai, an indigenous semi-nomadic tribe residing mainly in Kenya and Northern Tanzania. One of the most recognized rituals of the Maasai is the 'jumping ceremony', in which young men demonstrate their prowess, strength and agility through a series of jumps. Unlike many other tribes, women have a strong voice in their community. Their shaved heads, bright colourful clothing, beads, and the removal of one of the lower teeth identify them. Women also stretch their earlobe, in which a piece of jewellery or a bead is inserted to keep the gap open (Gupta, 2005). The unique red robe, worn by men and women alike, has become a material and visual sign of the tribe (Klumpp, 1981).

The Ethiopian Karo is a small tribe residing on the banks of the river Omo. Scars adorning the chest identify the women of the tribe, symbolizing their strength and endurance to bear and grow children and maintain the household. The unique hairstyle of the women resembles a helmet. Men also scar themselves, usually symbolizing a victory or a successful hunt, by cutting their flesh and rubbing the wound with ash (DeMello, 2014).

The Surma tribe, also residing in Ethiopia, was once nomadic but are now famers. The men of the tribe are famous for their stick ritual in which young unmarried men face each other with long stick (*donga* sticks). The winner, knocking down his opponents, is carried on a wooden platform to a group of young women, one of which will choose him as her husband. The women of the tribe are known for their lip discs. Six months prior to the young woman's marriage, her mother gives her a small ceramic disc that will keep changing in size, gradually broadening, and enlarging the opening in the girl's lower lip. The longer the disc, the larger the dowry the young bride will receive. Once more, the capacity to endure pain, as well as the patience and commitment shown through the lip disc are socio-cultural standards of femininity and worth (Russell, 2010). The Surma are also known as masters of body paint, using their whole bodies as canvas. Every morning during the courtship season, the men, women, and children of the Surma go to the riverbank and gather white chalk. Mixed with water the chalk becomes white paint used to paint their bodies (Beckwith, 2012).

The result of this unique fusion between the platform of the *kokeshi* and the three African tribes as cultural context is extremely interesting. While keeping the proportion of traditional *kokeshi*, the graphics of each doll was excavated from one of the three tribes. The main part of each doll was made of maple, keeping the whitish color of the traditional *kokeshi*. Additional choice of wood was made according to the visual attributes of each tribe.

The Maasai *kokeshi* is recognizable by the lengthy earlobes and the typical red robe, both of which are made of the reddish Bubinga wood. The Karo *kokeshi* is recognizable by the women's unique hairstyle, resembling a reddish helmet, along a V-shaped pattern symbolizing the tribe's scarring rituals. Both the hair, as well as the scar pattern are made of light Wenge wood. Finally, the Surma *kokeshi* is recognizable by the lower lip disc, along a unique tunic worn on one shoulder, leaving a single breast bare. Both lip disc as well as the Surma tunic are made of dark Wenge wood. As we have seen, building upon seemingly alien cultures, through a material common platform (the *kokeshi* dolls) allows students to confront and challenge their own beliefs and convention, triggering a deeper sense of professional empathy. These empathy-driven narratives are found not only through cultural objects, bus also through common myths and legends, as we shall see next.

2.6.6 Red riding hood netsuke

Once upon a time there was a little girl who was so sweet and kind that everyone loved her. Her grandmother, who loved her more than anyone, gave her a little cap made of red velvet, which suited her so well that she wanted to wear it all the time. Because of that, everyone took to calling her red riding hood (Pullman, 2012:137).

As is the case with visual or material culture, folk tales are a vessel to identify and distribute global and local norms, values, and identities. Anthropologists were adamant in their understanding of the correlation between myths, tales and socio-cultural systems (Lévi-Strauss, 1955; Bidney, 1953; Leach, 2013, among many others). The various types of myths, tales, legends, and other culturally imbued narratives surpass the modest scope of this volume. However, suffice to say that narratives in their many types, highlighting or neglecting a moral punch, are a globally accepted form of narration (Eller, 2014).

Famously, folk stories and myths mirror the social group's needs, fears and moral constraints, much in the Freudian tradition, which explains the number of dreams and surreal parts in various famous folk stories found in cultures around the world (Ashliman, 2004). In other words, coined by anthropologist Edward Sapir, our socio-cultural raison d'être is mirrored in the structure and sounds of our language (Mannheim, 2015).

The centrality of the Red Riding Hood tale can be seen in its varying versions found in Korea and Japan (Dundes, 1989).

The Aarne-Thompson Index is a well-known classification system for folktales was developed in 1910 by Finnish Antti Aarne and was translated and enlarged by American Stith Thompson in 1928. In this system, tales are classified according to themes, heroes, and other variants, in a manner enabling researchers to compare various folktales in different cultures. The categories include animal tales, ordinary folk tales, jokes and anecdotes, formula tales and unspecified tales (Zipes, 2000). The story of Little Red Riding Hood, numbered 333 in the Aarne-Thompson index, holds global attributes of good and evil, nature and culture, youth and adulthood, wisdom, and naiveté and more (Ashliman, 1996). It was through this fable that the designer of these specialized netsuke attempted to rephrase the role of folktales in contemporary lives and their points of correlation between three different cultures – Africa, the US, and Japan (Figure 2.6.7).

The first literary version of Little Red Riding Hood was written in 1697 by Frenchman Charles Perrault and titled 'Le Petit Chaperon Rouge'. Perrault probably knew an older oral version of the tail in which an unnamed peasant girl meets a werewolf:

Figure 2.6.7 Red riding hood netsuke, Japanese (upper row), Western (middle row), and African (bottom row).

The wolf asks her whether she is taking the path of pins or needles. She indicates that she is on her way to becoming a seamstress by taking the path of the needles. The werewolf quickly departs and arrives at the grandmother's house, where he devours the old lady and places some of her flesh in a bowl and some of her blood in a bottle. After the peasant girl arrives, the werewolf invites her to eat some meat and drink some wine before getting into bed with him. Once in bed, she asks several questions until the werewolf is about to eat her. At this point, she insists that she must go outside to relieve herself. The werewolf ties a rope around her leg and sends her through a window. In the garden, the girl unties the rope and wraps it around a fruit tree. Then she escapes and leaves the werewolf holding the rope.

(Zipes, 2000: 301–2)

Contrary to the original version, in which the girl fends for herself, Perrault turned her into a spoiled, naïve, and helpless peasant wearing a red cape, symbolizing her sinful nature. After waging a bet with the wolf, she loses both the bet and her grandmother. Upon arriving at the grandmother's house, she climbs into bet with the wolf and after asking him several questions is devoured in much the same manner as her grandmother. The morale of the tale is simple, dictating the fate of immoral girls, showing interest in unknown men. Only in 1812, the brothers Grimm added to their version the *Jäger* (gamekeeper) who saves both the girl and her grandmother (Zipes, 2000; Haase, 2007). Following Tehrani (2013), the echoes of this classic tale in other cultures brought to fruition the materialization of three sets of *netsuke* depicting this epic tale of dichotomous forces embedded in a simple depiction of good and evil.

The series, designed by Gal Berlinka, include nine *netsuke* figurines, three from each culture, representing a good character, a bad character, and a neutral one. While the role of the good and bad characters is easily understood as a depiction of social norms, the neutral character's role in to create a certain ambiguity necessary to further highlight the differentiation between good and evil. In structuralism, a mediator whose role is to highlight the differences, as well as the relation between the two binds the two binary oppositions (Mascia-Lees, 2010).

In the Western tradition, the good character, symbolizing morality is the huntsman, re-enacting order in the chaotic situation caused by evil. The bad character, symbolizing nature, chaos, ferocity, baser instincts (following the classic structuralist anthropology), is obviously the wolf. The neutral character, symbolizing naiveté, childhood, John Locke's *tabula rasa* and the amenability of youth, is obviously Little Red Riding Hood. We can clearly see these three netsuke crafted from a combination of maple, Wenge, and red Eucalyptus via stack lamination.

In the African (Yoruba) tradition, the good character, symbolizing order and safety, is Zangbeto, the keeper of the forests, fighting witches, and thieves. The bad character, symbolizing nature, chaos, ferocity, baser instincts, is Anansi, the mischievous spider. Unlike other trickster gods, Anansi serves as mediator between his father, the sky God Nyame and the human realm. The neutral character symbolizing naiveté, neutrality, helplessness facing nature, and an innate ability to get in trouble is man (Haase, 2007; Abiodun, 2014). The three Yoruba *netsuke* are made of Wenge wood and crafted via a classic African technique of derogation.

In the Japanese (Shinto in this case) tradition, the good character, the discoverer of the Japanese archipelago, the chief of earthly kami, *Sarutahiko-Okami*, usually depicted sporting a long black beard. The bad character, symbolizing fear, loneliness, ghostly death, and the unknown is *Yurei*, usually depicted as a pale woman with long black hair. The neutral character, symbolizing mediation and intelligence and bewitchment, is the likeable messenger of the *kami*, the fox spirit of *Kitsune* (Williams, 2009; Picken, 2010; Davisson, 2015). The three Japanese *netsuke* are crafted from a combination of Wenge and maple via joinery.

The depiction of good, evil, and neutrality differ in various cultures, leading to differing style and techniques. Therefore, empathy derives not only from variations in cultural values, body imagery or religious beliefs, but of the very definitions of good and bad as well. Through learning about the Other, as well as understanding and accepting differences, designers hone their craft in a positive direction to include and cater for excluded communities and reflect about the very essence of industrial design – the standard.

2.6.7 Of culture and empathy

> The word material is the result of the Romans' attempt to translate the Greek term hyle into Latin. Hyle originally meant 'wood'.
>
> (Flusser, 2013: 22)

As we have seen, our main pedagogic approach lies in two parallel layers. First, we wish to present the students with the importance of theoretical and methodological understandings of different cultures, not only as an academic necessity, but as a key factor to trigger practice-based innovation. By starting a project with a deeper socio-cultural understanding of a situation, scenario or concept, the designer opens an almost endless array of possibilities. Furthermore, through this research one identifying with and understanding the other, thus triggering empathy through an unexpected path.

Going back to Henry Dreyfuss' influential work *Designing for People* (2003 [1955]), brings to our attention that while function, assembly and aesthetics are important, the deeper meaning of an object is as equally important. Flusser, in his mischievous fashion, calls to attention the linguistic-semiotic relation between the essence of craft and wood as a cultural matter. The final phrase of the above quote illustrates the importance and deep relation between design, materials, and culture, as we have seen throughout this article. Wood, culture (philosophy, religion, norms, rituals and objects and technology, as we have seen) and design have become so integrated in the works presented here they have become almost synonyms. In this last section let us frame these works in contemporary design issues redefining the designer and the design process.

The Bauhaus and its offspring offered a universal mode of design. Integrating new technologies, modes of production and a global semiotic material and visual language, their legacy transcends time. The late 20th century and the early years of the 21st century brought forth global design characterized by the strength of brand identity, advanced ergonomics and the rise of the software (could-computing, autonomous cars, apps, etc.). While German designers redefine Korean car brands, Japanese designers' rule over the Italian Mecca of design, and we all use the same apps, view the same streams of data and buy the same designed goods, global design is here to stay. However, facing political turmoil, amidst eco-social movements, anti-branding and economic crises, the importance of the local has never been more acute. Rather than turning to the not-so-fresh concept of the 'glocal', we wish to suggest a different route, turning to two well-known concepts – *design culture* and *craft*.

As the number of cultures far surpasses the number of states throughout the world, the importance of integrating cultural aspects to the design process is crucial, more so with current immigrant/refugee streams across Europe. However, as we have seen in this article, the issue is not to imitate or copy, but rather understand and create a different interpretation, or rather of fusion of another culture's principles and one's own. Throughout these pages, we followed the footsteps of design culture researchers such as Guy Julier (2013) and Glenn Adamson (2013), highlighting the cultural aspects of design and the interdisciplinary aspects of the discipline. However, our focus and scope are different. Through understanding different cultures, under the material umbrella of design, we define design anthropology not just as a theoretical or applied addition to design practice, but as an inherent addition, one that should become a crucial part of the practice. Penny Sparke (2013) discussed at length the intricacies of the cultural and historical elements of the evolution of design, bringing forth a culture of consumption and rise of the era of design. Julier (2013) describes 'design culture' as something designers *do*, but also as part of the flows of global culture, embedded in communication, media, and social norms. As Julier rightfully explains, design culture deals not only with ways products are produced and planned, but also used, lived, perceived, and understood in our daily surroundings. Julier describes design culture as consisting of several layers: process, context-informed practice, organizational or attitudinal, as agency (the ability to generate social change) and finally, as pervasive but differentiated value. As a result, Julier (2006) calls for a shift of attention from 'visual culture' to 'the culture of design', focusing on the broader issues and layers of design as a socio-cultural agent. Indeed, our deeper pedagogic raison d'être lies in the notion that empathy in relation to design practice resolves not through the obvious and 'easy' choices, such as design for inclusivity or design and disability but rather through the unobtrusive title of 'design and culture'. Through a deeper understanding on another's culture, the designer questions their own belief system, norms, and conventions.

In this chapter, we highlighted a different approach towards design that in many ways contradicts many of the design pedagogical approaches used by designers. Rather than start by defining the design brief according to a problem, functional issues, or target population, we started by defining a cultural, social, ritualistic, religious, or philosophical attribute of a specific social group. We then move to redefining this abstract feature vis-a-vis visual and material features and finally proceed to defining an object and manufacturing technologies. In a way, our design process went as follows: by highlighting cultural attributes, we believe designers will take their place as Renaissance humanistic agents, materializing socio-cultural concepts and influencing society. In doing so, using wood as a major material, the idea of 'makers' will be imbued with an extra layer of relevance and meaning. While the term 'craft' suffers from a multitude of trendy new names (neo-craft, new-craft, post-craft, etc.), Glenn Adamson (2007; 2013) calls for integrating the term 'makers'. As we can see throughout this chapter, craft should not be reduced to material folk product or historic leftovers, but rather a mode of contemporary innovation. However, we believe that this term far transcends its William Morris-imbued meaning to conclude contemporary designers integrating new (3D printing for instance) and traditional (cabinetmaking) technological ways of production. As the production and use of materials derive from socio-cultural traits, the designer focuses on the design situation rather than just the product, rendering the designer's work to a much broader vista of a cultural interpreter.

As we believe design anthropology is crucial for young designers, as an innovative creative tool, this article helps redefine the possibilities of this sub-discipline not only as a supplier of theoretical or methodological venues, but also as a practice-based trigger of innovation and interpretive empathy towards the other. Understanding and identifying with others will not only help us better understand our place in the world but more importantly, redefine the very role of designers in their communities, as well as their impact in society.

Notes

1 A good starting point, among many, would be the volume *Embodies Practices* (1997) edited by Kathy Davis.
2 Famously, in medieval Japan, feminine beauty also included shaved and painted eyebrows and blackened teeth, along whitened face, and small, red-painted lips.
3 http://www.state.gov/j/drl/rls/irf/2006/71342.htm
4 By that we are referring to Bourdieu's famous theory of mimicking a ruling cultural field to shift the power balance between reform and orthodox powers (see, Bourdieu, 1984; 1993).

References

Abiodun, R. (2014). *Yoruba art and language: Seeking the African in African art*. Cambridge: Cambridge University Press.

Adamson, G. (2007). *Thinking through craft*. Oxford: Berg.

Adamson, G. (2013). *The invention of craft*. London: Bloomsbury Academic/V&A Publishing.

Andrews, E. D., & Andrews, F. (1964). *Shaker furniture: The craftsmanship of an American communal sect*. Mineola NY: Dover Publications.

Andrews, E. D. (2012). *The people called shakers*. Mineola, NY: Courier Corporation.

Ashliman, D. L. (ed.). (1996). *Folklore and mythology*. Pittsburgh: University of Pittsburgh Press.

Ashliman, D. L. (2004). *Folk and fairy tales: A handbook*. Westport Conn: Greenwood Publishing Group.

Avella, N. (2004). *Graphic Japan: From Woodblock and Zen to Manga and Kawaii*. Hove: RotoVision.

Becksvoort, C., & Sheldon, J. (2000). *The Shaker legacy: Perspectives on an enduring furniture style*. Newton CT: Taunton Press.

Beckwith, C. (2012). *Painted bodies: African body painting, tattoos, and scarification*. New York: Rizzoli International Publications

Bidney, D. (1953). *Theoretical anthropology*. New York: Columbia University Press.

Bourdieu, P. (1984). *Distinction: A social critique of the judgement of taste*. Cambridge MA: Harvard University Press.

Bourdieu, P. (1993). *The field of cultural production: Essays on art and literature*. Cambridge: Polity.

Breen, J., & Teeuwen, M. (2010). *A new history of Shinto*. Oxford: John Wiley & Sons.

Brown, T. (2009). *Change by design: How design thinking transforms organizations and inspires innovation*. New York: Harper Collins.

Buchanan, R. (1992). Wicked problems in design thinking. *Design Issues*, *8*(2), 5–21.

Davisson, Z. (2015). *Yurei: The Japanese ghost*. Seattle WA: Chin Music Press Inc.

DeMello, M. (2014). *Inked: Tattoos and body art around the world*. Santa Barbara CA: Greenwood Press.

Dreyfuss, H. (2003 [1955]). *Designing for people*. New York: Skyhorse Publishing Inc.

Dundes, A. (1989). Interpreting "Little Red Riding Hood" psychoanalytically. In: A. Dundes (ed.), *Little Red Riding Hood: A casebook*, pp. 192–238. Madison WI: University of Wisconsin Press.

Eller, J. D. (2014). *Introducing anthropology of religion: Culture to the ultimate*. Abingdon: Routledge.

Fine Woodworking. (2001). *In the Shaker style: Building furniture inspired by the Shaker tradition*. Newton CT: The Taunton Press.

Flusser, V. (2013). *Shape of things: A philosophy of design*. New York: Reaktion Books.

Foster, L. (1991). *Women, family, and Utopia: Communal experiments of the shakers, the Oneida Community, and the Mormons*. Syracuse, NY: Syracuse University Press.

Geertz, C. (1973). *The interpretation of cultures*. Basic Books.

Geertz, C. (2012). *Available light: Anthropolcgical reflections on philosophical topics*. Princeton NJ: Princeton University Press.

Gimlin, D. L. (2002). *Body work: Beauty and self-image in American culture*. Berkley: University of California Press.

Grogan, S. (2007). *Body image: Understanding body dissatisfaction in men, women and children*. Abingdon: Routledge

Gunn, W., & Donovan, J. (2012). Design anthropology: An introduction. In W. Gunn, Donovan, J. (eds.), *Design and anthropology*, pp. 1–16. Farnham: Ashgate.

Gunn, W. O. T., & Charlotte Smith, R. (2013). Preface. In W. Gunn, T. Otto & R. C. Smith (eds.), *Design anthropology: Theory and practice*, pp. xiii–xiv. London and New York: Taylor & Francis.

Gupta, A. (2005). *Human rights of indigenous peoples: Protecting the rights of indigenous peoples* (Vol. 1). New Delhi: Gyan Publishing House

Haase, D. (2007). *The Greenwood encyclopedia of folktales and fairy tales*. Westport Conn: Greenwood Publishing Group.

Harris, D. (2001). *Cute, quaint, hungry and romantic: The aesthetics of consumerism*. Cambridge MA: Da Capo Press.

Isomae, J. I. (2014). *Religious discourse in modern Japan: Religion, state, and Shintō*. Boston, MA: Brill.

James, P. (2017). Culture. In K. A. Reinert (ed.), *Handbook of globalization and development*, pp. 209–225. Edward Elgar.

Julier, G. (2006). From visual culture to design culture. *Design Issues*, *22*(1), 64–76.

Julier, G. (2013). *The culture of design*. London: Sage.

Kassay, J. (1980). *The book of Shaker furniture*. Amherst MA: University of Massachusetts Press.

Kirkland, R. (2004). *Taoism: The enduring tradition*. Abingdon: Routledge.

Klumpp, D. R. (1981). An historical overview of Maasai dress. *Dress*, *7*(1), 95–102.

Koizumi, K. (1986). *Traditional Japanese furniture*. Tokyo: Kodansha international.

Leach, E. (2013). *The structural study of myth and totemism* (Vol. 4). Abingdon: Routledge.

Lifshitz, F. (2009). *An African journey through its art*. AuthorHouse.

Lévi-Strauss, C. (1955). The structural study of myth. *Journal of American folklore*, 428–444.

Mannheim, B. (2015). Introduction: Intertwined Traditions. In: N. Bonvillain (ed.), *The Routledge Handbook of Linguistic Anthropology*, pp. 1–10. Abingdon: Routledge.

Manzini, E. (2016). Design culture and dialogic design. *Design Issues*, *32*(1), 52–59.

Mascia-Lees, F. (2010). *Gender and difference in a globalizing world: Twenty-first-century anthropology*. Long Grove IL: Waveland Press.

Matsuo, K. (2007). *A history of Japanese Buddhism*. Folkestone: Global Oriental.

McDaniel, R. (2014). *Zen masters of Japan: The second step east*. Rutland VT: Tuttle Publishing

Nosco, P. (1996). Keeping the faith: Bakuhan policy towards religions in 17th century Japan. In P. F. Kornicki, & I. J. McMullen (eds.), *Religion in Japan: Arrows to heaven and earth*, pp. 136–155. Cambridge: Cambridge University Press.

O'Sickey, I. M. (1994). Barbie magazine and the aesthetic commodification of girls' bodies. In: S. Benstock & S. Ferriss (eds.), *On fashion*, pp. 21–40. New Brunswick NJ: Rutgers University Press.

Okazaki, M., & Johnson, G. (2013). *Kawaii!: Japan's culture of cute*. Munich, London and New York: Prestel.

Oldstone-Moore, J. (2003). *Taoism: Origins, beliefs, practices, holy texts, sacred places*. New York: Oxford University Press

Otto, T., & Charlotte Smith, R. (2013). Design anthropology: A distinct way of knowing. In W. Gunn, T. Otto, & R. C. Smith (eds.), *Design anthropology: Theory and practice*, pp. 1–32. London and New York: Taylor & Francis.

Parker, R., & Pollock, G. (2013). *Old mistresses: Women, art, and ideology*. London: IB Tauris.
Pevsner, N. (2005). *Pioneers of modern design: From William Morris to Walter Gropius*. New Haven Conn: Yale University Press.
Picken, S. D. (2010). *Historical dictionary of Shinto*. Lanham MD: Scarecrow Press.
Pollock, G. (1988). *Vision and difference: Femininity, feminism, and the histories of art*. Abingdon: Routledge.
Pullman, P. (2012). *Fairy tales from the Brothers Grimm: A new English version*. New York: Penguin.
Russell, R. R. (2010). *Gender and jewelry: A feminist analysis*. Rebecca Ross Russell.
Segal, K., & Ventura, J. (2019). *In short*: Design. Tel Aviv: Books in The Attic [in Hebrew].
Segal, K., & Ventura, J. (2019a). From material to abstract: Contemporary Israeli design manifested through the culture of wood. *Israel Affairs*, *25*(5), 826–853.
Shakarov, A. & Senatorova, L. (2015). *Traditional African art: An iIllustrated study*. Jefferson NC: McFarland.
Sieber, R., & Rubin, A. (1970). *Sculpture of black Africa: The Paul Tishman collection*. Los Angeles: Los Angeles County Museum of Art.
Sparke, P. (2013). *An introduction to design and culture: 1900 to the present*. Abbingdon: Routledge.
Spielvogel, L. (2003). *Working out in Japan: Shaping the female body in Tokyo fitness clubs*. Durham NC: Duke University Press.
Sprigg, J., Larkin, D., & Freeman, M. (1991). *Shaker: Life, work, and art*. London: Cassell.
Tehrani, J. (2013). The phylogeny of Little Red Riding Hood. *PLoS ONE* 8(11): e78871.
Toffoletti, K. (2007). *Cyborgs and Barbie dolls: Feminism, popular culture and the posthuman body*. New York: IB Tauris.
Veblen, T. (2007 [1899]). *The theory of the leisure class*. Oxford: Oxford University Press.
Ventura, G., & Ventura, J. (2017). Milk, rubber, white coats and glass: The history and design of the modern French feeding bottle. *Design for Health*, *1*, 1–21
Ventura, G., & Ventura, J. (2018). From classic content analysis to multi-faceted design research, *SAGE Research Case Studies* (online).
Ventura, J. (2016). Design situation, in *The Bloomsbury Design Library* (online).
Ventura, J., & Bichard, J. A. (2017). Design anthropology or anthropological design? Towards 'social design'. *International Journal of Design Creativity and Innovation*, *5*(3-4), 222–234.
Ventura, J., & Shvo, G. (2018). I just cut my finger in a Ninja fight: The semiotics and hermeneutics of the band-aid. *Punctum*, *4*(1), 179–201.
Williams, G. (2009). *Shinto*. New York: Infobase Publishing.
Wolf, N. (2002). *The beauty myth: How images of beauty are used against women*. New York: Harper-Collins.
Worobey, J., & Worobey, H. S. (2014). Body-size stigmatization by preschool girls: In a doll's world, it is good to be "Barbie". *Body Image*, *11*(2), 171–174.
Zipes, J. (ed.). (2000). *The Oxford companion to fairy tales*. Oxford: Oxford University Press.

Part 3

Design thinking in practice and professional training

DOI: 10.1201/b22630-16

3.1 Designing for change in complex socio-technical systems

Lina Rylander[a,c] *and Magnus Eneberg*[b]

[a]*Integrated Transport Research Lab, KTH Royal Institute of Technology, Stockholm, Sweden;* [b]*KTH Royal Institute of Technology, Integrated Product Development and Design, Stockholm;* [c]*Scania CV AB*

3.1.1 Introduction

The design industry has moved into new domains with an expanded offering, such as service design (Eneberg, 2015), and the number of designers working in the service design field has been growing fast. The purpose of this paper is to increase our understanding of how designers perceive past and future development in their discipline. The level of involvement by different stakeholders in the design process in relation to an expanding design offering is investigated. Further on, we explore the designer's role in working with system transformation and transition and how design methods may contribute to this.

Public organizations and business organizations are faced with complex and ambiguous challenges, such as developing environmental, economic, and socially sustainable solutions. At the same time, possibilities arise with, for instance, digital technology. New technology and large societal changes create a need and ability to handle the transformation of complex socio-technical systems, such as the health care system and transportation system. Complex socio-technical systems exist in many different contexts, and the conversation regarding changes and challenges in such systems has been discussed within the design field (Vink, 2019; Norman & Stappers, 2015; Sangiorgi, 2009).

Designers often work with changes in different contexts, and the attitude toward changes has developed over time. For example, as Irwin (2015) describes, "Historically, change has been viewed as something that can be "managed" through centralized, top-down design processes that produce clear, predictable outcomes."; however, when approaching complex challenges, work with large-scale changes and complexity might need to evolve, and we believe that designers have the possibility to contribute to this.

Lately, the interest in system thinking in a "designerly" context has been emphasized for sustainable challenges (Sevaldson et al., 2010), but design and systems thinking have differences in their approaches (Pourdehnad et al., 2011). Nevertheless, designers have started to apply methods and tools from the systems thinking field (Jones, 2014), and Costa et al. (2019) have identified a need to integrate design methods and processes into systems thinking to manage complex system challenges. One term that can be used to describe the integration of design thinking and systems thinking is systemic design (Ryan, 2014) Moreover, when designing complex systems, new approaches are needed to understand changes within such systems (Irwin, 2015), and Irwin (2015) suggests that the designer needs to be aware of its role in the system. However, as an actor in a system, you cannot experience the whole system you are part of; you can only experience your way through the system (Buchanan, 2001). Thus, the challenge in integrating systems thinking with design will affect the designer, and its effects on the design field are still little explored.

DOI: 10.1201/b22630-17

3.1.2 Research methodology

3.1.2.1 Overview of the data collection process

This paper builds on two explorative studies. The first study is part of a larger project we performed at Scania where we are involved as design researchers. Scania provides transport solutions to its customers, and the focus of the project is the development of a vehicle health management system for self-driving vehicles at the company. Our role as researchers is to apply a human-centered design methodology. The curiosity and openness to design methods at the company made us interested in the expectations of design and how design could contribute to this context. Furthermore, trying to increase our understanding of the designer's role in transforming large-scale socio-technical systems, such as moving toward self-driving vehicles, led us to the second study in which we interviewed designers regarding their experiences and reflections on the issue.

3.1.2.2 Interviews, observations, and workshop at Scania

The research project we have been working on took place at Scania, which offers transportation solutions. The company is undergoing a transformation, from producing heavy vehicles (trucks) to developing electrified and self-driving vehicles and delivering transportation services. In other words, the company is working on disruptive innovation. During recent years, the company has started to employ design methods to develop services and digital tools and been interested in increasing, for instance, co-creation.

The purpose of the project has been to explore how the vehicle health management system will change when the vehicles become driverless, and there is no longer a driver present in the system. The system consists of, for instance, humans, technical artifacts, and social frames. Thus, a socio-technical system representation can be applied to describe the system. Our role in the project has been to apply design methodology to support and increase understanding of the socio-technical system and interactions and relations between humans and non-human actors. An illustration of the method process applied in this study is shown in Figure 3.1.1.

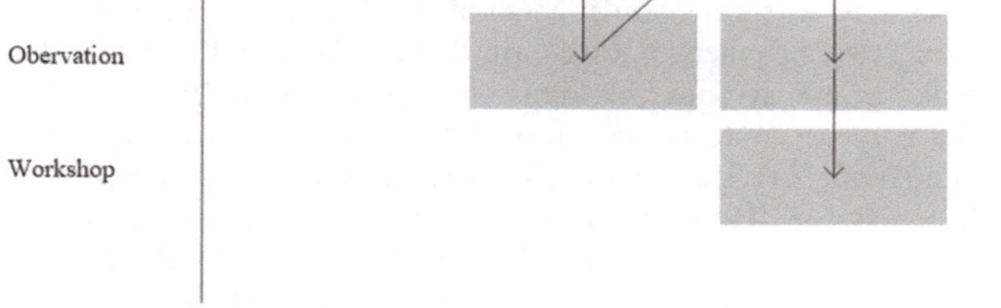

Figure 3.1.1 The method process.

We have performed an exploratory study where we had the role of design process facilitators and conducted two field trips and 11 semi-structured interviews with drivers and service center personnel. Moreover, we have performed three semi-structured interviews to derive fault scenarios with internal vehicle health management experts. The faults were elaborated on from the first interviews. Later, we used the scenarios during a workshop with internal experts at Scania to understand further the technological perspectives and challenges. Additionally, we have had discussions with internal experts regarding our concepts, for example, concerning the technical aspects of self-driving vehicles

We performed the interview analysis by identifying keywords, and the notes were coded and analyzed. Moreover, through synthesizing, we developed themes regarding the driver's role. From the interview analysis, several fault scenarios were generated and used during the workshop. The scenarios were applied as discussion triggers concerning new system design aspects. The study has resulted in a system map describing key actors in the vehicle health management process and information flow. The system map tool has also been used to visualize different concepts, such as one where the driver is represented and another representing a new system design concept. Moreover, we have identified social aspects important for the system, such as relations, interactions, professional experiences, and unique competencies.

From our experiences in the project, when adapting a service design methodology, we have developed insights that have raised questions regarding the designer's role and the design process. Therefore, a second study was performed to deepen the discussion.

3.1.2.3 An interview study with design consultants

In the second study, we conducted semi-structured interviews with six designers working at different firms and with varying backgrounds, as presented in Table 3.1.1. The interviews aimed to obtain perspectives on the designer's role in the design process, the involvement of actors in the design process, and the design industry's change. Initially, from what we learned in the design research project, we identified five core themes used to formulate a questionnaire. The themes were the following: (1) Is design education important, (2) Reflections on the concept of design thinking, (3) Teams practicing design, (4) The designer's role in the design process, and (5) Involvement of non-designers in the design process. The interviews were conducted over Microsoft Teams and later transcribed.

Table 3.1.1 Background of the interviewees

No.	*Working experience*	*Education*	*Current working role*
1	10 years	Industrial and product design	Design consultant, Customer experience strategist
2	15 years	Industrial and product design	Design consultant, Service designer
3	20 years	Visual Communication	Design consultant, Senior Product Designer in UX
4	7 years	Business administration and product design	Design consultant, Service designer
5	1.5 years at Scania	Computer science and engineering, and interaction design	Scania, UX designer
6	18 years at Scania	Industrial and product design	Scania, Senior product and service designer

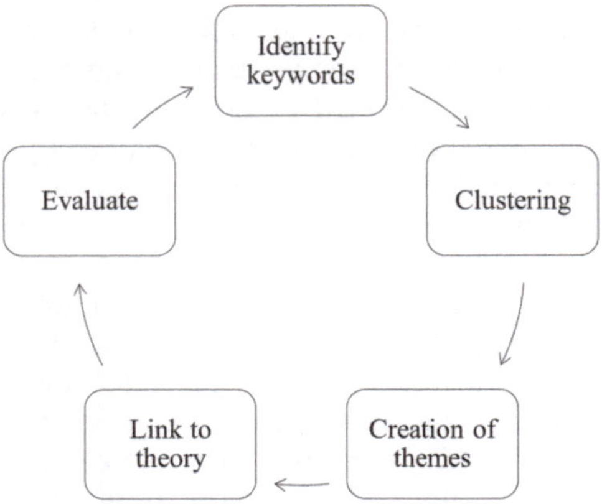

Figure 3.1.2 The data analysis loop inspired by Gioia (Gioia et al., 2013).

A digital board called Miro was used as we clustered the data. The data analysis was inspired by the Gioia Methodology (Gioia et al., 2013), and the analysis and synthesis of the material were performed iteratively. First, we wrote quotes and keywords on digital post-it notes and created clusters to analyze the data. Then, in the synthesis, we developed themes. Later, the themes were related to literature to conceptualize the findings. An illustration of the analysis process is presented in Figure 3.1.2.

The research method we have applied in this study has been limited by the Covid-19 pandemic, and integrative methods could have been used. Such examples are focus group interviews with designers or collaborative workshops among different stakeholders at Scania. For example, a focus group interview with designers could have been valuable to discuss the different perspectives and experiences among the respondents. However, to overcome the limitations of non-physical meetings, we decided to spend much time on the interviews and gather respondents with different backgrounds. Moreover, the literature study was an essential part of contextualizing and bringing relevant insights to the study.

3.1.3 Moving from intradisciplinarity toward transdisciplinarity

3.1.3.1 A pursuit of having a strategic impact

The application of intradisciplinary work can be described as work performed within one discipline (Stember, 1991). In a "designerly" context, it can be exemplified by design work that does not require a high level of involvement from other stakeholders. The designer mainly focuses on reaching out to, empathizing with, and acting as the end-users' spokesperson. Here, visualization tools are mainly used to communicate the results of the design process.

During the industrial era, the offering of the designer was limited to a sequential value chain logic in which designers were commissioned to add value at the end of the product development process (Eneberg, 2015). Morelli (2002) argues that "Designers activities usually have focused on material artifacts (…) rather than on systemic solutions including services." At the same time, the design industry struggled to move into new domains beyond working with functional and aesthetically appealing products (Eneberg, 2015). The aim was to move from an operative role and contribute to the strategic development of client firms.

When applying design in the strategic development of client firms, the client must understand how design as a process can contribute beyond the aesthetically appealing product. This was discussed by one of the design consultants:

> When you start to see design as a way of working, you can start designing anything. Design an organization, a product or a service, an information campaign, a business strategy. (2)

Through our studies at Scania, we have identified that the main focus in the design process has been on ergonomics, functionality, and aesthetics. In addition, the truck driver has often been involved, giving input on the design process to create an environment based on the user group's needs. During the interviews with designers at Scania, it was mentioned that there had been a shift in the design department regarding the level of involvement of end-users, which is related to different design disciplines. In several departments at Scania, the designer's role has not changed; for example, designing the truck's appearance is still an important task. Nevertheless, in other departments, the design tasks have changed focus, and designers work more strategically, closer to other disciplines.

3.1.3.2 Increased level of involvement

For the designer to have a greater strategic impact in client firms, the designer must be involved earlier in the product development process and work closer to other competencies inside the client firm; hence, multidisciplinary work is needed to engage several disciplines. Choi & Pak (2006) states, "Multidisciplinary draws on knowledge from different disciplines but stays within the boundaries of those fields." In a "designerly" context, multidisciplinary work could be exemplified with methods like interviews, observations, and exploratory workshops with participants from different departments in the client firm. The aim is knowledge acquisition and combining different perspectives to understand the problem to be solved, such as understanding activities and interactions in client firms to add new insights into the design process (Buchanan, 2001). This was exemplified in the interviews:

> it is about getting in the statistician's competence and perspective into the design work. (2)
>
> But before these meetings, you must explain to them that everyone should go into their roles, you should not pretend to be a designer, so they feel comfortable in such a context. (3)
>
> If the client has a customer service where they receive many comments about the service or product. (…) I often try to access customer service. (3)

To achieve multidisciplinary involvement in the design process, the clients need to comprehend how design can contribute beyond the product, focusing on the process and how it means working in a "designerly way" (Cross, 2006). As design thinking has received increased attention in management, it also paves the way for the design industry to raise awareness among clients about how design methods and processes may enhance interpretation and collaboration between multiple stakeholders. In our study, collaboration was described as nourishing to the design process, and the design consultants described that multidisciplinary teams support innovations and give flavor to the solutions:

> Secondly, you minimize risks and solve problems to a greater extent, and solutions can be improved by becoming more 'bulletproof' and/or more 'flavorful.' (1)

Nevertheless, working in multidisciplinary teams can sometimes be challenging, and developing approaches to work together may be difficult (Choi & Pak, 2006), implying a need to find tools that facilitate the discussions. One design consultant exemplified this:

> The most challenging projects are where people with different competencies work together. And there can be tug-of-war and long discussions. But when you manage to get all the competencies together, it gets so much better. (2)

We found that visualization is an essential tool in the design process to communicate with other disciplines participating in the design process. Furthermore, with the help of trigger material, visualization can also be used to explore new conceptual ideas and develop new knowledge and perspectives.

One design competence that has received increased attention at Scania is UX design, focusing on digital support systems and technical tools that support tasks at the truck workshop. The designers involve system developers and workshop personnel in the design process to understand their perspectives and increase the understanding among developers of how technical systems can be developed to satisfy users' needs. The UX designer we interviewed at Scania describes the collaboration as follows:

> If there are larger functionalities [to be implemented in the technical system], you need to involve developers, and of course, product owners to see that you have thought right when it comes to the process. (…) For us, it is obvious, but you have to ask those who will code it in the end so that it is feasible. So, it is much juggling between us designers but between developers and product owners as well. (5)

Exploring end-user needs and experiences is not only a method to support the designer with knowledge but it also mediates knowledge and perspectives among other disciplines. As UX design has gained more focus in the company, it is notable that UX designers collaborate with more disciplines within the company than designers have traditionally done at Scania. Surprisingly, even if a focus on the end-user has always been important in the design process, the study shows that the product designers working at Scania continue to work more independently with design within the design team. In contrast, UX designers are part of the development team and involve the end-users and other disciplines in the design work more frequently, or as a senior designer at Scania expressed it:

> The big change I experience, what I mentioned with UX, is how UX has grown so that all the micro-interactions have become a larger part of our interaction with the users than before. (6)

Moreover, the UX designer at Scania pointed out that the UX designer's role may need to increase their knowledge in business, as they work closely with the market department:

> In general, I also think, as a UX designer, I must also be able to keep track of how business works as well. Before, it may not have been so, but it is also a trait that is becoming important, as I understand it. That you understand business and not just the users. (5)

We, as researchers, are studying systems and actors involved in the vehicle health management process of self-driving vehicles. Our study started with a multidisciplinary approach that aimed to increase our understanding of the actors involved in the system, such as truck drivers and workshop personnel. We performed interviews, and the purpose was to provide us with insights into our design process and gain different perspectives on the vehicle health management process, such as problem-solving, interactions, and the fault handling process. When analyzing the data, we realized that the vehicle health management process is much more complex than we first thought. The company offers many services, and the network surrounding the vehicle health management system consists of many actors, including human actors but also different kinds of artifacts and technical systems. Therefore, we realized a need for more integrative methods such as organizing co-creation workshops to increase our and the company's understanding of the problem.

The increased collaboration in the UX field is one example of how designers today are involved throughout the development process and work as a glue between different disciplines. The increased level of multidisciplinary collaboration has also raised a desire among management and other disciplines in the company to know more about the design thinking process.

3.1.3.3 Management consultancies acquiring design consultancies

Lately, large-scale management firms have acquired design agencies, and design has been identified as a tool to work with a high level of complexity. Such examples are disruptive innovation and sustainability challenges. Thus, management firms have identified the design discipline's contribution in this area. However, two of the design consultants expressed a worry about what effects the development of being part of management consultancies will have on the design industry and the designer's role in the future.

> Innovation and new ways of working can be lost. That you continuously push the [design] industry forward. Many small independent agencies stand for what makes us unique, and they are more pioneers. I think that will be lost. (1)
>
> If the smaller studios stop existing, it will be a little more boring, if you understand what I mean. We need both. We need diversity. (3)

In transition design, designers act as agents for change (Irwin, 2015), and management firms may have the power and resources to enable new ways of designing for change. Two examples where the support from management firms may contribute are disruptive innovation and digitalization.

> But it depends on what the future needs. But the future may need a larger holistic approach sometimes. Larger budgets so you can build bigger platforms. (1)

Being part of a large management firm may enable designers to engage in large-scale projects with many diverse participants, which can be challenging to offer as a smaller designer agency. In addition, the "power" of belonging to a large organization may support operating design projects from the beginning to implementation. One respondent, however, questioned if belonging to a larger organization will lead to the desired effect:

> Disruptive is good, but it is difficult to establish it in the large organization (1)

3.1.3.4 The designer as facilitator, teacher, and coach

Designing has moved from being understood as a relieving service, that is, that the client outsources the aesthetics aspect of a product, to an enabling service (Eneberg, 2015). Enabling entails, in the context of service-dominant logic, shared learning activities with the aim to perform a task in a new and better way (Normann, 1992, 2001; Vargo & Lusch, 2008). Although an enabling design service may consequently result in organizational learning and change in the client company, it may also result in a need for a higher degree of involvement where participants from the client organization actively participate in the design process (Eneberg, 2015).

> They [the clients] are much more involved in the design process these days than they were 5–10 years ago when we saw ourselves as the design experts. (2)

A service can be regarded as an interactive process that needs planning and organization where all actors in the system are considered (Morelli, 2002). Therefore, several parts of an organization may be affected by the design process, not least due to the different touchpoints between the customer and service provider, which were exemplified by one of the design consultants:

> (...) and look into how several channels are connected. We can not only solve it on the web, but the web is connected to perhaps a physical product that is connected to a store that is connected to an information letter. (2)

As mentioned previously, exploratory workshops in a multidisciplinary context may support the designer in defining the problem space or exploring new conceptual ideas. Workshops can also be used as a method to enable co-creation among different disciplines in the client firm. Sanders & Stappers (2008) argue that co-designing teams will increase in diversity, for instance, involving different stakeholders in the design process. This type of involvement can be described as interdisciplinary or, as Choi & Pak (2006) illustrate it: "Interdisciplinarity analyzes, synthesizes and harmonizes links between disciplines into a coordinated and coherent whole," and design methods, such as workshops, may enhance sensemaking and collaboration between the actors involved (Eneberg, 2015; Jahnke, 2013; Verganti, 2009). By creating conditions for affordance, that is, supporting an environment that allows the individual to perform actions in shared activities (Greeno, 1994), the designer may facilitate the opportunity for different thought networks to merge and be reshaped, such as changes in mental models (Vink et al., 2019).

As the client offers more complex services to their customers, a desire to integrate the design process into their development processes has emerged. This can be exemplified by Scania, which is undergoing a transformation in the technical area, namely developing electric and self-driving vehicles, which also leads to a need to transform how they are organized and with whom they collaborate. We were asked to facilitate a design project focusing on one of the services that will be affected by the transforming technology, and service design is seen as a tool that can be useful to tackle this transformation. When the project we participated in at Scania started, we experienced a desire to learn more about service design. We were asked to arrange courses to share our design competence with engineers within the research and development department.

Designers are more often commissioned to take the role of a facilitator to enable inter-organizational collaboration, co-creation, and co-designing, such as in workshops. Moreover, the designer's role is also moving toward being a coach and teaching the design process and methods to the employees in the client organizations, as mentioned in the Scania example and the interviews with design consultants:

> I am there as a coach and say that now you should hold interviews, and then you should do this. And it is hard. I think. It is challenging. (4)

There is a significant difference between working with aesthetics and functionality in product design and taking a facilitating, teaching, or coaching role in client firms. As in every profession, designers are individuals with different motivations and driving forces.

> Some designers work as experts, and some work as facilitators. (2)
>
> A designer is, for me, someone who can collaborate and give nourishment to the collaboration. (1)
>
> It is difficult to tar every designer with the same brush. Some designers do not like to work with mixing competencies. (1)

Hence, the design profession and the different expertise designers provide are increasingly becoming more heterogeneous as the clients have become aware of the strength of the design process and its methods.

A possible effect of introducing design thinking to the management field is that people who do not have a formal design education can start practicing design. Respondents in our study had different opinions regarding this development. One positive consequence is that it may create a cumulative understanding of design. However, it can be challenging to learn the design mindset, which two design consultants discussed:

> It cannot be done by someone who does not have any kind of experience or background or education or understanding or has practiced any kind of design. (4)
>
> It has been a great challenge with people who do not have design skills, not always, but idea generation often has higher quality with those who are trained in it. To be careless and to be daring. It is sometimes challenging to focus on 'the new' with people who are in a comfort zone. (1)

Another issue raised by the designers was that when educating others in design, the designer will need to compromise the design process and the results.

> if it is to be both a learning project and the result is important. Then there will be a huge conflict within me (...) if you have a person who has done 40 projects, design projects, and if you have three people who have never heard of design before, it will not be as good. (4)
>
> It takes more time to include non-designers. It is a higher investment. It will be a compromise on which results you want. (1)

Design offerings have moved from physical and digital artifacts to experiences, environments, and systems. Here, visualization is still an essential competence used by the designer; however, its application has widened when used to facilitate co-creation and co-design in an interdisciplinary context. Visualization tools are used as mediating artifacts that support intra- and inter-organizational collaboration and explore different perspectives, combinations of problems, contexts, and solutions.

> You may have prototypes and visualizations that you engage with and facilitate the discussion with. (4)
>
> To visualize and create a shared image was very important in order to unite the organization and focus their skills or expertise around a common goal; It was all about the visitor's perspective. (2)
>
> The visualization supports us in communication. It gives consensus on what one can create. (1)

We have identified a need for a systemic approach in the Scania project. Today, when performing fault handling and the troubleshooting of trucks, many actors participate in the process, and several parts of the organization are involved in developing tools, instructions, and technical functionalities for the vehicles. In addition, the system's complexity has increased with the development of self-driving vehicles, and the system will need to adapt to new needs when the system changes. As discussed in the interviews with the design consultants, visualizations are used as a communication tool to enhance discussions, create reactions, and present insights. In our research project at Scania, visualization has been a vital tool to co-create and gain a common understanding of the system and actors involved. For example, a system map may include an organizational presentation of actors and their interactions. However, a significant contribution by the designer is adding and illustrating the experience, needs, driving forces, and motives among participating actors. Different scenarios have been used to illustrate current needs in the research project. During a workshop, these scenarios were used to trigger discussions and enhance understanding among different disciplines.

3.1.3.5 Ever-increasing complexity

Not many years ago, companies stated that self-driving vehicles, such as trucks or taxis, would soon be a reality. However, it is obvious that it will take much longer than they first thought. The industry has realized a need for collaboration between system actors, such as policymakers and companies. Two examples of such collaborations are Scania and TuSimple[1] or Volvo with Aurora[2]. The need to extend partnership collaboration beyond industry is an essential strategy to evolve from being a provider of products to being a service provider within integrated mobility (McKinsey, 2016). As more complex functionalities are added in driverless and self-driving vehicles, troubleshooting becomes more challenging (Lanigan et al., 2011). Also, business models and services will change according to new business solutions and customer needs. Consequently, complexity increases with new touchpoints, and more actors are involved, for instance, policymakers. A holistic view is needed since the change will transform the transportation system with immense effects on society.

The concept of *socio-technical system* highlights the need to consider both technical and social aspects of the system. With inspiration from (Geels, 2005), the socio-technical system of goods-delivery transportation can be described as a system that consists of technology, such as technical systems and vehicles; regulations, policies and working regulations for drivers; user-practices and markets, such as requirements from the driver and haulage contractor; maintenance networks including workshops and service network; and cultural aspects such as working culture, norms, and institutional arrangements. Furthermore, understanding and supporting large-scale systems changes the different contexts and perspectives affected by the changes that need to be considered, which can be exemplified by making ethical considerations concerning disruptive innovation, such as self-driving vehicles (Sparrow & Howard, 2020).

The increased complexity in designing systems may need a new approach, where the level of collaboration, co-creation, and co-designing changes from that of being interdisciplinary to transdisciplinary. Transdisciplinarity is a key concept eliciting non-traditional perspectives and cross-fertilizing assumptions, forming new holistic approaches. For example, Choi & Pak (2006) describe transdisciplinarity as follows: "In transdisciplinary teams, not only goals but skills are shared (…) through role release and role expansion, transcends (hence "trans") the disciplinary boundaries to look at the dynamics of whole systems in a holistic way." In a design context, this could be described as disciplines working beyond existing boundaries.

> But in the future, maybe design thinking, or this thing to involve users and customers continuously in the development process is done by more people. Not just those who are educated designers. I hope it will be a natural part of any development process. (2)

The concepts of transition and transformation are often used to describe a large-scale change in complex socio-technical systems, and in both cases, the agency is described as a multi-actor process enabling innovation, learning, collaboration, and knowledge integration (Hölscher et al., 2018). However, there are contextual differences in which the two concepts are used, and Hölscher et al. (2018) propose a description of the differences concerning how the concepts are applied.

Transformation as a concept is often used when the focus is on change processes in society, intending to identify emergent change patterns and their consequences. Individuals' motives, driving forces, and values supporting a transformation focus on avoiding undesirable system change.

On the other hand, transition is used when describing how changes in complex adaptive systems occur with a stronger focus on change in socio-technical aspects and disruptive innovation to enable a sustainable transition in the subsystem.

3.1.3.6 Design enabling transition and transformation

In the study, we identified that the system focus has been accentuated in importance for the design industry and applying a system approach in design might influence how the designer works. One design consultant discussed how the involvement of actors outside the organization might be one way of handling the complexity and avoiding the development of unsustainable products and services:

> I think it is very important to include other actors outside the organization. (…) it can be other companies or public organizations (…) if you just stare blindly at user needs, you can create quite unsustainable solutions (…) So, this is where systems thinking comes in more and more. (4)

Irwin (2015) argues that new approaches are needed to understand changes within complex systems. Inspired by the work of Hölscher et al. (2018), we have identified essential aspects in the design competence that may support transition and transformation in complex socio-technical systems. Designing for transformation would imply a need to identify emergent change patterns. Bottom-up approaches targeting the micro-level perspectives are often used in design practices, not the least to understand the end-user. Searching for patterns is an essential tool when working with conceptualization (Vink, 2019). Hence, using a micro-level perspective would support our understanding of motives, driving forces, and values of individuals and emergent patterns, which may lead to undesirable changes when the system is transformed.

However, only focusing on the micro-level could lead to the loss of a holistic perspective and overview of the socio-technical system. Users in a system cannot experience the whole system they are part of; they only experience their pathway in the system (Buchanan, 2001). A proposal is a framework that includes methods and tools that support designers both from a micro and macro perspective (Vink et al., 2020). The design of socio-technical systems demands multi-directional processes, which are often conflicting, and the different perspectives must be zoomed in and out (ibid). Disruptive design interventions involving actors affected by large-scale system changes may also lead to changes in the mental models among the individuals and teams participating in the joint activity (Vink et al., 2019). This may also result in new transdisciplinary knowledge in the sense that assumptions are cross-fertilized and form new holistic approaches that can lead to disruptive innovation. The design intervention itself is essential; therefore, we argue that the designer performs actions in the design process that correlate with the characteristics of working with transition and may enable a sustainable transition in a subsystem.

The designer's future role is difficult to anticipate. However, in the study, we identified a desire among designers to work with disruptive innovation and sustainable transitions:

> We have to start adjusting. We may not have to produce new ones all the time. Look at removing or replacing. This can be difficult to convince sometimes. (1)
>
> I think it is good if the tools and methods can help organizations as forward and create better experiences and sustainable experiences. (4)

The sustainable aspect is vital as the long-term effects of the design process influence a sustainable mindset, not only from an environmental perspective but also from the system effects of new designs. Moreover, there are still many areas where design is not applied today, enabling further exploration of the design approach and design applications related to systems thinking. One respondent said:

> Where we are not is in policymaking, politics, and laws, or agriculture. There is still a lot of unexplored lands. (1)

The design discipline will need to collaborate with other disciplines not previously involved in the design process, such as system engineers. Combining design methods with methods used in other disciplines may support the designer in their pursuit to contribute to a sustainable transition in society. Sustainable challenges have enhanced interest in systems thinking within the design context (Sevaldson et al., 2010), and designers have borrowed tools and methods from systems thinking, for example, integrated methods and meta-design frameworks (Jones, 2014). In a sense, it can be deliberated upon whether the process and methods used in transforming large-scale transitions should be called design thinking or if it is even relevant.

> Design is a collective name for something you do. If you have a word for it, you can talk about it. Maybe it is good that there is a name for it. You can put an epithet on the activity. But if it should be called design or user-centered development or design doing or design thinking … I do not know … (2)

Systemic design is often described as the integration of a designerly way of working with methods from systems thinking, and it has also been discussed if systemic design should be its own field (Ryan, 2014). Another point of view could be to understand systemic design as an offering or problem space where the designer's competence can contribute to system changes. Also, the integration of systems thinking and design thinking might affect the designer's role, and Pourdehnad et al. (2011) accentuate a difference in systems thinking and design thinking, where stakeholders in design thinking are involved in enabling the designers to learn from the stakeholders when in systems thinking, that is, the designer is a stakeholder. Being a stakeholder rather than understanding other actors may contribute to a search for identity within the design discipline.

3.1.4 Conclusion

The designer role has evolved, and today designers are also engaged in facilitating, teaching, and coaching the design process in client firms. This shift is in many ways related to an increased need for sharing perspectives and methods between disciplines and the importance of collaboration, co-creation, and co-designing of new knowledge to solve ambiguous problems.

One field that is receiving much interest from the design discipline is systems thinking. We argue that designers would benefit from collaborating with disciplines as system engineers. Combined methods used by designers and system engineers may have the capacity to support and reach for change in complex socio-technical systems. Moreover, the need for a system perspective in design relates to system change and how it is described in transformation and transition literature. Methods used in the design process are relevant for both systemic transformation and transition, and it has been accentuated that designers need to iterate between macro and micro perspectives. In the context of system transformation, the designer may contribute by identifying patterns in the interaction among the actors involved in a system and their relations, motives, experiences, and driving forces. This may contribute to avoiding undesirable change.

Further on, disruptive design interventions may be used to provoke reactions and create conditions for a sensemaking process. The aim is to increase the understanding of mental models that may hinder a sustainable system transition on a macro level.

Today, designers work with tasks that demand different levels of involvement and collaboration, which can be put on a scale extending from intradisciplinarity to transdisciplinarity. Moreover, the emergence of systemic design has shown a search for identity in the design discipline, and as systems thinking has received more attention in the design discipline, the designer's role has changed. In intra-, multi-, and interdisciplinary involvement, the designer engages with others to understand them, but when engaging in transdisciplinary work, the designer's role changes from engaging with others to becoming a part of the system themselves. The relevance of the search for identity should not be diminished or disregarded; on the contrary, we need to continue the discussion about how the design discipline can or should evolve. Questions such as "Who is a designer?" and "What is design?" are highly significant as the designer becomes an agent and co-creator in complex socio-technical systems. However, the focus needs to be broadened and include how design can contribute to sustainable system transformations and transitions, which the study has shown that the designer both wants and already participates in today.

Acknowledgment

This work is supported by the Strategic Vehicle Research and Innovation Program (FFI) through Sweden's Innovation Agency (Vinnova).

Notes

1 https://www.scania.com/group/en/home/newsroom/press-releases/press-release-detail-page.html/3880923-scania-tests-self-driving-trucks-in-motorway-traffic
2 https://www.volvogroup.com/en/news-and-media/news/2021/mar/news-3932948.html

References

Buchanan, R. (2001). Design research and the new learning. *Design Issues*, *17*(4), 3–23.

Choi, B. C. K., & Pak, A. W. P. (2006). Multidisciplinarity, interdisciplinarity and transdisciplinarity in health research, services, education and policy: 1. Definitions, objectives, and evidence of effectiveness. *Clinical and Investigative Medicine*, *29*(6), 351–364.

Costa, J., Diehl, J. C., & Snelders, D. (2019). A framework for a systems design approach to complex societal problems. *Design Science*, *5*(e2), 1–32.

Cross, N. (2006). Designerly ways of knowing. *Design Studies* 3(4), 221–227.

Eneberg, M. (2015). *Beyond the product - Enabling design services in small and medium sized enterprises*. Lund University.

Geels, F. W. (2005). The dynamics of transitions in socio-technical systems: A multi-level analysis of the transition pathway from horse-drawn carriages to automobiles (1860-1930). *Technology Analysis and Strategic Management*, *17*(4), 445–476.

Gioia, D. A., Corley, K. G., & Hamilton, A. L. (2013). Seeking qualitative rigor in inductive research: notes on the Gioia Methodology. *Organizational Research Methods*, *16*(1), 15–31.

Greeno, J. G. (1994). Gibson's Affordances. *Psychological Review*, *101*(2), 336–342.

Hölscher, K., Wittmayer, J. M., & Loorbach, D. (2018). Transition versus transformation: What's the difference? *Environmental Innovation and Societal Transitions*, *27*, 1–3.

Irwin, T. (2015). Transition design: A proposal for a New Area of Design Practice, Study, and Research, Design and Culture. *Design and Culture*, *7*(2), 229–246.

Jahnke, M. (2013). *Meaning in the making - Introducing a hermeneutic perspective on the contribution of design practice to innovation*. University of Gothenburg.

Jones, P. H. (2014). Systemic design principles for complex social systems. In G. S. Metcalf (ed.), *Social systems and design* (Vol. 1, pp. 91–128). Springer Japan.

Lanigan, P. E., Kavulya, S., Narasimhan, P., Fuhrman, T. E., & Salman, M. A. (2011). *Diagnosis in automotive systems: A survey*. Parallel Data Lab, Carnegie Mellon University.

McKinsey. (2016). *McKinsey report automotive revolution & perspective towards 2030*. McKinsey & Company.

Morelli, N. (2002). Designing product/service systems: A methodological exploration. *Design Issues*, *18*(3), 3–17.

Norman, D. A., & Stappers, P. J. (2015). DesignX: Complex sociotechnical systems. *She Ji*, *1*(2), 83–106.

Normann, R. (1992). *Service management - Strategy and leadership in service business*. John Wiley Sons Inc.

Normann, R. (2001). *Reframing business: When the map changes the landscape* (1st ed.). John Wiley Sons Inc.

Pourdehnad, J., Wexler, E. R., & Wilson, D. V. (2011). Systems & design thinking: A conceptual framework for their integration. *55th Annual Meeting of the International Society for the Systems Sciences 2011*, 807–821.

Ryan, A. J. (2014). A framework for systemic design. *FormAkademisk - Forskningstidsskrift for Design Og Designdidaktikk*, *7*(4).

Sanders, E. B.-N., & Stappers, P. J. (2008). Co-creation and the new landscapes of design. *Co-Design*, *4*(1), 5–18.

Sangiorgi, D. (2009). Building up a framework for service design research. *8th European Academy Of Design Conference, Aberdeen, Scotland, April*, 415–420.

Sevaldson, B., Hensel, M., & Frostell, B. (2010). Systems-oriented design and sustainability. *Sustainability in Design Now! Challenges and Opportunities in Design Research, Education and Practice in the XXI Century*, 465–474.

Sparrow, R., & Howard, M. (2020). Make way for the wealthy? Autonomous vehicles, markets in mobility, and social justice. *Mobilities*, *15*(4), 514–526.

Stember, M. (1991). Advancing the social sciences through the interdisciplinary enterprise. *The Social Science Journal*, *28*(1), 1–14.

Vargo, S. L., & Lusch, R. F. (2008). Why "service"? *Journal of the Academy of Marketing Science*, *36*(1), 25–38.

Verganti, R. (2009). *Design driven innovation: Changing the rules of competition by radically innovating what things mean*. Harvard Business Press.

Vink, J. (2019). *In/visible - Conceptualizing service ecosystem design*. Karlstad University.

Vink, J., Edvardsson, B., Wetter-Edman, K., & Tronvoll, B. (2019). Reshaping mental models – enabling innovation through service design. *Journal of Service Management*, *30*(1), 75–104.

Vink, J., Koskela-Huotari, K., Tronvoll, B., Edvardsson, B., & Wetter-Edman, K. (2020). Service ecosystem design: Propositions, process model, and future research agenda. *Journal of Service Research*, *20*(10), 1–19.

3.2 Design Thinking meets innovation management

A case study examining how *Diplomatic Rebels* manage design in organizations

Hanna Lena Bogner and Anna-Sophia Rabe
Copenhagen Business School, Denmark

3.2.1 Introduction

Approximately 70% of innovation initiatives in organizations fail (Bucy, Finlayson, Kelly & Moye, 2016). While Design Thinking was brought to the business world as a new way to drive creativity and increase a firm's level of innovativeness, its success remains unclear. In fact, empirical studies indicate that there are many difficulties related to implementing Design Thinking in organizational settings (Carlgren, Rauth & Elmquist, 2016). Thus, this chapter aims to investigate how to successfully manage design and innovation in organizations. The case study of Diplomatic Rebels is presented – a company that seems to propose a different approach through introducing the concept of the *Diplomatic Rebel*[1] as an innovation manager. Focusing on the role of the individual *Diplomatic Rebel*, this research aims to contribute to the academic conversation by answering the following research questions:

- How do *Diplomatic Rebels* manage challenges of driving design and innovation in their organizations, based on their mindsets and skills?
- How do the diplomat and the rebel interrelate in the innovation process from ideation to implementation?
- How can the concept be understood in the light of literature, taking Design Thinking as a starting point?

The chapter contributes to the academic conversation on organizational innovation through Design Thinking by combining empirical findings with a thorough literature analysis. Through eight in-depth interviews with experienced innovation managers, the unique concept of the *Diplomatic Rebel* is introduced. Combining the characteristics of a rebel (Design Thinker) with those of a diplomat (manager), they have a unique entry for driving innovation. Studying the *Diplomatic Rebel* through the lens of literature, this chapter advances Design Thinking with theories from management literature. In doing so, the gap in the innovation process is filled: Design Thinking reflects the idea generation phase and is complemented with aspects from management literature for the implementation stage.

The chapter is structured to first give a literature review of the theoretical conversation in Design Thinking, in management and ambidexterity literature. Secondly, the method is explained. Thereafter, the empirical findings from the case study are presented. The main body of the chapter – the analysis and discussion – combines both empirical findings and

DOI: 10.1201/b22630-18

literature to answer the research questions. The chapter concludes with a summary, limitations and future research.

3.2.2 Literature review

The innovation process can be understood as two main processes: idea generation and implementation (Amabile, 1988; Baer, 2012; Rosing, Frese & Bausch, 2011) (Figure 3.2.1). Design Thinking has been brought to the business world as a new way to spark creativity in the idea generation phase and involve customers throughout the solution development. However, the implementation in the organization, defined as a socio-political process, is often neglected in Design Thinking (Carlgren et al., 2016; Rylander, 2009; Rosing et al., 2011). To fill this gap, management theory is reviewed.

3.2.2.1 Design thinking literature

Contrary to *designerly thinking*, which defines the designer's way of working, this study focuses on *Design Thinking* in a business context (Johansson-Sköldberg, Woodilla & Cetinkaya, 2013). While many associate Design Thinking with a workshop model (Hassi & Laakso, 2011; Brenner & Uebernickel, 2016), this study builds on Johansson-Sköldberg et al.'s (2013) holistic definition of Design Thinking as a general thinking style and way to approach complex innovation problems in organizations.

Taking a practice-based view, Design Thinking literature differentiates between *mindsets* and *practices* (Hassi & Laakso, 2011; Brenner, Uebernickel & Abrell, 2016). The mindset is defined through abductive reasoning (Dunne & Martin, 2006; Kimbell, 2011) and reflective reframing (Dew, 2007; Boland & Collopy, 2004). Further, Design Thinkers show an experimental and explorative mindset (Fraser, 2009; Brown, 2008; Holloway, 2009) combined with a tolerance for ambiguity (Drews, 2009; Brown, 2008; Dunne & Martin, 2006). They embrace a holistic 360° view to develop a full understanding of the customers' needs (Hassi & Laakso, 2011; Holloway, 2009). Lastly, Design Thinkers are optimistic (Brown, 2008) and future oriented (Dunne & Martin, 2006; Drews, 2009, Kimbell, 2011). The practices are thinking by doing (Brown, 2008), a human-centred approach (Brown, 2008; Dunne & Martin, 2006), collaborative work styles (Holloway, 2009; Kimbell, 2011) and visualizing (Brown, 2008; Rylander, 2009). Thus, Design Thinkers bring the mindsets and practices to think beyond boundaries, come up with novel ideas and involve customers to validate and develop solutions.

While Design Thinking has been widely praised, some authors express critique. Most criticism is directed towards a lack of measurable success and its application in the business world. Nussbaum (2011) criticizes that Design Thinking has turned into a linear-by-the-book process to fit organizational contexts. Kolko (2015) similarly argues that when managers become aware of the power of design, they see it as a solution to all their problems. Other authors see a gap between organizational values and those of Design

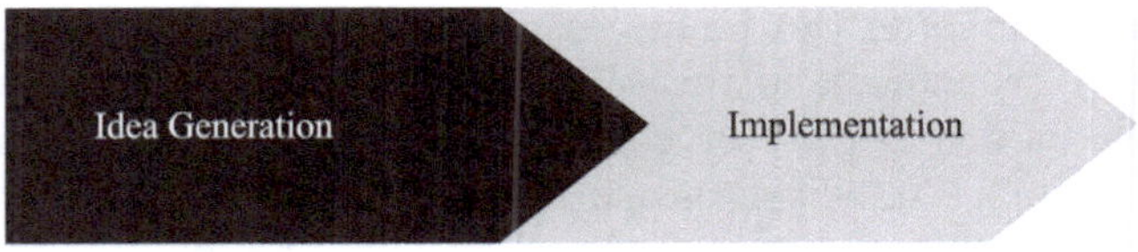

Figure 3.2.1 The innovation process: idea generation & implementation.

Thinking. For example, the playful and exploratory aspect of Design Thinking is often perceived as non-serious, and Design Thinking can cause political conflicts as it questions established business standards (Carlgren et al., 2014; Rauth et al., 2015). Overall, while critique has been expressed, few offer solutions towards successful implementation of Design Thinking in a business context.

3.2.2.2 Management literature

Large corporates often fail to innovate, and approximately 70% of all implemented change initiatives in organizations are not successful (Nikolaou, Gouras, Vakola & Bourantas, 2008; Altringer, 2013; Bucy, et al., 2016). Looking at the reasons why large firms fail to develop disruptive innovation, Assink (2006) focuses on innovation barriers within organizations. The *adoption barrier* is seen in "organizational rigidity and the existence of a dominant design carried on through successful concepts from the past" (Assink, 2006, p. 227), which results in organizations being unable to adapt to new structures and introduce disruptive innovation. The *mindset barrier* reflects a firm's inability to unlearn, the lack of distinctive competencies necessary for disruptive innovation, and obsolete mental models (Assink, 2006; Carlgren et al., 2016). Closely linked is the *risk barrier*, which describes a general risk averse corporate attitude. Sandberg and Aarikka-Stenroos (2014) find the mindset barrier to have the biggest impact on implementing innovation. Due to a fear of change and failure, conservative decision-making, and restricted organizational culture, employees are averse towards change. These negative feelings often lead to resistance (ibid.).

Here, Maurer (2009) identifies three different levels of resistance. The first level, *I don't get it,* stems from a lack of relevant information or knowledge, disagreement over the interpretation of data, or confusion over what it all means. The second level, *I don't like it*, is rooted in employees' fear of losing face, status, control, or the job through the initiated change. Lastly, the third level, *I don't like you*, reflects a lack of trust or confidence in the leader. Maurer (2009) further points out that many companies treat all resistance as first level resistance, whereas more than often the cause is emotional.

To deal with resistance, diverse streams in management literature acknowledge that a one-sided competence set is not sufficient and call for a change manager's mindset in two realms, spanning managerial aspects and social competences (Rezvani et al., 2016; Fisher, 2011; London, 1999). Building on that, Vakola et al. (2004) define that the personality traits of a manager should be strategic, achievement-oriented, orderly and persistent (*conscientiousness*) as well as emotional-diplomatic, cooperative, trustful and likeable (*agreeableness*) to better cope with change. Across several studies it was shown that extraversion and openness to experiment are not enough to manage change and that agreeableness and conscientiousness build a positive correlation with successful change implementation (Judge, Thoresen, Pucik & Welbourne, 1999; Vakola, 2004; Nikolaou et al., 2008).

Regarding the skills of a change manager, Van der Walt, du Plessis and Barker (2006) differentiate between *rational intelligence* (the ability to base decisions on a determinate logic), *emotional intelligence* (the ability to listen to employees, adapt and learn from each other) and *spiritual intelligence.* The latter can be linked to *cultural awareness* (Fisher, 2011), as it describes the ability to understand the values and meaning-seeking needs of team members to create a unified culture of change. Rezvani et al. (2016) and Vakola et al. (2004) find that emotional intelligence is particularly crucial in situations of change. Managers, who are able to perceive, understand and manage emotions of oneself and others, can turn these emotions into positive directions, and foster motivation for creativity and change (ibid.). Further, Fisher (2011) emphasizes that leading and influencing others, creating trust, being able to manage conflict as well as understand others' behaviour and cultural beliefs are crucial skills. Overall, communication skills are a key element in change management literature to educate and convince reluctant employees of why change is needed (Kotter & Schlesinger, 2008; Ford & Ford, 2009; Maurer, 2009; Turell & Earle, 2018).

3.2.2.3 Ambidexterity literature

Ambidexterity is widely discussed in the context of change and innovation, describing a manager's ability to combine both *exploration* and *exploitation* (Paliokaite & Pacesa, 2015; Rosing et al., 2011). Exploitation is the ability to deploy already existing competencies and structures in an organization, whereas exploration captures the discovery of new concepts (Paliokaite & Pacesa, 2015). Both competencies are needed for organizations aiming to change. Exploitation abilities are required to sustain efficiency, whereas exploration skills are essential for novelty (ibid.). Ambidexterity literature does not find common ground whether exploration and exploitation are to be embodied in one manager-person: Hunter, Cushenbery and Jayne (2017) claim that exploration and exploitation are a fundamental paradox and can thus not be incorporated in one person. Other authors find that these two personalities can co-exist as they come into play for different situational and personal factors (Bodwell & Chermack, 2010; Rosing et al., 2011; Keller & Weibler, 2014). While exploration is mostly encompassed in the idea generation phase, exploitation is mainly evident in the implementation process (Rosing et al., 2011).

3.2.3 Method

Studying the management of innovation processes in organizations, this research is based on a qualitative case study. As a case, the network-like company Diplomatic Rebels was chosen for two reasons. One, the case provided insights into different innovation processes in various large companies. Two, wanting to investigate Design Thinking, the company's term for an innovation manager, the *Diplomatic Rebel*, seemed a promising and novel concept to research.

The company Diplomatic Rebels was founded in 2017 in Copenhagen to support intrapreneurs – individuals who drive innovation in large international organizations (1000+ employees). The company's concept is a fellowship program, where intrapreneurs from different organizations and industries meet on a regular basis and exchange the challenges they face in their work as innovation managers. In group sessions, they discuss challenging topics and support each other. In addition, they get consultation and sparring from professional advisors. The fellows of the program are called *Diplomatic Rebels*. They are selected for the program based on having a position in innovation within a big corporation, as well as bringing a certain way of thinking and doing that is diplomatic and rebellious.

Following an inductive research approach, this study started out by exploring the context of innovation managers, where diverse perspectives were narrowed down as the research progressed (Saunders, Lewis & Thornhill, 2009; Qu & Dumay, 2011). Therefore, an unstructured group interview was conducted to explore diverse open-ended questions as a first step. The concept of the *Diplomatic Rebel* was subsequently set as the focus of the research, which was further narrowed down throughout seven individual semi-structured interviews with each innovation manager. Here it is important to highlight that we as researchers did not start out knowing what constitutes a *Diplomatic Rebel*. Rather we saw it as an interesting concept to explore and research Design Thinking empirically and theoretically. Throughout the research process the systems and characteristics of the *Diplomatic Rebel* were uncovered. Aiming to contribute to the field with new understandings of a social phenomenon, the focus was set on the interviewees' perceptions of their social interactions and their way of working. More specifically, factors for driving innovation were investigated through the real-life experiences of *Diplomatic Rebels*.

Table 3.2.1 provides an overview of all interviews. As far as possible, the interviews were held in person to observe behaviour, attitudes and environmental settings. However, three interviews were conducted via virtual calls, given the geological distance of respondents. The primary data collected through these eight interviews builds the empirical foundation of the study. Besides interviews, the research is influenced by secondary data consisting of documents published by the case company as well as non-text materials such as photos and artefacts used by the case company.

In a second step, an iterative data analysis was performed, alternating between empirical findings and literature. To mitigate bias and misinterpretations, all interviews were conducted and coded as a team. Following the inductive approach, all interviews were analysed through open coding. Thus, codes were formulated unsystematically at first and emerged progressively throughout the data analysis (Emerson, Fretz & Shaw, 2011; Miles, Huberman & Saldana, 2013). In the first coding cycle, interviewees' direct answers and behaviour were analysed line-by-line and assigned with predominantly *descriptive* and *in-vivo coding*, in some cases *process coding* (Miles et al., 2013). From the first coding cycle a list of unstructured codes emerged, which were again clustered and summarized into themes through second cycle codes. Guided by literature, higher cycles were developed, completing the final coding tree.

It is important to mention that a hybrid form was used, where empirical findings and literature findings are directly connected in the data analysis. Lastly, the inductive research approach proves particularly suitable for constructing new theory by studying contextual aspects of small samples (Saunders et al., 2009).

3.2.4 Findings

All interviewees faced several challenges in their jobs as intrapreneurs, with the role of driving innovation and disrupting status quo in their companies. To understand these challenges and how *Diplomatic Rebels* manage them, their cases[2] are presented here:

Thomas: transformed Flux, an international analytics company, from an internal process-focused into a customer-centric organization. He introduced a front-end innovation board and implemented entrepreneurial principles into the core culture. That new way of thinking triggered resistance amongst employees: "We do things ourselves. […] We design ourselves. It's much better. So, we don't want to let too many people in. This whole notion about open innovation and strategic partnerships is very alien to us" (Thomas, Interview). Likewise, C-management did not buy into Thomas' idea at first, as they did not understand the urgency of the innovation project over other strategic company matters. Through the fellowship program, Thomas learned to pitch the innovation project better by applying empathy and selling it at the right time to C-level and colleagues: "Maybe I didn't pitch it well enough […] but it seemed they had other things on their plate. […] At least I've become

Table 3.2.1 Overview of conducted interviews

Interviewee	*Matthew & Benjamin*	*Matthew*	*Benjamin*	*Greg*	*William*	*Lincoln*	*Alex*	*Thomas*
Company	Diplomatic Rebels	Diplomatic Rebels	Diplomatic Rebels	Spiel	News	Medant	E.G.	Flux
Position during time of interview		Managing Partner	Country Manager, Germany	Head of Capital Venture	Head of Analytics	Discovery Lead	Head of Innovation	Head of Global Quality
Interview Conduction	Unstructured Group Interview (Personal)	Semi- structured Interview (Personal)	Semi- structured Interview (Personal)	Semi- structured Interview (virtual)	Semi- structured Interview (Personal)	Semi- structured Interview (Personal)	Semi- structured (virtual)	Semi- structured (virtual)
Duration of interview	2:22:00 h	1:31:02 h	1:00:17 h	1:07:27 h	1:19:58 h	01:33:02	00:59:14 h	1:13:44 h

more mindful of sometimes you have a window of opportunity, because things are going on behind the scenes where they are up to here with worries" (Thomas, Interview).

Alex: changed the overall structure of E.G., a large international holding company. Like Thomas, Alex recognized that even though innovation is important, it is a lot of times not the most urgent matter in an organization. In the holding, all sub-companies previously ran innovation individually and in parallel with product development efforts. Alex broke those silos down. Instead, she introduced a holding-wide innovation board to focus on radical innovation, while the sub-companies were to focus on continuous development. Introducing this new organizational structure, Alex faced a lot of resistance, which stemmed from directors not feeling comfortable with the high risk associated with restructuring. To overcome that resistance, Alex appointed one change manager with a mandate to drive change as well as the sensemaking skills to address the strategic goals of the company.

Greg: initiated a radical innovation project in Spiel, a traditional toy manufacturer. His goal was to revolutionise the toy industry by transforming traditional toy bricks into digital worlds. As he did not get the buy-in from top management, Greg started to strategically build a "tribe" around him with colleagues who - in their free time - would develop the project. Greg emphasized the resistance his team faced from colleagues who did not accept the change of the company direction; "This is not Spiel. Spiel would never do this" (Greg, Interview). Greg understood it was important to counteract that conformity. "For people in big corporations, what happens is that they grow up through the system. [...] And then slowly they fall asleep, and they start getting more and more conform and they start thinking like everybody else 'oh we can't do this because the system says we can't'" (Greg, Interview). His team continued the project and eventually succeeded. At first, they started making T-shirts with the project's slogan to create awareness across the entire company. Further, the team learned to provide more information about the project, so people would understand "what it is all about". Greg emphasized the importance of being kind, empathic and approachable when talking to people about change. Over time, colleagues became curious and eventually, a hype was created bottom-up until C-level also bought into the idea.

William: wanted to change the culture and people dynamics at News, a national newspaper company. He stressed, that to succeed with new future strategies, the organization and its employees would need to become comfortable with change; "You have to figure out how to get people to change. And you will figure out that people hate change, and they only like change if it doesn't affect them" (William, Interview). With resistance being a natural human response to disliking change, William pursued a tailor-made approach to convince people. For some people he used numbers and statistics, for others he explained empathically how they could benefit from change. He spent a lot of time talking to people individually to meet their personal meaning-seeking needs; "You cannot go in and try to create a culture in an organization. Culture stems from the people, and you need one person to do that, and then that could influence others" (William, Interview).

Lincoln: led a digitization process in Medant, a world-leading healthcare group, by implementing a machine learning model. Initially, Lincoln was fully backed by C-level support, including necessary time, financial funds and human resources to develop the project. Just before go-to-market, Lincoln got a new supervisor. The new manager had little knowledge of the project and did not see any value in it. In the end, the new supervisor did not support the launch of the project, Lincoln changed position internally, and the digitization process was never implemented. Lincoln was the only case in this research study, who was not successful with his project as he did not gain the final supervisor support. Lincoln mentioned two reasons: Lack of ownership of his new manager and clashing mindsets. First, Lincoln said, he would have needed to involve his new manager more and bring him onboard from the beginning. Second, Lincoln's experimental, trial-and-error mindset conflicted with his supervisor's conventional theoretical approach. "I mean to be very honest; it's never going to work with him. [...] He is a different school" (Lincoln, Interview). Thus, in comparison to the other *Diplomatic Rebels*, the findings show a lack of empathy and communication for Lincoln to gain buy-in from his supervisor.

Amongst all five cases, the key challenge of resistance is observed – either from colleagues or supervisors. Two common reasons are found: Firstly, employees do not *understand* the need for change as it does not fit into the existing organizational traditions. As Greg exemplified: "[Even] many of the lead designers [...] were telling us 'No, no, no, this is not Spiel. Spiel would never do that.' So, we had a lot of resistance there" (Greg, Interview). Alex was challenged by a management board who did not understand the urgency for change; "[I was] facing a lot of resistance, having a hard time making people understand what we were doing and why we were doing it" (Alex, Interview). Secondly, employees *fear* the consequences of change, as it could threaten their jobs or competences; "All of them [the employees] they were a little bit afraid. Because it's innovation projects, or let's call it radical innovation, that maybe could disrupt the company or our products. Or new things that are too far from what they're working on" (Alex, Interview). William highlights that fear is a natural response to change; "There will be some people that [...] are afraid of the change to a degree where they just hate everything you are doing. It will feel like it's personal but it's just because they are super afraid. So, you will figure out that people hate change [...] It's a very natural thing." The findings reveal that resistance is one of the most common challenges when introducing change in organizations. Therefore, the key fundamental principle of *Diplomatic Rebels* is to acknowledge that resistance always occurs in corporate innovation. Thomas explained: "it is about empathy. Sort of understanding that resistance is typically due to fear of the unknown and maybe not understanding or having been heard. I would always start with that, figuring that out. You have a responsibility to at least understand people are not doing this to be cruel to you but because they don't understand."

A main finding of the chapter is that innovation cannot be implemented successfully without the organization's support. Supervisor support is inevitable to obtain the required resources for the project and gain the final approval for implementation. Colleagues' support is crucial as they are the ones delivering and implementing the change initiatives.

3.2.5 Data analysis

In line with the inductive research approach, the data analysis interlinks findings from the cases and literature directly. Hence, this section dives deeper into the mindsets and skills of the *Diplomatic Rebels* required to overcome resistance and innovation barriers; and how these can be understood through existing theory. As summarized in Figure 3.2.2, the analysis shows that the principles represented by Design Thinking fall in line with the mindsets and skills of the rebel. The diplomat with strong interpersonal competences and a strategic focus could not be understood through Design Thinking. Here, management literature is used. The grey-highlighted fields represent a gap in the respective literature field where management (diplomat) complements Design Thinking literature (rebel) and vice versa.

3.2.5.1 The Rebel

As one key result, the rebel's mindsets and skills are found to fall in line with the key principles of Design Thinking. Thus, the rebel "embodies" the Design Thinker below.

Mindset

Like Design Thinkers, the findings show the rebel to have a mindset of **abductive reasoning**. Instead of accepting norms (What is?), the rebel inherently questions the status quo, breaks up existing structures and comes up with new ideas (What could be?) (Hassi & Laakso, 2011). Alex exemplified she "didn't want to follow the same path as others" and showed out-of-the-box thinking by claiming she "was very curious about finding new

Skills	• Thinking by Doing • Human-Centered Approach • Visualizing	• Rational Intelligence • Communication Skills
	• Collaborative Workstyles (towards external stakeholders, e.g., consumers)	• Emotional Intelligence & Cultural Awareness (towards internal stakeholders)
Mindsets	• Future-Oriented • Optimistic • Holistic View	• Strategic
	• Abductive Reasoning • Experimental & Explorative • Risk-Taking	
		• Agreeable
	Rebel Design Thinking Literature	**Diplomat** Management Literature

Figure 3.2.2 Overview of the *Diplomatic Rebel*'s mindset and skills, reflected through academia, incl. differences and gaps between literature streams. Grey areas are aspects not covered in the opposite literature.

ways [...]". Through their **experimental and explorative** mindset the rebels pursue unknown fields and try out new things. The interviewees often referred to this as a trial-and-error and "try early, fail fast" mindset. Rather than viewing challenges as limitations, the rebel **embraces risks** and dares to disrupt existing thinking models (Fraser, 2009; Brown, 2008). This was reflected in comments such as "You don't ask permission; you just try to do things" and "willing to go the full way and if you get fired, then be it" (Lincoln, Interview). For example, Lincoln and Greg dared to proceed with their projects and introduced radical digitization processes in their companies, even though they did not get management support at first. This mindset enables the rebel to solve wicked problems and arrive at radical innovation (Dunne & Martin, 2006; Brown, 2008; Kimbell, 2011). Same as a Design Thinker, the rebel exhibits a **future-oriented mindset** that aims to truly "change [the company] for the better" (Drews, 2009). The *Diplomatic Rebels* explain it as: "We are all the time looking into 'How can we develop it better or how can we do things that might have a bigger impact?'" (Matthew, Interview). In addition, the rebels present a **holistic view** towards customer understanding (Hassi & Laakso, 2011). For example, Thomas restructured the company towards a customer-centric organization. Other interviewees expressed that the "mindset is about understanding, [...] have empathy basically to read a situation, understand what is going on here" (Greg, Interview). Lastly, an **optimistic mindset** was found. Optimism relates to passion and the intrinsic motivation and the drive to keep going where others would give up (Dunne & Martin, 2006; Gloppen, 2009). In this case study, optimism is also reflected by a strong self-confidence and self-determinism; "[T]he rebel type, they love to shine themselves. [...] someone who can be a little dominating, who wants to fill out the room" (Greg, Interview).

Two mindsets postulated in Design Thinking are not found in the cases, namely ambiguity tolerance and reflective reframing (Hassi & Laakso, 2011; Kimbell, 2011). Most *Diplomatic Rebels* pushed their initial idea through to the end, instead of accepting outside opinions that might prompt them to diverge; "I have to push against the resistance, because I believe it" (Benjamin, Interview). Thereby, the rebels can create dislike and stir resistance amongst colleagues; "If they stay in the rebel-mode, then people will find [...] them hard to work with, annoying. They will be people that you want to avoid" (Thomas, Interview).

Skills

Based on the findings, four predominant skills emerged for the rebel, similarly to the Design Thinker. Firstly, the skill of **thinking by doing** is found in the rebel's proactive way of working, recognizing trends early on and staying ahead in the innovative field. The rebel's actions seem to be led by intuition, a practice of "first doing, then thinking". Interviewees referred to it as "thinking in prototyping, thinking in testing" before extensive investments have been made (Bohemia, Liedtka & Rieple, 2012, p.157; Brenner & Uebernickel, 2016, p.8). Lincoln stressed the need to launch the digitization project as soon as possible to test it on the market: "it needed to go to market - so I launched that" (Lincoln, Interview). Secondly, rebels are skilled in a **human-centred approach**. Having a dialogue with users from early on is a core skill mentioned frequently by the interviewees. Especially Thomas highlighted this: "So instead of these developers being half-time product developers, I wanted them to have more strategic focus on the customer needs and pains, being out there with the customers". This highlights a crucial finding of the study: Whereas Design Thinking mainly focuses on the external user (Hassi & Laakso,

2011), the *Diplomatic Rebels'* understanding of human-centeredness also includes internal stakeholders, how they perceive the project and how to get them onboard. Thirdly, the findings confirm **collaborative work styles** as a key skill (Hassi & Laakso, 2011; Holloway, 2009; Kimbell, 2011). The interviewees showed strong cross-team collaboration skills. However, working across departments simultaneously brings certain challenges; As Lincoln highlighted: "[I]magine 10 people that have different calendars and you have to figure out how to get the right people to meet together and bring a project forward". These points of strategic stakeholder management and project planning are neglected in Design Thinking literature. Lastly, the interviewees were found to express their ideas through **visualising** (Hassi & Laakso, 2011), for example by using artefacts, LEGO Serious Play, or prototyping tools.

3.2.5.2 The Diplomat

The results show that the diplomat's mindsets and skills are described in management literature.

3.2.5.2.1 Mindset

The data analysis derives two core mindsets of the diplomat, being strategic and agreeable. The findings show that the **strategic diplomat** would approach tasks analytically in a holistic and long-term manner to find the most efficient and realistic solutions. These features can be linked to conscientiousness and orderliness, characterizing a person as being planful and organized (Judge et al., 1999b). The interviewees had a clear goal in mind while being aware of "the strategic targets of the company: where we're going, how to make money", as well as the internal values and meaning-seeking needs: "If you want to take an organization in a direction, […] you need to create change that goes along those values" (Matthew, Interview). The findings further show that the diplomat has a very people-focused mindset, coined as **agreeableness** – a mindset based on emotional intelligence that is deeply rooted in empathy (Judge et al., 1999b). The agreeable diplomat is a likeable person, who is "very patient [and] a good-people-person. So preferably someone who's able to have a conversation with colleagues on their level, so that they feel comfortable" (Greg, Interview). This element of "whole-hearted openness", "soft approach" and "trusted leadership" is a topic much discussed in Diplomatic Rebel'sfellow group. It constitutes a core principle of *Diplomatic Rebels*, described as "writing love letters" and as "making other people shine". As William put it: "Writing love letters is a great one. You have to make sure that people are encouraged and acknowledged because that's what drives human behaviour". Management literature supports these findings with agreeableness and sociological understanding being the core competencies of a project manager to create trust and bond with others (Meng & Boyd, 2017; Fisher, 2011).

3.2.5.2.2 Skills

In line with the strategic mindset above, the diplomat has the skill of *rational intelligence*, the ability to **think and act analytically** (Van der Walt et al., 2006). In the interviews this is reflected in presenting precise action plans, charts and numbers to convince stakeholders; "I made a project plan and I actually did it sometime prior to the project start", "I had just presented my findings that show that we could potentially improve our sales rep's revenue somewhere between 30–40%." (Lincoln, Interview). Beyond that, all interviewees demonstrated strong skills in *emotional intelligence* and *cultural awareness* (ibid., Fisher, 2011). Here, the diplomat is analysed as **understanding and supportive** towards their innovation teams. *Diplomatic Rebels* booked frequent touchpoints with their teams to understand their points of departure and feelings towards the organizational change; "You really need to make sure that everybody is feeling comfortable. If someone gets stuck, you need to maybe help them" (Benjamin, Interview). Being able to understand one's own as well as others' behaviours, emotions and values proves to be a crucial basis to create a culture of change (Rezvani et al., 2016). In addition, the interviewees highlighted the need to **involve and empower** team members from early on, which is reflected in the concept of "building a tribe" – a crucial skill of the *Diplomatic Rebels*. Despite many colleagues being hesitant towards the innovation project, all *Diplomatic Rebels* started out with people in their companies who believed in the project and then scaled it by engaging more and more colleagues. "You have to make sure that people are encouraged and acknowledged from early on because [...] then they are more willing to change". This goes back to empathy and speaking to colleagues on an "eye-to-eye level". As Matthew said: "You need to be able to tell a story that people can relate to, and they want to be part of. Because that's how you get people behind you". Engaging and involving is a crucial leadership skill to motivate and inspire others for the innovation project (Fisher, 2011). Therefore, strong communication skills are needed to bring people onboard (Kotter & Schlesinger, 2008; Maurer, 2009; Ford & Ford, 2009). The interviewees emphasized the importance of being able to **influence and convince** others through storytelling and explaining why change is needed. William for example emphasized how he persuaded colleagues for the innovation project by tailoring his communication approach to individual's needs. He would present different charts and statistics to strategically convince different stakeholders. Explaining the importance of a project, its benefit for individuals and why they should change, can unlock people's resistance towards change (Fisher, 2011).

3.2.5.3 The interrelation of Diplomatic Rebels

All interviewees referred to the *Diplomatic Rebel* as a unique concept and an equilibrated combination of competences; "I was amazed by this combination [...]. I felt very connected with it because I went through that process – being too much rebel that wasn't that good, but when I was too diplomatic, it didn't work either" (Alex, Interview). Throughout the fellowship, they learned that both competences of the rebel and the diplomat were needed to drive organizational innovation : "If you're only a diplomat, [...] radical innovation projects don't get approved because you're not pushing hard enough [...] because you don't have the power-through, the will of the rebel for radical change, to go the extra-mile" (Thomas, Interview). On the other side, being merely a rebel, one would oftentimes be disliked in the organization and stir resistance. For example, Alex realized she could not bring change forward in the E.G. holding group if her approach was too "pushy" for the board of managers; "I was a rebel, you know. So, I was very, very pushy, and in the process, I harmed a lot of relationships. [...] It was very hard in the company, because [...] their investments were big, so they were very reluctant of creating different things [...] so I realized I needed to be more diplomatic." Analysing the interrelation of the two, the findings show that the rebel and diplomat are not contradictory; "I don't think they need to be opposing in any way. [...] Because that would make it a bit schizophrenic [...]" (Greg, Interview). Instead, the rebel and diplomat are found to emerge interactively - depending on the situation and the person being addressed; "You need those two to go hand in hand. So even when you're doing something very rebellious, like getting up on stage and telling everybody that we need to think differently - We know the company is about to be disrupted and needs to do something. Then you need the diplomat at the same time, telling the story in a way that makes everybody believe in it and makes everybody understand that they can be part of it" (Greg, Interview).

3.2.6 Discussion

The following section aims to discuss in depth how *Diplomatic Rebels* manage the outlined challenges of corporate innovation, namely innovation barriers and resistance. Further is discussed how the rebel and diplomat interrelate at the two stages of the innovation process: the ideation and implementation phase (Amabile, 1988; Baer, 2012; Rosing et al., 2011). As outlined in the findings, it is argued that the rebel (Design Thinking) comes up with ideas and introduces new thinking models. For successful implementation, the diplomat (management literature) comes into play to overcome resistance and manage stakeholders. For simplicity reasons, this chapter refers to stereotypes of the rebel as the Design Thinker and the diplomat as a strategic and agreeable manager. However, this should not be seen as unequivocal, as the two sides are part of one person with overlaps occurring.

3.2.6.1 Innovation barriers

Literature has vastly shown that large organizations fail to innovate as they struggle to overcome adoption, mindset and risk barriers (Assink, 2006; Bucy et al., 2016). After all, that is the exact reason why Design Thinking has been brought to the business field: To help organizations innovate. This research argues that it is the rebel-side that helps organizations to break out of obsolete systems (Carlgren et al., 2016). In the following, it is discussed how the rebel manages the *adoption barrier, mindset barrier* and *risk barrier* (Assink, 2006).

Wanting to drive organizations forward in a future-oriented way, rebels strive to create change for the better (Lockwood, 2010). Through abductive reasoning and an "what if"-approach, the rebel questions existing norms and triggers disruptive thinking (Hassi & Laakso, 2011). Thereby, the rebel has the capabilities to overcome *adoption barriers* and break out of obsolete systems. That was demonstrated by all interviewees aiming to "stir things up and shake everybody up" in their elsewise traditional companies (Benjamin, Interview). Lincoln expressed the striving for disruption as "You don't ask for permission; you just try to do new things". Based on their designerly way of thinking, rebels do not only challenge existing norms, but also help companies find new ideas and solutions (Dunne & Martin, 2006). Through their creative vision, rebels come up with solutions that "break new ground, develop new models, and pioneer new approaches" (Turell & Earle, 2018, p. 173). The rebel's experimental and explorative mindset is a key capability to overcome the *mindset barrier* and come up with new solutions - where other employees would stay within existing mental frameworks such as "This is not Spiel. Spiel would never do this" (Greg, Interview). This challenger mindset was presented in all interviewees introducing radically new approaches in their companies, from digitization, restructuring and introducing customer-centric thinking models. The rebel type pushes radical ideas forward and engaged in a process of testing them from an early stage and improving them in iterations (Hassi & Laakso, 2011), which the interviewees referred to as "try early, fail fast" approach. Closely connected to the mindset barrier is the *risk barrier*. Perceiving ambiguous situations as desirable and feeling comfortable with taking risks, the rebel is set to cope better with change and thereby overcome the risk barrier (Judge et al., 1999a; Vakola et al., 2004). While their organizations fear the risks of innovation, the rebels accept the possibility of failing (Holloway, 2009; Plattner & Meinel, 2014). Both Lincoln and Greg proceeded with their projects in their free time, despite the explicit project refusal from C-level, going as far as risking their jobs.

While the rebel is a key enabler to break through these barriers and drive forward new solutions, there are drawbacks. By pursuing novelty and innovation in organizations, the rebel naturally stirs resistance in the company – as "people hate change" (William, Interview). Through their risk-taking mindset and willingness to drive change, the rebel can appear "too pushy, and harm a lot of relationships" (Alex, Interview). Thus, the rebel's disruptive capacity can come at the cost of sensitivity and empathy towards others. Introducing radical change without being mindful about internal stakeholder perceptions, the intrapreneur might not be able to implement the innovation project successfully. It was particularly Lincoln, who struggled to build trust with his new supervisor and ultimately failed to gain buy-in. This is an aspect neglected in Design Thinking theory. This chapter bridges the gap and adds to the academic conversation: Reviewing Design Thinking literature, some authors do acknowledge the importance to function as a visionary leader (Boland et al., 2008; Bohemia et al., 2012; Johansson-Sköldberg et al., 2013) and build a safe culture for employees (Kolko, 2015). In addition, most Design Thinking literature emphasises team collaboration and orchestrating diverse inputs from team members (Hassi & Laakso, 2011). However, there is little conversation in Design Thinking literature on how to lead teams, manage people and their emotions – and how to deal with fear around change, an aspect essential to get people's buy-in for innovation projects. Accordingly, this chapter brings the diplomat into play as a novel contribution to Design Thinking - a main contribution of this chapter (see Figure 3.2.2).

3.2.6.2 Resistance

The previous findings show that resistance is one of the most common challenges in innovation (Assink, 2006; Kotter & Schlesinger, 2008; Maurer, 2009). In the following, it is discussed how the diplomatic side enables the *Diplomatic Rebels* to manage resistance through their unique competences of agreeableness and strategic thinking.

3.2.6.2.1 Agreeableness

Agreeableness, being cooperative, trustful, and likeable to others, is an essential competence of the diplomat to help employees cope with change (Judge et al., 1999b; Vakola et al., 2004). While the rebel often creates friction with people, it is the diplomat who resonates well with people by understanding human behaviour (London, 1999). Based on the skill of emotional intelligence, *Diplomatic Rebels* understand their colleagues' standpoint and are able to bring them onboard for the project; "You need to be able to tell a story that people can relate to, and they want to be part of. And that again has to do with values: […] to connect to something that they truly believe is important" (Matthew, Interview). The interviewees expressed how they first struggled to bring people onboard, as they faced two different types of resistance (as outlined in the findings): Either people not understanding why change was needed *(I don't get it)* or people being afraid of the consequences of change *(I don't like it)* (Maurer, 2009). By involving their colleagues from early on and engaging them from the start and explaining why change is needed, the diplomat can overcome first-level resistance *(I don't get it)* (Maurer, 2009). For example, Lincoln and Greg "built a tribe", starting out by seeking support from few people and then scaling the team through followership. As Greg described; "You have to make sure that people are encouraged and acknowledged from early on because that's what drives human behaviour [...] then they are more willing to change". Here customized communication and rhetoric prove successful, to be responsive to multiple audiences (Kotter & Schlesinger, 2008; Maurer, 2009; Turell & Earle, 2018). For example, William excelled at making people understand why change is needed, by highlighting the individual benefits for each stakeholder. Based on cultural awareness and their emotional sensitivity, the diplomat can understand different systems of beliefs and can specifically address the meaning-seeking needs and motivations of employees (Fisher, 2011; Van der Walt et al., 2006). This becomes crucial to manage second level resistance *(I don't like it)* which often stems from fear (Maurer, 2009). By developing trusted relationships with colleagues, the diplomat provides a "safe house" to address fears and concerns within the team (Baer, 2012; Rezvani et al., 2016). Thereby, the diplomat can create a shared understanding and a positive perception of change by supporting the team members emotionally (ibid.; Vakola et al., 2004). For example, frequent touchpoints with their teams helped the *Diplomatic Rebels* to understand their colleagues' standpoint and address potential feelings of mistrust or fear towards the change project. The strong sensibility towards colleagues, demonstrated in the concept of "writing love letters" and "making other people shine", is found as a unique competence of the *Diplomatic Rebels*. Their empathy, care and sincere interest to understand other people, helps them in being exceptionally successful leaders in situations of change (ibid.).

It can be argued that the rebel/Design Thinker, is also able to build empathy with others (Holloway, 2009). Sato (2009) even claims that Design Thinkers must not only empathize with the customers' practical needs, but also with their emotions and social and cultural environment. The difference between this chapter and Design Thinking literature is, that Design Thinking focuses on the meaning-seeking needs of users and customers, but it does not cover internal stakeholder management. Yet, intrapreneurs do need to be agreeable and empathic with their colleagues to help them cope with their inner resistance to change. This internal diplomatic management aspect represents an essential contribution to Design Thinking, required to successfully manage design and innovation in organizations.

Lastly, it can be discussed whether too close emotional bonds can lead to underperformance and inertia as team members fear to express criticism (Boschma, 2005; Turell & Earle, 2018). This dilemma is often termed as over-embeddedness; a lack of diverse perspectives, impeding the innovation performance (Lowik et al., 2012; Andersen, 2013). Accordingly, strong buy-in ties should be seen with caution. For example, it could be questioned whether innovation is always required, in the first place. In the observed cases, the *Diplomatic Rebels* were merely concerned with effectively building trust to bring colleagues on board, whereas voices about concerns might have been unheard. For example, in Greg's case, many employees expressed that entering the digital gaming industry would betray Spiel's tradition in manufacturing physical toy bricks. Whereas the interviewees did not deviate from their initial path, innovation managers should see resistance as valuable feedback to continually reflect if the project is going in the right direction (Ford & Ford, 2009; Turell & Earle, 2018). Interestingly, Design Thinking provides perspectives here and puts importance on reflective reframing and ambiguity tolerance (Holloway, 2009; Dunne & Martin, 2006) - two elements found as shortfalls in the *Diplomatic Rebels* cases. More specifically, it is argued that Lincoln might have been able to succeed with his project, if he had acknowledged the perspectives of his new supervisor and integrated those into his project approach. The supervisor had strategic insights from a top management perspective that probably implied valuable reasons for not accepting the project.

3.2.6.2.2 Strategic thinking

The strategic and rational side to the diplomat is especially needed when it comes to convincing supervisors and overcoming their resistance - a second aspect disregarded in Design Thinking. Supervisor support is vital for successful innovation implementation as the project is not further scalable without the buy-in from higher management (Černe et al., 2016; Škerlavaj et al., 2014). Being a conscientious decision-maker who acts in line with the commercial goals and values of the organization, the diplomat can negotiate legitimacy for a project towards C-level management and supervisors (Turell & Earle, 2018). Whereas the rebel/Design Thinker focuses mainly on desirability and makes decisions based on external user insights and creative intuition (Hassi & Laakso, 2011), the diplomat can demonstrate the business viability of a project towards top management (Van der Walt et al., 2006; Turell & Earle, 2018). Especially William excelled at gaining his supervisors' backing through his tailor-made strategic approach and by presenting sales numbers according to whom he would talk to on C-level. Also Thomas underlined the importance to engage top-management with the right arguments, at the right time. Adding the strategic business perspective to Design Thinking, is a second contribution of this chapter.

The question could be raised whether the diplomat manipulates colleagues through customized rhetoric (Von der Walt et al., 2006; London, 1999). This research paper, however, argues that the diplomat merely communicates the rebel's true motivation and honest belief in the project in a way that others can resonate with. Whereas the rebel likes to shine on stage alone, the diplomat would "tell the story in a way that makes everybody believe in it and makes everybody understand that they can be part of it." (Greg, Interview).

Concluding on how the *Diplomatic Rebels* manage the innovation process, the rebel is predominant for the *idea generation phase* to overcome organizational *innovation barriers* and to push for change. For the successful *implementation phase* of change and innovation, one needs to manage *resistance* and bring stakeholders onboard – a core competence, where the diplomat comes into play. Wheras Design Thinking provides strong insights for the idea generation phase, the implementation phase in large organizational settings is vastly neglected in Design Thinking – including aspects like resistance management, internal stakeholder management, and strategic business viability.

3.2.6.3 Interrelation between Diplomat and Rebel

With the *Diplomatic Rebel*, this study introduces a new holistic concept, bridging the gap between Design Thinking (the rebel) and management literature (the strategic and agreeable diplomat). Questioning why those streams have not been combined before and whether conflicts might arise, this section aims to discuss if and how the two modes can be combined in one concept (one person) and how they interrelate during the innovation process. Here it is important to mention that the intrapreneur is taken as a starting point for this study, instead of innovation teams with different competences distributed across people (Design Thinkers and managers). Thus, the discussion is not rooted in the question, if different people could take on the different roles, but whether one person could occupy both – and how those roles interrelate.

Looking for literature that discusses Design Thinking and management jointly, ambidexterity literature prompts a first attempt to combine the two fields. Ambidexterity constitutes the dual roles in terms of *exploration* and *exploitation* (Paliokaite & Pacesa, 2015; Rosing et al., 2011). Here, exploration would represent the rebel, as it is defined as challenging existing systems and discovering new concepts. Exploitation would reflect the *strategic diplomat* as it is expressed in resource-sustaining and analytical approaches (Paliokaite & Pačesa, 2015). While authors in ambidexterity literature do not reach a consensus whether exploration and exploitation can be combined in one person or should be split between different roles, this chapter argues that the two modes can be unified – namely, in the unique concept of the *Diplomatic Rebel.* The empirical data shows that the interviewees do not experience an inner bias to reconcile the rebel and diplomat in one person. All interviewees agreed that the two modes complement rather than conflict each other. "I think in that sense they're not separated. Because that would make it a bit … . […] schizophrenic. […] In that sense one just enforces the other. I think it's a perfectly balanced mix" (Greg, Interview). Thus, in line with Bodwell and Chermack (2010), Rosing et al. (2011) and Keller and Weibler (2014), the findings show that the two roles can co-exist. Hereby, it is found that the rebel and diplomat come into play for different tasks and situations. The *Diplomatic Rebels* mentioned it was intuitive to them due to their high level of empathy and situation-understanding, in which situations to push for innovation and challenge norms, and when it was crucial to speak to people and engage them for implementation. Given the mindsets and skills of the rebel to constantly challenge status quo, come up with new solutions and inspire others for change, the rebel is mainly predominant in the ideation phase (Amabile, 1988). However, the rebel naturally stirs resistance as "people hate change" and if neglected, the innovation implementation would fail (Maurer, 2009). Thus, the diplomat is inevitable for implementation – yet, neglected in Design Thinking. With agreeableness and a strategic focus, the diplomat can overcome managers' and employees' resistance and gain their buy-in to successfully implement the innovation project. By introducing the unique concept of the *Diplomatic Rebel,* this chapter creates a comprehensive view on the innovation process, as it reconciles the two main activities into one role and makes the innovation process complete: spanning both idea generation (creativity) and implementation (innovation) (Amabile, 1988; Baer, 2012; Rosing, et al., 2011).

While ambidexterity focuses on exploration and exploitation capabilities, this chapter calls for a new holistic theory. The concept of *Diplomatic Rebels* enhances ambidexterity literature in two realms. Firstly, the combination of exploration *(rebel)* and exploitation *(strategic diplomat)* is advanced with the emotionally intelligent diplomat *(agreeable diplomat)*, which is crucial for successful implementation and bringing people onboard. Here a unique concept is introduced, which has not been explored in literature before. Secondly, this study enhances the academic conversation by proving, through empirical findings, that in practice the combination of the rebel and the diplomat in one role is possible, and that they interrelate situational at different stages of the innovation process. As this dual concept is unique, rare to find and difficult to imitate (Paliokaite & Pacesa, 2015), organizations looking to innovate[3] should recruit staff with the mindsets and skills of a *Diplomatic Rebel.*

3.2.7 Conclusion

This study examines the innovation process in large organizations – from idea generation to implementation – by studying *Diplomatic Rebels*. The rebel's mindset and skills are thereby analysed through Design Thinking literature. Confirming Design Thinking's purpose in the management field, the rebel proves crucial to overcome innovation barriers and bring an increased level of innovativeness to the business world. However, by challenging the status quo and introducing change, the rebel naturally stirs resistance within the organization – an aspect not regarded in Design Thinking so far. Here the chapter adds to Design Thinking by bringing the diplomat – discussed through management literature – into play as a crucial factor to turn creativity (idea generation) into innovation (implementation). With the combination of strategic and agreeable mindsets and skills, the diplomat can mitigate resistance and bring employees from different levels on board. If neglected, the innovation implementation is prone to fail.

Concluding, this chapter adds novelty to the Design Thinking field on an empirical and theoretical level. Firstly, the academic conversation is advanced with empirical evidence through a practice-based deep dive into a case study of multiple different companies In doing so, the mindsets and skills of a rebel (Design Thinker) are complemented with the agreeableness and strategic thinking of a diplomat. Secondly, this chapter contributes on a theoretical level. Design Thinking literature was brought to the management field because it introduced a new way of thinking and developing solutions, namely abductive reasoning. However, when looking at corporate innovation, the design way of working does not regard the organizational settings and the implications to manage people and resistance. Hence, this paper bridges the academic gap by bringingmanagement literature to Design Thinking. In particular, as a main contribution to literature, the chapter advances the innovation managers' competences with agreeableness and a high level of emotional intelligence and cultural awareness. Uniting the two realms, the *Diplomatic Rebel* proves a powerful concept to advance literature with a new holistic approach that bridges Design Thinking and management theory. With more companies aiming to innovate, the *Diplomatic Rebel* can provide a competitive advantage and can serve as an attractive profile to recruit.

3.2.8 Limitations & future research

Ambidexterity literature takes a first step at integrating exploration (rebel) and exploitation (strategic diplomat). However, the emotional element (agreeable diplomat) is missing. That's where this study takes off and introduces a novel and holistic concept. Taking a first practice-based view, future research and theory is needed to explore the concept further and go more into depth.

This study focuses on a one-sided view through the eyes of the *Diplomatic Rebels*, merely based on interviews. For future research, a more in-depth analysis could follow including more cases as well as ethnographic observations on how change is executed from the organization's view. The case-study indicates that there are further factors influencing the success of innovation managers, beyond the personal level (mindsets and skills). It would be interesting to investigate *Diplomatic Rebels* in relation to already established theories on success factors – for example, at the job level, team level, organizational level (Černe et al., 2016). Further, this study is built on a small sample size which could have influenced the results. Hence, conducting further experimental studies with a control group could prove valuable.

Lastly, the distinction made between mindsets and skills was mainly influenced by literature (Hassi & Laakso, 2011). However, there were no further consequences drawn from this distinction. Theory defines the mindset as being genetically determined (Judge, 1999a) and skills being a developable trait (Vakola, 2004). The interviewees suggest that the rebel is more intuitive and therefore more difficult to learn whereas the diplomat could be developed more easily. So, could the rebel be defined as a mindset and the diplomat a skill? Researching how far the diplomat and the rebel can be learnt from an empirical *and* theoretical standpoint would yield further insights into how organizations can increase their innovativeness by recruiting and training intrapreneurs with the characteristics of a rebel and diplomat. To reconcile and enhance the two competence streams of employees, emotional intelligence trainings could be a starting point to understand differing mental and emotional conditions and the ability to turn them into constructive directions at different stages of the innovation process (Rezvani et al., 2016).

All in all, this chapter calls for an integrative theory-building that extends Design Thinking with management literature and merges all sides of the mindsets, skills and tools of the *Diplomatic Rebel* – thereby advancing academia for Design Thinking and innovation in an organizational context.

Notes

1 We indicate the company named Diplomatic Rebels in non-italic font and introduce the concept of *Diplomatic Rebels* in italic font.
2 Both the interviewees' names and their respective companies are anonymised.
3 Annotation: We focus on companies aiming to drive organizational innovation. Thus, for other company setups, it might be desirable to look at the rebel or diplomat alone.

References

Altringer, B. (2013). A new model for innovation in big companies. *Harvard Business Review*. Retrieved from https://hbr.org/2013/11/a-new-model-for-innovation-in-big-companies

Amabile, T. M. (1988). A model of creativity and innovation in organizations. *Research in Organizational Behavior*, *10*, 123–167.

Andersen, K. V. (2013). The problem of embeddedness revisited: Collaboration and market types. *Research Policy*, *42*(1), 139–148.

Assink, M. (2006). Inhibitors of disruptive innovation capability: A conceptual model. *European Journal of Innovation Management*, *9*(2), 215–233.

Baer, M. (2012). Putting creativity to work: The implementation of creative ideas in organizations. *The Academy of Management Journal*, *55*(5), 1102–1119.

Bodwell, W., & Chermack, T. J. (2010). Organizational ambidexterity: Integrating deliberate and emergent strategy with scenario planning. *Technological Forecasting and Social Change*, *77*, 193–202.

Bohemia, E., Liedtka, J., & Rieple, A. (2012). Leading innovation through design. InE. Bohemia, J. Liedtka, & A. Rieple (eds.), *Proceeding of the DMI 2012 International Research Conference*. Boston.

Boland, R. J., & Collopy, F. (2004). *Managing as designing*. Redwood City, CA: Stanford University Press.

Boland Jr, R. J., Collopy, F., Lyytinen, K., & Yoo, Y. (2008). Managing as designing: Lessons for organization leaders from the design practice of Frank O. Gehry. *Design issues*, *24*(1), 10–25.

Boschma, R. (2005). Proximity and innovation: A critical assessment. *Regional Studies*, *39*(1), 61–74.

Brenner, W., & Uebernickel, F. (2016). *Design thinking for innovation*. Springer.

Brenner, W., Uebernickel, F., & Abrell, T. (2016). Design thinking as mindset, skills, and toolbox. In *Design Thinking for Innovation* (pp. 3–21). Springer.

Brown, T. (2008). Design thinking. *Harvard Business Review*, *86*(6), 84–92.

Bucy, M., Finlayson, A., Kelly, G., & Moye, C. (2016). The 'how' of transformation. *McKinsey & Company*. Retrieved from https://www.mckinsey.com/~/media/McKinsey/Industries/Retail/Our%20Insights/The%20how%20of%20transformation/The-how-of-transformation.pdf

Carlgren, L., Elmquist, M., & Rauth, I. (2014). Design thinking: Exploring values and effects from an innovation capability perspective. *The Design Journal*, *17*(3), 403–424.

Carlgren, L., Elmqvist, M., & Rauth, I. (2016). Exploring the use of design thinking in large organizations: Towards a research agenda. *Swedish Design Research Journal*, *11*(1), 23–32.

Carlgren, L., Rauth, I., & Elmquist, M. (2016). Framing design thinking: The concept in idea and enactment. *Creativity and Innovation Management*, *25*(1), 38–57.

Carlgren, L., Elmquist, M., & Rauth, I. (2016). The challenges of using design thinking in industry – experiences from five large firms. *Creativity and Innovation Management*, *25*(3), 344–362.

Černe, M., Hernaus, T., Dysvik, A., & Škerlavaj, M. (2016). A bird's eye view of the creativity-innovation nexus: The moderating role of supervisor support and decision autonomy. In *Human Resource Management, Innovation and Performance* (pp. 163–178). London: Palgrave Macmillan.

Dew, N. (2007). Abduction: A pre-condition for the intelligent design of strategy. *Journal of Business Strategy*, *28*(4), 38–45.

Drews, C. (2009) Unleashing the full potential of design thinking as a business method. *Design Management Review*, *20*, 38–44.

Dunne, D., & Martin, R. (2006). Design thinking and how it will change management education: An interview and discussion. *Academy of Management Learning & Education*, *5*(4), 512–523.

Emerson, R., Fretz, R. & Shaw, L. (2011). *Writing ethnographic fieldnotes*. Chapter 6. University of Chicago Press.

Fisher, E. (2011). What practitioners consider to be the skills and behaviours of an effective people project manager. *International Journal of Project Management*, *29*(8), 994–1002.

Ford, B. Y. J. D., & Ford, L. W. (2009). Decoding resistance to change. *Harvard Business Review*, (April), 99–104.

Fraser, H. (2009) Designing business: New models for success. *Design Management Review*, *20*, 56–65.

Gloppen, J. (2009) Perspectives on design leadership and design thinking and how they relate to European service industries. *Design Management Journal*, *4*, 33–47.

Hassi, L., & Laakso, M. (2011). Design thinking in the management discourse: Defining the elements of the concept. In *18th international product development management conference*.

Holloway, M. (2009). How tangible is your strategy? How design thinking can turn your strategy into reality. *Journal of Business Strategy*, *30*(2/3), 50–56.

Hunter, S. T., Cushenbery, L. D., & Jayne, B. (2017). Why dual leaders will drive innovation: Resolving the exploration and exploitation dilemma with a conservation of resources solution. *Journal of Organizational Behavior*, *38*(8), 1183–1195.

Johansson-Sköldberg, U., Woodilla, J., & Cetinkaya, M. (2013). Design thinking: Past, present and possible futures. *Creativity & Innovation Management*, *22*(2), 121–147.

Judge, T. A., Higgins, C. A., Thoresen, C. J., & Barrick, M. R. (1999a). The big five personality traits, general mental ability, and career success across the life span. *Personnel Psychology*, *52*, 621–652.

Judge, T. A., Thoresen, C. J., Pucik, V., & Welbourne, T. M. (1999b). Managerial coping with organizational change: A dispositional perspective. *Journal of Applied Psychology*, *84*(1), 107–122.

Keller, T., & Weibler, J. (2014). Behind managers' ambidexterity – Studying personality traits, leadership, and environmental conditions associated with exploration and exploitation. *Schmalenbach Business Review*, *66*(3), 309–333.

Kimbell, L. (2011). Rethinking design thinking: Part I. *Design and Culture*, *3*(3), 285–306.

Kolko, J. (2015). Design thinking comes of age. *Harvard Business Review*, (September).

Kotter, J. P., & Schlesinger, L. A. (2008). Choosing strategies for change. *Harvard Business Review*, (August), 130–139.

Lockwood, T. (2010). The bridge between design and business. *Design Management Review*, *21*(3), 5.

London, M. (1999). Principled leadership and business diplomacy. *Journal of Management Development*, *18*(2), 170–192.
Lowik, S., Rossum, D. Van, Kraaijenbrink, J., & Groen, A. (2012). How small firms innovate through bridging capabilities. *Journal of Small Business Management*, *50*(2), 239–256.
Maurer, R. (2009). Introduction to change without migraines. *Maurer and Associates*, *3*, 1–45.
Meng, X., & Boyd, P. (2017). The role of the project manager in relationship management. *International Journal of Project Management*, *35*(5), 717–728.
Miles, M. B., Huberman, A. M., & Saldana, J. (2013). Fundamentals of qualitative data analysis. In *Qualitative data analysis - A methods sourcebook* (3rd ed.). Los Angeles: Sage.
Nikolaou, I., Gouras, A., Vakola, M., & Bourantas, D. (2008). Selecting change agents: Exploring traits and skills in a simulated environment. *Journal of Change Management*, *7*(3–4), 291–313.
Nussbaum, B. (2011). Design thinking is a failed experiment. So what's next? Retrieved from https://www.fastcompany.com/1663558/design-thinking-is-a-failed-experiment-so-whats-next
Paliokaite, A., & Pačesa, N. (2015). The relationship between organisational foresight and organisational ambidexterity. *Technological Forecasting and Social Change*, *101*, 165–181.
Plattner, H., & Meinel, C. (2014). *Design thinking research: Building innovators* L. Leifer (ed.). Springer International Publishing.
Qu, S. Q., & Dumay, J. (2011). The qualitative research interview. *Qualitative Research in Accounting and Management*, *8*(3), 238–264.
Rauth, I., Carlgren, L., & Elmquist, M. (2015). Making it happen: Legitimizing design thinking in large organizations. *Design Management Journal*, *9*(1), 47–60.
Rezvani, A., Chang, A., Wiewiora, A., Ashkanasy, N. M., Jordan, P. J., & Zolin, R. (2016). Manager emotional intelligence and project success: The mediating role of job satisfaction and trust. *International Journal of Project Management*, *34*(7), 1112–1122.
Rosing, K., Frese, M., & Bausch, A. (2011). Explaining the heterogeneity of the leadership-innovation relationship: Ambidextrous leadership. *Leadership Quarterly*, *22*(5), 956–974.
Rylander, A. (2009). Design thinking as knowledge work: Epistemological foundations and practical implications. *Design Management Journal*, *4*(1), 7–19.
Sandberg, B., & Aarikka-Stenroos, L. (2014). Industrial marketing management: What makes it so difficult? A systematic review on barriers to radical innovation. *Industrial Marketing Management*, *43*(8), 1293–1305.
Sato, S. (2009) Beyond Good: Great innovations through design. *Journal of Business Strategy*, *30*, 40–49.
Saunders, M. N. K., Lewis, P., & Thornhill, A. (2009). *Research methods for business students* (5th ed.). Harlow: Pearson.
Škerlavaj, M., Černe, M., & Dysvik, A. (2014). I get by with a little help from my supervisor: Creative-idea generation, idea implementation, and perceived supervisor support. *Leadership Quarterly*, *25*(5), 987–1000.
Turell, Y. W., & Earle, A. G. (2018). Social entrepreneurs and field-level change: an institutional process model of social entrepreneurship. In *Social Innovation and Sustainable Entrepreneurship* (pp. 153–189).
Vakola, M., Tsaousis, I., & Nikolaou, I. (2004). The role of emotional intelligence and personality variables on attitudes toward organisational change. *Journal of Managerial Psychology*, *19*(2), 88–110.
van der Walt, E. A., du Plessis, D., & Barker, R. (2006). Exploring a spiritual intelligence (SQ) model of communication to recontextualise differences between management and employees. *Communication*, *32*(2), 264–290.

3.3 Design thinking and innovation in SMEs

A systematic literature review of barriers and best practices

Fatma Demir, Irina Saur-Amaral, and Daniel Ferreira Polónia
University of Aveiro

3.3.1 Introduction

Developing technologies and changing markets are creating new challenges and opportunities for companies (Goerge & Schoemaker, 2016). In the search for new solutions, the importance of corporate innovation management is increasing – even for small- and medium-sized enterprises (SMEs). The latter often lack resources to face these new challenges, even though they play a key role in the economic fabric of nations and will continue to do so in the expected near future (van de Vrande, de Jong, Vanhaverbek & de Rochemont, 2009). Especially for SMEs, the standard tools or approaches to innovation management are unsuitable and outdated (Assink, 2006). Design Thinking (DT) meets the needs of SMEs in terms of innovation capacity by promoting user-centricity and creativity, as well as uncovering unknown potentials. Thus, for SMEs, a creative approach to innovation development, such as DT, is even more pivotable and profitable (Assink, 2006).

Three main motivations guide the choice in a literature review: the kind of research interest, the characteristics and coverage of previous reviews, and the information explosion (Williams, Clark, Clark & Raffo, 2020). Maintaining an overview in specialized fields proves difficult, given today's information explosion; necessary information is often tricky to find. The identified research gap addresses a growing interest, yet a lack of overview of the field of DT in SMEs, as well as the innovation capacity (van de Vrande, de Jong, Vanhaverbek & de Rochemont, 2009). A research interest in this regard can be answered in principle with literature reviews, as they create an orientation service, show connections and disputes of previous works, and foster an understanding of the theoretical phenomena and terminology of the field (Williams, Clark, Clark & Raffo, 2020). Previous literature reviews do not fill the research gap sufficiently, as demonstrated by the corresponding systematic literature review results in the following section.

The selection of a specific systematic review methodology, especially in terms of its complexity, depends on the principal research objective and the general restrictions of the research context (Crossan & Apaydin, 2010).

This insight will help researchers to have more realistic expectations of DT and innovation processes in SMEs through a systematic approach and will help mentors guide newcomers through the phases of planning, implementation, and documentation (Carver, 2013). The most important initial considerations for the preparation of the systematic literature review relate to the formulation of a clear, focused research question, as all subsequent processes depend on its formulation explicitly (Okoli &

DOI: 10.1201/b22530-19

Schabram, 2010). A broad research question can be chosen for the review, which corresponds to the overall aim of the review. Because this work is invested in capturing the full scope of the research field, as well as specifically the developments in innovation strategy, the research questions are intentionally broad.

- *Which external sources of knowledge in research papers contribute to the definition of a DT innovation strategy and how can they be systematized?*
- *Which obstacles and barriers lead to the failure of implementing systematized innovation management systems in companies?*
- *How has the scientific output of the DT innovation strategy evolved in relation to SMEs?*

To answer these questions, the paper is structured as follows. First, we introduce the concept of DT in SMEs. Thereafter, we present the methodology and results of the systematic literature review. Finally, we discuss the results and present the conclusions.

3.3.2 DT as a creative development approach in SMEs

In SMEs, support of the entrepreneur(s) is the basis for the implementation of DT. This aspect is more crucial than would be the case in large companies or corporations, because it can be assumed that in SMEs, the chance is smaller that DT can be introduced in a single department without the knowledge and approval of the entrepreneur. The entrepreneur in an SME is responsible for the successful implementation of innovation strategies, like DT (Mortati & Cruickshank, 2011).

This is not possible without assistance and responsibility from the entrepreneur. While in a large company an implementation, especially at the beginning of the project, can be done on a small scale and the support of top management can be gained already with initial project results (Judy & Savatore, 2011), the strong focus and involvement of the entrepreneur in operations in SMEs is a prerequisite for the implementation of DT (Geldermann, Lerche & Sepulveda, 2018).

If, in addition to the operational business, little time for the creation and use of reflective Design Spaces is available (e.g., as a relocation of the DT space into a cloud-based virtual space) (Lim, Kim, Kim & Kim, 2019), it can be assumed that the willingness to deal with an approach that is probably unknown to the entrepreneur and does not correspond to the regular business procedure is rather small (Moultrie, Clarkson & Probert, 2006).

For the successful implementation of DT in SMEs, further success factors specifically tailored to SMEs are required. Time resources and budget are essential for successful innovation implementation (Acklin, 2013). However, here too, the prerequisites of SMEs and large companies differ: In large enterprises, there are often specific research and development departments (also creative spaces) as well as innovation managers or DT experts and thus places and people whose tasks include evaluating new approaches such as DT (Acklin, 2013).

In SMEs, employees tend to do this alongside their daily activities at work, and there are fewer specialized departments with experts. Also, SMEs are usually cautious about implementing innovative strategies such as DT without knowing the exact benefits beforehand (Acklin, 2013). Large companies, on the other hand, are more likely to give the search for optimization and innovation the necessary space.

3.3.3 Systematic literature review: Innovation strategy DT in SMEs

3.3.3.1 Research design

The basis for this work is a systematic literature review, a methodology that differs from classical reviews in that it aims to synthesize research in a systematic, transparent, and iterative process (Crossan & Apaydin, 2010). The idea behind systematic reviews is characterized as follows: They are transparent, focused, equal, and accessible, provide concreteness, perform homogeneity of research and practice communities. All of these conjoined lead to a synthesis. It provides a manifesto for science and for practice, as the compilation of knowledge research is managed systematically.

The goal of our systematic review is to structure the field of research on DT and innovations in the context of SMEs, to identify researched topics, to list the most important research gaps, and thus to contribute to theory development.

A systematic review includes both a quantitative, bibliographic analysis, and a more qualitative, thematic analysis (Saur-Amaral, Reis Soares & Proenca, 2018). Although systematic reviews can include other types of publications, we followed other researchers and focused on peer-reviewed academic journal articles in English, to ensure quality and reduce the sample to a manageable amount. In terms of the time frame covered, we argue that DT has received considerable attention and a strong interest in implementation from practitioners and international researchers since Brown published "Design Thinking" in the Harvard Business Review (Brown, 2008). In order to illustrate this trend, data from previous years were included in the analysis. Therefore, the review covers academic papers from 1998 to 2021.

3.3.3.2 Data collection

To identify the relevant innovation literature on DT in SMEs, a systematic literature review was conducted. The first step was to define the search terms for this literature search of abstracts and keywords. The exact search terms are listed in a structured review. The large database Web of Science for the years 1998–2021 was searched with the linked search term from the exact word sequences "Design Thinking" AND "Innov*" in title (see Table 3.3.1). The language "English" was selected for the search. To ensure the quality of the literature contributions, only ISI/WoS indexed journals were included in the analysis.

Search queries to the Web of Science database and initial hit lists were used to refine or, if necessary, extend the search terms used (see Table 3.3.2). Initially, this served to obtain a rough overview of the amount of literature available. Central works in the selected subject area and specific journals with complete table of contents and bibliographic information were then used as points of reference. Consequently, the search terms were evaluated based on a cursory scan of the results.

With a sufficient specification of the search queries, the result area could already be narrowed down to a manageable scope. However, the results still proved to be too broad,

Table 3.3.1 Exact search term for the systematic literature review

Database	*Exact search term*	*Number of articles*
Web of Science – Current Contents Connect	"Design Thinking" AND "Innov*" in Topic	308

Note: All searches were conducted on4 April 2021.

Table 3.3.2 Preliminary search terms for the systematic literature review

	Search term	*Results*	*Purpose*	*Criticism*
Surface search	"Innov*" (Current Contents Connect)	169,307	Edition of all works that contain a combination of words all about "innovation", "innovate", "innovative", etc. included	Too many hits, not very specific, usually wrong context
	"Innov*" AND "SME*" (Current Contents Connect)	2,804	Restriction of the works from the first query to those containing the term "SME*"	Too general, context continues to vary greatly
	"Design Thinking" (Current Contents Connect)	602	Overview of the hits on the topic DT in general	Consolidation of the topic of DT without direct reference to the topic of innovation
Advanced search	"Design Thinking" AND "Innov*" (Current Contents Connect)	308	Specification of the previously made request. Limitations to works that are explicitly engaged in DT in the field of innovation management.	Strongly limited search space
	"Design Thinking" AND "Innov*" – related to specific research areas	70	Sample of selected articles to analyze and import to NVivo	Very strongly limited search space, for importing it to NVivo

which is why the queries were subsequently specified once more. The investigation was finally limited to the results obtained by linking the terms "Design Thinking" and "Innov*". After eliminating the duplicates, 308 articles remained.

In the following phase of analysis, the collected works were exported to Endnote and NVivo and examined. In addition to extracting potential knowledge sources, specific content for SMEs was filtered. Likewise, barriers or failures of innovation were selected to show a powerful tool for elimination through failure analysis. These instruments were extracted from the articles using content-based data analysis.

The aim of this systematization was to synthesize the results and to identify differences, similarities, and failure factors of the innovations, and its need for research. This was used to identify all 308 literature entries and transfer them to Endnote and create an Excel list.

Consequently, the entries in Endnote were searched for duplicates and those were removed. Of the 308 papers, the title and abstract were checked for their content relevance and relevance to DT and innovation, and entries that did not match were removed.

The following exclusion criteria were applied: (1) contributions that considered Design as human-centered problem solving; (2) those that dealt with the specific design of things, methods, or products; (3) those that were purely located in Design Science, for example, advancing specific methods or dealing with the mindset of designers, and thus not concerned with the application of Design Methods for innovation; and (4) those that applied the term "Design Thinking" generically and especially in SMEs. The final sample in the analysis included 70 literature entries.

3.3.3.3 Descriptive data analysis

A descriptive analysis of the 308 articles revealed that the first article from the sample that linked DT and innovation appeared in 2007 (see Figure 3.3.1). From 2008 onwards, an increase in the number of articles can be observed, which might be attributed, among other things, to the Special Issue of the *Harvard Business Review* from 2008 with the much-cited (1,060 citations) article *"Design Thinking: Thinking like a designer can transform the way you develop products, services, processes - and even strategy"* by IDEO's CEO Tim Brown and

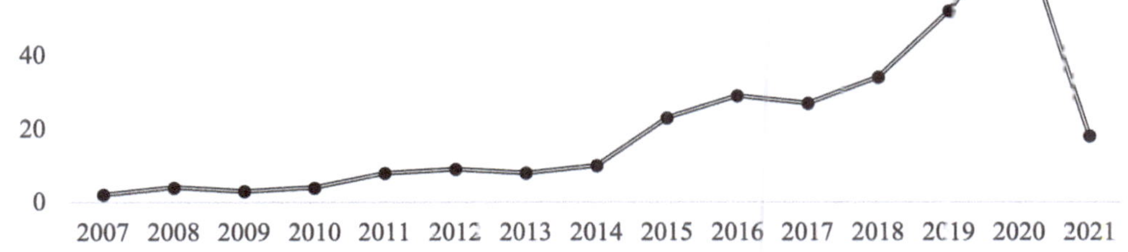

Figure 3.3.1 Publications per year.

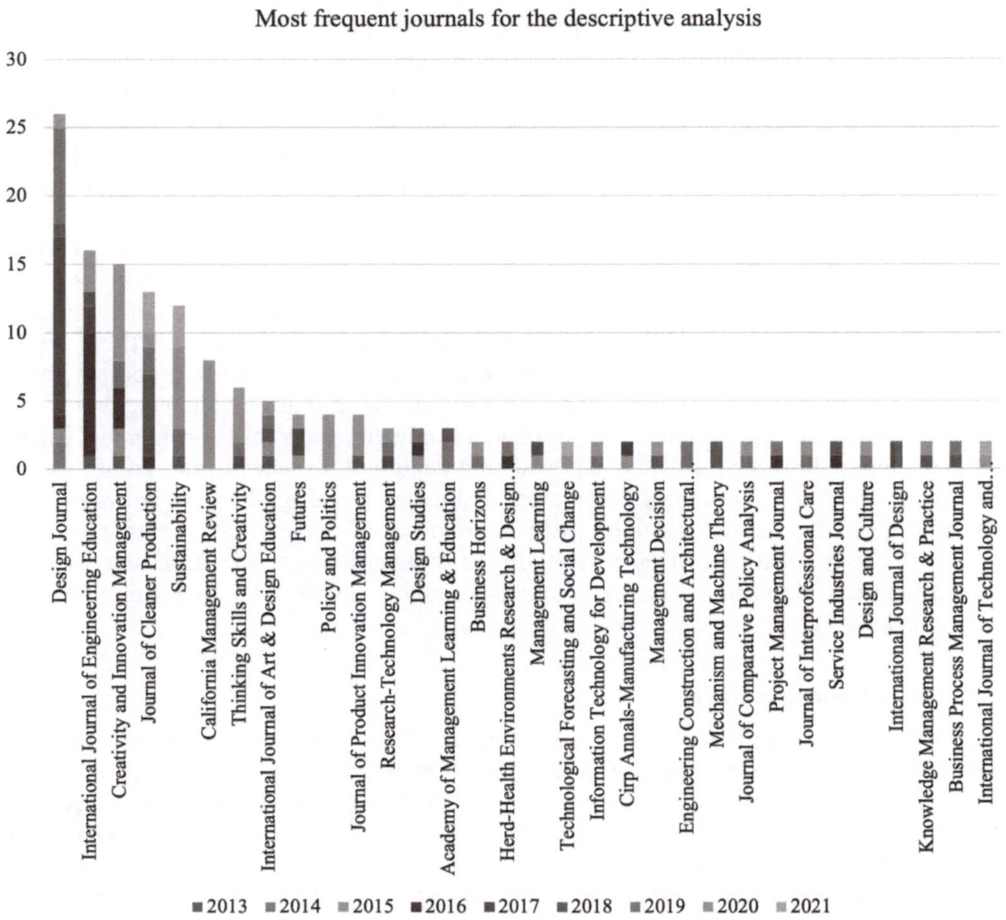

Figure 3.3.2 Most frequent journals for the descriptive analysis.

the resulting increased interest in the topic (Brown, 2008). 2020 showed a peak of 77 contributions.

Figure 3.3.2 considers the 152 most frequently occurring journals. The journal with the most frequent publications is Design Journal, by far a regular in the publications associated with DT and innovation, followed by International Journal of Engineering Education, Creativity and Innovation Management, Journal of Cleaner Production and Sustainability. As the year 2021 is not yet complete with research and its publications, the data in the graph is decreasing.

3.3.3.4 Content analysis

We read a total of 308 abstracts and excluded 238 papers that included the topics used in the search but did not address the innovation approach in terms of feasibility in different sizes of firms as a central topic area or as part of their theoretical innovation aspects.

The elaborated final selection contains 70 articles published between 1993 and 2021. Detailed content analysis in NVivo was conducted based on the final sample, which related to DT and innovation directly. The process included thematic coding to determine the main attributes of DT, looking in detail at SMEs.

We chose DT criteria for the development of the framework. First, we assessed the criteria and how they related to the topic, based on ambiguity. Our research team came to a common understanding that there was little or no talk about failure factors or failure of general innovation and DT projects. Therefore, criteria related to these aspects specifically were considered further. We then identified two groups: Failure of DT interventions as a parameter related to direct and indirect shortcomings in the articles (e.g., lack of understanding, lack of flexibility) and those that can be crystallized by notable successes (e.g., management support, customer-centric).

The factors for analysis were specifically adapted to SME characteristics. Attention was paid to the following SME-typical characteristics:

1 SMEs are characterized by the personality of the entrepreneur or the entrepreneurial family, which often combines management and entrepreneurship. Accordingly, SMEs are more dependent on the leader or the management body than a large company is on the board of executives (Kammerlander, Burger, Fust & Fueglistaller, 2015).
2 In SMEs, it can be observed more frequently than in large companies that the managing directors carry out operational activities and are thus less concentrated on management activities. Strategic instruments are rarely used in SMEs compared to corporations. The less intensive involvement with strategic activities is not necessarily a conscious choice. This poses a major challenge, as time must be found for innovative topics or the intensive examination of the future of the company, in addition to the varied tasks in day-to-day business (Gasda & Fueglistaller, 2015).
3 Many employees in SMEs are skilled workers who are trained within the company itself. Both entrepreneurs and employees in SMEs are highly involved and tend to be busy with their own tasks. Accordingly, innovations require additional capacities, which are rarely available in SMEs. But DT, which ideally involves people from several departments due to its interdisciplinarity and also challenges usual analytical business thinking, could be particularly difficult to implement in light of this SME characteristic (Liedtka & Ogilvie, 2011).
4 Faster product lifecycles and constantly changing technologies require an agile approach that sometimes exceeds the flexibility granted to SMEs. As a result, little time is available for implementing ideas alongside day-to-day business (Sarooghi, Sunny, Hornsby & Fernhaber, 2019).
5 In most cases, SMEs are more limited in their financial possibilities than large companies, which is especially noticeable in investment projects. SMEs are heavily dependent on the invested capital of the business owner (Kammerlander, Burger, Fust & Fueglistaller, 2015).
6 In SMEs, the entrepreneur often knows all employees personally. The company structure is lean, the hierarchy levels are usually flat, and the degree of formalization is low (Durst & Edvardsson, 2012).

From this selection of SME characteristics, the criteria of DT in SMEs are formulated and extracted in research papers.

In the context of the study, the specific key factors in Table 3.3.3 are divided into “Mandatory components”, “Beneficial components”, “Specific to SME” and “General factors”.

Table 3.3.3 Specific needs for DT in SMEs

No.	*Mandatory Necessary*	*Explanation*	*References*
1.	Management support for the DT team	In SMEs, the chances are smaller that DT can be introduced in a single department without the knowledge and approval of the entrepreneur. It is imperative to have the commitment and support of the entrepreneur for the project to be successful.	Appleyard, Enders, and Velazquez (2020) (Bason & Austin, 2019) (Crites & Rye, 2020) (Eppler & Kernbach, 2016; Lee & Ma, 2020; Snyder, Ingelsson & Bäckström, 2018) (Vagal et al., 2020)
2.	Time Resources	In SMEs, employee involvement in the innovative process is usually developed concurrently at most. There are fewer specialized departments. SMEs have to think twice whether they start a DT project without being able to calculate the exact benefits in advance.	(Agogino et al., 2016) (Ahn, Rundall, Shortell, Blodgett & Reponen, 2021; Beltagui, 2018) (Bicen & Johnson, 2015) (Cagnin, 2018) (Carlgren, Elmquist & Rauth, 2014) (Coco, Calcagno & Lusiani, 2020) (Crites & Rye, 2020; de Carvalho, da Hora & Fernandes, 2021) (Ghajargar, Mangano, De Marco & Giannantonio, 2017) (Kimbell, 2011; Nagaraj, Berente, Lyytinen & Gaskin, 2020; Yu, Qu & Hu, 2015)
3.	Budget for DT projects	Here the statement for Budget is the same as in “time resources”	(Baldassarre et al., 2020)
Beneficial			
4.	Small projects and small teams	In SMEs, the implementation of innovations looks different than in large companies. Here, even small projects can achieve greater success.	(Baldassarre et al., 2020) (Manzini & Rizzo, 2011)
5.	DT team strategically well positioned (team selection)	SMEs are less able to give the search for optimization and innovation the necessary space.	(Beckman & Barry, 2007) (Brown, 2008) (Fleury, Stabile & Carvalho, 2016) (Hölzle & Rhinow, 2019) (Kurtmollaiev, Pedersen, Fjuk & Kvale, 2018) (Nagaraj et al., 2020) (Seidel & Fixson, 2013) (Vagal et al., 2020)

			(Yang & Hsu, 2020) (Yeoman & Carvalho, 2019)
6.	Customer or benefit orientation is strategically anchored	SMEs will think twice before starting a DT project without knowing the exact benefits beforehand.	(Diepenmaat, Kemp & Velter, 2020) (Hankammer, Brenk, Fabry, Nordemann & Piller, 2019) (Hölzle & Rhinow, 2019) (E. Knight, Daymond & Paroutis, 2020) (Vetterli, Uebernickel, Brenner, Petrie & Stermann, 2016) (Yan, 2018)
Specific to SME			
7.	Potential benefits of DT project known in advance	Only if it can be seen early on that the initial investment has paid off will others follow.	(Carlgren et al., 2014)
8.	DT steps well defined	SMEs lack the financial resources to take bigger risks. Therefore, the steps for innovation should be taken in smaller but safe steps.	(Bairaktarova, Bernstein, Reid & Ramani, 2016) (Carmel-Gilfilen & Portillo, 2016) (Hookway, Johansson, Svensson & Heiden, 2019) (Pluchinotta, Kazakci, Giordano & Tsoukias, 2019) (Shafiee, Haug, Kristensen & Hvam, 2021)
9.	External DT experts to support the project	Due to the high involvement of employees in day-to-day business, it is usually unrealistic for SMEs to have the time, space and knowledge to implement DT internally, without external coaches.	(Ghajargar et al., 2017) (Jun, Morrison & Clarkson, 2014) (Kozlowski, Searcy & Bardecki, 2018) (Kulick, 2017) (Na, Cho & Harrison, 2017) (Shapira, Ketchie & Nehe, 2017)
10.	DT success is measured differently	Small steps lead to greater success.	(Brassett & O'Reilly, 2015) (Chandler & Ward, 2019) (Conforto, Amaral, da Silva, Di Felippo & Kamikawachi, 2016) (J. Knight, Fitton, Phillips & Price, 2019)
11.	Cross-departmental projects with DT	Due to the flat hierarchical levels in SMEs, communication within the team is much faster than in others.	(Seidel & Fixson, 2013)
12.	DT is also integrated and applied to existing projects	In SMEs, innovation approaches are often unconsciously applied in projects without being defined as such.	(Liem & Brangier, 2012) (Shafiee et al., 2021)

(Continued)

Table 3.3.3 (Continued)

No.	*Mandatory Necessary*	*Explanation*	*References*
13.	First application to design affinity areas	The advantage for SMEs is that DT can be introduced even with a small budget.	(Agogino et al., 2016) (Beckman & Barry, 2007) (Chandler & Ward, 2019) (Lande, 2016) (Lim, Kim, Kim & Kim, 2019)
14.	External experts are involved (almost indispensable for SMEs, but professionals may be available for corporations)	Outsourcing allows employees to focus on their own day-to-day work.	(Buhl et al., 2019) (Carlgren, Elmquist & Rauth, 2016) (Eppler & Kernbach, 2016) (Fleury et al., 2016) (Glen, Suciu & Baughn, 2014) (Kim & Strimel, 2020) (Liem & Brangier, 2012) (Olsen, 2015) (Yu et al., 2015)
General success factors independent of SMEs			
15.	DT is visually represented and communicated	Visual representation of DT is possible in both SMEs and large enterprises.	(Na et al., 2017)
16.	DT involved persons are professionally trained	Professional coaches can also be applied to all sizes of enterprises. However, this factor makes more sense and is more effective in SMEs.	(Na et al., 2017) (Shapira et al., 2017)
17.	Physical space for DT projects is available	This area appeals more to large companies than to SMEs.	(McGann, Blomkamp & Lewis, 2018) (McGann, Wells & Blomkamp, 2021)
18.	DT is easy to learn and use for company employees	DT is more practical to use and remains straightforward if you stick to the core steps. It is feasible in all sizes of companies.	(Souza et al., 2020)
19.	Internal employees involved in the project. Usually, no external experts necessary.	Large companies have their own innovation departments, so outsiders tend to be less represented.	(Brown, 2008) (Carlgren et al., 2014) (Roberts & Palmer, 2012) (Snyder et al., 2018)

As part of the systematic literature review, content-related criteria that can be tailored specifically to SMEs were analyzed. Therefore, these benefit criteria were once again weighed and divided. The first criteria under the category "Mandatory Necessary" are directly related and are more difficult to apply in SMEs due to their context than is the case for larger companies.

The supporting or beneficial factors are assumed to be directly related to their underlying SME characteristics. They also have an SME-specific background.

The third category contains those benefits that do not seem to be influenced by SME characteristics and are generally applicable to all company sizes. Finally, some general factors that are independent of SMEs but also interesting were developed.

However, this is only one perspective of successfully introducing innovations into SMEs using different approaches.

All the while, it is hard to guess how many SMEs are doomed to fail. Innovative products, processes, or services often fail before their "inventors" can even think about entering the market. Even products, for example, that meet customer needs in a completely new way often remain as a shadow of their potentials.

When developing an innovation, its developers and teams tend to face extreme challenges. A multitude of potential mistakes lurk in the innovation process that should be avoided. Only when you start pointing out situations that address failures can you draw lessons from them. Every mistake can become a costly lesson for a company. However, such costs can be reduced enormously by avoiding situations in which others have failed before. Therefore, Table 3.3.4 aims to present the failures of innovation in companies. These could be filtered out from the selected research papers.

Failures that are not accepted and shared lose their importance in the innovation world. As a result, oodles of time, money, and health are lost in the wake of repeatedly making the same mistakes – because no one talks about them (Rhaiem & Nabil, 2019) Interest in failure of innovation projects is shown by entrepreneurs and managers to bring about the prevention of problems and risks. Table 3.3.4 highlights the failures elaborated from the articles. These were listed from direct and indirect context. For the reasons mentioned above, SMEs cannot afford to make mistakes when implementing innovations in projects, it is advisable to take smaller steps and smaller projects.

It should also be kept in mind that SMEs often apply innovation approaches without labeling them as such. The integration of direct innovation strategies will lead to greater success.

3.3.4 Findings and discussion

The results were divided into several sections: First, we provided an overview of the research agenda and literature on DT and innovation. Consequently, the papers were analyzed descriptively and the availability of DT and innovation in the scientific field was examined. The analysis has shown that DT research in the innovation context is still a relatively novel field that has shown a steady development for about 15 years based on publications. A thematic analysis was then conducted, to present an overview of the DT parameter in relation to SME innovation practices.

Overall, the results demonstrate that our sample on DT and innovation is powerful, with 308 filtered articles explicitly addressing the topic. Narrowing it down even further, 70 articles dealt directly with the topic of DT and innovation and its key figures.

Table 3.3.4 Failure factors based on the nature of SMEs

No.	*Failure factors*	*Explanation (based on the nature of the SMEs)*	*Literature*
1.	Lack of customer input in customer analysis (customer goals, personas, customer journeys, etc.) and prevalent misconceptions of already knowing customer needs and expectations	Understanding customer goals and the current customer experience as perceived by the customer form the central basis for the subsequent active design of the customer experience. Companies must develop a deep understanding of the customer's goals, needs, perceptions, and interactions	(Liu & Lu, 2020) (Rau, Zbiek & Jonas, 2017) (Sohaib, Solanki, Dhaliwa, Hussain & Asif, 2019) (Snyder, Ingelsson & Backstrom, 2018)
2.	No comprehensive or incorrect survey of actual customer needs and expectations	A holistic and correct recording of the customer's needs and expectations require comprehensive ethnographic, qualitative and quantitative data collection and evaluation of internal and external information sources. Personal points of contact should be used for in-depth insights and feedback.	(Hankammer, Brenk, Fabry, Nordemann & Piller, 2019)
3.	Qualitative survey methods with the customer (surveys, interviews, etc.) are designed to create negative customer experiences	The instrument of direct collection of customers input is overused or misused. Data collection must be prevented from provoking negative reactions from the customer and creating survey fatigue, which undermines the willingness to provide critical feedback.	(Andreassan, Kristensson, Lervik-Olsen, Parasuraman, McColl-Kennedy & Edvardsson, 2016) (Lim, Kim, Kim & Kim, 2019) (Pande & Bharathi, 2020)
4.	No complete coverage of all touch points and no Evaluation of the importance of individual touchpoints from the customer's point of view	The identification of relevant touchpoints for the customer serves as a basis for the subsequent design. All touchpoints must be captured, and an assessment of relevance made from the customer's perspective.	(Hankammer, Brenk, Fabry, Nordemann & Piller, 2019)
5.	No linkage of the assessment of relevant touch points with entrepreneurial significance	The touchpoints deemed relevant by customers must also be assessed in terms of their economic significance for the company so that design priorities can be defined later.	(Martin, 2011)
6.	Insufficient analysis (resulting in a lack of understanding) of the customer's value-adding processes	Capturing the value-creating processes are a key step for the subsequent alignment of the company's own processes and value proposition. A deep understanding of how the customer creates value for its own customers are therefore needed.	(Appleyard, Enders & Velazquez, 2020) (Holzle & Rhinow, 2019)

7.	No identification of relevant stakeholders and decision makers as well as their holistic customer journeys, goals, and pain points	To create optimal experiences for all relevant customer stakeholder groups, the individual goals, pain points, and customer journeys must be captured holistically from start to finish.	(Bas & Guillo, 2015) (Geldermann, Lerche & Sepulveda, 2018) (Magistretti, Dell'Era & Doppio, 2020) (Pluchinotta, Kazakci, Giordano & Tsoukias, 2019)
8.	Insufficient analysis of a customer's interactions with other service providers in the ecosystem as well as with other customers	A holistic understanding of the customer experience requires capturing the value-creating ecosystem. Companies should analyze the experience ecosystem and understand how it generates added value and usage value for customers and how it shapes their expectations of the customer.	(Sorice & Donlan, 2015)
9.	No adequate recording of own value-adding processes as well as the current value proposition to customers	A shared understanding of the current value proposition and internal processes form the basis for subsequent design decisions and for aligning processes to deliver value for use to customers.	(Hankammer, Brenk, Fabry, Nordemann & Piller, 2019) (Hirano, Ishizuka & Sakaguchi, 2013)
10.	No situation analysis of the competencies and resources currently available in the company, related to customer experience and organizational performance	The subsequent implementation, monitoring, and control of the customer experiences requires dedicated competencies and resources. Only based on an analysis can any gaps be closed.	(Nagaraj, Berente, Lyytinen & Gaskin, 2020)

The most important articles for DT research and innovation are those with a design focus, with the Design Journal playing a dominant role, as already mentioned. In addition, articles in mainstream innovation journals such as International Journal of Engineering Education, Creativity and Innovation Management, Journal of Cleaner Production, Sustainability and California Management Review are found, which are also well represented in this analyzed topic area. This shows which subject areas also deal intensively with the innovation approach DT.

The purpose of the thematic content analysis was to identify essential elements and dimensions of DT practices in an SME context (focus on needs and failures – Tables 3.3.3 and 3.3.4). These are concrete measures of the success and failure of DT and innovation approaches, so as to increase innovation capability. As can also be seen in the above-mentioned Tables, the identified two essential building blocks (needs and failures) of DT in companies have been identified and explained.

The analysis reveals that the scientific literature has dealt with the implementation of thinking approaches in SMEs, but not entirely and exhaustively. Therefore, the topic is not very well known even in SMEs (Gasda & Fueglistaller, 2015). Because SMEs are not frequent users of DT, academia is only marginally invested in analyzing the subject (Kammerlander, Burger, Fust & Fueglistaller, 2015). Only a few SMEs do not share this approach. Moreover, SMEs are preoccupied with everyday business, which is why they are slow to integrate new approaches (Magistretti, Dell'Era & Doppio, 2020). For this reason, SMEs do not provide themselves with the capacity to develop "new things" and thus run the risk of losing some of their competitive advantages in the medium to long term. It seems to be the case that DT is not suitable for SMEs because of their characteristics (Magistretti, Dell'Era & Doppio, 2020). However, as already mentioned, there are also aspects that indicate a sensitivity of SMEs to DT elements. The receptiveness for DT in small and medium-sized enterprises is greater than in large companies (Shapira, Ketchie & Nehe, 2017), and due to flat hierarchies, SMEs can adapt changes more quickly than large companies.

As a basis for further research approaches, these factors can be utilized to further expand and strengthen the strategic sustainability behavior of SMEs and the innovation approaches DT is pursuing (Magistretti, Dell'Era & Doppio, 2020).

3.3.5 Conclusions

The aim of the study was to identify the needs and failures of DT and innovation approaches in SMEs by conducting a systematic literature review. Here, we focused on positive and negative implementation competencies. We contributed to the literature by showing that innovation approaches are, in principle, built on some unique DT parameters to deal with uncertainty at the management level, adaptability, flexibility and understanding, speed, and integration. Our finding implies that DT approaches related to well-known innovation approaches in our literature list and associated company size can be used to also test and improve innovation approaches in SMEs to respond to uncertainty and risks. The study is helpful for companies in revealing the limitations in exploring and comparing different solutions with a well-defined list of design approaches and principles. It is a guideline and a comparative first step to learn from mistakes of the past and work more efficiently in terms of time to implement innovations. It could also be a crucial research step in the current literature and practice in DT.

Our findings suggest a different theoretical path for the field. Stakeholders can now merge theoretical insight by using our categorized DT needs, not only to develop new solutions but also to compare different solutions in terms of performance in managing uncertainty in SMEs.

In addition, our text provides a fundamental comparison of management practices to advance the research field. In developing and elaborating solutions, researchers and practitioners can now utilize our research as an initial and preliminary reference. The theoretical way to design new hybrid approaches can provide more options for dealing with uncertainty. Researchers and scientists can test, continuously elaborate, and improve our DT solution approaches for SMEs in different business situations (e.g., SMEs and start-ups). They may also apply such DT criteria and principles in different phases of the DT approach, with different practices and techniques, and study the impact on innovation performance.

One promising avenue of research is to examine the design principles in the innovation management literature in relation to SMEs. Based on this, further studies and research could address other resource opportunities and their complexities. Moreover, the research team noticed that the presence of failures is seldom referenced in the literature.

The study has some limitations. First, we provide a preliminary list of failures and needs (Tables 3.3.3 and 3.3.4 are tailored to SMEs; however, we did not test these elements empirically, which is a future research direction).

Second, as this is a systematic literature review, an inherent limitation is related to the impossibility to generalize the results, or to provide managerial or policy implications.

Thirdly, the definition of SME encompasses a broad range of firms, from start-ups to incumbent firms, and failures and needs for these firms can be quite different. A specific analysis that considers the size, the age, the industry, and other characteristics as control variables may account for the differences among SMEs, is recommendable in future studies.

Also, the failure factors, based on indirect and direct representations regarding the DT criteria, were identified. Furthermore, we realized that the developed list based on the systematic literature review is not exhaustive. Ultimately, an opportunity arises to elaborate and investigate in further research, and additional DT criteria and principles tailored to SMEs can be added in the future.

Further research can be prepared, continuously improved, and supplemented in upcoming studies, in implementation of the DT project, joining an iterative process, which is so typical to DT.

References

Acklin, C. (2013). Design management absorption model: A framework to describe and measure the absorption process of design knowledge by SMEs with little or no prior design experience.

Andreassan, T., Kristensson, P., Lervik-Olsen, L., Parasuraman, A., McColl-Kennedy, J., & Edvardsson, B. (2016). Linking service design to value creation and service research. *Journal of Service Management*, 21–29.

Appleyard, M., Enders, A., & Velazquez, H. (2020). Regaining R&D Leadership: The Role of Design Thinking and Creative Forbearance. *California Management Review*, 12–29.

Assink, M. (2006, 04 01). Inhibitors of disruptive innovation capability: A conceptual model. *European Journal of Innovation Management*, 215–233.

Bas, E., & Guillo, M. (2015). Participatory foresight for social innovation. FLUX-3D method (Forward Looking User Experience), a tool for evaluating innovations. *Technological Forecasting and Social Change*, 275–290.

Beckman, S. L., & Barry, M. (2007). Innovation as a learning process: Embedding design thinking. *California Management Review*, *50*(1),25–56.

Brassett, J., & O'Reilly, J. (2015). Styling the future. A philosophical approach to design and scenarios. *Futures*, *74*, 37–48.

Brown, T. (2008, 06 01). Design Thinking. *Harvard Business Review*.

Carmel-Gilfilen, C., & Portillo, M. (2016). Designing with empathy: Humanizing narratives for inspired healthcare experiences. *Herd-Health Environments Research & Design Journal*, *9*(2), 130–146.

Carver, J. C. H. E. (2013). Identifying Barriers to the Systematic Literature Review Process. *IEEE International Symposium on Empirical Software Engineering and Measurement*, 203–212.

Crossan, M., & Apaydin, M. (2010, 07 20). A Multi-Dimensional Framework of Organizational Innovation: A Systematic Review of the Literature. *Journal of Management studies*, 1154–1191.

Durst, S., & Edvardsson, R. (2012, 05 01). Knowledge management in SMEs: A literature review. *Journal of Knowledge Management*, 879–903.

Gasda, J., & Fueglistaller, U. (2015, 01 01). Fostering individual-level ambidexterity in SMEs: A relational-contract perspective on informal external drivers of employees' ambidextrous behaviour. *International Journal of Entrepreneurial Venturing*, 217–236.

Geldermann, J., Lerche, N., & Sepulveda, J. (2018). Combining multi-criteria decision analysis and design thinking. *European Journal of Industrial Engineering*, 708–739.

Geldermann, J., Lerche, N., & Sepulveda, J. (2018). Combining multi-criteria desicion analysis and design thinking. *European Journal of Industrial Engineering*, 708–739.

Goerge, S., & Schoemaker, P. (2016, 08 01). Adapting to Fast-Changing Markets and Technologies. *California Management Review*, 59–77.

Goldschmidt, G., & Tarazi, E., eds. (2022). Expanding the Frontiers of Design: A Blessing or a Curse? *13th Design Thinking Research Symposium* (pp. 547–564). Haifa: DesignTech, Faculty of Architecture & Town Planning.

Hankammer, S., Brenk, S., Fabry, H., Nordemann, A., & Piller, F. (2019). Towards circular business models: Identifying consumer needs based on the jobs-to-be-done theory. *Journal of Cleaner Production*, 341–358.

Hirano, T., Ishizuka, A., & Sakaguchi, K. (2013). Innovation Activities by Co-creation Process. *Fujitsu Scientific & Technical Journal*, 391–396.

Holzle, K., & Rhinow, H. (2019). The Dilemma of Design Thinking in Innovation Projects. *Project Management Journal*, 418–430.

Hookway, S., Johansson, M. F., Svensson, A., & Heiden, B. (2019). The problem with problems: Reframing and cognitive bias in healthcare innovation. *Design Journal*, *22*, 553–574.

Judy, M., & Savatore, B. (2011). Continuous innovation in SMEs: How design innovation shapes business performance through doing more with less. *Proceedings of the 12th International CINet Conference - Continuous Innovation*, 696–708.

Kammerlander, N., Burger, D., Fust, A., & Fueglistaller, U. (2015, 07 01). Exploration and Exploitation in Established Small and Medium-sized Enterprises: The Effect of CEOs' Regulatory Focus. *Journal of Business Venturing*, 582–602.

Liedtka, J., & Ogilvie, T. (2011, 07 27). Design for growth: A design thinking tool kit for managers. *Columbia Univers. Press*, 227.

Lim, C., Kim, K., Kim, M., & Kim, K. (2019). Multi-factor service design - identification and consideration of multiple factors of the service in its design process. *Service Business*, 51–74.

Lim, C., Kim, K., Kim, M., & Kim, K. (2019). Multi-factor service design: Identification and consideration of multiple factors of the service in its design process. *Service Business*, 51–74.

Liu, A., & Lu, S. (2020). Functional design framework for innovation design thinking in product development. *Cirp Journal of Manufacturing Science and Technology*, 105–117.

Magistretti, S., Dell'Era, C., & Doppio, N. (2020). Design sprint for SMEs: An organizational taxonomy based on configuration theory. *Management Decision*, 1803–1817.

Martin, R. (2011). The Innovation Catalysts. *Harvard Business Review*, 82.

Mortati, M., & Cruickshank, L. (2011). Design and SMEs: The trigger of creative ecosystems. *Conference on Designing Pleasurable Products and Interfaces*, 1–8.

Moultrie, J., Clarkson, J., & Probert, D. (2006). A tool to evaluate design performance in SMEs. *International Journal of Productivity and Performance Management*.

Na, J., Choi, Y., & Harrison, D. (2017). The Design Innovation Spectrum: An Overview of Design Influences on Innovation for Manufacturing Companies. *International Journal of Design*, 13–24.

Okoli, C., & Schabram, K. (2010, 01 02). A Guide to Conducting a Systematic Literature Review of Information Systems Research. *Sprouts: Working Papers on Information Systems*, 10–26.

Pande, M., & Bharathi, S. (2020). Theoretical foundations of design thinking - A constructivism learning approach to design thinking. *Thinking Skills and Creativity*, 637-637.

Pluchinotta, I., Kazakci, A., Giordano, R., & Tsoukias, A. (2019). Design Theory for Generating Alternatives in Public Decision Making Processes. *Group Decision and Negotiation*, 341–375.

Proenca, J., & Jiménez-Sáez, F. (2020). Design of Services for the Incremental Innovation Management in SMEs. *Universidad & Empresa*, 1–20.

Rau, C., Zbiek, A., & Jonas, J. (2017). Creating Competitive Advantage from Services A Design Thinking Case Study from the Commodities Industry. *Research-Technology Management*, 48–56.

Rhaiem, K., & Nabil, A. (2019, 03 28). Learning from innovation failures: A systematic review of the literature and research agenda. *Review of Managerial Science*, 189–234.

Sarooghi, H., Sunny, S., Hornsby, J., & Fernhaber, S. (2019, 07 22). Design Thinking and Entrepreneurship Education: Where Are We, and What Are the Possibilities? *Journal of Small Business Management*.

Saur-Amaral, I., Reis Soares, R., & Proenca, J. (2018). Business model innovation: Towards a conceptual framework. *Tourism & Management Studies*, 80–93.

Shapira, H., Ketchie, A., & Nehe, M. (2017). The Integration of Design Thinking and Strategic Sustainable Development. *Journal of Cleaner Production*, 277–287.

Snyder, K., Ingelsson, P., & Backstrom, I. (2018). Using design thinking to support value-based leadership for sustainable quality development. *Business Process Management*, 1289–1301.

Sohaib, O., Solanki, H., Dhaliwa, N., Hussain, N., & Asif, M. (2019). Integrating design thinking into extreme programming. *Journal of Ambient Intelligence and Humanized Computing*, 2497–2504.

Sorice, M., & Donlan, C. (2015). A human-centered framework for innovation in conservation incentive programs. *Ambio*, 788–792.

van de Vrande, V., de Jong, J., Vanhaverbeke, W., & de Rochemont, R. (2009). Open innovation in SMEs: Trends, motives and management challenges. *Technovation*, 423–437.

Wardati, N. K., & M.., E. (2019). The Impact of Social Media Usage on the Sales Process in Small and Medium Enterprises (SMEs): A Systematic Literature Review. *Procedia Computer Science*, 976–983.

Williams, R., Clark, L., Clark, W., & Raffo, D. (2020, 09 17). Re-examining systematic literature review in management research: Additional benefits and execution protocols. *European Management Journal*, 13.

Appendix – List of References used for the analysis

Aagaard, A., Saari, U. A., & Makinen, S. J. (2021). Mapping the types of business experimentation in creating sustainable value: A case study of cleantech start-ups. *Journal of Cleaner Production*, *279*, 23182-23182.

Abookire, S., Plover, C., Frasso, R., & Ku, B. (2020). Health design thinking: An innovative approach in public health to defining problems and finding solutions. *Frontiers in Public Health*, *8*, 459-459.

Agogino, A. M., Beckman, S. L., Castanos, C., Kramer, J., Roschuni, C., & Yang, M. (2016). Design practitioners' perspectives on methods for ideation and prototyping. *International Journal of Engineering Education*, *32*(3), 1428–1437.

Ahn, C., Rundall, T. G., Shortell, S. M., Blodgett, J. C., & Reponen, E. (2021). Lean management and breakthrough performance improvement in health care. *Quality Management in Health Care*, *30*(1), 6–12.

Andreassen, T. W., Kristensson, P., Lervik-Olsen, L., Parasuraman, A., McColl-Kennedy, J. R., Edvardsson, B., & Colurcio, M. (2016). Linking service design to value creation and service research. *Journal of Service Management*, *27*(1), 21–29.

Appleyard, M. M., Enders, A. H., & Velazquez, H. (2020). Regaining R&D leadership: The role of design thinking and creative forbearance. *California Management Review*, *62*(2), 12–29.

Bairaktarova, D., Bernstein, W. Z., Reid, T., & Ramani, K. (2016). Beyond surface knowledge: An exploration of how empathic design techniques enhances engineers understanding of users' needs. *International Journal of Engineering Education*, *32*(1), 111–122.

Baldassarre, B., Konietzko, J., Brown, P., Calabretta, G., Bocken, N., Karpen, I. O., & Hultink, E. J. (2020). Addressing the design-implementation gap of sustainable business models by prototyping: A tool for planning and executing small-scale pilots. *Journal of Cleaner Production*, *255*, 20295-20295.

Ballie, J. (2019). Design thinking for progress. Initial insights from an evolving design-led business support programme for Scotland. *Design Journal*, *22*, 981–995. doi:10.1080/14606925.2019.1595412

Bas, E., & Guillo, M. (2015). Participatory foresight for social innovation. FLUX-3D method (Forward Looking User Experience), a tool for evaluating innovations. *Technological Forecasting and Social Change*, *101*, 275–290.

Bason, C., & Austin, R. D. (2019). The right way to lead design thinking. *Harvard Business Review*, *97*(2), 82-+.

Beltagui, A. (2018). A design-thinking perspective on capability development: The case of new product development for a service business model. *International Journal of Operations & Production Management*, *38*(4), 1041–1060.

Bicen, P., & Johnson, W. H. A. (2015). Radical innovation with limited resources in high-turbulent markets: The role of lean innovation capability. *Creativity and Innovation Management*, *24*(2), 278–299.

Buhl, A., Schmidt-Keilich, M., Muster, V., Blazejewski, S., Schrader, U., Harrach, C.,… … Sussbauer, E. (2019). Design thinking for sustainability: Why and how design thinking can foster sustainability-oriented innovation development. *Journal of Cleaner Production*, *231*, 1248–1257.

Cagnin, C. (2018). Developing a transformative business strategy through the combination of design thinking and futures literacy. *Technology Analysis & Strategic Management*, *30*(5), 524–539.

Carlgren, L., Elmquist, M., & Rauth, I. (2014). Design thinking: Exploring values and effects from an innovation capability perspective. *Design Journal*, *17*(3), 403–423.

Carlgren, L., Rauth, I., & Elmquist, M. (2016). Framing design thinking: The concept in idea and enactment. *Creativity and Innovation Management*, *25*(1), 38–57.

Chandler, L., & Ward, A. (2019). Immersed in design: Using an immersive teaching space to visualise design solutions. *International Journal of Art & Design Education*, *38*(2), 314–327.

Cheah, Y. H., Chai, C. S., & Toh, Y. (2019). Traversing the context of professional learning communities: Development and implementation of Technological Pedagogical Content Knowledge of a primary science teacher. *Research in Science & Technological Education*, *37*(2), 147–167.

Coco, N., Calcagno, M., & Lusiani, M. (2020). Struggles as triggers in a design-thinking journey. *Creativity and Innovation Management*, *29*, 103–115.

Colombo, S., Cautela, C., & Rampino, L. (2017). New design thinking tools for the next generation of designer-entrepreneurs. *Design Journal*, *20*, S566–S580.

Conforto, E. C., Amaral, D. C., da Silva, S. L., Di Felippo, A., & Kamikawachi, D. S. L. (2016). The agility construct on project management theory. *International Journal of Project Management*, *34*(4), 660–674.

Crites, K., & Rye, E. (2020). Innovating language curriculum design through design thinking: A case study of a blended learning course at a Colombian university. *System*, *94*, 2334-2334.

de Carvalho, R. A., da Hora, H., & Fernandes, R. (2021). A process for designing innovative mechatronic products. *International Journal of Production Economics, 231*, 7887-7887.
Diepenmaat, H., Kemp, R., & Velter, M. (2020). Why sustainable development requires societal innovation and cannot be achieved without this. *Sustainability, 12*(3), 1270-1270.
Eines, T. F., & Vatne, S. (2018). Nurses and nurse assistants' experiences with using a design thinking approach to innovation in a nursing home. *Journal of Nursing Management, 26*(4), 425–431.
Eppler, M. J., & Kernbach, S. (2016). Dynagrams: Enhancing design thinking through dynamic diagrams. *Design Studies, 47*, 91–117.
Fleury, A. L., Stabile, H., & Carvalho, M. M. (2016). An overview of the literature on design thinking: Trends and contributions. *International Journal of Engineering Education, 32*(4), 1704–1718.
Geissdoerfer, M., Bocken, N. M. P., & Hultink, E. J. (2016). Design thinking to enhance the sustainable business modelling process - A workshop based on a value mapping process. *Journal of Cleaner Production, 135*, 1218–1232.
Geldermann, J., Lerche, N., & Sepulveda, J. D. (2018). Combining multi-criteria decision analysis and design thinking. *European Journal of Industrial Engineering, 12*(5), 708–739.
Ghajargar, M., Mangano, G., De Marco, A., & Giannantonio, R. (2017). Design thinking applied to data storage innovation: A case study. *Design Journal, 20*, S3776–S3788.
Glen, R., Suciu, C., & Baughn, C. (2014). The need for design thinking in business schools. *Academy of Management Learning & Education, 13*(4), 653–667.
Hankammer, S., Brenk, S., Fabry, H., Nordemann, A., & Piller, F. T. (2019). Towards circular business models: Identifying consumer needs based on the jobs-to-be-done theory. *Journal of Cleaner Production, 231*, 341–358.
Hirano, T., Ishizuka, A., & Sakaguchi, K. (2013). Innovation activities by co-creation process. *Fujitsu Scientific & Technical Journal, 49*(4), 391–396.
Holzle, K., & Rhinow, H. (2019). The dilemmas of design thinking in innovation projects. *Project Management Journal, 50*(4), 418–430.
Jun, G. T., Morrison, C., & Clarkson, P. J. (2014). Articulating current service development practices: A qualitative analysis of eleven mental health projects. *Bmc Health Services Research, 14*, 20-20.
Kim, Y. S., Jeong, J., Hong, Y., & Hong, S. J. (2020). A Schema for systematic service imagining: Context-based activity modeling. *Sustainability, 12*(22), 9558-9558.
Kim, E., & Strimel, G. J. (2020). The influence of entrepreneurial mindsets on student design problem framing. *IEEE Transactions on Education, 63*(2), 126–135.
Kimbell, L. (2011). Rethinking design thinking: Part I. *Design and Culture, 3*(3), 285–306.
Knight, J., Fitton, D., Phillips, C., & Price, D. (2019). Design thinking for innovation. Stress testing human factors in ideation sessions. *Design Journal, 22*, 1929–1939.
Knight, E., Daymond, J., & Paroutis, S. (2020). Design-led strategy: How to bring design thinking into the art of strategic management. *California Management Review, 62*(2), 30–52.
Ko, Y. T., Chen, M. S., Yang, C. C., & Zheng, M. C. (2015). Modelling a contradiction-oriented design approach for innovative product design. *Proceedings of the Institution of Mechanical Engineers Part B-Journal of Engineering Manufacture, 229*, 199–211.
Kozlowski, A., Searcy, C., & Bardecki, M. (2018). The reDesign canvas: Fashion design as a tool for sustainability. *Journal of Cleaner Production, 183*, 194–207.
Kulick, G. (2017). Adding plus value to development aid projects through design strategy: Experiences from Pakistan. *Design Journal, 20*, S411–S423.
Kummitha, R. K. R. (2019). Design thinking in social organizations: Understanding the role of user engagement. *Creativity and Innovation Management, 28*(1), 101–112.
Kurtmollaiev, S., Pedersen, P. E., Fjuk, A., & Kvale, K. (2018). Developing managerial dynamic capabilities: A quasi-experimental field study of the effects of Design Thinking training. *Academy of Management Learning & Education, 17*(2), 184–202.
Lande, M. (2016). Catalysts for design thinking and engineering thinking: Fostering ambidextrous mindsets for innovation. *Journal of Engineering Education, 32*(3), 1356–1363.

Lee, C., & Ma, L. (2020). The role of policy labs in policy experiment and knowledge transfer: A comparison across the UK, Denmark, and Singapore. *Journal of Comparative Policy Analysis*, *22*(4), 281–297.

Lim, C., Kim, K. H., Kim, M. J., & Kim, K. J. (2019). Multi-factor service design: identification and consideration of multiple factors of the service in its design process. *Service Business*, *13*(1), 51–74.

Liem, A., & Brangier, E. (2012). Innovation and design approaches within prospective ergonomics. *Work-a Journal of Prevention Assessment & Rehabilitation*, *41*, 5243–5250.

Liu, A., & Lu, S. (2020). Functional design framework for innovative design thinking in product development. *Cirp Journal of Manufacturing Science and Technology*, *30*, 105–117.

Magistretti, S., Dell'Era, C., & Doppio, N. (2020). Design sprint for SMEs: An organizational taxonomy based on configuration theory. *Management Decision*, *58*(9), 1803–1817.

Manzini, E., & Rizzo, F. (2011). Small projects/large changes: Participatory design as an open participated process. *Codesign-International Journal of Cocreation in Design and the Arts*, *7*(3-4), 199–215.

Martin, R. L. (2011). The innovation catalysts. *Harvard Business Review*, *89*(6), 82-+.

McGann, M., Blomkamp, E., & Lewis, J. M. (2018). The rise of public sector innovation labs: Experiments in design thinking for policy. *Policy Sciences*, *51*(3), 249–267.

McGann, M., Wells, T., & Blomkamp, E. (2021). Innovation labs and co-production in public problem solving. *Public Management Review*, *23*(2), 297–316.

Na, J. H., Choi, Y., & Harrison, D. (2017). The design innovation spectrum: An overview of design influences on innovation for manufacturing companies. *International Journal of Design*, *11*(2), 13–24.

Nagaraj, V., Berente, N., Lyytinen, K., & Gaskin, J. (2020). Team design thinking, product innovativeness, and the moderating role of problem unfamiliarity. *Journal of Product Innovation Management*, *37*(4), 297–323.

Olsen, N. V. (2015). Design thinking and food innovation. *Trends in Food Science & Technology*, *41*(2), 182–187.

Pande, M., & Bharathi, S. V. (2020). Theoretical foundations of design thinking - A constructivism learning approach to design thinking. *Thinking Skills and Creativity*, *36*, 637-637.

Pluchinotta, I., Kazakci, A. O., Giordano, R., & Tsoukias, A. (2019). Design theory for generating alternatives in public decision making processes. *Group Decision and Negotiation*, *28*(2), 341–375.

Rau, C., Zbiek, A., & Jonas, J. M. (2017). Creating competitive advantage from services: A design thinking case study from the commodities industry. *Research-Technology Management*, *60*(3), 48–56.

Roberts, D. L., & Palmer, R. (2012). Developing a visceral market learning capability for new product development. *International Journal of Market Research*, *54*(2), 199–220.

Seidel, V. P., & Fixson, S. K. (2013). Adopting design thinking in novice multidisciplinary teams: The application and limits of design methods and reflexive practices. *Journal of Product Innovation Management*, *30*, 19–33.

Shafiee, S., Haug, A., Kristensen, S. S., & Hvam, L. (2021). Application of design thinking to product-configuration projects. *Journal of Manufacturing Technology Management*, *32*(1), 219–241.

Shapira, H., Ketchie, A., & Nehe, M. (2017). The integration of design thinking and strategic sustainable development. *Journal of Cleaner Production*, *140*, 277–287.

Snyder, K., Ingelsson, P., & Backstrom, I. (2018). Using design thinking to support value-based leadership for sustainable quality development. *Business Process Management Journal*, *24*(6), 1289–1301.

Sohaib, O., Solanki, H., Dhaliwa, N., Hussain, W., & Asif, M. (2019). Integrating design thinking into extreme programming. *Journal of Ambient Intelligence and Humanized Computing*, *10*(6), 2497–2504.

Sorice, M. G., & Donlan, C. J. (2015). A human-centered framework for innovation in conservation incentive programs. *Ambio*, *44*(8), 788–792.

Souza, A., Ferreira, B., Valentim, N., Correa, L., Marczak, S., & Conte, T. (2020). Supporting the teaching of design thinking techniques for requirements elicitation through a recommendation tool. *IET Software*, *14*(6), 693–701.

Vagal, A., Wahab, S. A., Butcher, B., Zettel, N., Kemper, E., Vogel, C., & Mahoney, M. (2020). Human-centered design thinking in radiology. *Journal of the American College of Radiology*, *17*(5), 662–667.

Vetterli, C., Uebernickel, F., Brenner, W., Petrie, C., & Stermann, D. (2016). How Deutsche Bank's IT division used design thinking to achieve customer proximity. *Mis Quarterly Executive*, *15*(1), 37–53.

Yan, M. R. (2018). Improving entrepreneurial knowledge and business innovations by simulation-based strategic decision support system. *Knowledge Management Research & Practice*, *16*(2), 173–182.

Yang, C. M., & Hsu, T. F. (2020). Integrating design thinking into a packaging design course to improve students' creative self-efficacy and flow experience. *Sustainability*, *12*(15), 5929-5929.

Yeoman, P., & Carvalho, L. (2019). Moving between material and conceptual structure: Developing a card-based method to support design for learning. *Design Studies*, *64*, 64–89.

Yu, M. J., Qu, M., & Hu, J. (2015). A research method to capture design state based on multi-fuzzy cognitive mapping. *Journal of Intelligent & Fuzzy Systems*, *29*(6), 2669–2575.

3.4 The impact of outsourcing and collaboration on the use of intuition and deliberation

A study of site analysis in the context of architectural design

Diana Osmólska and Alan Lewis
University of Manchester

3.4.1 Introduction

There is evidence that architects use intuition in their work, particularly in the design process (Akin, 1986, 1990; Canter, 1974; Lawson, 2004b; Schön, 1991), and also in site analysis (Burns & Kahn, 2005; Darke, 1979; Leoto & Lizarralde, 2019; Lewis, 2017; Rohde et al., 2020). Previous studies have focused on individual designers' use of intuition. However, architectural design and site analysis are not practised individually (Imrie & Street, 2011). A project involves external consultants, internal team members, regulators and Local Planning Authorities who all affect outcomes (Sinclair, 2020). This is especially apparent during site analysis, where information from multiple sources is gathered, synthesised and applied to test the feasibility of proposed construction projects. Collaboration and outsourcing are essential to site analysis, yet how they impact architects' thinking has received little attention.

During site analysis, a proposed development is assessed to determine its viability in a particular location. The process includes evaluating existing conditions and exploring design options (LaGro Jr, 2013). It is during those initial stages that ideas are generated that can shape the entire project (Darke, 1979; Lawson, 2004a). It is also during site analysis when there is the greatest potential for optimisation of the project (Kovacic & Zoller, 2015; Rohde et al., 2020). Such optimisation can happen through architects' detailed analysis, outsourced information or collaboration. The decisions taken during those stages can influence up to 80% of the project's environmental and operational costs (Bogenstätter, 2000) and have off-site impact on biodiversity, flooding and traffic (LaGro Jr, 2013).

Arguably, the site-analysis literature provides an inaccurate description of architects' processes, implying the use of linear procedures, in which information is collected and synthesised before design can begin (e.g., LaGro Jr (2013) or Sinclair (2020)). The design-studies literature, however, suggests that architects often start by forming design solutions, which are then evaluated as information is gathered (Akin, 1990; Darke, 1979; Lawson, 2006; Schön, 1991). Such studies have focused on architectural design, but much could be learned by considering the related activity of site analysis. Specifically, there is scope for exploration of the impact of solution-focused approaches on how architects gather and synthesise information, tasks intrinsic to site analysis. In this way, studying architects' use of intuition in site analysis expands design thinking to encompass processes of information gathering and evaluation, that is, non-design activities that are integral to design. Those activities are particularly important as they affect design-quality and performance, as noted above.

DOI: 10.1201/b22630-20

We have an understanding that expert architects work using pattern recognition, where they automatically recall past-situations based on their experience and design habits, and apply those to current situations (Akin, 1986, 1990; Canter, 1974; Lawson, 2004b). If, however, we acknowledge the co-productive nature of architectural practice, our current understanding of architects' intuitive processes can change. This may open new fields of discussion and expand current understanding of design thinking. To explore the use of intuition and deliberation in a collaborative environment, this paper will draw on dual processing theory (DPT). DPT recognises two types of thinking, Type 1 which is automatic and intuitive, and Type 2 which is deliberate and slow (Evans & Stanovich, 2013). The concept of fast and slow thinking is widespread both in philosophical and psychological writing (Frankish & Evans, 2009) and has been applied in multiple fields of study such as economics (Kahneman, 2011; Rand et al., 2012), medicine (Brink et al., 2019; Brush et al., 2017) or judicial practices (Wissler et al., 2012).

DPT assumes that we are all born with essential Type 1 abilities and develop others through learning. A prolonged use of a technique can allow an individual to become an expert and perform it intuitively. An example of such phenomena in architectural practice was described by Lewis (2017), who discussed how his interviewees could look at an architectural drawing, and intuitively judge whether a proposed building would have adequate daylight. However, not all intuitions are based on expertise (Evans & Stanovich, 2013), because Type 1 is constantly "on" generating ideas and suggestions (Kahneman, 2011). Hence, intuition can be used even if there is not enough skill to support the accuracy of judgments (Stanovich, 2018).

Type 2 is associated with complex decisions and problems. It requires concentration for optimal performance, for example, when calculating 126×17 (unless you are an expert in mental arithmetic), because it is bound by the working memory and its capacity. Type 2 is used for rationalisation of intuitions, hypothetical thinking, argumentation, safety and reduction of errors (De Neys, 2017; Evans, 2019). Deliberation needs motivation, relevant knowledge and cognitive capacity to perform optimally.

The way both thinking-types combine is that Type 1 provides Type 2 with information. Type 2 tends to approve these suggestions immediately, especially when there is an ingrained feeling of rightness, which is a characteristic of Type 1 (Ackerman & Thompson, 2017; Thompson et al., 2011). Alternatively, Type 2 justifies suggested intuitions or deliberates and reasons for alternative solutions (Evans, 2019).

To look beyond intuition and deliberation as aspects that affect an individual, Hemmatian and Sloman (2020) discussed how outsourcing and collaboration can be explored through DPT. The process of outsourcing might be linked to the use of intuition, as it hinges on information one does not possess, but which can be relied on, and supplied by another more informed individual (Hemmatian & Sloman, 2020). In those situations, Type 2 processes can become outsourced to reduce effort and complexity. It is possible to confuse outsourced information with individual knowledge and in turn, overestimate one's understanding of the process. Just knowing that there is access available to experts who understand a particular process is enough to provide a non-expert with the confidence needed to proceed with a particular task (Barbey et al., 2021; Hemmatian & Sloman, 2020; Rabb et al., 2019). Possibly, architects become dependent on other experts as a result of outsourcing, for its convenience and reduced effort.

Collaboration can encourage deliberation especially when there is a common goal (Barbey et al., 2021; Hemmatian & Sloman, 2020). This is not to say that intuition cannot emerge and be prompted by collaboration, but it is possible that deliberation takes more of an active role in collaboration. This could be due to the need for clarity and understanding between team members, which lends itself to a more structured process of working and explaining. In collaborations, however, members of the same team can also share a common language and sense of understanding (Lloyd, 2000; Lloyd & Busby, 2001), potentially reducing the need for deliberation.

By studying collaboration and outsourcing, this paper will explore the involvement of different co-operators and their extended effect on-site-analysis-processes and, in turn, design. Also, considering how architects think, when working in co-production, will enhance our understanding of practice processes.

3.4.2 Background

3.4.2.1 Site analysis

The current literature suggests that site analysis involves following instructions, theories, set standards and guidelines on how to analyse a site successfully (De Chiara & Koppelman, 1984; LaGro Jr, 2013; Lynch & Hack, 1984; Makstutis, 2018; Rubenstein, 1987; Tait, 2018). It incorporates scientific methods to predict, for example, daylight (Mavromatidis et al., 2014), sunlight (Nault et al., 2015), ventilation levels (Ko et al., 2018) or multiple physical properties at the same time (Leidi & Schlüter, 2013). Site analysis processes described in the current literature are arguably Type-2 dependent, that is, they assume the use of deliberation.

There is a possibility that site analysis also involves the use of intuition, which is yet to be acknowledged in the site-analysis literature. The literature on design expertise suggests that architects utilise automatic pattern recognition (Akin, 1986, 1990; Canter, 1974; Lawson, 2004b). Researchers who discussed architects' site analysis process (Darke, 1979; Lewis, 2017) and scientific methods for site analysis (Rohde et al., 2020) also indicated reliance on intuition, although without specifically outlining how intuition was used.

Despite the cruciality of site analysis in building development, it has received little attention until now. There is a potential for this research to influence the current site analysis and design literature.

3.4.2.2 Collaboration

There have been studies on collaboration on forms of design outside architectural practice, that suggest disagreement and challenging of ideas can lead to more careful analysis of solutions (Stempfle & Badke-Schaub, 2002). Moreover, collaboration can prompt a team to explore more factors before a decision is made. In collaboration, the knowledge of a team is larger than that of any individual (Ottosson, 2003). There are suggestions outside the area of design that groups perform to the standard of their best member (Moshman, 1998). The design-studies literature also acknowledges how critical networks are in the transfer of knowledge and their application in action (Cross et al., 2006; Lloyd & Busby, 2001; Luck, 2007; Schön, 1991). As suggested by Hemmatian & Sloman (2020) DPT can provide the means to explore those networks.

Current literature also indicates how verbalisation of ideas, collective problem solving (Cross et al., 1996; Ding & Ng, 2010; Gill et al., 2000; Lawson, 2006; Murphy, 2005) and pitching ideas to a team (Lloyd & Busby, 2001) can lead to the use of deliberation. It is not necessarily the case, however, that collaboration always leads to deliberation. For example, storytelling can act as a form of discourse within a design team. Stories can begin to shape ideas during design processes (Lloyd, 2000; Lloyd & Busby, 2001). Designers' use of storytelling can be accompanied by "selective interpretation" (Brereton et al., 1996), where the story being told is based on incomplete knowledge of a situation. This could be linked to Type 1's need for coherence which is privileged over the need to full and accurate information (Kahneman, 2011).

The research on collaboration and working in practice is still limited (Ding & Ng, 2010). It focuses primarily on team meetings or project discussions. In architectural practice, other players are involved. For example, site analysis would include the client (and the client's advisors), lead designer (who gathers input from each team member), principal designer (managing health and safety), designers (architect, landscape-architect, interior designer), engineers (civil, structural, mechanical and electrical [M&E], fire and others) and other specialists (such as cost, acoustic or daylight consultants) (Sinclair, 2020). There is a need for research that explores architects' thinking processes when working with those professionals towards a shared goal.

3.4.2.3 Outsourcing

The concept of outsourcing was defined by Hemmatian and Sloman in the field of cognitive science as information "one does not know oneself but assumes that someone else (whose identity may be unknown) can supply" (Hemmatian & Sloman, 2020:103). Hemmatian and Sloman in their description of outsourcing mostly present examples of non-professionals utilising knowledge from an expert. However, it is possible the implications of outsourcing change when it is used between professionals. Such changes might impact on how outsourcing is defined.

Research outside of the area of design suggests that people might believe they know more than they do, by attributing outsourced knowledge as their own (Ward, 2021). This raises questions regarding who takes the credit for the outsourced actions (Barbey et al., 2021).

By recognising that architects work with multiple players, there is an indirect acceptance that those actors provide information. An architect needs to consider the data from all those actors to shape a project, even when not in the position of a lead consultant. Outsourcing can also happen between colleagues working for the same company. Although current design-studies literature explores collaboration, there is little discussion of implications of outsourcing on how professionals work. This research will begin to fill that gap, by exploring how outsourced information is applied, when it is used, its links to intuition and how it differs from collaboration.

3.4.3 Research aims

This paper explores how outsourcing and collaboration impacts on the use of intuition and deliberation within site analysis, focusing on the following questions: Is collaboration more likely to result in deliberation or intuition? And similarly, is outsourcing more likely to result in the use of deliberation or intuition?

3.4.4 Methods

3.4.4.1 Participants

To investigate the research aims, the article draws on semi-structured interviews with practising architects and architectural assistants. For the purposes of this study, 21 participants were interviewed. Participants varied in their experience, positions within the company, types of work undertaken and practice sizes. The researchers decided to have a varied representation of practitioners undertaking site analysis in their daily practice to explore differences between their approaches. Interviews were conducted until researchers decided the data had reached saturation.

Purposive sampling was used, with potential participants identified through architectural-practices' websites (Etikan et al., 2016). Additional participants were recruited through snowball sampling (Biernacki & Waldorf, 1981). All of the participants were employed in the United Kingdom and undertook site analysis in their work. Participation in the study was voluntary. Table 3.4.1 provides more specific participant detail.

All interviews were included in the analysis of data. Longer interviews were richer in contextual detail, but shorter interviews allowed researchers to have a broader understanding of what is happening in practice, and to reach saturation.

3.4.4.2 Data collection

The interview questions were intended to facilitate the researchers in understanding the site analysis process. The questions were open-ended and focused on asking architects to explain from the beginning how they go about analysing a site. Prior to each interview, interviewees were asked to be prepared to discuss a specific past project. When it was necessary to gain more detail about a particular response or process a closed question was asked. However, those were generally avoided to allow interviewees to describe site analysis processes in their own words (Fisher & Geiselman, 1992).

It was possible that due to "desire for esteem" (Proctor, 2006) participants would present themselves and their processes in the best possible ways, not necessarily in line with their actual ways of working or thinking. Hence, the researchers decided to build a rapport with the interviewees (Fielding, 1993; Phoenix, 1994). The interviewing researcher had experience of working in architectural studio. This allowed a familiarity bond to be built with the interviewees, restricting participants' feelings of being examined or under threat, and allowing the interviews to proceed more openly. The interviewing researcher was careful to not let their preconceptions based on their architectural-experience lead the interview (Chew-Graham et al., 2002). As interviewees were expected to present themselves in the best possible light, statements that showed interviewees' culpability in making errors were judged to be reasonably accurate. Additionally, to overcome the limitation of architects' self-presentation, the interviewer sometimes followed open-ended question with questions that required explicit answers.

All interviews were conducted by one researcher via telephone or video link (Zoom). During the interviews, some participants shared their screens to talk through plans, drawings, diagrams, presentations, competition submissions and planning application documents such as full or outline design and access statements.

This research was conducted during COVID-19 pandemic, where there was a need to adapt from face-to-face interviews to accommodate for the restrictions in place (Chew-Graham, 2020). Research suggests that interviewing via video-link carries similar strengths to telephone-interviews, such as convenience and accessibility (Block & Erskine, 2012; Gray et al., 2020). However, different to telephone-interviews, commonality of life during pandemic allowed researcher to easily build a rapport with the interviewees reducing some of the weaknesses associated with distance-interviews.

The interviews were digitally audio-recorded and transcribed verbatim. This study was approved by the University of Manchester Research Ethics Committee.

Table 3.4.1 Participants' information

Interviewee number	*Experience (years qualified as an architect [Q] or years without qualifying [WQ])*	*Position within the company*	*Type of work undertaken "Mixed" involved in 3 or more different types of work*	*Practice size Large: more than 50 employees Medium: between 11–50 employees small: less than 11 employees*	*Length of the interview (min)*
A1	15 Q	Senior Architect	Healthcare	Large	94
A2	3 Q + 11 WQ	Architect	Healthcare, Education	Large	32
A3	3 WQ	Arch-Assistant	Inhouse healthcare team	Unknown	19
A4	26 Q	Director	Residential, commercial	Medium	80
A5	24 Q	Director	Residential, commercial	Small	62
A6	7 WQ	Arch-Assistant	Mixed	Large	64
A7	22 Q	Director	Mixed	Small	48
A8	25 Q	Director	Education, mixed	Medium	39
A9	1 Q	Architect	Mixed	Large	75
A10	20 Q	Director	Masterplanning, Mixed	Small [solo practitioner]	77
A11	4 WQ	Arch-Assistant	Residential/conversion	Medium	61
A12	24 Q	Director	Masterplanning, Mixed	Medium	32
A13	5 WQ	Arch-Assistant	Mixed	Medium	50
A14	2 Q + 3 WQ	Director	Residential, Commercial	Small [solo practitioner]	95
A15	22 Q	Director	Conversion, Education	Small [solo practitioner]	69
A16	16 Q	Director	Residential	Medium	99
A17	10 Q	Director	Mixed	Small [solo practitioner]	50
A18	1 WQ	Arch-Assistant	Large mixed projects	Large	53
A19	17 Q	Regional Director	Mixed	Large	51
A20	10 WQ + 0.5 Q	Architect	Mixed	Medium	47 + 92
A21	5 Q	Senior Architect	Large mixed projects	Large (Star practice)	109

3.4.4.3 Data analysis

Inductive thematic analysis was undertaken. Although the researchers had a pre-existing theoretical understanding of intuition through DPT, they did not anticipate the focus on collaboration and outsourcing; rather, this interest was prompted by the data, combined with an awareness of recent design-studies literature (Braun & Clarke, 2006).

Transcripts were analysed using NVivo 12 software. The themes were identified, by considering the extent to which aspects of the data captured "something important to the … research question[s]" (Braun & Clarke, 2006:82). The analysis, therefore, while inductive, was also informed by relevant literature. As Braun & Clarke have observed, "Researchers cannot free themselves of their theoretical and epistemological commitments, and data are not coded in an epistemological vacuum" (2006:84).

The use of deliberation or intuition cannot be directly observed, hence researchers had to make assumptions based on the interviews and transcripts. Sometimes it was possible to infer the use of intuition or deliberation from a specific quote, which was then provided in the findings section. On other occasions, the use of intuition or deliberation was only apparent through a number of interview questions and the flow of the conversation. For example, as requested, an interviewee would talk-through their project and only towards the end, or by being prompted by an emerging question, an aspect discussed earlier would reveal itself as intuitive or deliberate. Due to complexity of those situations, the researchers decided to paraphrase the key aspects to capture and explain the phenomena. When findings were paraphrased and reported, their prevalence in particular transcripts was noted.

3.4.5 Collaboration in architectural practice

Site analysis within the interviews was described as a collaborative process involving internal and sometimes external collaborators. The interviewees collaborated with clients [A1,16], contractors [A2,3] and users [A1]; with fellow architects and assistants [A10,2,20,21,6,14,19,4,9,15]; with consultants where there was an opportunity to involve them in the project at an early stage [A10,3,16,5,15] and planners only when there was a shared goal [A10].

A1's report can be used as an example of how such collaboration happened in practice. A1 discussed how when working on a new cancer centre, a retired head of radiotherapy helped to develop the brief and program with her extensive experience. A1 also involved other clinical staff and specialists to gain feedback on the project's proposed occupancy, purpose, dimensions, spatial connections and hierarchy. Then it was A1's role to synthesise information to explore design options. The proposals were then tested and reviewed, as A1 explains how *"I must have done a half a dozen presentations leading up to submitting the planning application to keep the Trust updated and the team updated with the progress of the design …" (A1).* The design proposals formed part of testing the feasibility of the site.

The process described by A1, where the information is gathered and repetitively communicated through the involvement of different experts is similar to what normative literature outlines for programming stages (LaGro Jr, 2013; Maxwell & Brown, 1993; Tusler et al., 1993), suggesting a deliberate process.

3.4.5.1 Architects' desire for collaboration

The interviews with solo-practising architects particularly stood out, as they emphasised how much they missed practice-based exchange of ideas [A17,10,15], potentially because they could compare their experience with when they worked at a multi-person office [A17]. The contrast could allow them to realise how collaboration can prompt discussions that do not occur when an individual is working alone (Wallace & Ahmed, 2003) in turn, improving proposals through deliberation (Mercier & Sperber, 2011; Trouche et al., 2014). This led one architect to *"set up a network of smaller businesses in [his city]"* made of construction industry specialists. They want to use the network as *"sounding board for free advice"* and *"do design reviews for each other and potentially check stuff over"* (A10).

Other interviewees also discussed how they appreciated forms of collaboration. A21, reported how his practice operates on pin-up bases, where they showcase their work every morning to the project team, and begin discussing and elaborating on ideas. A4 and A19 discussed having project presentations across the office to gain feedback.

The reliance on others in collaboration can be desirable for architects as it can allow them to work more efficiently. For example, A9 utilised the collaborative setting to work quickly on multiple things at a time, potentially intuitively. He could work in such a way because he *"... [relied] on the other staff in the office to pick up the detail ..."*, this was achievable due to common goals set in collaboration.

3.4.5.2 Joint problem solving

Collaborative problem-solving appeared to involve deliberation. Interviewees discussed how when mistakes emerged, the team tended to come together to jointly solve those. The mistake acted like a common goal that drove the team together. For example, A9 discussed how when they neglected to consider topography at the beginning of analysis, and developed an infeasible proposal *"everyone panicked and had to do a lot of work very quickly ... and just working very, very, very hard in the panic ... So it wasn't it wasn't a disaster by any means, but ... It was a pretty scary two weeks ..."* A similar situation was described by A6, A1 and A10, where the team had to come together to resolve site analysis issues they initially neglected to consider.

Sometimes a problem emerged because of a particular design concept chosen for the project or due to site constraints. That problem then became the focus of deliberation, an area where collaborative effort was placed to resolve the issue. For example, A20's project was near a motorway, and as a result the design team had to consider the pollution, how to dampen the sound, ventilate the building and the cost of the project. A16 wanted to discharge service water into a river that was protected at certain times of the year. Those problems led to careful considerations and involvement of specialists. However, by resolving these problems, architects brought improvement to the area of problem-focus. Resolving a problem is not equal to the betterment of the design as a whole (Chakrabarti, 2003). In that respect, placing effort in one area, does not result in deliberate reasoning on other aspects of the project. Additionally, after effort has been placed, practitioners might be less willing to engage in Type 2 for the next challenge, because effort is tiring (Baumeister et al., 2018; Vohs et al., 2014). There were no apparent reports within the transcripts of architects undergoing a complex process of resolving a problem, followed by another problem, or multiple problems at once, in relation to site analysis.

3.4.5.3 Negotiation as a result of collaboration

Although not directly considered as a form of collaboration, negotiation, where discussion is aimed at reaching an agreement between the parties involved, emerged as a theme when working in collaboration within site analysis [A16,9,15,1,2,21]. Based on reasoning literature we can assume that deliberation is responsible for the development of arguments and persuasion, which are present in negotiation (De Neys, 2017; Mercier & Sperber, 2011). Interviewees reportedly developed strategies to persuade other collaborators of their ideas. For example, A16 described how when working with a developer and surveyor he intentionally *"set the bar high [for the project] so they have something to cut. Almost like giving them a chance so they feel like they have something to do"* (A16). The process might have also involved hypothetical thinking, a feature of Type 2, where different possibilities and their consequences were explored prior to being shown to the collaborator (Evans, 2020; Evans, 2019).

There are often meetings arranged with the local planners to discuss larger projects prior to submitting planning applications (in the United Kingdom proposed buildings require planning permission before construction can begin). Those meetings are another example of where collaboration led to negotiation. For example, an interviewee explained how they prepared aspects to propose during the meeting to *"sweeten the deal"* (A9) for the planners to accept their design proposals, and overlook certain factors of site analysis that are not to the expected standard. A9 provides an example of such deal, where they are *"providing this amount of funding for this, which they [planners] really want, and you're gonna be ... redoing the public realm around the site ... and you'll be providing X amount more workspace, which is important, in their strategic plan ..., maybe they can overlook the rights [to light] issue if it's not a major concern"*. Other forms of negotiation emerged when architects had to communicate with Government funding bodies, Historic England, or Conservation Officers [A16 and 15]. For example, A16 describes how *"Conservation Officers [... want a building ...] to look old."* This is an illustration of a site constraint, where the design of a new building might be constrained by the need to make it resemble adjacent historic buildings.

3.4.5.4 Individual thinking in a team setting

When introducing the concept of collaboration, A1 was used as an example for his work with external actors. However, in the interview, A1 downplayed the importance of internal collaboration. Similarly, A15 emphasised his own role, where he was the Lead Architect on a project. This could be linked to taking such collaboration for granted, or architect's ego when presenting his project. Alternatively, it is due to the hierarchical setup of an office, as one interviewee compares his experience to *"... other offices I worked [in], the decision making is very hierarchical. So, it comes from the director, director thinks something, and everyone just does it ..."* (A21). Another interviewee discussed how when the Lead Architect wanted to develop an idea, the support or comments from fellow team members did not affect the design [A20]. In those situations, potentially the Lead Architect wants to have autocracy in decisions on design and the feasibility of a proposed building on a specific site. However, individuals who are thinking on their own will not be able to benefit from collaboration (Mercier & Sperber, 2011).

3.4.5.5 Collaborative overlooking

Although collaboration can lead to further deliberation, there were reports of collaborative overlooking of certain site analysis aspects. This phenomenon can be attributed to intuition, based on Kahneman's (2011:85) explanation of Type 1 characteristic "What you see is all there is (WYSIATI)". WYSIATI implies that Type 1 creates a story from the information an individual can see. In those circumstances, the quality and quantity of evidence supporting a particular story does not matter. Instead, it is the quality of the story that engenders confidence in an individual. For the sake of a coherent story, an individual might fail to recognise that critical evidence is missing. Arguably, it is not just subjective confidence that such stories can facilitate, they can also foster collective confidence. The interviewees A1, 2, 6, 9, 10 and 13 believed what they saw as a team, in their project stories, was all there was. They confidently proceeded with their projects, until they realised, they were collaboratively overlooking missing crucial evidence. The development of such stories was also discussed within A20 and 21's interviews but without mention of overlooking of evidence.

The following quote from A1 provides an example of the phenomena within a team setting:

"A1:	*We did a health centre in [City] in 2018 … it was ready to submit for planning at stage 3. And later on in the day, the client sent through the underground utilities information, which showed a high voltage electricity cable running right under the building, which meant that actually to divert that would cost the client 150,000 pounds. So, it blew the budget out of the water straightaway. […]*
Researcher:	*is there a reason why the client didn't send it earlier on?*
A1:	*I think it was a mistake. I think that the architects … have … just … basically forgotten. [And] the project manager forgot about it. … And it wasn't until I think it was a planning requirement as part of the submission that they wanted to see underground information."*

In this case, there was a collective acceptance of the project, its story and progression, which reduced the need to deliberate or look for additional information. A1's team had meetings, negotiations and discussions with the client, planners, and other collaborators. The project was ready to be submitted for planning, hence, site analysis was completed, and the design was coming to an end as well. However, it was not until A1's team was prompted by an external cue, in needing to supply information, that they realised they were missing critical evidence. The project had to be re-worked once they realised their mistake, which led to delays, loss of time and fees.

3.4.6 Outsourcing

There were instances of outsourcing in the interview data, which aligned with Hemmatian and Sloman's (2020) definition. For example, the interviewees outsourced internally through relying on information from architectural assistants, and in-house experts, data gatherers and analysts. External outsourcing was also prominent, where the data was provided by external consultants. In the description of findings below, however, the definition is extended to include outsourcing of decisions, where decisions are made by another individual.

3.4.6.1 Outsourcing effort for efficiency

Type 2 processes are dependent on cognitive resources (De Neys, 2017). Overall, overworking in forms of cognitive, emotional or physical effort can result in greater reliance on Type 1. After we expend energy we have a tendency to save what is left (Baumeister et al., 2018; Kahneman, 2011; Vohs et al., 2014). Hence, the interviewees appeared to outsource some site-analysis-related effort, to facilitate their deliberation in other places. For example, architectural assistants can become outsourced to produce site analysis for the project lead, as described by A16: *"… I didn't have time to sit there and do it … I say, "Here you go," to one of the Part 1 s (architectural assistants) [you do it] …" (A16).*

Similarly, instead of researching different aspects of a site, A19 described how they outsourced knowledge from fellow-office-architects with specific areas of expertise. Those architects could utilise their knowledge to very quickly provide information to be incorporated into site analysis. The process differs from collaboration as it does not involve engaging in constructive discussions.

Some practices reportedly employed staff who specialise in specific data-gathering methods for their site analysis. For example, A13's practice employs a data analyst, A19 has a team who works with GIS (Geographic Information System) and A21's practice has a research department that uses scientific methods to provide information to the design team, a workspace department that focuses on developing information about the future of the site and project, and a sustainability team. The material is then provided to the design team and utilised in the project.

Interviewees [A13,16,19,21] explained utilising outsourced information through forms of feeling of rightness. If there was a low feeling of rightness, deliberation would be engaged, and the particular area of information would be investigated, but otherwise it was intuitively accepted and applied to the project (Thompson, 2009; Thompson et al., 2011; Thompson & Morsanyi, 2012).

3.4.6.2 Information stored elsewhere

Sometimes an in-house expert can develop and share information to aid site analysis, such as a guide to planning applications, or compile certain regulations for ease-of-use. The members of the office might begin to utilise such information, but they can also assume they know more than they do. The actual knowledge and its in-depth implications are retained by the expert who supplied the information. It is also possible for the knowledge to be shared across other experts who understand the information provided. Utilising such aids can be attributed to outsourcing, and be based on intuition (Rabb et al., 2019; Sloman & Rabb, 2016). For example, A2 reported how they have access to *"these spreadsheets [with] all of those calculations [for example for occupancy and spatial areas which] … you don't need to be doing, because someone's already done them and they are tried and tested".*

Other examples described by the interviewees include pre-made documents and templates which contain ways of presenting site analysis, or examples of how to construct certain analytical diagrams or calculations [A1,20]. Those could then be applied to projects *"… from page to page. It was like, access, then it was sun path, then it was noise, potential noise, then it was … protected trees and greenery areas around … and then your little massing studies at the end"* (A20). A similar approach reportedly was taken with housing standards, where a *"variance of stuff we used before"* (A16) is outsourced to form new proposals. Another example was described by A11, where her practice developed a style that had to be applied when working on projects.

3.4.6.3 Outsourcing to reduce complexity

Some Type 2 processes can be outsourced to reduce complexity. Interviewees commonly avoided complex analysis by outsourcing it [A1,3,5,6,7,9,11,13,14,17,18,19,20]. Hence, if they needed to understand an aspect of a site, instead of utilising prescriptive or scientific methods they could rely on an expert specialised in the area of enquiry to deliver the knowledge (Hemmatian & Sloman, 2020).

This approach appears to reflect an attitude in which external consultants are responsible for providing technical knowledge, while the role of an *"architect is to actually bring in the right partners at the right time"* (A19) [similarly A2] (also discussed in Jamieson et al., 2011). As observed above, however, architects might not only be outsourcing the analytical aspects of site analysis but also consultants' understanding. For example, A17 discussed how he had *"a specialist involved to do those types of calculations"* but A17 *"didn't understand the numbers"*. Other interviewees also reported being unable to fully understand the reports [A3,7,13,14,17,18,19,20]. An example of such phenomena can be seen in the following conversation:

"Researcher:	*Did you do any calculations to make sure that there's enough light?*
A1:	*We didn't, no, because we knew that the volume of light coming from this atrium would be sufficient through what the M&E guys told us. … Likewise, natural lighting, artificial lighting and ventilation in-take and extract and a lot of other things utility, services is all the M&E consultant, we design the space and help deliver the brief.*
Researcher:	*Does what M&E consultant … inputs in, does that change the design quite often?*
A1:	*Not really.*" (A1) [similar conversation happened with A7]

It should be noted that in the above quote, the information given by the consultant did not lead to collaboration. The interviewee was also not involved in any analytical work prior to consultant's provision of information. This could be linked to interviewees' desire to reduce cognitive load and their arguably "lazy" Type 2, which prefers not to engage in a topic where lack of expertise might lead to increased effort and energy. Arguably there was no application of the findings to the project either, as the design was not affected (Baumeister et al., 2018; Hemmatian & Sloman, 2020; Kahneman, 2011; Vohs et al., 2014).

Sometimes to reduce complexity and consultancy costs, interviewees described how they did not engage in analytical aspects, such as *"daylight analysis"* (A18). They would only conduct such analysis if asked to do so by a planner in the submission process [A18,11]. However, at that stage, it can be too late to adjust the design to apply the information. Similarly, some interviewees reported how consultants' reports needed to be done, as a "tick-the-box-exercise", with no intention to synthesise such information into the project [A6,9], hence supporting the argument that reports did not affect the design.

Other interviewees discussed how they looked for pointers within report's summary pages, to highlight any aspects which needed consideration [A1,7,13]. Possibly in practice, information is accepted and set aside, unless there is a cue that triggers a low feeling of rightness and prompts deliberation.

When there was data that needed applying, interviewees looked for specific instructions they could swiftly employ. For example, if the report said there is *"arsenic in the ground, so you need to have a capping layer of 600 mm of new topsoil"*, the interviewee *"would be like fine, okay, so the ground levels of the site need to come up by 600 mm …"* (A5). Type 2 might be involved in reading and comprehending the report; arguably Type 2 is always present in some form when reading and formulating a response The DPT literature assumes that Type 2 is always "on" but in a low-effort mode (De Neys, 2017; Kahneman, 2011). However, the instructions of what needed to be done were outsourced and applied.

Another perspective on outsourcing is when the external consultant is content to oversee and guide the architect through the process to reduce the complexity of their task at a later stage. As A15 explains: *"[the environmental, M&E and structural engineer] want me to give them the right zones of space, so that when they are on the project at the building regs stage, they've got enough space to do what they need to do. They've learned over the past three years that if they don't help me early on, it's really difficult for them at the next stage because I haven't given them enough space ..."* (A15). However, such process of inter-reliance was achieved over-time.

Not all interviewees relied on consultants for information. A21 reported how it is important for him to use analytical software which can help shape the design. That way as the design progresses, he can make sure the proposal is optimised, for example, the proposal makes use of the orientation on site. Additionally, the software can highlight any parts which need further attention. By contrast, consultants' reports can be utilised for precise information, after the design has been set. A21's approach aligns with some of the scientific methods discussed earlier.

3.4.6.4 Outsourcing decisions

Sometimes, planners were outsourced by architects as design validators. For example, *"I think if ... it went to planning, and you received an approval straightaway. Then yeah, it's a good sign that you've done all your job correctly. And you've done the analysis, and it kind of justified the design process"* (A13) [also reported by A11, A6 and A10]. In those situations, possibly architects were substituting difficult questions such as whether the analysis was satisfactory, with easier ones, such as whether the planners liked the scheme. Substitution, where one question is answered instead of another, is a feature of Type 1 thinking (Kahneman, 2011:97). Substitution was also used to outsource what needs to be incorporated into site analysis based on *"what the planners want"* (A8) [also A18 and A8].

Reportedly once planning consent has been granted less consideration could be placed on site analysis. For example, A6 discussed how nearby buildings *"are being overshadowed by us"* but *"because this ... already been ... [through] ... planning, it is going to get done, so, it's fine"* (A6). The importance of decisions associated with site analysis diminished, implying that the initial process of site analysis was reliant on outsourced planners' needs. Overall, there is less need to engage in Type 2 when the decisions are insignificant (Evans, 2019).

The process of structuring and aligning a project to suit planners' needs can reportedly fail, when there is a change of staff, as explained by A16:*"We had many discussions with one planner, and then he left, another came, and they said this completely does not suit the requirements and it does not work for them, it does not make sense, the aesthetic does not match what we are after." (A16).*

Potentially, the reason why there is minimal collaboration between the planners and architects is because they are working to a different set of goals for the project and the wider built environment. If the goals aligned, it could become *"a team-based thing, rather than ... the architect's game"* as A14 reports, who has experience of working in collaboration with the local authorities and planners. Reportedly such collaboration is rather uncommon. However, it could lead to some benefits discussed in the previous section, such as joint discussion and improvement of ideas.

Difficult decisions can also be outsourced by architects to their clients. Hence, analysis would be resolved, when the client is satisfied [A1,19].

3.4.7 Discussion

By switching the focus from design processes to information gathering, which is central to site analysis, we can open new perspectives on design thinking. Specifically, exploring how collaboration and outsourcing are used in site analysis reveals their impact on intuition and deliberation.

3.4.7.1 Extending the frontiers of design thinking through the study of site analysis

Previous studies of design indicate that designers, including architects, often start by generating and evaluating potential solutions, rather than gathering and synthesising information before commencing design (Akin, 1990; Darke, 1979; Lawson, 2006; Schön, 1991). According to these studies, information-gathering and design happen simultaneously. Such studies, however, have focused on design; less attention has been paid to the ways in which designers collect information. This study expands design thinking by focusing on site analysis, a design-related activity that centres on the gathering and synthesis of information. Moreover, it considers information-gathering processes through the lens of DPT, which provides insights, especially in relation to architects' use of Type 1 thinking, with its capacity to overlook information.

Looking at site analysis through the lens of DPT challenges some accepted ideas around design expertise. In the current literature, much emphasis is placed on the ways in which designers, including architects, develop and deploy expertise in design processes (Akin, 1986, 1990; Canter, 1974; Lawson, 2004b; Schön, 1991). Specifically, there is a perception that architects learn over time, allowing them to intuitively recall situations by meaningful chunks when prompted by cues. However, it appears that architects can also take the default act of applying information without conscious reflection. Unlike pattern recognition the information is not based on consolidated knowledge (Akin & Akin, 1996). This raises the possibility that vital information will be overlooked in decisions on design or on the feasibility of a building on a given site. The implications of overlooking information are apparent from the literature on post-occupancy which highlights the severity of problems caused by design errors (Stevenson, 2019).

3.4.7.2 Co-production in architecture

Collaboration and outsourcing are essential to the information-gathering processes involved in site analysis. By investigating these, this study broadens our current understanding of collaboration (Cross et al., 2006; Ding & Ng, 2010; Gill et al., 2000; Goldschmidt, 1996; Lawson, 2006; Lloyd & Busby, 2001; Luck, 2007; Murphy, 2005; Ottosson, 2003; Schön, 1991; Stempfle & Badke-Schaub, 2002) by highlighting instances where team-work prompted deliberation or intuition. Current literature indicates that in collaboration, differences and challenges (Stempfle & Badke-Schaub, 2002) and forms of negotiation (Brereton et al., 1996; Lloyd & McDonnell, 2009; McDonnell, 2009) can be present, and those can lead to more deliberate ways of working, which was echoed in this study's findings.

A novel and critical finding is how intuition can be present in a team setting in the form of collaborative overlooking. This happens when missing critical evidence is overlooked for the sake of a coherent story about a project. Such, selective interpretation was also reported by Brereton et al., (1996). However, this study's findings indicate that those stories engendered confidence that was shared across the team. Arguably, this Type 1 characteristic, usually attributed to individuals' use of intuition (Kahneman, 2011), can manifest in team-based settings. This shared sense of confidence can undermine the benefits associated with collaboration, such as the challenging of ideas, which prompts deliberation.

Although there have been previous studies on collaboration in design, there have been far fewer studies on outsourcing. Within site analysis literature (LaGro Jr, 2013; Makstutis, 2018) there is an assumption that architects receive information from external parties but the process behind its application is not discussed. A key finding was that architects rely on outsourcing for efficiency and to reduce complexity. Due to outsourcing, architects can feel as if they have all the necessary information, if they receive reports on key issues. However, the automatic nature of outsourcing may mean that architects do not necessarily understand the reports or have read their content. Hence, they can have a misplaced sense of confidence. The findings indicate that outsourcing led to the use of deliberation only when the information received was accompanied by a low feeling of rightness. However, if the information is beyond architects' understanding, it might not be possible for them to recognise its relevance, and in turn, have an associated feeling of rightness, to engage Type 2 (Stanovich, 2018).

A related issue is some architects' apparent inability to understand consultants' reports. Potentially, because architects know they can avoid understanding complex structures by outsourcing them, they need never gain competence in the task in question (Hemmatian & Sloman, 2020). Furthermore, reflecting on consultants' reports would require architects to expend a considerable amount of energy (Baumeister et al., 2018; Vohs et al., 2014), especially if there is lack of competence in the area (Dreyfus & Dreyfus, 2005). However, this might be unfeasible in an environment where architects are under pressure to produce high-quality outputs within a limited time (Jamieson et al., 2011; Nelson, 2017; Serginson et al., 2013; Sinclair & Ashkanasy, 2016).

This study also begins to explore co-production and its effect on the design produced. Information delivered from an outsourced external actor, or fellow office staff member, did not appear to affect the design, unless there was an issue that emerged and needed resolving. The contribution of outsourced actors was mostly uncredited. There is an understanding that architects work collaboratively (Goldschmidt, 1996; Imrie & Street, 2011), which was shown in the findings, however, the lead architects also wanted to portray themselves as the sole authors.

3.4.7.3 Limitations and further work

Semi-structured interviews are post-rationalised reports of events. Trying to obtain data on intuition or deliberation through interviews is difficult, especially since it is dependent on interviewees' recollection of the event. Brief events which occurred some time ago can be forgotten (Ericsson & Simon, 1984; Paletz et al., 2011). Hence, further research is needed to explore key questions emerging from this research. For example, ethnographic work on initial and post-completion stages of a project would be beneficial to explore implications of architects' decisions, and experimental research on architects' competencies would show the basis of expertise in architects' intuitive judgments.

3.4.8 Conclusion

Through studying site analysis, and the integral activities of collaboration and outsourcing, this paper expands design thinking. Investigating design-related information-gathering processes yields new insights into how architects work, specifically on architects' use of intuition beyond pattern recognition.

The findings indicate that collaboration is more likely to result in deliberation when there is a desire between team members to work together, when faced with a difficult problem that needs resolving, and when it leads to negotiation. Intuition, however, might be present in collaboration, particularly where shared narratives cause the overlooking of information, which in turn can lead to re-works. Outsourcing allowed for efficiency and reduced projects' complexity but seemed more likely to lead to the use of intuition. In the interview data, deliberation was prompted during outsourcing only when there was a common goal between the information supplier and architect, or when outsourced information led to architects' low feelings of rightness. The prominence of outsourcing in this study raises questions regarding the type and level of expertise that architects require in order to gather and synthesise information, in undertaking site analysis.

References

Ackerman, R., & Thompson, V. A. (2017). Meta-Reasoning: Monitoring and control of thinking and reasoning. *Trends in Cognitive Sciences, 21*(8), 607–617.

Akin, Ö. (1986). *Psychology of architectural design*. Pion.

Akin, Ö. (1990). Necessary conditions for design expertise and creativity. *Design Studies, 11*(2), 107–113.

Akin, Ö., & Akin, C. (1996). Frames of reference in architectural design: Analysing the hyper acclamation (A-h-a-!). *Design Studies, 17*(4), 341–361.

Barbey, A. K., Patterson, R., & Sloman, S. A. (2021). *Cognitive neuroscience meets the community of knowledge* [Preprint]. PsyArXiv.

Baumeister, R. F., Tice, D. M., & Vohs, K. D. (2018). The strength model of self-regulation: Conclusions from the second decade of willpower research. *Perspectives on Psychological Science, 13*(2), 141–145.

Biernacki, P., & Waldorf, D. (1981). Snowball sampling: Problems and techniques of chain referral sampling. *Sociological Methods & Research, 10*(2), 141–163.

Block, E. S., & Erskine, L. (2012). Interviewing by telephone: Specific considerations, opportunities, and challenges. *International Journal of Qualitative Methods, 11*(4), 428–445.

Bogenstätter, U. (2000). Prediction and optimization of life-cycle costs in early design. *Building Research & Information, 28*(5–6), 376–386.

Braun, V., & Clarke, V. (2006). Using thematic analysis in psychology. *Qualitative Research in Psychology, 3*(2), 77–101.

Brereton, M. F., Cannon, D. M., Mabogunje, A., & Leifer, L. J. (1996). Collaboration in design teams: How social interaction shapes the product. *Analysing Design Activity*, 319–341.

Brink, N. V. den, Holbrechts, B., Brand, P. L. P., Stolper, E. C. F., & Royen, P. V. (2019). Role of intuitive knowledge in the diagnostic reasoning of hospital specialists: A focus group study. *BMJ Open, 9*(1), e022724.

Brush, J. E., Jr., Sherbino, J., & Norman, G. R. (2017). How expert clinicians intuitively recognize a medical diagnosis. *American Journal of Medicine, 130*(6), 629–634.

Burns, C., & Kahn, A. (2005). *Site matters: Design concepts, histories and strategies*. Routledge.

Canter, D. (1974). *Psychology for architects*. Applied Science.

Chakrabarti, A. (2003). Processes for effective satisfaction of requirements by individual designers and design teams. In *Human behaviour in design* (pp. 132–141). Springer.

Chew-Graham, C. A. (2020). Involving patients in research during a pandemic. *Health Expectations: An International Journal of Public Participation in Health Care and Health Policy*, *23*(3), 521–522.

Chew-Graham, C. A., May, C. R., & Perry, M. S. (2002). Qualitative research and the problem of judgement: Lessons from interviewing fellow professionals. *Family Practice*, *19*(3), 285–289.

Cross, N., Christiaans, H., & Dorst, K. (1996). *Analysing design activity*. Wiley.

Cross, R., Laseter, T., Parker, A., & Velasquez, G. (2006). Using social network analysis to improve communities of practice. *California Management Review*, *49*(1), 32–60.

Darke, J. (1979). The primary generator and the design process. *Design Studies*, *1*(1), 36–44.

De Chiara, J., & Koppelman, L. (1984). *Time-saver standards for site planning*. McGraw-Hill Book Company.

De Neys, W. (2017). Bias, conflict, and fast logic: Towards a hybrid dual process future? In *Dual process theory 2.0* (pp. 47–65). Routledge.

Ding, Z., & Ng, F. (2010). Knowledge sharing in architectural design institutes: A multiple-case study. *Construction Innovation*.

Dreyfus, H. L., & Dreyfus, S. E. (2005). Peripheral vision: Expertise in real world contexts. *Organization Studies*, *26*(5), 779–792.

Ericsson, K. A., & Simon, H. A. (1984). *Protocol analysis: Verbal reports as data*. The MIT Press.

Etikan, I., Musa, S. A., & Alkassim, R. S. (2016). Comparison of convenience sampling and purposive sampling. *American Journal of Theoretical and Applied Statistics*, *5*(1), 1–4.

Evans, J. S. B. (2020). *Hypothetical thinking: Dual processes in reasoning and judgement* (2nd ed.). Routledge.

Evans, J. S. B. T. (2019). Reflections on reflection: The nature and function of type 2 processes in dual-process theories of reasoning. *Thinking & Reasoning*, *25*(4), 383–415.

Evans, J. S. B. T., & Stanovich, K. E. (2013). Dual-process theories of higher cognition: Advancing the debate. *Perspectives on Psychological Science: A Journal of the Association for Psychological Science*, *8*(3), 223–241.

Fielding, N. (1993). Qualitative interviewing. In N. Gilbert (ed.), *Researching social life* (pp. 135–154). Sage Publications Ltd.

Fisher, R. P., & Geiselman, R. E. (1992). *Memory enhancing techniques for investigative interviewing: The cognitive interview*. Charles C Thomas, Publisher.

Frankish, K., & Evans, J. St. B. T. (2009). The duality of mind: An historical perspective. In J. Evans & K. Frankish (eds.), *In two minds: Dual processes and beyond* (pp. 1–30). Oxford University Press.

Gill, S. P., Kawamori, M., Katagiri, Y., & Shimojima, A. (2000). The role of body moves in dialogue. *International Journal of Language and Communication*, *12*, 89–114.

Goldschmidt, G. (1996). The designer as a team of one. In N. Cross, H. Christiaans, & K. Dorst (eds.), *Analysing design activity*. Elsevier.

Gray, L. M., Wong-Wylie, G., Rempel, G. R., & Cook, K. (2020). Expanding qualitative research interviewing strategies: Zoom video communications. *The Qualitative Report*, *25*(5), 1292–1301.

Hemmatian, B., & Sloman, S. A. (2020). Two systems for thinking with a community: Outsourcing versus collaboration. In *Logic and uncertainty in the human mind: A tribute to David E. Over* (pp. 102–115). Routledge.

Imrie, R., & Street, E. (2011). *Architectural design and regulation*. Wiley-Blackwell.

Jamieson, C., Robinson, D., Worthington, J., & Cole, C. (2011). *The future for architects?* Report. London: University of the Arts.

Kahneman, D. (2011). *Thinking, fast and slow*. Macmillan.

Ko, W. H., Schiavon, S., Brager, G., & Levitt, B. (2018). Ventilation, thermal and luminous autonomy metrics for an integrated design process. *Building and Environment*, *145*, 153–165.

Kovacic, I., & Zoller, V. (2015). Building life cycle optimization tools for early design phases. *Energy*, *92*, 409–419.

LaGro Jr. J. A. (2013). *Site analysis: Informing context-sensitive and sustainable site planning and design*. John Wiley & Sons.
Lawson, B. (2004a). *What designers know*. Elsevier/Architectural Press.
Lawson, B. (2004b). Schemata, gambits and precedent: Some factors in design expertise. *Design Studies*, *25*(5), 443–457.
Lawson, B. (2006). *How designers think: The design process demystified* (4th ed.). Elsevier/Architectural Press.
Leidi, M., & Schlüter, A. (2013). Exploring urban space: Volumetric site analysis for conceptual design in the urban context. *International Journal of Architectural Computing*, *11*(2), 157–182.
Leoto, R., & Lizarralde, G. (2019). Challenges in evaluating strategies for reducing a building's environmental impact through Integrated Design. *Building and Environment*, *155*, 34–46.
Lewis, A. (2017). The mathematisation of daylighting: A history of British architects' use of the daylight factor. *The Journal of Architecture*, *22*(7), 1155–1177.
Lloyd, P. (2000). Storytelling and the development of discourse in the engineering design process. *Design Studies*, *21*(4), 357–373.
Lloyd, P., & Busby, J. (2001). Softening up the facts: Engineers in design meetings. *Design Issues*, *17*(3), 67–82.
Lloyd, P., & McDonnell, J. (2009). *About: Designing-analysing design meetings*. Taylor and Francis Group.
Luck, R. (2007). Using artefacts to mediate understanding in design conversations. *Building Research & Information*, *35*(1), 28–41.
Lynch, K., & Hack, G. (1984). *Site planning*. MIT Press.
Makstutis, G. (2018). *Design process in architecture: From concept to completion*. Laurence King.
Mavromatidis, L. E., Marsault, X., & Lequay, H. (2014). Daylight factor estimation at an early design stage to reduce buildings' energy consumption due to artificial lighting: A numerical approach based on Doehlert and Box–Behnken designs. *Energy*, *65*, 488–502.
Maxwell, C. E., & Brown, D. R. (1993). Programming processes for military health care facilities *Professional Practice in Facility Programming*, 249–278.
McDonnell, J. (2009). Collaborative negotiation in design: A study of design conversations between architect and building users. *CoDesign*, *5*(1), 35–50.
Mercier, H., & Sperber, D. (2011). Why do humans reason? Arguments for an argumentative theory. *Behavioral and Brain Sciences*, *34*(2), 57–74; discussion 74-111.
Moshman, D. M. M. G. (1998). Collaborative reasoning: Evidence for collective rationality. *Thinking & Reasoning*, *4*(3), 231–248.
Murphy, K. M. (2005). Collaborative imagining: The interactive use of gestures, talk, and graphic representation in architectural practice *Journal of the International Association for Semiotic Studies*, *156*, 113–145.
Nault, E., Peronato, G., Rey, E., & Andersen, M. (2015). Review and critical analysis of early-design phase evaluation metrics for the solar potential of neighborhood designs. *Building and Environment*, *92*, 679–691.
Nelson, C. (2017). *Managing quality in architecture: Integrating BIM, risk and design process* (2nd ed.). Routledge.
Ottosson, S. (2003). Collaborative product development considerations. In U. Lindemann (ed.), *Human behaviour in design* (pp. 164–173). Springer.
Paletz, S. B. F., Schunn, C. D., & Kim, K. H. (2011). Intragroup conflict under the microscope: Micro-conflicts in naturalistic team discussions. *Negotiation and Conflict Management Research*, *4*(4), 314–351.
Phoenix, A. (1994). Practising feminist research: The intersection of gender and 'race' in the research process. In M. Maynard & J. Purvis (eds.), *Researching women's lives from a feminist perspective* (pp. 49–71). Taylor and Francis.
Proctor, R. (2006). The architect's intention: Interpreting post-war modernism through the architect interview. *Journal of Design History*, *19*(4), 295–307.

Rabb, N., Fernbach, P. M., & Sloman, S. A. (2019). Individual representation in a community of knowledge. *Trends in Cognitive Sciences*, *23*(10), 891–902.

Rand, D. G., Greene, J. D., & Nowak, M. A. (2012). Spontaneous giving and calculated greed. *Nature*, *489*(7416), 427–430.

Rohde, L., Jensen, R. L., Larsen, O. K., Jønsson, K. T., & Larsen, T. S. (2020). Holistic indoor environmental quality assessment as a driver in early building design. *Building Research and Information*, *49*(4), 460–481.

Rubenstein, H. M. (1987). *A guide to site and environmental planning* (3rd ed.). Wiley.

Schön, D. A. (1991). *The reflective practitioner: How professionals think in action*. Avebury.

Serginson, M., Ladinski, V., Giddings, B., & Messer, S. (2013). Assessing the effectiveness of architectural design communication through public participation methods. *International Journal of Design Management and Professional Practice*, *6*(1), 61–83.

Sinclair, D. (2020). *RIBA plan of work*. In A. Tait & L. Carmichael (eds.). RIBA. www.ribaplanofwork.com

Sinclair, M., & Ashkanasy, N. M. (2016). Intuition: Myth or a decision-making tool? *Management Learning*, *36*(3), 353–370.

Sloman, S. A., & Rabb, N. (2016). Your understanding is my understanding: Evidence for a community of knowledge. *Psychological Science*, *27*(11), 1451–1460.

Stanovich, K. E. (2018). Miserliness in human cognition: The interaction of detection, override and mindware. *Thinking & Reasoning*, *24*(4), 423–444.

Stempfle, J., & Badke-Schaub, P. (2002). Thinking in design teams-an analysis of team communication. *Design Studies*, *23*(5), 473–496.

Stevenson, F. (2019). *Housing fit for purpose: Performance, feedback and learning*. Routledge.

Tait, J. (2018). *The architecture concept book*. Thames & Hudson.

Thompson, V. A. (2009). *Dual-process theories: A metacognitive perspective*. Oxford University Press.

Thompson, V. A., Turner, J. A. P., & Pennycook, G. (2011). Intuition, reason, and metacognition. *Cognitive Psychology*, *63*(3), 107–140.

Thompson, V., & Morsanyi, K. (2012). Analytic thinking: Do you feel like it? *Mind & Society*, *11*(1), 93–105.

Trouche, E., Sander, E., & Mercier, H. (2014). Arguments, more than confidence, explain the good performance of reasoning groups. *Journal of Experimental Psychology: General*, *143*(5), 1958.

Tusler, W. H., Zilm, F., Hannon, J. T., & Newman, M. A. (1993). Programming: The third dimension. In W. F. E. Preiser (ed.), *Professional practice in facility programming* (pp. 227–247). New York: Van Nostrand Reinhold.

Vohs, K. D., Baumeister, R. F., Schmeichel, B. J., Twenge, J. M., Nelson, N. M., & Tice, D. M. (2014). Making choices impairs subsequent self-control: A limited-resource account of decision making, self-regulation, and active initiative. *Motivation Science*, *1*(S), 19–42.

Wallace, K., & Ahmed, S. (2003). How engineering designers obtain information. In U. Lindemann (ed.), *Human behaviour in design* (pp. 184–194). Springer.

Ward, A. F. (2021). People mistake the internet's knowledge for their own. *Proceedings of the National Academy of Sciences*, *118*(43).

Wissler, R. L., Williams, K. E., & Saks, M. J. (2012). Dual-processing models of admissibility: How legal tests for the admissibility of scientific evidence resemble cognitive science's system 1 and system 2. *Va. JL & Tech.*, *17*, 354.

3.5 Considering architect's and clients' interactions through the design thinking framework

Audrey Mertens, Yaprak Hamarat, and Catherine Elsen
Université de Liège

3.5.1 Context of the research project

3.5.1.1 Background of the research project

This research is part of a wider research project aiming to provide a better understanding of how architects and end-users interact nowadays and to suggest ways to renew these interactional models. As the intertwined relationship between designers and end-users constitutes a crucial part of the design process, involving users in the design process is more than ever essential to the project success (Lawson, 2005; Estevez & Léglise, 2015; Sarkar & Gero, 2017; Arboleda, 2020). Recently, researchers have put into question the traditional model of architectural design, seen as the result of a sole Master's artful persuasion (Prost & Chaslin, 2014). This model has been said to be completely outdated and no longer practicable (Siva & London, 2011; Albrecht, 1988; Macaire, 2009; McDonnell & Lloyd, 2014). As an alternative, this chapter leans towards the design thinking framework as an inspiration to help architectural routines evolve when it comes to users' and architects' interactions, while still being adapted to architects' needs and field constraints.

3.5.1.2 Belgian context

Considering the architectural practice, we specifically focus on housing in Belgium (in Wallonia & Brussels), where any construction work affecting the envelope or structure of a building requires an architect in charge. In the process of digging into existing interactional models, one of the first steps consists of getting a better grasp of architects' actual organizations and current practices.

One of the most well-known Belgian association for consumer protection ("Test-Achats") has shown that amongst 1330 respondents who built their personal homes between 2000 and 2013, only half felt satisfied with their architect(s) at the end of the process. Almost half of the participants moreover expressed having experienced stress and 27% confessed having suffered from irritability and anxiety throughout the process (Nauwelaers & Rossini, 2014). Another questionnaire broadcasted in Belgium showed that the client is among the « top 5 factors » that make architects' jobs more difficult on an everyday basis (Stals et al., 2018). This friction occurring in the housing design process between architects and end-users highlights the need to further question the social interactions underlying the design process, as these interactions seem to be a source of dissatisfaction.

DOI: 10.1201/b22630-21

3.5.2 Architectural design through the design thinking framework (a thematic literature review)

End-user low satisfaction levels constitute a key issue nowadays in architectural design (Siva & London, 2011). Discounting the importance of alarming reports, architectural design indeed missed several opportunities to systematize user-centered and participatory approaches, therefore lagging behind other design fields (Bacqué et al., 2011; Van der Linden et al., 2019). If housing architecture is a branch of the design practices, we question the influx of design thinking in that scale of practice.

We will develop this connection through two levels: conceptual and pragmatic. First, we touch upon the principles of design thinking as a reflective approach (section 2.1). Second, we tackle its use as a process and a set of tools for architects (section 2.2). This framework will be used as a lens to look into our empirical data, as to later frame how it might support user-centeredness in architectural practice.

3.5.2.1 Might design thinking be a mindset … for architects?

As Brenner and Uebernickel put it, "innovation is made by humans for humans" (2016, p. 8). This human-centered premise is presented as one of the fundamentals triggering all design thinking principles. Another fundamental aspect would be combining both divergent thinking – following unconventional paths – and convergent thinking to eventually grasp the details and stakes of the most promising path (Cooper, 2008; Brenner & Uebernickel, 2016). Those lines of thinking, by nature, require experimenting and iterating, prototyping, failing, implementing and enhancing prototypes, looping with this process until the output meets most expectations in a satisfactory manner. Of course, some architects can already be referred to as *design thinkers* in that regard, but one has to observe that most of those concepts are not systematically and/or explicitly taught in architecture schools, nor implemented in most architectural offices (Nishimura et al., 2017). Looking into architectural practices, another question arises: whose insights and user-centered data might drive such iterative processes in architecture, be they applied? Architectural offices functioning with design teams (instead of an individual architect working on his/her own project) might already be closer to the design thinking mindset, in the sense that during collective ideation, processes get challenged by other designers, all of them working together with creative confidence and optimism (Brown, 2009).

Nevertheless, in the architectural field, more often than not we see architects operating in the close, tiny, mono-disciplinary microcosms of their firms. This is especially true in Belgium, where research has shown that most architects work in structures of less than 10 employees, 42,7% of surveyed architects (N=572) even working alone, or with just one colleague (Stals et al., 2017). Historically and structurally speaking, architects thus do not tend to exit their microcosms and cultivate empathy towards end-users, at least not as often as designers from other design disciplines. Attention has been brought to this issue repeatedly (Tribout, 2012; McDonnell & Lloyd, 2014) and, as a consequence, participatory practices slowly gained more interest from architects, especially in public spaces' design or building projects driven by the public sector (Biau et al., 2012; Zetlaoui-Léger, 2013). If such citizens' input, needs and expectations are nowadays considered as enriching these divergent-convergent processes, it is not clear if such participatory practices might also percolate smaller scale projects, such as housing projects.

3.5.2.2 Might design thinking act as a process, a toolbox … for architects?

The fragmented nature of the architectural design field, as underlined by Jacob (2016), can be challenging to analyze. Based on Lawson (2005), Jacob provides a brief model of the architectural process, arguing that during the first step called "formulating", architects have to define, identify, understand, frame and explore highly complex, intricated and ill-structured design problems. Emphasis is put on the fact that architects, while navigating such problem space, should generate stories to reframe the issues from different points of view and perspectives. In this case, we argue that being in contact with the key actors eventually populating the artifact being designed, for example, the future end-users, might help open up some relevant perspectives, that architects while working between themselves might otherwise not open to. From where we stand, this first phase might well be the cornerstone, with the involvement of end-users profoundly changing the whole story. This aspect of Jacob's model that we defend is closer to matching Brenner and Uebernickel's (2016) "need finding" step (Figure 3.5.1).

In his second step, Jacob (2016) delineates the "representation phase", when architects externalize their ideas and thoughts to represent the problem through models or drawings, among other visual supports depending on the nature of the outcome. If we compare Jacob's model to Brenner and Uebernickel's (2016), this step also overlaps the "need finding" step mentioned earlier (Figure 3.5.1). This is the moment where architects create externalized, mediating supports and could further include other actors into the architectural conversation. In Jacob's third and final step, architects come up with one or more solutions – also referred to as "moving" by Lawson (2005) – in a form of divergent/convergent process, which connects well with design thinking principles of ideating, prototyping and testing (Figure 3.5.1). Still, according to Jacob (2016), designers should integrate objective/technical and subjective/aesthetic judgments in making choices and be able to reflect in action, and also be able to reflect on how they go about the design process itself, for example by keeping sketchbooks or collecting artifacts reflecting what they consider to be good design to learn from when considering future designs. Reflecting on Lawson and Jacobs's process descriptions, we realize those do not fall very far from Brenner and Uebernickel's design thinking description (2016). For both authors, the sequence of steps from "formulating" to "moving" rarely follows a linear logic (Lawson,

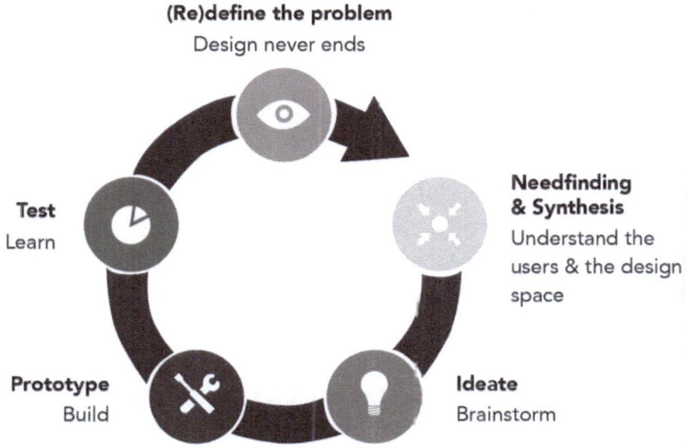

Figure 3.5.1 The design-thinking micro process, in Brenner & Uebernickel (2016, p. 11).

2005), but is rather continuous and iterative. However, whilst keeping Lawson's and Jacob's hypothesis and developments in mind, we will base our comparative analyze on Brenner and Uebernickel's model (Figure 3.5.1).

If the five steps forecasted in Figure 3.5.1 are typically implemented to aim for judicious innovation through design thinking, it is still unclear at this point how (or if) those five steps might be transposed or interpreted in the small scale of housing architecture practices, or if any of the tools Brenner and Uebernickel have popularized with their 2016 book – that is, Stakeholder Maps, Empathy Maps, 5-whys, AEIOU-Method, Personas, Observation and Storytelling – are taken upon by architects in the Belgium housing context. Several researches have indeed shown that, in small-scale architectural practices, most architects rarely go beyond early conversational interactions to reach out to users' needs and expectations (Norouzi et al., 2015; Van der Linden et al., 2017).

3.5.2.3 Focus

In this chapter, we specifically question the posture architects tend to have towards their end-users and we use design thinking as a framework through which we study these architectural practices. We first want to summarize the process and meeting points of architects and end-users during small-scale housing projects. We question the architects' postures toward the end-users and look for success stories in these relationships. Finally, we look for the patterns depicted by the interviewed architects in regard of the design thinking principles, steps and tools raised here above.

3.5.3 Research methodology

After building a theoretical understanding of our use of the design thinking model through international scientific literature, we put our focus on Belgian practices through empirical inquiry. We broadcasted a questionnaire through the Belgian Architects' Association, as to invite all Walloon and Brussels architects affiliated to this Association to take part in this study. We then conducted five semi-structured, in-depth interviews.

On that basis, we consider the data through the lens of abductive analysis (Timmermans & Tavory, 2012): with Brenner and Uebernickel's principles as a comparative model, we analyze and compare the phenomenon described by architects in terms of design methods and we let the data express unexpected trends, adding specific and original depth to the model regarding our field (e.g., housing architecture in the Walloon Context).

3.5.3.1 Online survey and profile of the participants

The recruitment questionnaire, aiming to profile different significant groups of architects active in the housing sector, got 96 respondents. Questions touched upon their location in Belgium, their experience, education, the scale of their agency, their level of responsibility, the percentage of housing projects overall, the type of clients they usually deal with (private individuals and families/investors …), if they feel impacted in their daily practice by their relationships with their clients, and eventually how often they interact with them. We grouped the respondents in view of their answers, and we selected an architect amongst each main group to conduct an in-depth interview and collect qualitative data on the actual day-to-day practices regarding the interactions with their clients. We added the profile of another architect who had not participated in the survey but was interviewed in the same conditions than the others. This last profile was a pragmatic choice yielded by an opportunity: a senior architect in contact with our research lab agreed to participate in the interviews even though he had not responded to the questionnaire. This allowed us to include the profile of someone who was not amongst the architects innately drawn to this problematic (and spontaneously taking part in the survey), that is, someone who was not necessarily predisposed to volunteer and participate in our research, but rather did it give favorable circumstances.

In this chapter, we thus focus and analyze interviews of five architects (two female, three male), situated in three different provinces (Brussel [2], Liege [2], Luxembourg [1]) and who mainly practice residential architecture. Only one is fully dedicated to individual clients with a single-family housing project, while others also work for residential projects carried by institutions (e.g., social housing and co-op housing), self-builders, or real estate investors. They have various professional experiences (less than six years [1], between 10–14 years [2], more than 25 years [2]). Only one of the participants is trained as an architectural engineer and four others are architects, one with a collective housing specialty. On a daily basis, they work in an environment welcoming between 3 and 14 collaborators, except for one of them who works alone. All of them hold leadership positions, three for their own companies and two for third parties. With the exception of one of them (for whom we could not collect data through the survey), the others declared that the relation with the client influences a lot their motivation and the quality of their daily life at work. To conclude, we identified that the way these architects meet their clients during the early design phases of architectural project is quite different, notably in terms of frequency. While one of them declares meeting the client only twice, others meet between three and five times, and another can meet with them up to eight times before reaching the building permit phase.

3.5.3.2 Semi-structured in-depth interview and analysis

These interviews aim to collect rich qualitative data on architects' relations with their clients, who are also the end-users of the building being designed, since we are here looking at housing projects cases. The objective is to understand multiple interactions that intertwine different issues related to communication, architectural culture, money management and time, among others. This complex situation, called a "wicked problem" by Horst W. J. Rittel (see [Buchanan, 1992] for a genealogy of the concept of design thinking), requires a global understanding of the phenomenon and a co-construction of some shared sense between the researcher and the interviewee (Imbert, 2010; Denzin & Lincoln, 2011). Therefore, a conversational place is necessary for architects to develop their point of view when it comes to the role given to the end-users throughout the architectural practice.

Semi-structured interviews are nowadays recognized for their added value when researching architectural or design processes. For example, four Japan-based designers were interviewed to study "their perception of Western and Japanese people's behaviour in co-design workshops" (Taoka et al., 2021, p. 2), while 20 interviews were realized to identify "perceptions, sources and tools, as well as perceived barriers and motivations for inclusive design" in Flemish architectural practice (Van der Linden et al., 2016, p. 33). Holopainen (2010) studied a service design project in order to understand how designers and architects design their services, with which particularities. Part of a broader ethnographic approach, they conducted eight scheduled, semi-structured interviews to "provide[d] a more structured way of discovering the opinions of the informants" (Holopainen, 2010, p. 602). As we see in these examples, even though such interviews should be completed with other methodological approaches to deeply understand architects' practices (see 5. Limits and Perspectives), semi-structured interviews allow to reveal architects' perspectives, at least in an exploratory approach.

Through general and specific questions, both on residential architecture and relations with clients, the semi-structured interviews created a reflexive conversation and revealed the experiences, opinions, feelings and particular situations experienced by the architects during their careers. A set of 20 questions organized into six thematic categories thus guided the conversation on the specific topic of interactions between architects and end-users, while leaving the possibility for the participants to explore other subjects more freely when relevant and when the allocated time allowed. In brief, five one-to-one recorded semi-structured interviews were realized through Zoom (3), in the interviewee house (1), or hosted at a university office (1) and lasted between one and two hours. We focused only on discursive content on basis of a written transcription of all of the audio recordings (h = 8:58:57). After such a transcription, we did a comparative analysis of the interview data, using an abductive approach. To do so, we used our research topics operationalized through the questions asked to interviewees. We annotated common and singular patterns meaningful to our research object, that is, interactions between architects and end-users. We aim to reveal the plurality of the data without generalizing it (Beaud & Weber, 2010; Dodier & Baszanger, 1997). To this end, we differentiate the interviewees from one another using pseudonyms to preserve their anonymity. We compared these results to the design thinking framework, thus revealing similarities and differences between architectural practices and design thinking and discussing the added aspects of the architect's narrative to enrich the reference model. First, we conduct analysis using a collective lens, pointing out the big overlaps in the interviewees' narratives that echo the design thinking model. Then we zoom into an individual lens and match specific excerpts corresponding or diverging from the model to illustrate differences and surprising specificities as expressed through our interviews.

3.5.4 Findings and discussion

Looking into the data, we first focus on some insightful practices and comments raised during the interviews. Then, we delineate the similarities and gaps between the design thinking framework and the practices depicted by the interviewed practitioners.

In most narratives, we tried to grasp the elements that caused friction, or on the opposite were actually working really well. Thus, we acknowledge that the following anecdotes are not representative of all interactions that may occur between end-users and architects, but these examples can serve as a basis for further reflection on the variety of practices. We also want to drive the reader's attention to the fact that some of these short stories are anecdotal in the interviewee's practice, while some other narratives are more customary and thus representative of this single interviewee's routines.

3.5.4.1 Overlapping positions

The five interviewed architects told us stories about their practices when encountering clients that would reach out to them for their housing project. These narratives allow us to create a summary timeline on the most frequent meeting points occurring between architects and clients throughout the design phase of the project (at least in their contexts). This sequence of steps is a compilation of the main steps described by these five participants, practicing architecture in Belgium in offices employing less than 20 collaborators. The process described includes a first contact phase between the user and the architect, when they both "test each other" in order to decide whether or not they are going to be working together. The second phase, sometimes already forming in parallel with the first phase, is about creating or adjusting a brief, firstly through physical or virtual meetings and conversations, secondly by presenting reference images and/or bringing along the first sketches and drafts of the architect's ideas ("sketch", "*esquisse*" in French). When the architects sense there is a good enough grasp on the user's needs and expectations, they define and detail the layouts and enter the phase they call preliminary project or pre-project ("*avant-projet*" in French). Finally, when the architects consider all the variables as set, they undertake the building permit documentation, which is mandatory for any work on the building's envelope or structure, delineating a rather fixed version of the project ("*dossier de permis*" in French). Of course, this sequence appears as rarely linear, and does not do justice to the messiness and the complexity of an actual design process. This simplification is displayed as a support for the reader's comprehension of the main steps that are discussed in the rest of this chapter.

One of the main strains mentioned by every interviewed architect, apart from the usually tight budget to hold, is the little time they have for all day-to-day responsibilities. They are in charge of usually small teams of co-workers and do not have all the support staff they would dream of as to make more time for meta-cognitive thinking. This could be an explanation for the architects' lack of hindsight on their practices and lack of reflection when it comes to enhancing their interactions with their clients in general. This is a key point that has to be kept in mind for any future recommendations.

Another explanation that could be explored is also brought up by several interviewees. Indeed, Ken, Charly and Richard raise the question of the architectural educational background compared to the actual practice. Whereas handling these complex social interactions require huge amounts of social skills from the architect's part, all three of them mention that being an architect sometimes feels like having to be a psychologist, while never receiving any training in such matter. Architectural training often focuses on the form, function and techniques of the end-product design (e.g., the building, the public space …) but lacks coaching on the overall process of designing, especially in terms of social interactions with colleagues (Calixte, 2021) but also with clients. If empathy is often presented as key to a relevant design (Heylighen & Dong, 2019), these architects regret not having an actual psychology course. Whether this might have been recently implemented in architectural trainings around Belgium should be looked into; yet we wonder: would it be possible to accompany senior architects with further continued training to bridge this gap, in order to enhance their current practices, whilst still fitting their tight schedule? As the leader of his architectural firm, Ken does mention that he called an expert in psychology to give all of his associates a course on collaborating and working as a team. This kind of courses could be a great inspiration to adapt a format of classes tackling the psychology behind interactions with users.

Overall, they all seem to be generally in favor of user-centered design in their discourse, but still often depict a vision of the "wise" architect facing the "naive" end-users. Richard, while emphasizing the importance of being at the user's service, still adopts a paternalistic position, explicitly saying he knows what to do for the good of the occupants, that he "offers" his high-quality viewpoints and "help them take a step back to re-examine their needs". Whilst this hindsight coaching is non-rebuttably welcome and rich, we advocate that any user should be considered as the main "expert" when it comes to their needs, desires and uses.

Richard does mention the difference between designing for private mono-family housing (with very specific, individual choices), and larger scale housing when the design is thought to be suiting "everyone". In that regard, he says he enjoys larger projects, public housing for instance, but still designs some residential projects from time to time. Lola also expresses her preference for clients who let themselves be taken by the hand, who listen to what she suggests, without challenging her too much. Still regarding user expertise, Charly addresses the difference of confidence she might put in a user, depending on whether such a user holds certain skills, background or knowledge in construction-related topics. She values clients that come with certain craftsmanship skills and show themselves to be open to debate and exchange with her. This again expresses a certain reluctance to seriously considering every single end-user's input as expertise. All five of the interviewees state that they refuse clients who arrive with ready-made plans to be signed, but we were not able to collect information on whether those layouts were copied or stolen from another project, or if some or all of the mentioned cases might have actually been drafted by the clients themselves. However, this still highlights the reluctance these architects express about not being the initiator of the design process and design concept.

3.5.4.2 Individual excerpts and anecdotes

Another possible explanation for this reluctance to accept ready-made layouts is brought up by Ken during his interview:

> You don't expect to get the gift [of recognition] but to get it, to feel a real thank you and that your work was appreciated, that is clearly the reason I'm in this business. (…) I could earn more, and stress less [in another job] but I would like it less because I wouldn't have the fulfillment that my job brings me. Recognition is part of that. The fulfillment is to be able to please, to bring projects to life for the people trusting us, bringing our touch to it for people to live in.

From his point of view, the actual passion in his job comes from the fact that he is able to create and gift the project and get full recognition for it; in that regard, we can understand the reluctance he shows when it comes to sharing credit on any design ideation process. However, couldn't he get the recognition he craves by including the end-users more and thus co-creating together with them an even more relevant project, fitting their wishes and needs? The shift in authorship touches a delicate cultural preconception, stating that only the artist is the creator of an art piece he/she offers to the world.

For Leo, the ideal mission for an architect could be summarized as designing living spaces that allow people to blossom. In contrast with this utopia, he says, is the reality of the architect's task: to respect and apply norms, regulations and to take responsibility when things go wrong. On the one hand, he puts the user's wellbeing as the main target, but then on the other hand, it seems like the practice and real-life constraints keep him from actually fulfilling this goal. He also mentions the complexity of communicating with people who do not have an architectural background or sensitivity to the design culture. According to him, people do not have sufficient general skills in terms of reading a blueprint or harmonizing elements and materials. Again, the architect is placed as *the* expert.

When specifically asked about their thoughts on participation practices, in general, all the interviewees pause for a short while. Whilst Ken is not quite sure what participation means, and while Lola shuts down the conversation saying firmly "it's good to listen to all the people whose lives will be impacted by the project", Richard praises it for public projects, but would not go there for smaller scale houses. However, he does detail a participation process he was involved in, in the context of a school project, where all the potential future users were invited to a consultation committee and were invited to speak. They visited the site, room by room, noting down everything that was said, listening to and reframing the debates between stakeholders. He also mentions some work groups with teachers and maintenance staff, of children needs being collected, and even the presentation of a model to the children as to get feedback in an iterative process. The design was then presented to the parents and another feedback loop was implemented to take into account their comments and opinions.

His enthusiasm in narrating this memorable project surely contrasts with his reluctance in including the end-user in the very first steps of the design process when it comes to private housing for instance. Once again, the cultural architectural background of these architects does not include any sort of training in participatory approaches and co-creation methods, which might explain such reluctance.

Regarding Lola's case, even though she did not manifest a particular interest in per-say participation, she does have an interesting posture when it comes to client communication. She qualifies her mission *prima* facie as to have to put all of what she has learned to ensure the comfort and wellbeing of the occupants. She prefers private housing projects and puts a great emphasis on a human connection with her clients, introducing herself not only as an architect but reminding that she is a human being, a woman, a mother too. She wants her clients to be able to empathize with her and tells them "We are going to live together for a few months" before signing the contact. She also mentions what she calls her "feedback methods":

> I sent a card every year (…) I take a picture of one of the year's projects and send it as a postcard to each client at the end of the year, to tell them 'Hi, I'm still here. I hope all is well'(…) I'll publish out images from previous projects. I make a post on Instagram, (…) saying: Here it's already been that many months since the project was finished, or things like that. And [feedback] works through replies to those kinds of posts or replies to stories for me. (…) I've also had clients call me (…) they give me a nice little comment. Or a client who was pregnant at the same time as me, who wrote to me when her baby was born. Yes, little things like that, but always positive (…) without wanting to sound pretentious. And I also think it is because I'm filtering my clients.

However, she still admits that two of her stories are not that easygoing. She addresses her vision on what went wrong in these relationships:

> It happened two times, and the two times the clients were people who came to me, not because they were convinced that what I was doing was great, or that it was what they wanted. No, they came because they needed an architect, already had been kicked out by another one. And because somebody had told them that they knew somebody … that was me. And I was kind of as a last resort.

Leo also has a success story hidden in an anecdote recalled in his interview, that could be taken as an inspiration for similar cases. In this project, he was in charge of redesigning a small house where his clients experienced difficulties in the daily use. He suggested setting up a small "simple" solution, like a prototype, to test what it could be like to partition the living room, and thus created a small entrance hall with a wardrobe. Eventually, the prototype met the expectations of the end-users so well that it was sustained, and the project did not need to be taken further since the issue had been solved. This can be seen as a smart and effective application of the prototyping principle in a small-scale housing architecture context.

Another isolated success story happened in Richard's career, about 20 years ago. He could not recall who initiated this method, but for a specific client, he had to narrate the space. He did have blueprints and sketches but did not take them out at first. Instead, he did some storytelling, he interpreted a daily journey taking place in the designed spaces. This scenario of life, staging the space in the tale while involving the listener, was a success says Richard, a great moment shared with his client, who then became and has since stayed a friend of his. The project did not evolve much afterwards, from what he recalls, as if this was a crucial step when the end-user could project himself into the immaterial design at this time and was content with what he heard. Unfortunately, this was a one-time experience for Richard, thus nothing indicates that this could be an unfailing method to follow. However, the fact that this experience had a positive outcome can be interpreted as a positive signal to test this method further.

3.5.4.3 Patterns of design thinking in architectural practices

Putting together these five narratives and all of the stories of project within, we delineate the practices that can be considered as complete or incomplete steps of a design thinking-like process, discussing potential add-ons specific to our field (e.g., housing architectural practices in Wallonia).

3.5.4.3.1 Need finding and synthesizing

The collection of user data is very variable, and as mentioned earlier, can even vary for a same architect depending on the client he/she meets. As harsh as this elitist way of doing may seem, some architects did confess this practice: depending on the budget of the client and the scale of the project, the architect will bring a different approach to the user's needs, such as higher attention to detail, visiting their homes and ensuring a better client support overall.

Every architect interviewed did use reference images, most of them with a procedure encouraging clients to come up with images they really like and images they dislike and having them explain why. This photo elicitation-like technique (Clark et al., 2013) has proven to be really effective in bringing up reactions and insightful information about the wishes of the client.

Ken also has an original method for collecting precious information. He asks the clients to have a notebook or a large piece of paper that will lie on the kitchen counter or anywhere in the current living space to note down whatever comes to their minds regarding the project, even the smallest thing, even if they think it might seem ridiculous or irrelevant. This allows him to collect some more detailed and intimate information and also trains the clients to actually think and write down what they really need. Although this method sounds really promising, it relies first on the discipline and rigorousness that the clients put into writing things down, and also on their ability to reflect on what they do and what they need. Although this method might miss the actual information that could be observed in action (Zahle, 2012), it still can be a great start to collect further information on uses, expectations, needs and desires of the clients.

Charlie's method is more conventional and consists of a questionnaire of preferences that she sends to her new clients to collect broad information for the project.

Lola had the opportunity to meet with clients initiating new ways of presenting themselves. In one case, it was a family and they had prepared a slide presentation to present all the members of the family, their usages of the rooms, their needs and expectations. "They premade the job for me", she recalls. In the second case, a couple handed her a mood board for the project, which she also welcomed warmly. Although these two cases could be a coincidence, they might also be evidence that the sense of humanity she brings to her communication practices is paying off, as the clients feel freer to share intimate information with her. On another level, this could be another thread to pull on for a shift in the relationship: a thread that could not be pulled by architects, but for once could be pulled by the clients; a way to empower and encourage them to take the lead on unraveling their habits and uses to their architect. As we see it, we conclude that there is no apparent consensus on the methods or tools to collect user data, nor to constitute a brief in small-scale housing projects. Everyone seems to do things in their own way.

3.5.4.3.2 Ideating

Sometimes, it seems that the architects will skip right to the ideation step, collecting only some basic information on the brief and then hurtling to come up with sketches and drafts already. We might want to hear the user's side of this story in order to frame if this might be perceived as an issue.

This ideation step is not so well documented in the interviewees' narrative and still seems to be cultivated as the mystery of artistic creation. Ken does mention that he often goes to reference books for inspirational images, that can also serve later on to communicate with the client.

Only one thing appears quite clearly in the architects' narratives, and somehow goes against the conclusions of our thematic literature review: none of them mention ideas brought by the users when it comes to the design of volumes, layouts, organization, light … . Sometimes, they mention that the clients should chose some of the materials, at best, or sometimes just the interior finishing or some of the equipment, but do not really seem ready to let go with, nor share their creative processes.

3.5.4.3.3 Prototyping

As previously mentioned, some actual tools come in handy when it comes to giving end-users a sense of the future project. Mood boards are often brought up as a useful way to "draft a vision of the space" quite early in the design process. Leo preaches for physical scale models but recognizes that it is slowly fading out of the architects' common practices, as they have less and less time and budget to work with nowadays. He also mentions Virtual Reality (VR) headsets that can allow end-users to virtually move around a model of the project. However, this kind of technology is not supported by Richard. These are interesting tools, he says, but the time to invest in regard to what we want to show is not yet profitable for small projects. If the model is there and well-built for construction or design purposes, it is a blessing to be able to show it to the client as a plus. But if the model is not accomplished enough, it can be detrimental to the project. Building the model is very time consuming and not every architect is trained to model virtually their drawings. Richard and Charlie agree that to be able to showcase a sensory aspect of the chosen materiality, it is easier to work with a still image of an outlook that can be further worked with rendering mode. Lola also creates these kinds of sensory previews but uses a *collage* method to style her prospective views on the project. She adds a layer to the use of these images by suggesting taking the end-users to the room or place that is modeled, in order to better visualize the changes and discuss the designs on actual site.

She also did once some "*prototyping*" with tape and cardboard on site to help end-users get a better sense of the spaces delineated in the project. She is also interested in VR and augmented reality, which would allow clients to virtually walk through the project but insists, just as Richard, on the lack of time to make such models, as well as the lack of funds for this kind of equipment. She also worries that it might lead people to think that you can "easily" move an element while not realizing all of the consequences implied by small changes and thus not realizing how long it can take to actually implement these small kinds of changes.

3.5.4.3.4 Testing and redefining

All of the techniques described here above are supposed to be implemented at some point between the "sketch" phase and the end of the "preliminary project" phase. As a design project is rarely a linear process but rather an iterative, messy process (Brenner & Uebernickel, 2016; Jacob, 2016), often when the client is consulted, there is an update and a change to apply. In this case, it might be important to highlight the fact that it is often not *per se* a test, but rather the architect reaching out to ask for some validation from their clients.

Architects are inclined to accept the changes requested by their clients up to a certain point. Indeed, one of the interviewees insists on the fact that challenging the first drafts is common and clients are welcomed to do so. But as the architect's time is very precious and counted, the interviewed architects have a policy of accepting up to 4 or 5 modifications throughout the "sketch" and "preliminary project" phases and will often even try to aim for a maximum of 3 major modifications.

To that end, Ken and Richard for instance present two rough sketches at the very beginning of the project, in order to rule out the unfollowed paths without having to explore them further. This resonates with Brenner's "fail often and early" principle (2016, p. 8).

3.5.4.3.5 Issues with feedbacks

The first four design thinking steps can be somehow referred to, or at least reinterpreted through the architects' narratives. Commonly emphasizing the importance of empathy and pointing to the iterative process of working through variants of the project, the definition and ideation steps yet vary the most from one profile to the other.

The fifth step however seems to be the most critical. While listening to architects depicting their practices, what stood out the most is the limited feedback of the user on the project being designed, and even more the lack of user's feedback on the overall collaboration process. Indeed, none of the interviewed profiles had an actual feedback method in place to grasp the possible difficulties or dissatisfactions their clients might have experienced during the design and construction process. This particular issue resonates with the design thinking framework, where feedback is an essential step to understand and improve both the output and the process.

In the architectural field, the post-occupancy evaluation (POE) literature generally tackles this feedback issue. While indeed being addressed (see for instance: Love and Holt, 2000; Ng et al., 2011), this study sheds light on the fact that evaluation still appears to be rarely applied in practice, thus revealing a gap between the scientific and the architects' worlds, at least in view of the cases studied. Furthermore, we argue that POE methods undertake the satisfaction issue behind time, after the harm is already done. Even if it brings valuable information for architects to enhance their ongoing practices, the unsatisfied end-users are left with their frustrations and maintenance issues. We advocate in favor of a shift in architectural practice, and for an evaluation and feedback loop as soon as in the early phases of the design process, to aim for better end-user's satisfaction upstream rather than downstream. In that regard, we also want to drive the attention on the loop of *"sketch" ("esquisse" in French), "preliminary project" ("avant-projet")* and "*consultation of the clients/end-user*" described by the interviewed architects in regard of their design processes and emphasize the need for an actual user input during these phases, rather than just allowing them a light comment or validation.

3.5.5 Limitations and future research

If the research points out possibilities for improvement when it comes to architects' and end-users' interactions, five qualitative in-depth interviews cannot lead to generalities nor imply representativeness of all Belgian architects. However, the chapter still highlights that design thinking seems to be hardly diffusing in housing architectural practices, which raises the question of how architectural practices could benefit from broader inspiration from other design practices.

Even though interviews are limited to the "self-reporting issue" (Van der Linden et al., 2016); are restrained to the gap between reflexivity and practice (Boltanski & Thevenot, 1991); remain limited to what architects say they are doing or think they are doing and are influenced by what architects are open to share with the researcher vs. what they really do in practice, they remain a powerful exploratory method to study a professional community. Interviews reveal "the shared vocabulary, attitudes, and values of architects", as well as their relations to users (Van der Linden et al., 2016).

Worth to mention, this study is part of a broader set of interviews (n = 15) which will be soon completed and will be nurtured by observations conducted with an ethnographic approach. Therefore, such interviews constitute an excellent basis to build a solid step-by-step research method to tackle this complex topic of users' place in architectural practice. The complementarity between exploratory questionnaires, interviews and observations in the field is crucial to understand this specific research object (Van der Linden et al., 2016). Moreover, this research is part of a broader research program, including a study on participatory approaches developed in the design field (service, social, graphic, product). The aim is to understand contemporary architectural practices while identifying and adapting strategies developed in design to improve interactions between architects and end-users.

During autumn 2021, we conduct interviews with designers, begin observations with architects, and design upcoming workshops. Then, in winter 2022, we shift to participatory activities through workshops, develop prototypes and probes, and experiment deeply the levers and obstacles for a more collaborative approach between architects and end-users.

3.5.6 Conclusions

Through in-depth interviews, we addressed the main social encounters architects describe when describing the design phases of some of their small-scale housing projects. Then this chapter outlined some pain points and success stories extracted from those processes. Finally, we delineated and discussed some patterns depicted by the interviewed architects in regard of the design thinking principles, steps and tools raised here above, highlighting the (lack of) feedback processes at stake in these narratives. We advocate that an improvement in the interactions between end-users and architects is possible and argue for an architectural practice learning further from the design thinking principles. Moreover, we stand in favor of further involving users into the design process, the quality of the relationship between designers and users still constituting a key issue in architecture. Such an involvement and such a shift in the architect's posture towards the end-user seems to be encountering resistances in the architects' practices and has to be tackled keeping in mind the architects' real-life constraints such as budgets and timing. It nevertheless remains more than ever crucial to the success of any architectural project.

References

Albrecht, J. (1988). Towards a theory of participation in architecture: An examination of humanistic planning theories. *Journal of Architectural Education*, *42*(1), 24–31.

Arboleda, G. (2020). Beyond participation: Rethinking social design. *Journal of Architectural Education*, *74*, 15–25.

Bacqué, M., & Gauthier, M. (2011). Participation, urbanisme et études urbaines: Quatre décennies de débats et d'expériences depuis « A ladder of citizen participation » de S. R. Arnstein. *Participations*, *1*, 36–66.

Beaud, S., & Weber, F. (2010). *Guide de l'enquête de terrain: Produire et analyser des données ethnographiques* (4th ed.). Paris: La Découverte.

Biau, V., Fenker, M., & Macaire, E. (2012). Les métiers de l'architecture et de l'urbanisme à l'épreuve de l'implication des habitants et des usagers. In *Cahiers Ramau 6, L'implication des habitants dans la fabrication de la ville. Métiers et pratiques en question*, 11–28.

Brenner, W., & Uebernickel, F. (eds.) (2016). *Design thinking for innovation: Research and Practice*. Springer International Publishing.

Brown, T. (2009). *Change by design: How design thinking transforms organizations and inspires innovation*. New York: Harper Collins.

Boltanski L., & Thevenot, L. (1991). *De la justification. Les économies de la grandeur*. Paris: Gallimard.

Buchanan, R. (1992). Wicked problems in design thinking. *Design Issues*, *8*(2), 5–21.

Calixte, X. (2021). *Les outils dans l'activité collective médiatisée en conception: Traçabilité des usages au sein du processus de conception architecturale*. Doctoral dissertation, Liège, Belgique: Université de Liège.

Clark, J., Laing, K., Tiplady, L., & Woolner, P. (2013). *Making connections: Theory and practice of using visual methods to aid participation in research*. Research Centre for Learning and Teaching, Newcastle University.

Cooper, R. G. (2008). Perspective: The Stage-Gate® idea-to-launch process-update, what's new, and NexGen systems*. *Journal of Product Innovation Management* 25(3), 213–232.

Denzin, N. K., & Lincoln, Y. S. (eds.) (2011). *The Sage handbook of qualitative research*. London: Sage.

Dodier, N., & Baszanger, I. (1997). Totalisation et altérité dans l'enquête ethnographique. *Revue Française de Sociologie*, *38*(1), 37–66.

Estevez, D., & Léglise, M. (2015). Accountability and design. *International Journal of Design Sciences and Technology*, *20*(2) 99–102.

Heylighen, A., & Dong, A. (2019). To empathise or not to empathise? Empathy and its limits in design. *Design Studies*, *65*, 107–124.

Holopainen, M. (2010). Exploring service design in the context of architecture. *The Service Industries Journal*, *30*(4), 597–608.

Imbert, G. (2010). L'entretien semi-directif: à la frontière de la santé publique et de l'anthropologie. *Recherche en Soins Infirmiers*, *102*, 23–34.

Jacob, C. D. (2016). "Making is thinking": The design practice of crafting strategy. In W. Brenner & F. Uebernickel (eds.), *Design thinking for innovation. Research and Practice*, (131–140). Springer International Publishing.

Lawson, B. (2005). *How designers think: The design process demystified* (4th ed.) Routledge.

Love, P. E. D., & Holt, G. D. (2000). Construction business performance measurement: The SPM alternative. *Business Process Management Journal*, *6*(5), 408–416.

Macaire, E. (2009). Des architectes à l'épreuve de la participation. In F. De Coninck & J.-F. Deroubaix (eds), *Ville éphémère, ville durable – Nouveaux usages, nouveaux pouvoirs* (135–147). Paris: Édition L'Oeil d'Or.

McDonnell, J., & Lloyd, P. (2014). Beyond specification: A study of architect and client interaction. *Design Studies*, *35*(4), 327–352.

Nauwelaers, I., & Rossini, C. (2014). Construire, une brique dans le ventre pas toujours digeste. *Magazine Test-Achats*, *584*, 10–16.

Ng, S. T., Palaneeswaran, E., & Kumaraswamy, M. M. (2011). Satisfaction of residents on public housings built before and after implementation of ISO9000. *Habitat International*, *35*(1), 50–56.

Nishimura, S.-Y., Boda, S., & Sakurai, N. (2017). A Project based learning through international collaboration with students, inhabitants and local professionals. *7th World Engineering Education Forum* (WEEF), 589–592.

Norouzi, N., Shabak, M., Embi, M. R. B., & Khan, T. H. (2015). The architect, the client and effective communication in architectural design practice. *Procedia - Social and Behavioral Sciences*, *172*, 635–642.

Prost, R., & Chaslin, F. (2014). Pratiques de projet en architecture. Le tournant silencieux: Essai. In *Folio*, 250.

Sarkar, S., & Gero, J. (2017). The Topology of social influence and the dynamics of design product adoption. *Design Computing and Cognition DCC'16* (653–665). Springer.

Siva, J. P. S., & London, K. (2011). Investigating the role of client learning for successful architect-client relationships on private single dwelling projects. *Architectural Engineering and Design Management*, *7*(3), 177–189.

Stals, A., Elsen, C., & Jancart, S. (2017). Practical trajectories of parametric tools in small and medium architectural firms. In G. Çağdaş, M. Özkar, L. F. Gül & E. Gürer (eds.), *Proceedings of 17th International Conference, CAAD Futures: Future Trajectories of Computation in Design*, 461–473.

Stals, A., Jancart, S., & Elsen, C. (2018). Influence of parametric tools on the complexity of architectural design in everyday work of SME's. *Archnet-IJAR*, *12*(3), 206–227.

Taoka, Y., Kagohashi, K. & Mougenot, C. (2021). A cross-cultural study of co-design: the impact of power distance on group dynamics in Japan. *CoDesign*, *17*(1), 22–49.

Timmermans, S. & Tavory, I. (2012). Theory construction in qualitative research: From grounded theory to abductive analysis. *Sociological Theory*, *30*(3), 167–186.

Tribout, S. (2012). Quels freins à la participation des habitants du point de vue des concepteurs? In *Cahiers Ramau 6, L'implication des habitants dans la fabrication de la ville. Métiers et pratiques en question*, 199–216.

Van der Linden, V., Dong, H., & Heylighen, A. (2016). From accessibility to experience: Opportunities for inclusive design in architectural practice. *Nordic Journal of Architectural Research*, *28*(2), 33– 58.

Van der Linden, V., Dong, H., & Heylighen, A. (2017). The good client: How architect-client dynamics mediate attention for users. *Proceedings of Professional Practices in the Built Environment*, University of Reading, 174–183.

Van der Linden, V., Dong, H., & Heylighen, A. (2019). Tracing architects' fragile knowing about users in the socio-material environment of design practice. *Design Studies*, *63*, 65–91.

Zahle, J. (2012). Practical knowledge and participant observation. *Inquiry*, 55(1), 50–65.

Zetlaoui-Léger, J. (2013). Urbanisme participatif. In I. Casillo, R. Barbier, L. Blondiaux, F. Chateauraynaud, J-M. Fourniau, R. Lefebvre, C. Neveu, & D. Salles (eds.), *Dictionnaire critique et interdisciplinaire de la participation, GIS Démocratie et Participation*, Paris.

3.6 Design Thinking workshops for social workers

Joana Alves dos Santos[a], *Joana Moreira*[a], *and Katja Tschimmel*[b]

[a]*Mindshake;* [b]*Mindshake/University of Porto, ID+*

3.6.1 Introduction

Design was always a catalyst for innovation, chiefly in the industrial, architectural and graphic fields. Over the last decade, with numerous publications about Design Thinking (DT) (Brown, 2009; Martin, 2009; Tschimmel, 2012; Mootee, 2013; Uebernickel et al., 2020) and several case studies (Liedka & Ogilvie, 2011; Kimbell, 2014; https://thisisdesignthinking.net), the term has gained popularity in business and management media and became a label for the awareness that any kind of organization can benefit from designers' way of thinking and working, including within social institutions and initiatives (Brown & Wyatt, 2010; IDEO, 2015; Docherty, 2017).

Before becoming a popular concept for innovation, design thinking (at that time written in lowercase) had been defined and studied by the international research community, solely as the cognitive process of designers (Cross, Dorst & Roozenburg, 1992; Eastman, McCracken & Newstetter, 2001; Lloyd & Christiaans, 2001). In the last 15 years, the concept of design thinking has been stretched, and has broken free from its domain limits. Today, DT (now written in uppercase) is understood as a human-centered, collaborative and transdisciplinary innovation method, expressing the introduction of design culture and its tools into fields such as service innovation and organizational or social innovation (Brown, 2009; Uebernickel et al., 2020; Tschimmel, 2022). This move from design thinking as the way in which designers think, to Design Thinking as an innovation method embraced by the business world, offers many new challenges and opportunities to the design community. As many DT projects are lacking professional designers in their multidisciplinary teams, one of the questions the 13th DTR Symposium aimed to discuss is how useful the expansion of Design's boundaries is for the evolution of the design field and for society in general, and where the limits are (Goldschmidt & Tarazi, 2022).

This paper provides a theoretical and a practical approach to DT as a method for Social Innovation. The starting point of this research is based on the hypothesis that DT as a method for innovation can make a significant contribution to social workers' mindset and skills for developing innovative projects, while benefiting from a set of tools that design methodology offers. The goal of this paper is to demonstrate the potential the DT process and mindset have to change the way social workers collaborate creatively.

3.6.2 Contextual background

Since the beginning of the 21st century, social innovation has been high on the European agenda (Simon et al., 2014; Manzini, 2015). Social innovation and transformation is

DOI: 10.1201/b22630-22

generally seen as one way of contributing to economic growth, while at the same time promoting sustainability and addressing social challenges (Mulgan et al., 2007; Murray et al., 2010). Across Europe's member states, there is a range of funds, foundations, labs, research centers and networks (Moulaert et al., 2017), dedicated to supporting social transformation through Design and Knowledge, such as the DESIS network (Design for Social Innovation and Sustainability), SIKE (Social Innovation through Knowledge Exchange) or Social Innovation Academy. The role of designers in this scenario of fundamental social change toward sustainability includes supporting social organizations and their agents in collective projects where the culture of design and the methods designers use can be an important contribution to transformational processes. Besides their possible role as design activists, or triggers that initiate social projects, designers can act as facilitators, supporting ongoing processes and initiatives through their design-specific process knowledge. Projects such as the *Human Centred Design* model and the respective toolkit developed by IDEO (2015), in response to a request from the Bill & Melinda Gates Foundation, show that DT is an emerging method capable of responding to Social Innovation challenges.

Although DT for Social Innovation is proven to contribute to addressing user needs and to finding innovative and useful solutions for social projects, there is a lack of understanding about the usefulness of DT for social workers in supporting them over the process of finding innovative solutions for their communities. Most research in the domain of social innovation is focused on the contribution of DT to sustaining social innovation within the field of social organizations. Kummitha (2019), for example, studied how users persuade social organizations to adopt design thinking. But the focus of this research was the designers' roles in social organizations when adopting DT. We couldn't find any research project which addressed the DT training of social workers, with the goal of equipping them with supportive tools for their collaborative creative processes.

As professionals, we perceive the use of DT for social workers as adding value to the creation of knowledge within the innovation process by (1) being inclusive, (2) allowing collaborative workflow, (3) being motivating and empowering, and (4) having the potential for co-creation (Moreira, Santos, Palma & Tschimmel, 2020). According to Docherty (2017), the application of Design Thinking within social projects provides a "safe" space for different perspectives to be shared openly, for new insights to emerge and to empower participants in cooperating and creating shared visions.

This investigation emerged as an opportunity for us, as our team was refining a DT model for Social Innovation (which will be described in more depth below). In August 2018, the National Commission for the Promotion of Children and Youth Rights and Protection (CNPDPCJ) was looking for a partner to carry out a set of activities to support a project called "Adélia". This project intends to build Local Plans for the Promotion and Protection of the Rights of Children and Youth based on a local diagnosis and develop a set of activities that are part of this plan. However, the CNPDPCJ felt that these activities could benefit from being designed using a set of tools different from those that have so far been used by social worker teams. This partnership, therefore, enabled a line of research using the new DT Social Innovation Model. The research opportunity was particularly attractive, as the participants were mainly social workers who intervene with children at risk, and with their families. But perhaps above all, within the context of the Adélia project, in the area of risk prevention and parental skills improvement involving the broader context of the community, and not only families already on the radar of social services. In order to take this research further, we designed a DT workshop, directed at social workers (and other professionals), with our main research question being: Can social workers be DT practitioners?

3.6.3 Methodological approach

3.6.3.1 Research method

The methodological approach in this study integrates the social-critical paradigm of Research *through* Design, which according to Pedgley and Wormald (2007, pp. 72–73) means "research with a practical design element" or "research incorporating a design project". In our case, the design project was carried out by the three of us as DT researchers and trainers with more than a decade of experience and backgrounds in product design, research, and education. That means, in the words of Pedgley & Wormald, that the integration of the design activity was a means to an end, and not an end in itself. The principal aim of this research project was to verify through practical experimentation if social workers would get the necessary motivation and skills to become Design Thinking practitioners, as a result of our intervention. However, supplementary questions then arose:

- How can we make the DT process for social workers more comprehensible?
- Which are the DT tools from our Social Evolution 6 model that social workers perceive as the most useful for their work environment?
- How motivated to innovate do social workers become, after going through DT training?
- Which elements of design culture and which kind of DT tools, will be easy for social workers to integrate into their daily life?
- What is the impact of the workshop on social workers' mindset regarding awareness, motivation, and self-appraisal of their ability to apply DT?

Based on research by Chesson (2017) and by Schweitzer, Groeger and Sobel (2016), we conceived a framework for a questionnaire to help us to understand and measure social workers' Design Thinking knowledge and mindset. Our framework has 4 different levels:

- *DT Principles*: (1) Human Centered Approach, (2) Collaboration, (3) Openness to Risk/Permission to fail, (4) Holistic Approach, (5) Register of ideas, knowledge and process, and (6) Iteration.
- *Practices*: Work approaches, such as Observation, Questioning, Identification of patterns, Experimentation, Visualization of knowledge and ideas, etc.
- *Techniques*: Methods and tools supporting practices such as Mind Maps, Stakeholder Maps, Insight Maps, Brainstorming, etc.
- *Mindset*: Attitudes and ways of thinking (being empathetic, being divergent, not choosing the obvious, being playful, embracing failure, etc.).

In our research approach, we used the following techniques:

- Pre- and post-questionnaire to all the participants of the workshops.
- Field observation during the training.
- In-depth interviews with one random participant from each region at least 6 months after the training.

The pre and post-questionnaire considered the six principles from our framework and included a total of 27 questions divided into three sections (in the first section the participants had to indicate the level of agreement with a sentence, on a scale from totally disagree to totally agree; the second section participants had to choose what most closely reflected their positions, from opposing concepts; the third section had three open questions for feedback). With these questionnaires we evaluated their awareness, motivation, and capability to apply these principles. During the workshop training the facilitators took occasional notes of the most relevant insights. The in-depth interviews took place in June 2021 via Zoom, using a semi-structured guide with five open questions.

3.6.3.2 The DT model "Social Evolution 6"

In 2019, we developed a DT model for Social Innovation, called "Social Evolution 6" (Moreira et al., 2020), designed with the goal of helping companies to work on corporate social responsibility, and potentially, to collaborate with public services for their own modernization. Regarding the workshop's context, the model was the relevant DT model to use as a base for social workers' training.

The Social Evolution 6 model (SE6) (Figure 3.6.1), is a natural evolution of the Portuguese Design Thinking model Evolution 6^2, developed between 2012 and 2015 (Tschimmel, 2012, 2018), and registered under Creative Commons, combined with the Social Innovation Spiral from NESTA.

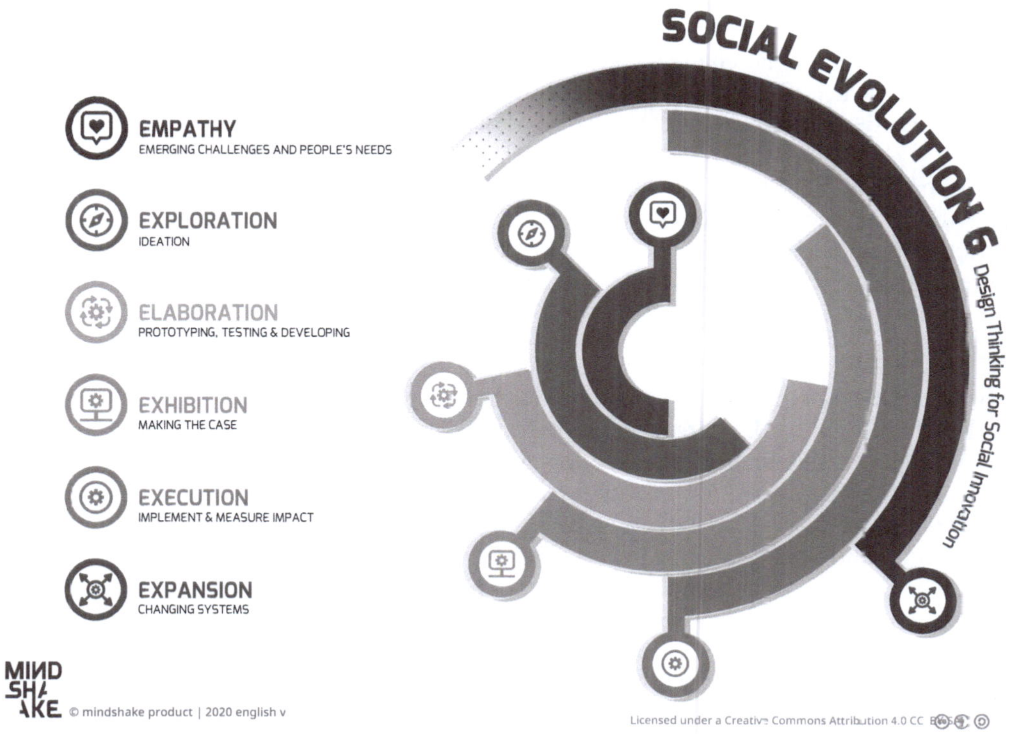

Figure 3.6.1 The Social Evolution 6 model - Design Thinking for Social Innovation.

Taking into account these two "mother" models on which Social Evolution 6 is based, it also comprises six phases, and maintains the rationale of the letter "E," existing in the E6, both for Portuguese and English. There are many other DT process models developed in academic or business context, which consider 3, 4, or 5 phases, depending on the organization's approach. The major difference between all these models does not lie in the number of phases (which are frequently related to the ease of memorizing the model's name), but in the visual representation of the process and the integration of several DT tools.

Our Social Evolution 6 model consists of the following 6 phases (Moreira et al., 2020):

- **Empathy** – Emerging challenges and social needs: Explores challenges and needs that emerge directly from people and their circumstances, and the goal is to identify a problem to solve.
- **Exploration** – Generation of ideas: This phase is about exploring wild ideas and thinking in analogies in order to find ideas that will possibly be an answer to the challenge.
- **Elaboration** – Prototyping, testing and development: Materializing ideas and testing helps develop an idea further, transforming an idea into a concept.
- **Exhibition** – Presentation of the project: It is necessary to present the idea properly in order to garner the support on which it depends for its successful development. This presentation is at the level of content and form, which means that not only are the necessary tools developed for the idea to receive economic support, but also that the narrative supports the idea.
- **Execution** – Implementation and impact assessment: Associated with the implementation and impact assessment, are sets of theories and tools that emerge exclusively from the social context, and which are used in practice.
- **Expansion** – Systemic change: The final phase corresponds to the ultimate desirable impact, which is Systemic Change and it requires a complex interaction of culture, consumer behavior, business practice, legislation, and public policy.

For each phase of SE6, we also propose a set of tools as a way to support the development of a project. The workshop program was designed according to each phase and respective proposed tools. The Expansion phase was the only phase that wasn't considered, due to the research purpose and the training context.

3.6.4 Workshop

In March 2020, the authors were ready to start a program of Design Thinking for Social Workers to be replicated in three regions of Portugal: The North, the Center, and the South. The promoter of this initiative as mentioned above, was the *National Commission for the Promotion of Children and Youth Rights and Protection* – CNPDPCJ – in the context of a project dedicated to Positive Parenting. This initiative was aimed at practitioners in the field of childhood and youth, integrated into several entities, both public and private.

The designed program comprised a 5 day-workshop for Social Workers (7 hour/day) and a public presentation to relevant stakeholders (Government representatives, Social Entrepreneurs, Managers of Social Innovation Hubs) in each region with the aim of developing local projects addressing children's rights and positive parenting (Figure 3.6.2).

The first group (25 social workers) from the Northern region was scheduled for March 2020. Unfortunately, given the pandemic outbreak that occurred in the meantime, the last two days of training as well as the Pitch Day were postponed to September of the same

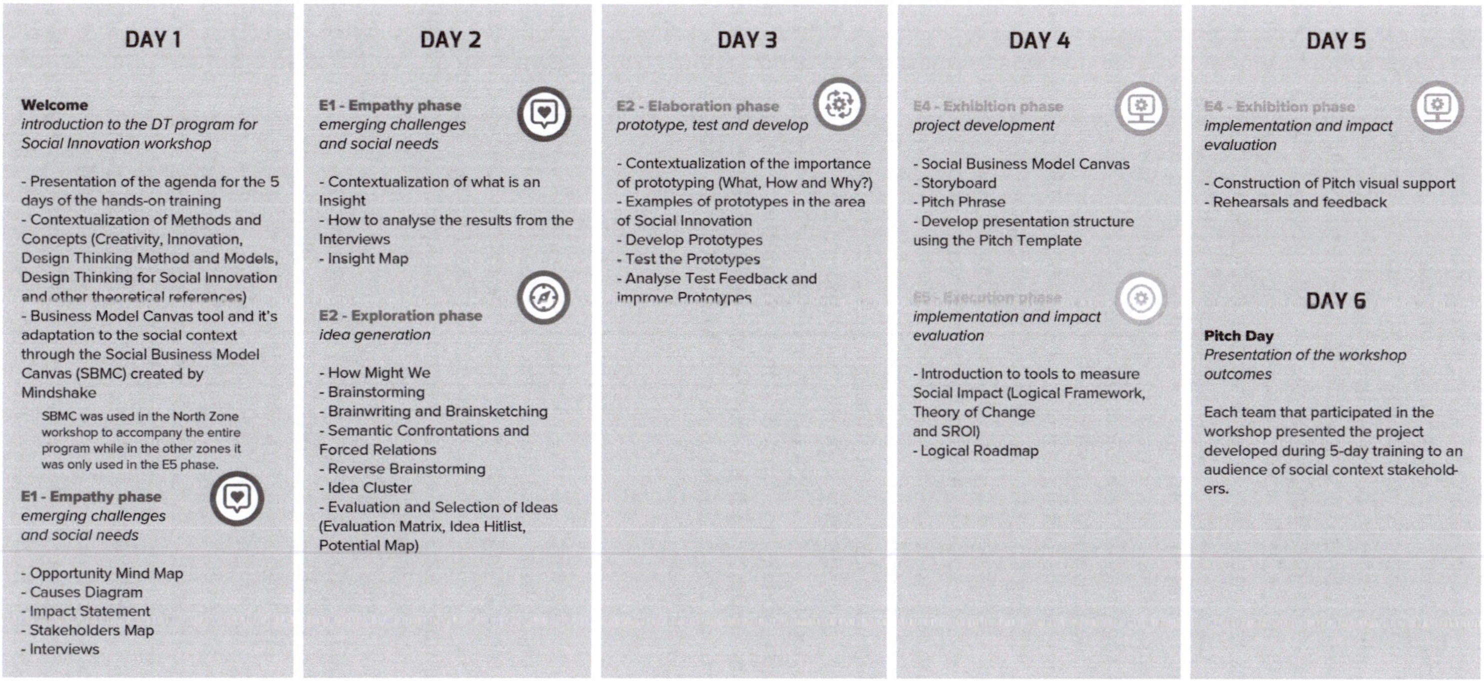

Figure 3.6.2 Program of the 5-day workshop.

year. This interruption led to the withdrawal of two of the participating teams. The remaining sessions were conducted online, using Zoom and the collaborative platform Miro. This hybrid format was unexpected and made it difficult to familiarize all the participants with collaborative digital tools. All participants were unanimous in considering that these tools allowed a better visualization of the contents than a mere Zoom call, even though it was not possible to use the Miro tool to its full potential.

The training sessions for the Central region were also scheduled for March and were rescheduled for October 2020, when it was possible to carry out the entire program in person, as well as the Pitch Day. The total number of social workers was 22.

The Southern region sessions, initially planned for November 2020, took place in an exclusively online format in April 2021. A total of 15 social workers were involved. The Presentation Event – Pitch Day – took place in a webinar format.

The design of the program was determined by the Social Evolution 6 model's phases and consisted of the following activities:

At the beginning of the training, participants, who applied in teams, were posed with the challenge of building an initiative or social business that would fit the Local Plan for the Promotion and Protection of the Rights of Children and Youth of their city.

Accordingly, in each session, after a more theoretical contextualization, we urged the participants to use several tools and templates, following the phases defined in the model. In the 1st phase, the teams designed an Opportunity Mind Map to explore the knowledge about positive parenting and all information gathered by local diagnosis of their region/city, so that they could identify a problem or opportunity to explore. From there, the participants resorted to several techniques, mainly using templates designed or adapted specifically for this model, that guided them through the 5 initial stages of social transformation. In this journey, we highlight the use of Stakeholder Maps, Interviews, "How Might We" questions and a Social Business Model Canvas that supported a Pitch presentation on the final day.

3.6.5 Findings

The findings of our Research through Design come from two different perspectives: the social workers' perception (Questionnaires, Feedback tool "I like/I wish" and Deep Interviews) and the authors' analysis of their observations made during the workshops combined with the results from the questionnaires, informal feedback, and interviews. The findings resulting from the several research techniques are complementary and added value to our final conclusions.

Our observations were focused on the collaborative dynamics and engagement of the participants in the DT process, as well as the way participants applied the proposed tools. We perceived the social workers initially as very passive and expecting a traditional teaching format. As these social workers were not used to practical training, they had problems getting out of their comfort zone and adopting a more hands-on approach. Furthermore, we observed the participants had various difficulties in understanding and using the tools according to our instructions, and also understanding the context in which they could apply them. However, the tools presented were perceived by the participants as beneficial in helping them solve complex and challenging societal problems.

The initial questionnaire was answered by 40 social workers, but the final questionnaire was only answered by 21. The main reason was clearly the difficulties thrown up by the pandemic situation. In the following text, we describe the findings from these sources:

At the beginning of the training, 27,5% of the participants were not familiar with the DT process, 40% claimed that they had heard about it but didn't really know what it was about, and 10% % stated that they already used some of its tools. The remaining 9 people (22,5%) said they were familiar with the method.

From the initial questionnaire, it is possible to infer that the participants' expectations about this 5-day training were learning "new work methodologies" and tools, having contact with new perspectives, and applying the method in projects with children and their families in the future. They also believed this was a method "to look for innovative solutions to common/old problems" and, as well, one that supplied much-needed support for generating ideas and finding solutions for social problems. Some participants also mentioned the need to "think outside the box", to be more creative and that new tools could also be a way of motivating teams and beneficiaries of their organizations. In summary "to be able to draw my ideas, in order to make them happen!"

The findings from the questionnaires applied both in the beginning and at the end of the training are divided into the 6 DT principles (1. Human Centered Approach, 2. Collaboration, 3. Openness to Risk/Permission to fail, 4. Holistic Approach, 5. Register of ideas, knowledge and process, and 6. Iteration), identified in the framework built for the questionnaire as mentioned above, and sub-divided into *awareness*, *motivation,* and *capability*. There were also synthesis questions for each one of the principles.

3.6.5.1 Human centered

Awareness: Although discreetly, the results show an inclination toward the understanding that knowing how a problem affects people is one of the fundamental elements for solving it. At the end of the training, more than 50% of the respondents totally agreed with the statement "The most important step in solving a problem is to understand how it affects people".

Motivation: The results show no particular improvement in terms of motivation for empathizing with people affected by a given problem, and most of the answers "agreed" or "totally agreed" with the claim "When I'm solving an issue, I'm curious to understand how people are affected by this situation".

Capability: At the end of the five DT sessions, while answering the question "When someone tells me how they are affected by a given problem, I can readily look at the situation from their point of view", trainees seem to be more moderate in self-appraisal of their capacity to put themselves in someone else's shoes, since some answers shifted from "totally agree" to "agree" or "partially agree". Although not particularly positive, we understand, also according to our observation during the training, that this type of evolution in answers may be due to a greater awareness of how their preconceptions about how a problem affected people may not be real. There is a realization that it is necessary to go through a process of empathy to observe and question specific people affected by specific problems.

However, trainees conclude overwhelmingly, in this part of the questionnaire dedicated to the Human Centered principle, that understanding how people are impacted by a problem is important or very important, and there has been a great evolution in terms of responses (Figure 3.6.3).

3.6.5.2 Collaboration

Awareness: From the question "Collaborating with others is essential to create the best solutions", we conclude that trainees are more aware of the importance of collaborating with others, and the percentage that "totally agrees" that only in this way it is possible to

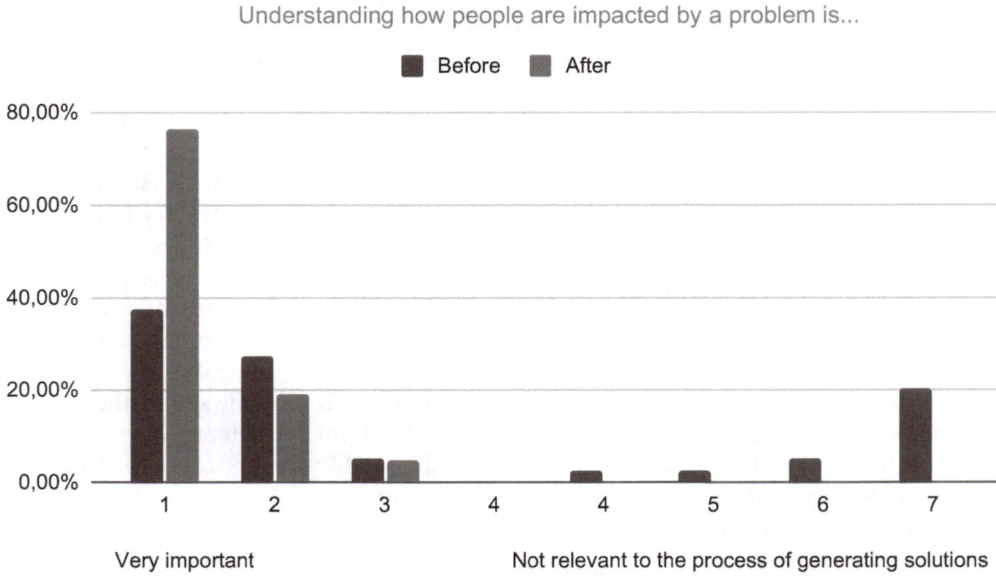

Figure 3.6.3 Graphic result of the opinions before and after the training on the statement "Understanding how people are impacted by a problem is …".

create the best solutions, rises visibly, to more than 75%. Besides the Questionnaire, this perception is reinforced in the Deep Interviews. All interviewees underlined their higher awareness of the importance of collaborative work in innovation processes.

Motivation: Likewise, the motivation to collaborate has significantly improved. In the question "When solving problems, I actively seek other people's involvement", people that "partially disagreed" disappeared and those who "partially agree" diminished. The percentage of people that "totally agreed" rose to more than 45%.

Capability: When answering the question "My abilities to create new solutions stand out more as a team than when working individually", as in the case of the theme related to Human-Centered, there seems to have arisen a more acute awareness, self-analysis and moderation regarding the ability to put collaboration into practice and obtain better results. The number of people that "totally agreed" to the statement diminished and was replaced with people that "agree" or "partially agree".

Trainees learnt from this formative process that working with others has more value than the opposite, placing all answers in the range from 6 to 8 (Figure 3.6.4).

3.6.5.3 Openness to risk taking/permission to fail (creative thinking)

Awareness: To the question "Failure is part of the problem-solving process" the results are slightly better in terms of the awareness, but they are not so expressive of change that definitive conclusions can be drawn. At the end of the formative process, the majority of people agree (38%) or totally agree (43%) with the statement.

Motivation: The group reveals greater motivation for experimentation in the question "I always want to try new things". Those who "disagreed" and "partially disagreed" prior to the training disappeared, and the percentage of trainees at the end of the training that "agreed" or "totally agreed" rose to 90%.

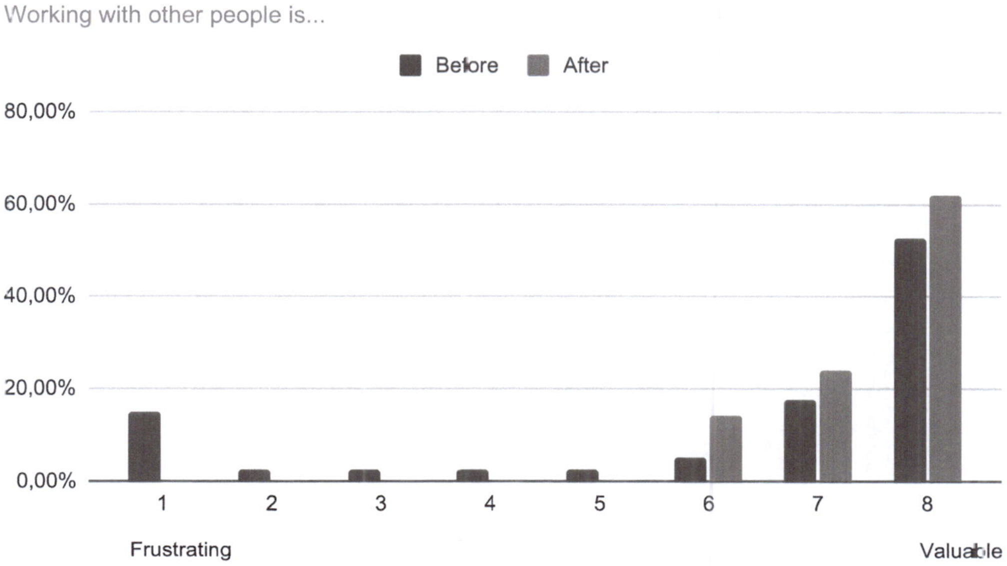

Figure 3.6.4 Graphic result of the opinions before and after the training on the statement "Working with other people is ...".

Capability: Similarly, to previous themes, with regards to the ability to accept risk, there is a decrease in the number of people who say that they "totally agree" with the statement, but there is also a decrease in the number of people who "partially disagree". The results look more judicious and realistic to the question "I'm very comfortable with the risk inherent in trying something completely new" and more than 70% of the answers fell in the "agree" or "partially agree" area.

In the summary question, there is a clear indication that most trainees understand that failure is a form of learning (Figure 3.6.5).

3.6.5.4 Holistic approach

Awareness: Included in the Holistic Approach principle, the statement "I think it's important to look at an issue from different perspectives" demonstrates that the group grasped the concept that different perspectives are advantageous when analyzing an issue. All disagreement with the statement disappeared at the end of the training, 39% of the people agreed and 61% totally agreed.

Motivation: The answer to the question "When I work on a given problem, I look for complementary or contradictory information" does not show significant improvements regarding tolerance and/or the search for ambiguity. Most of the answers are similar to those at the beginning of the training and are in the partially to totally agree areas.

Capability: The largest shift in responses was among those who "fully agreed" with the statement to "partially agree" (Figure 3.6.6). The insight we have, from our observation, is that the work practices of this particular group of people is not very inclusive of divergent opinions and that these five full days of training are not enough to change that attitude.

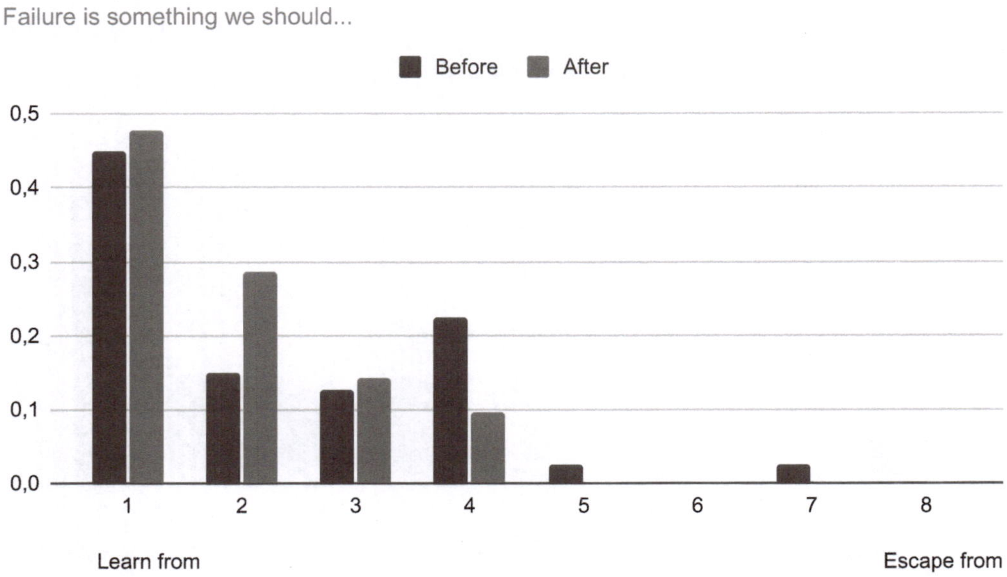

Figure 3.6.5 Graphic result of the opinions before and after the training on the statement "Failure is something we should ...".

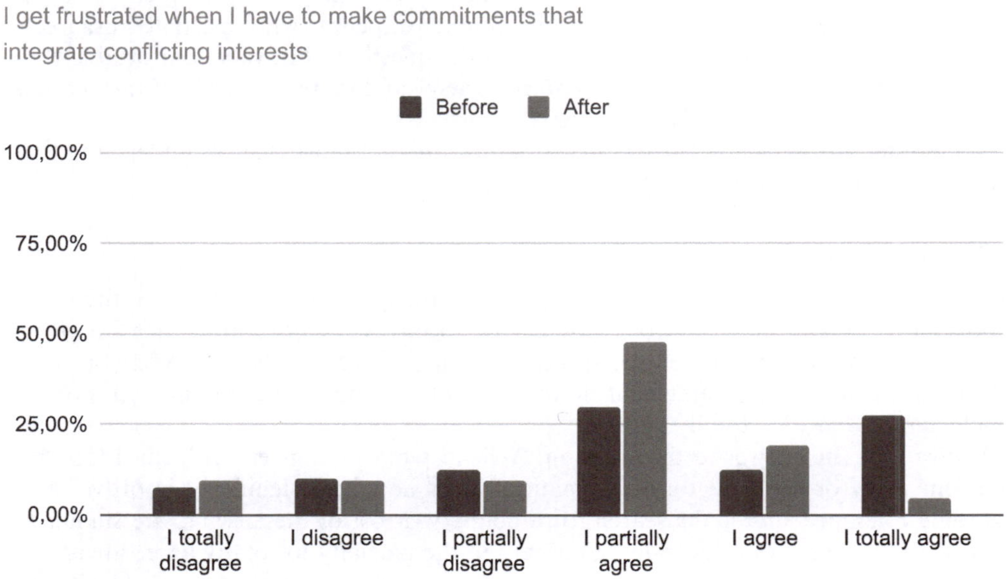

Figure 3.6.6 Graphic result of the opinions before and after the training on the statement "I get frustrated when I have to make commitments that integrate conflicting interests".

3.6.5.5 Register of ideas, knowledge, and the process (making things tangible)

Awareness: The graphic (Figure 3.6.7) shows clearly that people became more aware of the power of images and drawings to communicate.

Motivation: At the end of the training there are more people (38%) who disagree with the statement that "Images and drawings distract me when I'm trying to find a solution to a problem".

Capability: There doesn't seem to be a particularly positive evolution in the ability to think through design/sketches and prototypes, which obviously reveals that nobody can develop visualization skills in a 5-day workshop. The principle of registering ideas and knowledge (making things tangible) is probably one of the most difficult skills to develop in non-designers (Figure 3.6.8).

3.6.5.6 Iteration

Awareness: In this statement "Reflecting on the opinion of others only delays the search for a solution", trainees evolved in their position. At the end of the training, 85% of the group "disagrees" or "totally disagrees" with the idea that considering others' opinion delays solutions versus 60% prior to training.

Motivation: Likewise, the number of people who "totally agree" with the statement "In a creative process, I believe in taking two steps backward to take three steps forward" has increased from 30% to 48%.

Capability: The percentage of people who fully agree with this statement after training has increased from less than 10% to almost 40% (Figure 3.6.9). In fact, we believe, based on our

Figure 3.6.7 Graphic result of the opinions before and after the training on the statement "Images and drawings are a great way to explain ideas and solutions to others".

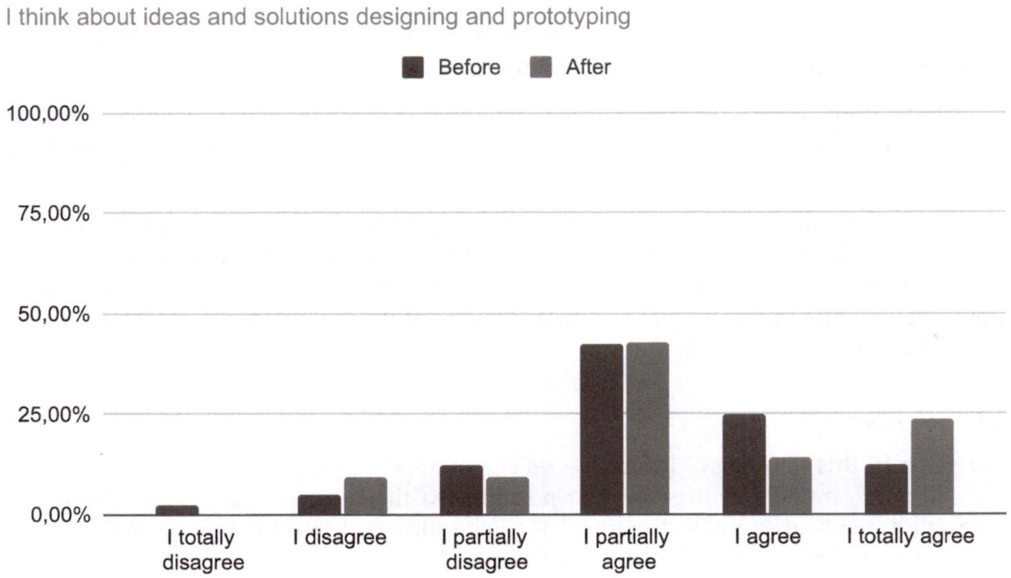

Figure 3.6.8 Graphic result of the opinions before and after the training on the statement "I think about ideas and solutions and prototyping".

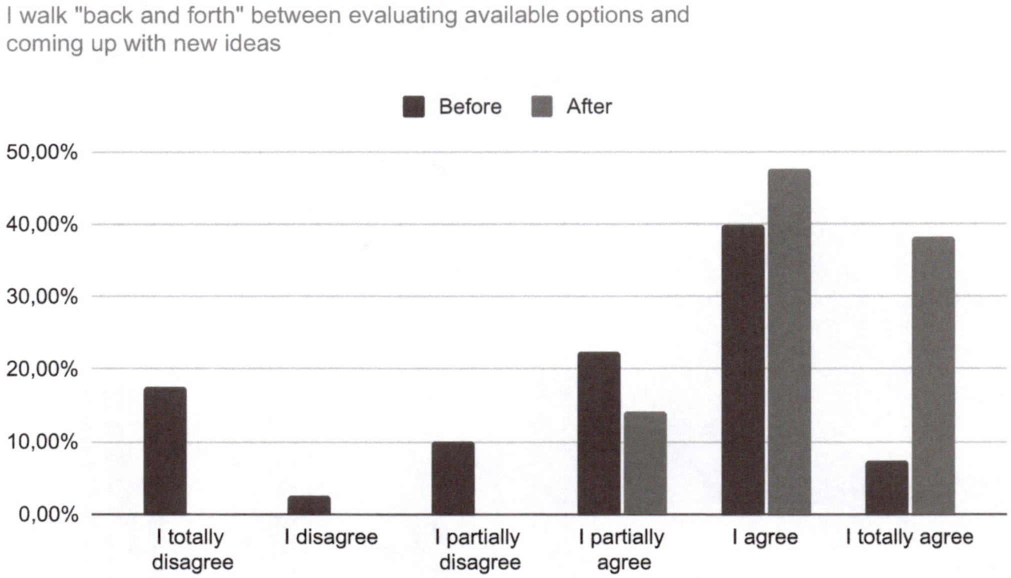

Figure 3.6.9 Graphic result of the opinions before and after the training on the statement "I walk "back and forth" between evaluating available options and coming up with new ideas".

observation, that people were able to verify that the DT process, having reached the goal of building a new project, often involved iterating between ideation and idea selection.

3.6.5.7 Findings from open questions

Regarding the open question, posed in the end Questionnaire (Q), "What advantages do you see in a process that uses Design Thinking in the work you develop in the area of intervention/participation/social work?", and also including some insights from the Deep Interviews (DI), the most interesting answers can be clustered as follows:

3.6.5.7.1 Human-centered

Q: "The advantage is that this process focuses on the human, from the perspective of who is the user, in social work it is important to know the problem that afflicts the context/individual and intervene directly in the problem in order to improve the individual's quality of life and consequently society."

Q: "Ability to give 'voice' to all stakeholders and create negotiated and shared solutions."

3.6.5.7.2 Collaboration

Q: "Teamwork is very beneficial and all the steps to be taken to reach a solution are very important."

Q: "It is very creative, an igniter of new ideas, actively involves a team, stimulating."

DI: "Now I tell people that DT is a 'glue' that helps to gather contributions from different people and perspectives."

3.6.5.7.3 Openness to risk taking/permission to fail (creative thinking)

Q: "It is a creative methodology that helps to think outside the box which is very necessary for social work that is too standardized for a certain way of working."

Q: "Creativity and new solutions."

Q: "It helps to 'get ideas', to look at the problem in a different way and in the process realize that it is possible to intervene using more creative approaches."

DI: "I enjoyed the challenge of looking at the same things I've been looking at for a long time, but in a different way, shaking them. Like grabbing the same thing but grabbing it in an innovative way. It's fun, and very difficult."

3.6.5.7.4 Holistic approach

Q: "We were able to see a given problem from different perspectives and then find more than one solution."

DI: "The Stakeholder Map is very important for us as we work with many different partners, and it is valuable to quantify and qualify relationships."

3.6.5.7.5 Register of ideas, knowledge, and the process (making things tangible)

Q: "A more realistic and promising process."

Q: "Helps to structure ideas, understand problems and create solutions."

Q: "I think it is important, as it makes us reflect, to have a concrete idea of the whole process and it is easier to find solutions."

Q: "Helps gather ideas to prepare a project, evaluating all the issues, not dispersing us, and focusing on the essential."

Q: "Assertiveness, better time and resource management."

DI: "You, for me, have become an equation: this structure works and causes change in the community or wherever we want to apply it."

3.6.5.7.6 Iteration

There was no relevant answer that would have led us to a meaningful conclusion.

Regarding the question about obstacles in using DT tools to develop ideas or solve problems in the area of intervention/participation/social work, we can summarize most of the comments with these testimonies:

Q: "The lack of knowledge of the existence of these tools by the vast majority of workers in the social area and families, prejudice toward the use of this method, and resistance to change, in my opinion, are the biggest obstacles."

DI: "In normal work processes, most social workers believe they already have a deep understanding of the problems, and thus act on preconceived ideas."

In short, since most of their peers did now have access to this training, many people saw ingrained work habits, and also lack of financing, as the biggest obstacles to implementing DT tools in their daily work routine.

Regarding the tool "I like/I wish" (Figure 3.6.10), used to gather feedback from the trainees, it was clear that the most common denominator in terms of the things that people appreciated were the "dynamics" of the training itself. Also mentioned were sharing among participants and the discussion moments, creating prototypes, and getting to a final presentation for the Pitch Day. Citing one of the social workers "I liked better understanding the importance of having a model that guides our intervention; to follow the steps and not rush to the solution; to work in a team; to prototype; … ".

As for what people would like to have had in this training, most answers referred to the lack of time to explore the support materials and tools, to incubate ideas, to iterate more using the prototypes and also more time to discuss with the project team with the dedicated attention of the DT facilitator. Nonetheless, one of the participants wrote: "I would like to have more time, but I also understand that this way forces us to focus on the goal".

3.6.6 Research limitations

This research was conducted during the Covid-19 pandemic and obviously, it had an impact on various aspects of the scheduled program. These series of workshops, initially designed to take place in person, at three different times of the year and in three different locations, had to be adjusted several times in order to follow the safety measures against the virus. What happened was that the first workshop was 3 days in face-to-face format and 2 days adapted for online, the second workshop was fully face-to-face and the third fully online.

Due to the uncertainty and the short time to adjust to the preventive measures imposed by the Portuguese Government, the number of participants was lower than expected, as was their availability to fully participate, due to remote work and the adjustments that each institution also suffered. It is important to take into account that many of the

Figure 3.6.10 Picture of the "I like/I wish" tool at the end of the first 3 days of training with the Northern region group.

participants work in institutions where there are several risk groups such as the elderly and children. Many of them also had to be in the preventative mode because abuse and risk situations for children and youth rose significantly during the lockdown. This situation, unfortunately, also affected the quantitative participation in our questionnaires (from 40 before the training to 21 after the training).

3.6.7 Conclusions

This paper provided a theoretical and practical approach to DT as a method to change the mindset of social workers, improving their way of collaboration and developing innovative projects through design culture.

Based on the Questionnaire answers, our conclusion to the main research question **"Can Social Workers be Design Thinking practitioners?"** is that there isn't a huge positive evolution in trainees as Design Thinkers. However, through our observation and informal dialogue with the participants, the feedback received in the "I like/I wish" template and the in-depth interviews demonstrate in some aspects *(awareness, motivation* and *learning*) many benefits for the social worker's mindset, regarding collaboration, empathy and creative thinking (divergence, playfulness, experimentation, etc.).

Regarding the supplementary question "What is the impact of the workshop on the social workers' mindset regarding awareness, motivation and self-appraisal of their ability to apply DT?" our conclusions are as follows: Concerning the "Human-Centered" principle, the trainees seem not to have gained much more motivation or ability to put themselves in the shoes of others, however, they clearly recognize that this is a fundamental step toward building social innovation solutions. There seems to be a better self-perception of each trainee's real competence in empathetic thinking. With regard to "Collaboration" there was a clear improvement in awareness and motivation of the advantages of working together and a moderation in the self-perception of competence. We believe that most of these professionals had never faced this kind of creative working environment and realized that their usual practices were not really collaborative (even when developed in a team) and that their perspectives about social problems do not correspond to the beneficiaries' perspectives (the people they work for). Nonetheless, we are convinced, giving the participants' motivation, that if institutional obstacles are overcome, these practices can become integrated and part of social workers' mindset.

On the subject of "Experimenting and accepting risk and failure", there seems to be a slight improvement in awareness, an improvement in motivation and a greater moderation in response as far as ability is concerned. However, the participants seem to have "learned" this concept. There doesn't seem to have been much learning about a holistic approach to problems that integrates different views or even conflict. The participants recognize that it is necessary to look at different perspectives but do not understand the need for uncertainty and opposing views. There seems to be more awareness and motivation for using visual tools to register ideas and knowledge during the creative process. However, there was no improvement in competence and learning of how to visualize ideas (drawing and prototyping).

It was unexpected for us that the evolution of self-appraisal of participants' capabilities in some skills related to the main principles of DT was negative. Nevertheless, upon critical reflection on the results, we concluded that somehow these results reflect a more judicious evaluation of their own capabilities. Often, in the learning process, trainees become aware of what they do not know and realize, in the process, the skills that they lack.

Regarding the question of how motivated social workers are to innovate, after going through DT training, the questionnaires and interviews lead us to the conclusion that the social workers are very motivated to learn more about the DT process (by expressing high enthusiasm and belief in the power of DT), but that they still do not see a context for the practical application in real projects. The obstacles they identified are related to the resistance to change of the social institutions, the lack of finance and the lack of DT knowledge/training of their peers. The motivation and desire to learn more from the DT method was demonstrated by three requests we received after the workshops from different social institutions to facilitate DT workshops in other contexts. We have already answered two of these requests by the realization of a workshop for the Red Cross and another for ASAS - Santo Tirso's Solidarity and Social Action Association.

Concluding with which elements of the design culture and which kind of DT tools, social workers will be able to integrate easily in their daily life, our findings have shown that the social workers evaluated very highly the dynamic and systematization of the DT process, the collaboration, human centeredness and the divergent thinking. The participants recognized clearly that Design Thinking focuses on establishing an innovation-friendly climate in their organizations, for the long term, and they are willing to advocate for it. Concerning the DT tools, the in-depth interviews revealed that in general, the social workers don't remember the name of the techniques they applied in the workshop, with the exception of the Opportunity Mind Map, Stakeholder Map, and the Cause Diagram, probably because these tools are in a way connected with their common ways of working and most pressing concerns.

This research was essentially conducted to give us an understanding of how to improve social projects through our DT knowledge. Thus, as a creativity and design consultancy we, of course, asked ourselves how we could make the DT process more comprehensible for social workers. By the end of this study phase, we conclude that our Social Evolution 6 model needs to be improved, above all in relation to the specific language and to the templates used in the training sessions. At the moment, one of the authors of this paper is involved in a series of follow-up events and is aware of the work some of the participants are developing. There are some promising signs, however, it is still too soon to evaluate the long-term impact.

In future workshop programs, in order to make the process more comprehensible, we will concentrate more on a learning-by-doing approach and leave theoretical explanations of DT principles, models, and tools behind. As a next step in our research, after improving some templates and the workshop program, we will run a workshop with the social and private sector in order to test the improvements and develop tools to follow up and measure the impact of the (1) learnings from the workshops, (2) projects created during the workshops, and (3) future projects developed by the participants.

Regarding the topic of the DTRS13 conference, the usefulness of the expansion of Design's boundaries, based on our training experience with social workers, we consider the expanding of design frontiers to clearly be a blessing for social institutions and also for all kinds of other organizations, as the DT process and tools help them to think more empathetically, collaboratively and creatively. For designers, the extended perception of Design and the recent DT boom, which goes beyond Social Innovation, is a great opportunity to export their process knowledge to other domains. Not only can and should the designer be a member of social innovation teams, but often he/she could take on the role of a facilitator in the development of innovative projects, guiding 'social innovation agents' or providing training on DT procedures and techniques for social environments.

References

Brown, T. (2009). *Change by design. How design thinking transforms organizations and inspires innovation*. New York: Harper Collins Publishers.

Brown, T., & Wyatt, J. (2010). Design thinking for social innovation. *Stanford Social Innovation Review*. Available at https://ssir.org/articles/entry/design_thinking_for_social_innovation, viewed at 20/06/2021.

Chesson, D. (2017). *Design thinker profile: Creating and validating a scale for measuring design thinking capabilities*. PhD Thesis. PhD Program in Leadership and Change, Antioch University. Available at https://aura.antioch.edu/etds/388, viewed at 20/06/2021.

Cross, N., Dorst, K., & Roozenburg, N. (eds.) (1992). *Research in design thinking*. Delft University Press.

Docherty, C. (2017). Perspectives on design thinking for social innovation. *The Design Journal, 20* (6), 719–724.

Eastman, C., McCracken, M., & Newstetter, W. (eds.) (2001). *Design knowing and learning: Cognition in design education*. Oxford: Elsevier Science Ltd.

Goldschmidt, G., & Tarazi, E. (eds.) (2022). *Expanding the frontiers of design: A blessing or a curse? Proceedings of 13th Design Thinking Research Symposia (DTRS)*. DesignTech, Faculty of Architecture & Town Planning Technion – Israel Institute of Technology, Haifa.

IDEO (2015). *The field guide to human-centered design.* IDEO.org Publisher. Available at https://www.designkit.org/, viewed at 20/06/2021.

Kimbell, L. (2014). *The service innovation handbook.* Amsterdam: BIS Publisher.

Kummitha, R. K. R. (2019). Design thinking in social organizations: Understanding the role of user engagement. *Creativity and Innovation Management, 28*(1), 101–112.

Liedka, J., & Ogilvie, T. (2011). *Designing for growth. A design thinking toolkit for managers.* New York: Columbia Business School.

Lloyd, P. & Christiaans, H. (eds.) (2001). *Designing in context: Proceedings of Design Thinking Research Symposium 5.* Delft: Delft University Press.

Manzini, E. (2015). *Design, when everybody designs: An introduction to design for social innovation.* Cambridge, London: The MIT Press.

Martin, R. (2009). *The design of business. Why design thinking is the next competitive advantage.* Boston, Massachusetts: Harvard Business Press.

Mootee, I. (2013). *Design thinking for strategic innovation.* New Jersey: Wiley & Sons.

Moreira, J., Santos, J., Palma, G., & Tschimmel, K. (2020). *Design thinking para a inovação social - desenvolvimento do modelo Social Evolution 6.* In Raposo, D., Neves, J., Silva, J., Castilho, L. C., & Dias, R. (eds.), *Investigação e ensino em* design *e* m*ús*ica, Vol. II (169–176), Coleção Convergências Research Books, Castelo Branco: Ed. IPCB.

Moulaert, F., Mehmood, A., MacCallum, D., & Leubolt, B. (eds.) (2017). *Social innovation as a trigger for transformations.* Brussels: European Commission.

Mulgan, G., Tucker, S., Rushanara, A., & Sanders, B. (2007). *Social Innovation. What it is, why it matters, and how it can be accelerated.* London: The Young Foundation.

Murray, R., Caulier-Grice, J., & Mulgan, G. (2010). *The open book of social innovations. Social Innovator Series: Ways to design, develop and grow social innovation.* The Young Foundation & NESTA.

Pedgley, O., & Wormald, P. (2007). Integration of design projects within a Ph.D. *Design Issues, 23*(3), 70–85.

Schweitzer, J., Groeger, L., & Sobel, L. (2016). The design thinking mindset: An assessment of what we know and what we see in practice. *Journal of Design, Business & Society, 2*(1), 71–94.

Simon, J., Millard, J., Keller Lauritzen, J. R., Carpenter, G., Schimpf, G., & Leszek, P. (2014). *Doing social innovation: A guide for practitioners. A deliverable of the project: "The theoretical, empirical and policy foundations for building social innovation in Europe" (TEPSIE).* European Commission – 7th Framework Programme. Brussels: European Commission, DG Research.

This is Design Thinking (s.d.). Case studies. available at https://thisisdesignthinking.net/category/cases/, viewed at 20/06/2021.

Tschimmel K. (2022). Creativity, Design and Design Thinking—A Human-Centred ménage à trois for Innovation. In: Raposo D., Neves J., Silva J. (eds.). *Perspectives on Design II. Springer Series in Design and Innovation*, vol 16. Cham: Springer.

Tschimmel, K. (2018). *Toolkit evolution 6^2. An e-handbook for practical design thinking for innovation.* Porto: Mindshake.

Tschimmel, K. (2012). Design thinking as an effective toolkit for innovation. In *Proceedings of the* XXIII ISPIM Conference: Action for Innovation: Innovating from Experience. Barcelona.

Uebernickel, F., Jiang, L., Brenner, W., Pukall, B., Naef, T., & Schindlholzer, B. (2020). *Design thinking: The handbook.* World Scientific Publishing.

3.7 Design process of live-action video instructions

Yvonne Eriksson[a], *Katrin Skagert*[b], *and Per-Erik Ekwall*[c]

[a]*School of Innovation, Design and Engineering, Mälardalen University, Sweden;* [b]*Division Material and Production, Department Method for Product Realization, RISE Research Institute of Sweden, Gothenburg, Sweden;* [c]*Faculty of Media Studies, Dalarna university, Falun, Sweden*

3.7.1 Introduction

Today's Sweden may be described as a "risk society", where communication about safety and risk prevention commonly takes the form of warnings and instructions. In this chapter, we present an innovative way of designing video instructions, namely co-designing them with their presumed users. In a professional safety control context, the creation of visual instructions is normally the affair of specialists that design the instructions in accordance with warning design standards such as, for example, the ANSI Z535 for safety signs and colors (Wogalter & Mayhorn, 2017). However, the description of design element properties that can alert workers to hazards is of questionable use if the actual hazards are partly unknown and exist in a new type of hazardous condition. A case in point is the Covid-19 pandemic. Therefore, new types of hazards in the workplace require innovative, designerly approaches with regard to the communication of hazard control and disease prevention procedures (Ekwall, Ädel & Nyström-Höög, 2021). Our instructional approach involves live-action videos, since this particular format allows for a quick-turnaround between design iterations, and functions very well when instructions rely heavily on human speech and human movement (Eriksson & Eriksson, 2019; Ganier & de Vries, 2016).

We have used a Design Thinking (DT) process (Cross, 2010; Badke-Schaub, 2022) and Research through design (RtD) (Giaccardi, 2019). DT was helpful in the problem-framing process, and for gaining an understanding of the user and empathising with them; namely personnel working in homecare for elderly people. In the iterative ideation phase, we used photography and videos. The aim of the project was not the end product but to gain a deeper understanding of what is required for the creation of instructional videos; in this case, instructional live-action videos. Therefore, the project could also be defined as a research through design process, an artefact-centred design approach, where two researchers (specialised respectively in design, and occupational safety and health management) involved stakeholders working with homecare in a municipality in the southern region of Sweden. The stakeholders involved held different occupations such as manager, nurse, homecare workers (with and without formal training), occupational therapist, and union representatives. The project started in July 2020 and lasted for one (1) year. It was financed by Afa Insurance, and the municipality contributed through the participating personnel.

3.7.1.1 Background

Elderly care services in Sweden have certain problems related to a shortage of educated workforce, which is also the case in many other countries (WHO, 2016). The pandemic

DOI: 10.1201/b22630-23

seems to have only made these problems more acute (Swedish Association of Local Authorities and Regions, 2021). Employees in the elderly care sector are often part-time employees, about a quarter have Swedish as their second language, and about the same number have attended zero or less than one year of nursing school (Strandell, 2019). To support these workers, and to prevent the spread of Covid-19, various how-to videos have been developed which show them how to use protective equipment and follow guidelines for basic hygiene in care environments. However, the videos in question do not adhere to the segmenting and contiguity design principles. They are too lengthy, and above all, they are inconsistent, for example, what is said and what is done in the videos is not synchronised. In addition, they were recorded in a hospital environment and did not relate to people working with homecare. It is important to keep in mind that this category of employees did not have a tradition of using personal protection equipment since they mainly assist with daily hygiene routines (where they use gloves), preparing food and cleaning up. Therefore, our instructional support was developed with this specific group of staff in mind.

3.7.1.2 Aim and research question

The aim of this design project is to explore ways of co-designing instructional videos, together with representatives from the elderly care sector, that show how to use Personal Protective Equipment (PPE) and how to follow basic hygiene routines. To do this, we ask this primary and basic research question: ***How can a co-design process improve understanding of the challenges involved in the design of instructional live-action videos?***

3.7.1.3 Research through design

By co-designing videos in accordance with a Research Through Design (RtD) framework, the focal point of this design project is not the end product per se, but a mutual knowledge creation which, hopefully, will be beneficial for both homecare workers, designers of the instructional videos, as well as the patients and elderly in nursing care facilities. This approach essentially means that we regard the end-product, i.e., the final video instruction, as a good enough, albeit improved, design solution that is based on realistic expectations, as opposed to (lofty) ideas of design perfection. The means by which to achieve such a good enough design solution is the RtD framework since RtD offers ways for the stakeholders to engage in deeper dialogic interaction with regards to how to communicate the proper use of personal protective equipment and basic hygiene routines.

Similar to RtD-researcher Giaccardi (2019), we consider digital communication objects as connected things capable of illuminating both hidden and obvious knowledge that are the result of dialogical interaction between stakeholders. In doing so, the communication object (the video instruction) reflects the role of the RtD artifact in participatory processes of co-design and co-reflection. At the heart of any RtD-project lies the activity of prototyping. The RtD prototype, indirectly helps form a sort of consensus between non-designers and designers. This is what Giaccardi (2019, p. 141) calls a "stabilizing character" of the RtD-artifact. Oftentimes, RtD-researchers use prototypes to support processes of reflection, in terms of what did not work and what did work in the design process. If used like this, the artifact is used primarily in its evaluative capacity. In the study presented in this paper, the video instruction is used in this way. But more importantly, it is also a vehicle for theory building (i.e., generate generalisable knowledge) as well as enabling stakeholders to simulate and "see" procedures and practices in a new light, thereby being able to identify what might be regarded best-practice routines.

3.7.2 Visual communication

Visuals are perceptually tangible to various degrees and can portray objects and spaces and demonstrate spatial relations. A visual can be an object by itself, or a representation of an object based on its similarity to what is represented, or it can be based on abstract conventional signs. Visual communication is part of every culture, even though we may sometimes use it in different ways. Interestingly, even a very simple symbol can carry a complex meaning, which is something that needs to be learned. Visual communication can be defined as the use of one or several pictures to visually represent and illustrate a concept for the explicit purpose of communication. We use visual communication to express ourselves with the aim of sharing thoughts, ideas, emotions, or information. Regardless of whether the content is complex or not, all communication is an externalisation of our "inner voice", our thoughts; and the purpose is to make these thoughts accessible and understandable by someone else. Whether we want to explain a concept or call attention to something, or entertain, it's all about sharing with others. It is not only socio-cultural aspects that matter in visual studies. We also need to take cognitive aspects into consideration. Our ability to imagine things influences the way we see and perceive the world, which ties into Mitchell's (2015) suggestion to distinguish images from pictures; where images are to be regarded as mental images or metaphors, and pictures as actual representations. This definition indirectly highlights the difference between two distinct modes, or types, of visual instructions. There are visual instructions that are meant to aid users in creating "mental images", that is, by being mentally engaging (Mayer & Clark, 2016) such as, for example, abstract, static, instructional drawings, as well as visual instructions that are more "naturalistic", such as live-action videos that display people who function in more direct ways since they offer humans a kind of mirror neuron advantage, being imitable (Eriksson & Eriksson, 2019). It is necessary that those who are learning a task or approach understand why something should be done in a specific way. Visual instructions need to be motivational. In addition, in useful visual instruction, the user should be able to easily identify what the displayed image or video really shows or represents (Eriksson & Eriksson, 2019). The videos we have designed are the live-action video format type, that is, recorded videos. The videos feature a care worker working at an elderly person's home.

3.7.2.1 Live-action video

In contrast to computer-generated animations, live-action videos are recorded with video cameras, showing events and procedures as they occur over time in real-life situations. The rapid adoption and development of new communication technologies such as powerful video-capable smartphones have revolutionised the face of instructional communication, in essence making it a lot easier and affordable to create and co-create various types of video instructions using human instructors, as well as offer simple ways to very quickly distribute content to the right type of audience/user. Hence, in recent years, studies investigating the usability and design implications of live-action how-to videos have increased. For example, Eriksson and Eriksson (2019) explored how direct affordances inform users' abilities to imitate construction procedures; Cheung et al. tested the effects of video instructions on procedural skills transfer in clinical practice (2020); Biard, Cojean and Jamet (2018) investigated the role of pacing and segmenting in procedural video instructions demonstrating how to make hand orthoses; Ganier and de Vries (2016) investigated the relationship between users' performance and static or live-action instructional design in suture workshops; and Yi et al. (2019) studied what role the live-action video-mediated presence of an instructor plays in students' learning when decoding instructions. The underlying reasons behind this research development have to do with ease-of-use editing and recording technologies (such as iPhones and various kinds of web cameras) which have now become commonplace, and online distribution venues (such as YouTube) that allow for a quick turnaround of recorded videos. However, at the heart of research efforts such as these, there is the issue, and research problem, of the role design plays and how designerly ways of producing instructional content relate to learning outcomes.

3.7.2.2 The cognitive load aspect

Defining and measuring learning outcomes associated with online video instructions are specialised interests in the field of cognition and learning, as most notably exemplified by Mayer (cf. 2019) and Sweller et al. (cf. 2019). According to Mayer's Cognitive Theory of Multimedia Learning (CTML), as well as Sweller's Cognitive Load Theory (CLT), learning outcomes are intrinsically linked to the interplay between the capacities of limited working memory and unlimited long-term memory. Over the years, this insight has indirectly warranted the employment of design guidelines (also known as "design principles") for visual instructions, which are meant to lessen the cognitive load of working memory. Such guidelines are also applicable to the live-action video format (Eriksson & Eriksson, 2019). Despite the growing number of so-called DIY videos (do-it-yourself videos), explainer-videos and how-to videos on the internet and in social media, in the field of information design, research on live-action visual instructions has been rare. Still, information design scholars Spinillo and Perozza (2015) recommend that animated procedural instructions, which may or may not be live-action, should be based on Mayer and Moreno's design guidelines for animations (2002) in order to support communication effectiveness:

1 Video content should include both pictorial elements and narration (text or voice).
2 On-screen text that refers to pictorial content should be presented near the pictorial content.
3 The video content and the narration must be presented simultaneously.
4 Only relevant and coherent elements of animation should be included in the video to avoid extraneous elements (including sounds).
5 Video content should be associated with audio narration rather than with on-screen narration.
6 If audio narration is employed with pictorial content, do not add on-screen text.
7 Verbal content (on-screen text or voice narration) should be presented in a conversational and informal manner.

Video is a time-based, or transient, medium. The transient effect refers to the negative effects associated with the constant flow of new information in videos that cause cognitive load. Unsurprisingly, according to some online education scholars, presentations that are too lengthy are the main problem with instructional videos in general and may possibly explain why the implementation of Mayer's multimedia design guidelines in lecture-style videos, which are partly live-action, have failed to generate positive learning outcomes (Lamb, 2015). None of Mayer and Moreno's design guidelines will aid in making instructional videos more useful for users, in the sense that they will make the videos in question shorter. In other words, none of Mayer and Moreno's design guidelines include the so-called segmentation guideline, which is especially geared towards combatting cognitive load via the means of breaking down multimedia lessons and, hence, decreasing the negative effects of the transient effect (Biard, Cojean & Jamet, 2018). However, with respect to the live-action video format, segmentation can, nowadays, be easily achieved via user interaction whereby the user uses control buttons integrated with the viewing interface, thereby essentially achieving self-paced lessons (Biard, Cojean & Jamet, 2018). This also allows for some level of user-interactivity.

In summary, then, the design guidelines proposed by Mayer and Moreno (2002) seem to overlook the interactivity-factor that has been proven to promote comprehension and learning (Koć-Januchta et al., 2020). The issue of interactivity, and the lack thereof, not only concerns the ability to control the pace of the instructional video content but is also connected to the type of modality (written explanatory text versus spoken narration). Empirically based studies have shown that spoken narration is dependent on the type of interactivity/pacing (cf. Koć-Januchta et al., 2020). If the pacing of the video is largely system-paced, an auditory narration leads to a better learning outcome than written text; but if the learner completely controls the pacing by themself, written text is better (Koć-Januchta et al., 2020).

3.7.2.3 The live-action advantage

Curiously, the transient effect, which is normally considered to be something extremely negative in instructional situations, may be advantageous in the live-action video context because "transience", simply put, is the normal state of the natural world. If one acknowledges that mediated communication may be understood as a kind of exchange in a natural environment, and that this environment is the basic and essential ingredient in the formation of messages, then we might have to accept that "natural communication" also involves things that are known without being overtly communicated and/or taught (Geary, 2008; Paas & Sweller, 2012). In other words, gestures, facial expressions and aspects of human speech do not require overt teaching methods since some communication exchanges take place, and become "meaningful", between bodies and in bodies, thereby bypassing working memory altogether. According to Paas and Sweller (2012), there is actually no evidence that basic problem-solving skills, planning skills and decision-making (i.e., means-end analyses) can be taught at all, since these skills are part of our predetermined repertoire of biologically primary knowledge which we are pre-programmed to know and be able to automatically relate to. Thus, biological primary knowledge cannot be taught, and therefore requires no explicit tuition (Geary, 2008). This approach to natural information processing systems provides the rationale behind labelling live-action videos as the perfect medial vehicle for exploiting humans' bodily architecture. Essentially, the rationale is that biologically primary knowledge is integral to live-action videos, as they often feature people who make actual human movements, are involved in basic problem solving, use human speech (often in one's mother tongue) and exhibit naturally occurring cues and signals (cf. Eriksson & Eriksson, 2019). Therefore, the explicit teaching part in some Live-action-video instructions can be omitted. Immersion can be a good enough approach (Paas & Sweller, 2012).

This presents the key advantage for designers using the live-action format. If some aspects of the overall instructional goal can be left without overtly attending to them, more effort can be placed into instructing about what is normally the primary goal of any instruction: to drive the acquisition of biologically secondary knowledge. Biologically secondary knowledge is culturally and socially acquired knowledge that is not preprogramed in humans, and that "resides" in the mind (Sweller, Ayres & Kalyuga, 2011). In other words, biologically primary knowledge may leverage instructional attempts to promote acquiring biologically secondary skills (Paas & Sweller, 2012). The reason for this is that when we convey information belonging to the biologically primary knowledge realm, very few cognitive recourses are required (cf. Eriksson & Eriksson, 2019). This frees up cognitive space for other, more demanding, mental activities which are the real important ones and the focal point of the instruction in quest of standard operational procedures to prevent the read of Covid-19.

Covid-19 is classified as an epidemic disease in Risk Class 3, which means that employers are responsible for ensuring staff members have access to personal protective equipment and that the equipment meets safety standards (WHO, 2021; Swedish Work Environment Authority, 2020). Safety standards and regulations for basic hygiene routines are often translated into written standard operational procedures (SOP) in terms of instructions, norms, and rules to manage and prevent the spread of diseases. Those SOPs can be described as a dimension of work as imagined (WAI), while work as done (WAD) is how it is performed in practice. WAI, in terms of SOPs, cannot fully reflect actual events and the performance of work, but is needed to create a framework for WAD. Yet, a continuous reconciliation is required through reflection and learning to adjust and harmonise between WAI and WAD (Carvalho et al., 2018). In this sense, the design process in this study worked as a continuous reconciliation.

3.7.3 Method

This study is a RtD study. According to Herriott (2019), the overall RtD process can be divided into four fundamental phases. The background research Phase 1 includes a general risk analysis of how, when, and why protective clothing should or should not be used. The method used in Phase 1 was a semi-structured interview where one interview was with a mixed group in the collaborative municipality, and the other was with a group of four homecare workers in another municipality. Participants in the municipality we collaborated with were union representatives for different professions working in elderly care (i.e., assistant nurses, registered nurses, physiotherapists and administrative staff), as well as healthcare managers and medical responsible nurses (MRN). In relation to the written hygiene routines and the result from the interviews regarding in what order the PPE is put on and taking off we created a storyboard. In Phase 2, the visual prototypes were designed and produced based on the results of Phase 1. A slideshow was created with photos and accompanying text in plain language, explaining what PPE to put on or take off, and how to do it. The slideshow was analysed together with the mixed group with focus placed on the order of instructions on how to dress and undress. The result from the analyses of the slideshow created a basis for Phase 2. The design of the live-action video prototypes started. The live-action video was recorded using an iPhone on a tripod in the hall entrance of one of the researchers' homes. This researcher also performed the role of the main actor/instructor in the video. An objective camera angle was used, and the tone of the verbal instruction was instructive, in plain language. It was an iterative process that generated seven (7) versions of the live-action video. The analyses of the videos were completed using visual methodology (Rose, 2016) and content analyses (Krippendorff, 2004) with focus on the hygiene routines. In this second step of the design process, when we started to videotape prototypes, we mainly met with two nurses who are responsible for hygiene routines. This was partly due to restrictions related to physical meetings, and because of the second wave of the pandemic which made the homecare workers very busy. The nurses were involved in the co-design of the prototypes by taking part in the analysis of all seven steps (the seven versions of the prototypes).

In Phase 3, video production and process analysis, the final video was produced based on dialogues related to the video prototypes. Digital workshops were held with homecare staff to capture how the visual instructions are perceived and understood. Ten focus group interviews were conducted online over two days, with a total of forty-two (42) homecare workers from two units in the municipality. On the first day, technical administration/support was undertaken by a staff member from the administration department. On the second day, the manager of the group took care of administering the meeting and participated in all interviews. The two final versions of the videos of how to put on and take off PPE were shown. The participants watched the videos for the first time. The videos were shown on a computer screen in the room where the participants met, while two members of the research team participated online from two different locations.

The discussions were recorded with the participants' permission, and the researcher wrote field notes during the process. We asked the following questions:

Before Covid-19, when was protective equipment used?
How did you learn how to put on and take off the protective equipment?

Then we showed them the videos with the instructions. First, the one that shows how to put on PPE, and then the one showing how to take off PPE. We asked for spontaneous thoughts:

What do you recognise and what is different compared to how you usually do this?
What are the advantages or disadvantages of visual instructions like this one that we have made?

We also asked about the value of the video in relation to work to be done.

In Phase 4, the comments and reflections were sorted after those questions, and a manifest content analysis was conducted (Graneheim & Lundman, 2004; Krippendorff, 2004).

3.7.4 Results of the research through design process

Here, we describe the first three (3) phases of the research through design process, and end up in a discussion in Phase 4, with new information and a modified theory.

3.7.4.1 Phase 1—risk analysis

Phase 1 started with an initial meeting with the strategic and operative management group for the caregiver staff in the local municipality. Ten people (10) attended, including union representatives for different professions (i.e., registered nurses, assistant nurses, physiotherapists, and administrative staff) and MRN for homecare. This first meeting we held was our only physical meeting. We met the group at their regular monthly one-hour meeting between August and December 2020. We started by presenting the project and our aim of designing visual instructions for homecare workers related to how to use PPE. Initially, we talked about whether it was only elderly care or whether care for disabled (personal assistants and supported housing) would be included. We started discussing specific work situations where protective equipment is used. Here, the representative from the union was careful to distinguish between "personal protective equipment" and "protective equipment". Protective equipment, such as a plastic apron, is used to prevent the contamination of the ordinary workwear and the spread of infection between patients or the elderly persons receiving care. PPE is used for the safety of the employee, and can include medical masks, goggles or face shields, and can differ depending on the situation.

The participants pointed out that the accessibility of PPE can differ locally; for example, if there is space in a nursing home to have the PPE standing in the corridor or inside the user's rooms. For homecare, the user may be required to change into their PPE in the stairwell. The MRN made it very clear that all the instructions and local risk assessments were in place and the staff had received clear written instructions (we interpreted a sub-meaning to mean that if they do not do it right, it is because of them—not because they did not receive good enough instructions).

During our dialogue about how it works in practice, it became clear that the dressing situation varies depending on how the care recipient lives—in an apartment or in a house—and how the home is designed, especially the entrance. The most critical moment, we learned, is during undressing and determining when and where the discarded PPE are to be discarded. We also noted that it was unclear when, where and how employees clean their face shields between care recipients' homes. The meeting led to a mutual decision to focus on how to put on and remove PPE. In our dialogue with the meeting participants, the group helped define what part of the instructions for PPE could benefit from visual instruction.

After the meeting with the management group, we took part of and read the municipality's basic written hygiene routines. We conducted a semi-structured interview online with four homecare workers in another municipality to gain a broader understanding of how basic hygiene routines are used in practice by different organisations and parts of the country. We asked them to describe, step by step, in what order they put on and take off the PPE. Two out of the four interviewees worked in a nursing home, while two of them worked in homecare services as homecare workers in people's homes.

3.7.4.2 Phase 2—visual prototypes

Based on the information gathered in Phase 1, we designed a prototype of the visualisation in the form of a slide show with still photography, including a short text on every slide explaining what to do—which means the slideshow had a high degree of redundancy. The photos were taken by a student, with another student acting as a homecare worker, who was instructed by the research team. The slideshow consisted of 25 slides. Many of them contained several pictures demonstrating every single step, aiming to visualise, step by step, how to for example, put on PPE in line with how we have understood the basic hygiene care instructions. The first challenge we faced in the early stage of

Figure 3.7.1 Example from the slideshow.

the design process was how to put on different elements of PPE (Figure 3.7.1). We only received information in this order:

1 Start by sanitising your hands.
2 Put on shoe covers without touching your shoes.
3 Put on the gloves.
4 Then put on the apron, mouth guard and face shield.

When starting the design process by just following these steps, we faced a lot of practical challenges and not only when it came to how to put on the different elements, such as lack of instructions regarding where to put the jacket and bag when entering the door of the care recipient's home. These details were under discussion during the whole process, from the making of the slideshow to the final video. However, the staff commented on the fact that in the video the surrogate caretaker was not wearing working clothes. The basic hygiene routines are not adapted to the private sphere and are not concretised either. They are rather on an abstract level and learned during nursing education. Then it is up to every co-worker to interpret the basic hygiene routines and apply them to the best of their ability. Before the pandemic, this was not a big issue. But because of the pandemic, it became a big problem, especially at the beginning of the virus' spread. The existing instructional videos were based on a hospital context and did not offer any guidance.

The second part of the slideshow contains instructions on how to take off the PPE in a correct way, without causing contamination, and where to put the used PPE. The instructions on the slides accompanying the pictures advised that the PPE be taken off in the hallway before leaving the care recipient's home and thrown in the trash can in the hallway, within a garbage bag. The homecare worker should start by removing the shoe covers and throwing them in the trash. Then remove the gloves (we asked how) by pulling them inside out so you do not touch the outside of the gloves. Throw them in the trash and sanitise your hands. Remove the face shield without touching the outside of it (what do you do with it then?) Place it on the floor while you take off your mouthguard and apron (our solution). Remove the mouthguard by holding only the loops and then sanitise your hands again. Remove the apron. Start by unloosing the apron at the back (our solution). Fold the apron inwards (our solution). Pull it up over your head. Roll up the apron with the outside inwards (our solution). Throw the apron in the trash. Finally, put the face shield in the backpack, or carry it in your hand (our solution). Sanitise your hands. The last thing to do is throw the garbage bag outside in the dustbin.

The design team and the management team met for a dialogue regarding the content of the slideshow. The discussion was mainly about the advice that homecare workers should start by sanitising all the way up to their elbows, before putting on the shoe covers. Most comments were related to the undressing part. Such as: *Do not lay the face shield on the floor! Wear gloves while removing the face shield. The gloves should be taken off after the face shield. Throw away gloves in bucket/bag at the door. Sanitise the face shield* (with spray bottle or plastic wrap for face shield cleaning). *Sanitise the hands. Lay the face shield down in a backpack* (all homecare workers wear a backpack with the PPE in it).

Even though we asked them to focus on how to put on and take off the PPE, several of the staff commented on the clothing since they were not work clothes but the student's own clothes—as well as the fact that she was wearing jewellery. We were told that home homecare workers in general leave the garbage in the care recipient's hallway, and never brings it with them. The garbage has been an issue throughout the whole project, and we will discuss it later in the paper.

Based on the comments on the slideshow, we started to make video prototypes using an iPhone. Since the video instructions were intended for homecare workers visiting people's homes, we chose an entrance hall to an apartment to frame the situation authentically. The camera was stationary, and the actor (a member of the design team) entered the entrance hall and put the bag on the floor before beginning the process by using hand sanitiser and putting on the PPE. The actor talked to the camera during the whole process. Every single step was commented upon, using straightforward language, with clear instructions and no motivations given for how to act, only for what to do; for example, "Sanitize your hands, wrists, and underarms".

In the first prototype, our team member acting as staff started with a mistake: she put on the shoe covers before sanitising her hands. That was probably an effect of the slide show, for which she was involved in the design of the pictures. The rest of the prototype video took into consideration the comments that we had gathered in the meeting with the management group. In the prototype video the representative from the design team that acted as a care worker wore private clothing, but in a neutral colour to avoid creating distracting information related to the clothing, and no jewellery. However, the staff commented on the fact that she was not wearing working clothes (Figure 3.7.2).

After finishing the first prototypes, one showing how to put on the PPE and one showing how to take it off, we showed them to the nurses. In a dialogue with them, we gained new comments and suggestions on how to improve the video instructions. They were mainly about details but, from the perspective of basic hygiene care, they were important. For example, the backpack should not be put on the floor, they said, but needs to be placed on a stool, chair, or table to avoid contamination. They also commented that the actor put the face shield on the table in the hallway before sanitising it. They suggested that the face shield, instead, be carried in a separate plastic bag which should be hung on the door handle when entering the care recipient's home. In a later prototype, we received the comment that the handle needs to be sanitised before hanging the plastic bag on it.

The undressing part was also a challenge since it has to do with details such as the quality of the apron plastic which affects the ability to tear away the apron, how to sanitise the face shield and where to put it when leaving the care recipient. The instructions from the nurses were that the apron should be wrapped up and, at the same time, the gloves should be taken off inside out with the apron inside the glove, together with the other glove. We were told that the face shield could be sanitised with spray and left to dry in a clean plastic bag.

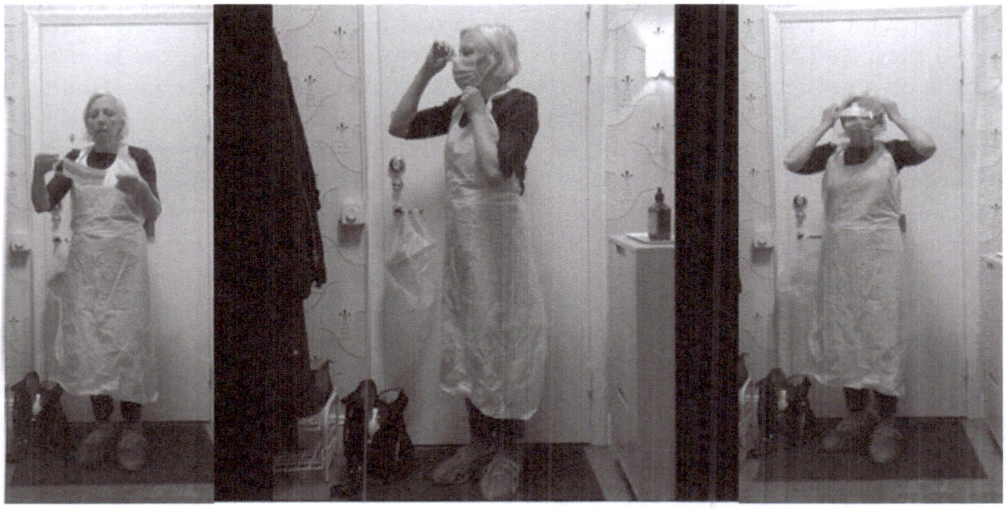

Figure 3.7.2 From the prototype video showing how to put on a mouth guard and face shield.

With every version of the prototype, new details that were not noticed in the previous video were commented on. All the meetings took place online. However, all seven of the prototypes were made in the same entrance hall with the same actor and video recorder. The design team interpreted this as the development of the instructions—the closer the video got to an acceptable result; the more details became visible.

From our dialogues, it seemed to us that the nurses had limited experience of homecare in practice. They did not support the design team regarding how to handle the basic hygiene routines in practice. Instead, they referred to the text document. One of them commented on the last prototype regarding hand sanitation, arguing that three pumps from the bottle are necessary and at least 6 ml should be used. We later used the advice of three pumps but ignored the instructions of how many millilitres should be used, since it is not possible to use that information in practice.

Once we had agreed upon the last version of the prototypes for putting on and taking off the PPE, with some small modifications, the final manuscript for the live-action-video was produced.

3.7.4.2.1 Result of Phase 2: Video prototypes

Using videos in the co-design process generated good insights among the involved partners—the designers and the nurses. This is because live-action videos are associated with visual realism, partly due to the fixed camera position (the iPhone) while recording. For viewers, the video content is directly recognisable, in terms of what is performed and how. The actor in the videos explains what to do using a plain and straightforward language, without giving any explanations of why (i.e., without verbal contextual information). This was a conscious decision since we wanted to keep the videos short; and during the prototype process, there were no comments on the verbal presentations from the nurses, only on the visual parts. The tangible nature of the video—it is imitable—made it possible for the nurses to reflect on how to put the hygiene routines into practice.

This gave the nurses the opportunity to reflect on the relationship between general roles for routines and the very concrete situation that occurs when a homecare worker enters a person's home. The gap between the written bullet points and the visualised situation became highly apparent. As we learned, the video format offers an advantage in one way, and a disadvantage in another. The advantage was the ability to discuss practical issues regarding how to put on parts of the equipment and in what order to put on the bag and the jacket. The disadvantage was the discussion on details such as what clothing the actor was wearing in the prototype, details that were irrelevant to the focus of the prototypes since the aim was to understand how to put on and how to take off PPE and in what order to do so. Every single one of the seven (7) video prototypes was rectified following the discussion with the nurses. To conclude, the prototype phase not only provided insights into the complexity of the design process of live-action videos but also highlighted the necessity of involving the users of instructions to gain a full understanding of how such instructions are likely to be perceived.

3.7.4.3 Phase 3—Video production and process analysis

In the last design step, we involved a professional filmmaker and a professional actress, which requires a specific skill. Since the professionals were not involved in the design process, they were not familiar with the project, aside from the brief they received. The video was made in the same entrance hall as in the prototypes, with a static camera. The manuscript was based on what was agreed upon during the last meeting with the management group and the representative from the staff group. The recording took place in one day, and several versions were made. In one version, for how to put on and take off PPE, the actor talks simultaneously to the camera, in the same way as in the prototypes, explaining what she was doing. In an additional version, she acted without narrating her steps, and afterwards the voice recording was taped in a closeup camera angle. Besides filming the full PPE routine, several versions were filmed showing how to use parts of the PPE. One with gloves, apron and face shield; another with gloves, apron and face mask; and a third with only gloves and apron. The aim was to create visual instructions that could be used even after the pandemic, and in more ordinary situations in which only basic hygiene routines are necessary.

The design of the final versions was completed together with the professional film maker who edited the films in relation to how much time one can expect a homecare worker to spend watching the video. This led to a mixture of full-picture views and close ups. The length of the video on how to put on PPE is 3.32 minutes, and the video on how to take it off is 1.42 minutes, and the actors describe what to do and demonstrate it simultaneously. There are no explanations in the video, only instructions. The instructions are segmented by the actor's gestures, and it is easy to follow even for staff that are not fluent in Swedish (https://play.mdh.se/media/t/0_vm6olg6n; https://play.mdh.se/media/t/0_2d3gbhev). In the final version of the live-action- video, we had access to working clothing that the actor wore, but they were borrowed from a different region of Sweden than the one we were collaborating with. Some of the staff did not recognise the clothes as working clothes and commented on that (Figure 3.7.3).

The final videos were shown to the management group. This was followed by a discussion about the content and details in the videos. The meetings took place online, due to the pandemic and related restrictions.

Figure 3.7.3 From the final live-action-video showing how-to put-on the mouth guard and face shield.

Photo: Magnus Göthlund.

3.7.4.3.1 Input from the target group

To gain information and input directly from the target group for the visual instruction, digital workshops were conducted. We learned from the staff that prior to Covid-19, PPE was seldom used, only in cases of calicivirus, where they used gloves, aprons and mouth guards. In cases of calicivirus, they always put on the PPE before entering the home of the care recipient. Other situations in which they used protective equipment such as gloves and aprons were close-contact care situations, such as helping with shower or bathroom visits. In these situations, they put the equipment in the care recipient's bedroom or bathroom. They said that prior to Covid-19, they had never used a visor before.

We asked the staff how and where they had learned about the regulations for how to put on and take off the PPE. The ones who had a formal education said that they had learned basic hygiene care in school. The others said they had learned from short instructions when they started working in elderly care. We found that there was great variety in the ways they handle basic hygiene care in practice. Much is dependent on the actual situation, and some of the staff mentioned that elderly people stand waiting for them just inside the front door in the entranceway, and it is difficult to put on the PPE before they get too close. Others mentioned that the living conditions and how the home is designed influences where you can put the jacket and/or backpack. Sometimes the floor is the only place. One of the comments was about the entrance hall in the video. They commented that it was too organised, with an empty chest of drawers and only a bottle of hand alcohol for sanitation, and a chair where the backpack could be placed. It is more common, they said, that the entrance hall is very narrow and full of things which makes it impossible to place the backpack in any other place except the floor. Even the jacket needs to be placed on the floor in some homes. Most of the groups also mentioned that it is not possible to leave hand sanitiser or gloves in the care recipient's home since they get stolen by friends and relatives (See Table 3.7.1).

After we had already conducted the prototyping sessions, and before we met with the ten staff groups, we realised that new directives and instructions for handling PPE before and after use had been rolled out by the municipality. Now, they have bags that contain aprons, gloves, and mouth guards, and before they leave the care recipient. they put it all back in the bag as garbage. We did not know this when we met the staff and, therefore, it is not included in their comments above. Most of the spontaneous comments concerned the gloves and details on sanitising hands and face shields (see Table 3.7.1). Yet, the

Table 3.7.1 Selection of video comments

Day 1	*Highlighted comments*	*Day 2*	*Highlighted comments*
Group 1: 7 persons	• We use double gloves, it's not necessary to sanitise that often. • Sanitise hands and put on mouth guard outside the apartment/house. • The visor must be sanitised and dried.	Group 1: 5 persons + manager	• The bag does not look like our backpacks. • She forgot the garbage.
Group 2: 7 persons	• She is not wearing same kind of apron as the one we use. • We use double gloves, it's not necessary to sanitise that often. • Sanitise hands and put on mouth guard outside the apartment/house. • The visor must be sanitised and dried. • The video shows the same order as we follow to put on and take off PPE.	Group 2: 4 persons + manager	• You always put the backpack/bag on the floor. • We don't sanitise the handle. • It is not always possible to take off the apron as suggested. • We dry the visor after sanitising it.
Group 3: 3 persons	• She is not wearing the same kind of apron as the one we use. • We have never sanitised the handle. • The visor must be sanitised and dried. • 'I don't sanitise my hands that frequently.' • We use double gloves.	Group 3: 4 persons + manager	• We take the garbage with us when we leave. • We don't put on gloves when sanitising the face shield. • We sanitise our hands between every "task". • We dry the face shield after sanitising and put it in the backpack.
Group 4: 3 persons	• We have never sanitised the handle. • The video is fine. It covers the most important parts.	Group 4–5: 4 persons + manager	• We have never sanitised the handle. • We wash our hands between every task. • We put on shoe covers before gloves.We undress in the following order:Apron, gloves, shoe covers, mouth guard and face shield.
Group 5: 4 persons	We use double gloves, it's not necessary to sanitise that often. • Sanitise hands and put on mouth guard outside the apartment/house. • Visor must be sanitised and dried.		

groups commented and described how they managed details differently, even though they followed the same instructions. That underscored our reflection during the earlier Phase 2, that the MRNs seemed to lack experience on how the work and basic hygiene routines are performed, and that implied the WAI in the written instructions is not easy to follow in the WAD. The interview participants from the first day had extensive working experience and had developed their own routines for how to handle PPE. When we mentioned the use of double gloves to the second-day staff group, they immediately referred to rules that state that the use of double gloves is not allowed. The presence of the manager of the group that we interviewed on the second day was unmistakable. She was very ambitious and commented on what was written in the basic hygiene routines and what she had taught the staff. However, the groups still commented on details that did not match the way they performed their job when visiting care recipients.

3.7.4.3.2 Challenges and benefits of the visual instructions

The actual possibilities of following SOPs to use PPE and prevent the spread of Covid-19 seemed to be influenced by homecare workers' earlier knowledge and experiences, the spatial design in the care recipient's home and their approach, and local routines and the praxis developed within the team. The results indicate that the live-action video format has both challenges and benefits when it comes to visual instructions that are connected to these themes as well. The main advantage of instructional live-action videos is, simply put, "transience", which reflects the normal situation in the natural world. The staff could therefore easily recognise the situation and follow the instructions, the length of the videos was significant, and all participants found it appropriate regarding the required information. As expected, the live-action videos freed a lot of cognitive capacities and working memory power to decode and process the overall meaning and message of the instructions for the users. In other words, this format functions very well as a point of departure for deep-probing and constructive discussions that concern the real issue here: what elderly homecare workers actually do in the given situation, and what they can do to improve their disease prevention work methods.

All ten groups found the videos to be very useful and easy to follow, regardless of earlier knowledge and experience. The overall reactions from the ten groups were that they found the videos to be instructive, with a good balance between information and performance. They thought that the length of the videos was adequate and that the spoken instructions fit well with the actors' performance. One comment from Group 2, on the first day, was that they wished they had seen the video when the pandemic started. Another comment from Group 3, on the same day, was that "all new staff should see it". On the second day, we received comments from Group 1, mentioning that the video works well for people with limited skills in Swedish.

We asked if they would have preferred an animated video, but all groups preferred the live-action video, since it makes the instructions more realistic and easier to follow. The live-action video was experienced as "natural communication" (Geary, 2008; Paas & Sweller, 2012), even though the interviewed staff found the hallway entrance to be "too perfect" and clean. The main disadvantage of the live-action video format is another side of its advantage, namely, that it is "naturalistic", and displays a wide range of specific, and distracting, visual details, such as the displayed entrance hallway in the videos. That invited comparisons to other known entrance hallways and what a "standard" entrance hallway should look like. In this context, this kind of discussion is not very fruitful.

Visuals have other conditions for statements than verbal language. Language can be imperative, while visuals are always compulsory since they are tangible and show what something looks like. Therefore, the challenge of designing visual instructions is to create generalisable instructions that work for various users in different situations and are still instructive for a specific purpose. The discussion regarding the aprons indicates how easy it is to build in visual noise into an instruction.

Visual communication of this kind becomes perceptually concrete since all the communication that is performed is visible in the form of gestures, facial expressions, and actions of various kinds. In the videos, the staff could follow what to do by observing in which order the different parts of the PPE were put on and how. It also became clear, when the actor deviated from their local routines and praxis of how the staff usually performs the tasks, that these deviations would become a basis of a discussion that elucidates that the staff could not follow the basic hygiene routines to the letter because of certain contexts and situations.

The ambition of this project was to conduct a continuous dialogue with employees working in elderly care, but because of the second wave of Covid-19 in the actual region, it was not possible. Instead, the dialogue was conducted with nurses responsible for basic hygiene routines. From the dialogues in the meetings with the ten groups, we learned that the project would have included too many ad hoc solutions and suffered from too many deviations from the basic hygiene routines. That sparks a discussion about when and how to use participatory design as an approach, and when to use an interactive design approach as we have done in this project.

3.7.4.4 Phase 4—new information and modified theory

The RtD process is about learning and gaining knowledge, both with respect to the process itself, and the learning exchange between the participants in the co-design process (Castro & Lima, 2017). Thus, in this RtD context, it is more relevant to talk about the creation of new knowledge, than to only point out that stakeholders, over the course of the project, gain new information about the problem area (i.e., new data on disease prevention in Swedish elderly care services). From the perspective of the researcher and prototype designer, one important type of knowledge is relatable to insights into how photographs and moving image sequences function as visual instructional prototypes. First and foremost, their "naturalism", invites the onlooker to look for anomalies that stand out in an assumed natural context. For example, in the first RtD phase, when looking at the slideshow made up of photographs presented in a sequential manner, the stakeholders noticed details that we as designers had not noticed, such as jewellery and odd (private) clothing choices. Abstract images, such as line drawings in the form of, for example, action diagrams, on the other hand, would not have been associated with such comments since they are highly abstract and altogether "fantasies" devoid of a natural context in which things may be noticed as anomalies. In hindsight, perhaps it would have been relevant to use instructional line drawings early on in the RtD process in order to get the routines right (how to put on the PPE and how to take it off) instead of getting into more complex discussions about both the process and details, such as what to wear and what not to wear at a certain instance, in a certain situation. On the other hand, if more abstract instructional imagery had been used as the primary RtD medial object at this point, perhaps we would not have gained important knowledge regarding how, for example, used hygiene articles should be disposed of at an elderly person's home (in a waste bucket, close to the exit door, with a plastic bag that is immediately discarded elsewhere after use).

All in all, the design of the live-action video prototypes, including the still photograph sequences, taught us that they are useful when it comes to concretising what should be done and how, step by step. However, the prototyping also taught us that it is extremely difficult to provide the final exact and correct video instruction on how to follow specific hygiene routines in a homecare setting since, simply put, as noticed by several nurses, there is no "real" and fixed homecare setting, they all vary.

Since the home homecare workers' working conditions differ in relation to spatial design in the care recipient's home and influence their ability to strictly follow SOPs, the videos do not function so much as instructions, but more like education. They create the opportunity to reflect on what it looks like in different homes, in various situations, and how one may adapt basic hygiene routines and the use of PPE to the surrounding conditions. During the research project, the researchers noticed that risk assessments, national and local guidelines and instructions regarding what, when and how to use different types of PPE have changed several times since the first wave of Covid-19 in spring 2020. This also affected the development of visual instructions and the risk that they would quickly become obsolete. This type of video instructions, however, obviously revealed details that are difficult to obtain in written instructions.

Consequently, we learned from the workshops with the homecare workers that the final versions of the live-action videos work as a starting point for a very constructive discussion about what to do, and what not to do. However, they do not show what homecare workers define as "a real situation". In other words, the videos show ideal situations, not the familiar but unique "messes" that homecare workers often meet in peoples' entrance halls. Furthermore, according to the hygiene specialist (a nurse), the videos failed to correctly communicate how much rubbing alcohol is to be used when disinfecting hands and arms. We, therefore, conclude that visual instructions that are based on specific regulations appear to be unsuitable for the live-action format. It is a kind of paradox: The specificity of the live-action format, on one hand, is useful since it invites critical perspectives and discussions on anomalies. On the other hand, it does not easily offer information that is "malleable" enough to be useful for the individual user, and that can be generally applied to diverse situations and contexts. In other words, the live-action information is too concrete to easily be turned into unique mental images that aid the individual user to better comprehend the particular instruction in question (cf. Tversky, Morrisson, & Betrancourt, 2002). In summary, then, the live-action video format functions very well as an ease-of-use medial object and catalyst for a valuable and constructive discussion about procedures and routines. However, the live-action video format is, probably, less valuable as an actual instructional tool when a user is performing a routine. This is to infer that, while the live-action format normally relieves pressure on working memory (Eriksson & Eriksson, 2019), it is also not mentally engaging enough to foster focus, selection and the re-organisation of information which are critical to being able to "think visually" (Arnheim, 1969) and perform a certain task. This insight partly modifies the theory about the live-action video as an instructional tool since it highlights the value of the live-action format's indirect effects in a pre-production phase of instructional content; namely, its capacity to generate more knowledge out of a certain instructional situation than the designer pre-inputs into it (cf. Schön, 1983). By visualising instructions and reflecting on what employees do in practice, the gap between written instructions on WHAT to do (WAI) and HOW it is performed in reality (WAD) became obvious.

Especially in the discussion with the responsible nurses who seemed to have experience from practical homecare. Carvalho et al. (2018) used simulation exercises and observations made by researchers to explore the gap between WAI and WAD, they emphasised that the participatory method in the analysis seemed to contribute to reflection and learning. Our results show that live-action videos can be a useful tool for adjusting and harmonising between WAI and WAD and can be easier to use for educational purposes when introducing work routines to newly recruited staff. It may also be a time and cost-effective complement to training exercises. This is also to point out that the employment of live-action video instructions is an enabling media technology that may support development work, in which the objective and goal of the instructional project are identified.

3.7.5 Conclusion

The aim of this design project was to explore ways of co-designing video instructions, to answer the question: How can a co-design process improve the understanding of challenges in the design of instructional live-action videos? We used a case study to answer the question where we, together with representative workers from elderly care sectors, show how to use Personal Protective Equipment (PPE) and how to follow basic hygiene routines, with regard to the implications of Covid-19 in an elderly care context. To use Schön's (1987) terminology, we consider the knowledge type in question to be "knowledge-in-action". The video instructions become visual, mediated manifestations of this type of knowledge, that Schön refers to as a passive form of "knowledge-in-action" (1987). Consequently, the video prototypes, when viewed and listened to by a user, are meant to support "reflection-in-action", that is, the capacity of homecare workers to consciously think about what they are doing while they are doing it.

Live-action videos may be useful for educational purposes and when introducing work routines to new staff. In particular, the results show that the main improvements derived from using a co-design process were input on the details needed to make the video more realistic and more reflective of real-world scenarios. In part, we think this is due to the fact that the live-action format allows users to use more cognitive capacities to discover and gain insights into key details (i.e., image and sound elements). These details sometimes convey just the correct kind of activity, while other times they distract or seem contrived. Thus, over the course of the project, the video prototypes became more and more nuanced and pedagogically straight to the point. Yet, at-home elderly care is complex work, with various conditions in every home and far from "Standard Operational Procedures". Hence, we think that the live-action instructional format of this kind can function very well as a basis for reflection on how to actually perform work tasks, even for experienced staff. However, the best fit for the proper use of PPE and basic hygiene routines is probably a combination of visual instructions and practical training. No instruction alone, without actual "reflection in action", will result in better and more sound practises and routines.

Acknowledgement

The authors would like to thank Afa Insurance for their financial support of this study.

References

Arnheim, R. (1969). *Visual Thinking*. University of California Press: Berkeley, CA, USA.

Badke-Schaub, P. (2022). Design Thinking between Myth and Methodology. In Eriksson, Y. (ed.), *Different perspective in design thinking*. CRC Press.

Biard, N., Cojean, S., & Jamet, E. (2018). Effects of segmentation and pacing on procedural learning by video. *Computers in Human Behavior*, 89, 411–417. 10.1016/j.chb.2017.12.002

Carvalho, P. V. R., Righi, A. W., Huber, G. J., Lemos, C. F., Jatoba, A., & Gomes, J. O. (2018). Reflections on work as done (WAD) and work as imagined (WAI) in an emergency response organization: A study on firefighters training exercises. *Applied Ergonomics*, 68, 28–41.

Castro, T., & Lima, D. (2017). *Designing for children using the RtD and HCD approaches*. (pp. 481–490). Springer International Publishing. 10.1007/978-3-319-58706-6_39

Cheung, J. J. H., Kulasegaram, K. M., Woods, N. N., Moulton, C., Ringsted, C. V., & Brydges, R. (2018). Knowing how and knowing why: Testing the effect of instruction designed for cognitive integration on procedural skills transfer. *Advances in Health Sciences Education: Theory and Practice*, 23(1), 61–74. 10.1007/s10459-017-9774-1

Cross, N. (2010). *Designerly ways of Knowing*, Springer Verlag.

Eriksson, P. E., & Eriksson, Y. (2019). Live-action communication design: A technical how-to video case study. *Technical Communication Quarterly*, 28(1), 69–91. 10.1080/10572252.2018.1528388

Ekwall, P-E., Ädel, A. & Nyström Höög, C. (2021). Towards a unified affordance approach: searching for congruent meaning making in COVID-19 warning designs, *Social Semiotics*, DOI: 10.1080/10350330.2021.1995306

Ganier, F., & de Vries, P. (2016). Are instructions in video format always better than photographs when learning manual techniques? the case of learning how to do sutures. *Learning and Instruction*, 44, 87–96. 10.1016/j.learninstruc.2016.03.004

Geary, D. C. (2008). An evolutionarily informed education science. *Educational Psychologist*, 43(4), 179–195. 10.1080/00461520802392133

Giaccardi, E. (2019). Histories and futures of research through design: From prototypes to connected things. *International Journal of Design*, 13(3), 139–155.

Graneheim, U. H., & Lundman, B. (2004). Qualitative content analysis in nursing research: concepts, procedures and measures to achieve trustworthiness. *Nurse Education Today*, 24(2), 105–112

Koć-Januchta, M. M., Höffler, T. N., Prechtl, H., & Leutner, D. (2020). Is too much help an obstacle? effects of interactivity and cognitive style on learning with dynamic versus non-dynamic visualizations with narrative explanations. *Educational Technology Research and Development*, 68(6), 2971–2990. 10.1007/s11423-020-09822-0

Krippendorff, K. (2004). *Content analysis: an introduction to its methodology*. Thousand Oaks, Calif.: Sage.

Lamb, R. A. (2015). A Makeover for the Captured Lecture: Applying Multimedia Learning Principles to Lecture Video. Doctoral dissertation. Nova Southeastern University. Retrieved from *NSUWorks. Graduate School of Computer and Information Sciences*, (36) h7p://nsuworks.nova.edu/gscis_etd/36

Mayer, R. E. (2019). Thirty years of research on online learning. *Applied Cognitive Psychology*, 33(2), 152–159. 10.1002/acp.3482

Mayer, R. E., & Moreno, R. (2002). Nine ways to reduce cognitive load in multimedia learning. *Educational Psychologist*, 38(1), 43–52.

Paas, F., & Sweller, J. (2012). An evolutionary upgrade of cognitive load theory: Using the human motor system and collaboration to support the learning of complex cognitive tasks. *Educational Psychology Review*, 24(1), 27–45. doi:10.1007/s10648-011-9179-2

Rose, G. (2016). *Visual Methodologies. An introduction to Research with Visual Material*. Sage Publishing.

Schön, D. (1983). *The Reflective Practitioner. How Professionals Thin in Action*. Farnham, Surrey: Ashgate.

Spinillo, C. G., & Perozza, R. (2015). *An interactive guide to design animated visual instructions in Brazil*. (pp. 374–381). Springer International Publishing. 10.1007/978-3-319-20898-5_36

Strandell, R. (2019). Homecare workers under pressure–A comparison of the work situation in Swedish homecare 2005 and 2015. *Health Soc Care Community*, 28, 137–147. 10.1111/hsc.12848

Swedish Work Environment Authority (Arbetmiljöverket). *Personal Protective equipment* (PPE). (https://www.av.se/en/health-and-safety/diseases-infection-and-microbiological-risks/risk-of-diseases-in-the-workplace/coronavirus-and-workplace-safety-and-health/personal-protective-equipment-ppe/ accessed 20 June 2021)

Swedish Association of Local Authorities and Regions (2021). *Fakta om äldreomsorgen i ljuset av coronapandemin* [Facts about elderly care in the light of the coronavirus pandemic]. (Fakta om äldreomsorgen (skr.se), accessed 23 June 2021).

Sweller, J., Ayres, P., & Kalyuga, S. (2011). *Cognitive load theory*. New York: Springer.

Sweller, J., van Merriënboer, J. J. G., & Paas, F. (2019). Cognitive architecture and instructional design: 20 Years later. *Educational Psychology Review*, 31(2), 261–292. 10.1007/s10648-019-09465-5

Tversky, B., Morrisson, J. B., & Betrancourt, M. (2002). Animation: Can It Facilitate? *International Journal of Human-Computer Studies*, 57(4), 247–262. doi:10.1006/ijhc.2002.1017.

Wogalter, M. S., & Mayhorn, C. B. (2017). Warning Design. In A. Black, P. Luna, O. Lund, & S. Walker (eds.), *Information Design. Research and Practice* (pp. 331–348). New York: Routledge.

World Health Organization (2016). *Global strategy on human resources for health: workforce 2030*. Geneva: World Health Organization. (Global strategy on human resources for health: Workforce 2030 (who.int), accessed 23 June 2021)

World Health Organization (2021). *COVID-19: Occupational health and safety for health workers*. Interim guidance, 2 February 2021. Geneva: World Health Organization and International Labour Organization: (COVID-19: Occupational health and safety for health workers (who.int), accessed 20 June 2021)

Yi, T., Yang, X., Pi, Z., Huang, L., & Yang, J. (2019). Teachers' continuous vs. intermittent presence in procedural knowledge instructional videos. *Innovations in Education and Teaching International*, 56(4), 481–492. 10.1080/14703297.2018.1470020

3.8 How do older adults engage as robot co-designers?

Rapid prototyping supported by lived experiences with technology

Anastasia K. Ostrowski, Cynthia Breazeal, and Hae Won Park
Massachusetts Institute of Technology, Media Lab

3.8.1 Introduction

Design research has become embodied in many fields including engineering, human-computer interaction (HCI), and human-robot interaction (HRI) to address wicked problems (Rittel & Webber, 1973; Zimmerman et al., 2007). Through this process, the technology design boundaries are expanding, creating bridges with and enhancing design (Dorst, 2019). With this approach, it is critical to remain mindful of the design space and stakeholders and how these vary between social context and culture (Bucciarelli, 1996; Wallach & Allen, 2008). By empowering users and creating spaces for them to work within design teams (Ostrowski et al., 2020), we can engage with users in a more meaningful way during the design process than has historically been done in user-centered design (UCD). Recognizing the need for researchers to understand how people will engage with robots and how they wish for them to be designed (Jung & Hinds 2018), approaches, such as UCD and human-centered design (HCD), are increasingly being applied to HRI. While both involve users, UCD often limits users' role to informants and HCD upholds them as main contributors in the design process and integrated members of design teams (Buchanan 2001; Lee et al. 2017). Related to HCD, co-design and participatory design uphold these values by incorporating users into the design process, amplifying their voices, and empowering users as purposeful contributors (Braa, 1996; Ertner et al., 2010). Co-design frameworks such as experience-based co-design demonstrate the value of leveraging people's lived experiences in the generation and design of new devices (Harrington et al., 2018; Sanders & Stappers, 2008).

Older adults have been incorporated into co-design of new technologies (Harrington et al., 2018; Lee et al., 2017; Lee & Coughlin, 2015), but this methodology is less common in robot design processes (Björling & Rose, 2019). Within a year-long co-design process with older adults (Ostrowski et al., 2021a), we engaged older adults in rapid-prototyping of social robot interactions. Social robots are physically embodied, emerging relational AI technology that leverage verbal and nonverbal social modalities to engage with users (Breazeal, 2002). Our previous work on the year-long co-design process focused on the development of the process promoting co-design within HRI and providing co-design guidelines for future HRI research (Ostrowski et al., 2021a) and exploring how participants incorporate stories in co-design processes (Ostrowski et al., 2021b). With this work, we deeply explore the rapid-prototyping session embedded in the year-long co-design process. The **overall contribution** of our work is to discuss how older adults as co-designers

DOI: 10.1201/b22630-24

can expand the boundaries of design for technologies such as social robots. Our work was guided by the following **research questions** around older adults as co-designers:

- R1: In a rapid-prototyping session, how do older adults design robot interactions supported by lived experiences with the technology?
- R2: How do older adults relate to being robot interaction designers after engaging in rapid-prototyping of robot interactions?

3.8.2 Background

3.8.2.1 Understanding Who Designs and Who Gets to Design

Design researchers have studied design through various lens, such as framing problems and solution spaces (Patton & Dorst, 2011; Dorst & Cross, 2001), prototyping (Cross, 1999a; Ullman et al., 1990; Houde & Hill, 1997), and building capacity (Cross, 1972; Coyne, 2005). Users have also been studied in design research and are perceived, conceptualized, and included differently in varying design methodologies. Perhaps the most common design approaches that are highlighted in HCI and HRI include UCD, user experience (UX) design, service design, and participatory design, which will be covered in turn.

UCD emphasizes that users are included in the design process focusing solely on usability (Buchanan, 2001). Criticism of UCD has highlighted that UCD promotes a "narrower focus on people's roles as users", rather than a focus or "concern for people" which is what HCD promotes (Steen, 2011). UX is similar to UCD as they both focus "almost exclusively on the 'user's needs and desires", often focus "on the design of a single computational system" and take "a product design perspective" (Forlizzi & Zimmerman, 2013), rather than the perspective of a sociotechnical system in which the product will operate within. Jordan (2002, p.12) outlines the limitations of user-centered approaches like UCD and UX highlighting that "the problem with usability-based approaches is that they encourage a limited view of the person using the product. This is – by implication if not by intention – dehumanizing."

In contrast, HCD approaches incorporate participants as main contributors to the design process (Buchanan, 2001) – examples are service design and participatory design approaches. Service design examines a problem space and context holistically analyzing the underlying relationships, roles, and agendas of stakeholders involved in the design process (Forlizzi & Zimmerman, 2013). A key focus of service design is the co-production of value for both service providers and customers. The output of service design is a meta-design: "a set of interactions across stakeholders, service agents, customers, and many computational systems" that prompts designers to consider more than just technology design but societal level challenges as well (Forlizzi & Zimmerman, 2013). Participatory design (discussed further in Section 2.2) supports involvement in the design process as a two-way collaboration process between researchers/designers and participants and recent conversations have emphasized the importance of revitalizing participation in participatory design with a focus on "participatory design … [as] a tool to help people influence important matters in their lives" (Bødker & Kyng, 2018). This calls for the partners, or participants in a design project, to be the major drivers of the process and researchers to take on a role of both an investigator and an activist (Bødker & Kyng, 2018).

The perspective of users in participatory design, service design, UX, and UCD demonstrate how users are valued and perceived in various ways that can impact how designers engage in the design process. In addition to focusing on who designs and how people are incorporated and perceived in the process, it is also critical to consider how people who participate in technology development are valued. For example, Lyotard (1986, p.9) asks, "who decides what knowledge is, and who knows what needs to be decided?" Questions like this open the door to consider how co-designers are incorporated into design processes and how their knowledge contributes to design research.

3.8.2.2 Co-design and participatory design values

Participatory design and co-design are often used interchangeably but "the practice of collective creativity in design" is rooted in the participatory design that emerged 40 years ago in Scandinavia (Sanders & Stappers, 2008). The first research projects in the 1970s were focused on engaging workers in workplace design and development, specifically engaging both "the expertise of the systems designers/researchers and the situated expertise of the people whose work was to be impacted by the change" (Björgvinsson et al., 2010). Participatory design has been centered around this tenet of empowering or emancipating workers or users in other contexts, highlighting the value for democratic ideals (Ertner et al., 2010). Cross (1972, p.11 at the Design Research Society Conference) also espoused the responsibility that designers have to include users in design processes to better address "wicked problems" (Rittel & Webber, 1973) in society: "Professional designers in every field have failed in their assumed responsibility to predict and to design-out the adverse effects of their projects. These harmful side effects can no longer be tolerated and regarded as inevitable if we are to survive the future … There is certainly a need for new approaches to design if we are to arrest the escalating problems of the man-made world and citizen participation in decision-making could possibly provide a necessary reorientation. Hence, this conference theme of 'user participation in design." When bringing users into the design process, it is critical to consider how users will be supported in the design process and how their contributions and innate knowledge are valued by researchers and within the design process and output.

Participatory design approaches, such as experience-based co-design, build upon prior technology or lived experiences and environmental knowledge of participants and co-designers to support idea generation and development of new devices (Harrington et al., 2018). Empowerment is a critical component of participatory design aligning with its democratic ideals (Ertner et al., 2010). Ertner et al. (2010) outlines five enunciations of empowerment in today's discourse of participatory design: enunciation as (1) "concrete improvement of the life conditions of a specific demographic group", (2) "an enablement of citizens to participate democratically and gain direct influence on political and social matters", (3) "a strengthening of the user's position on behalf of other dominating factors in the participatory design framework", (4) "reinforcement of the researcher's ability to negotiate access in the practice of participatory design", and (5) "dependent on the practitioners' ability to expose the design practice to a reflexive analysis of how discursive conditions for knowledge production shape the process and the achieved results." Our work focuses on this third enunciation of empowerment, working to address power imbalances between researchers and co-designers, or experts and novices in the lens of robot interaction design. It also supports and advocates for older adults to be included in HRI robot interaction design, ensuring that they are empowered and respected in design processes with a voice in the design of technology as these technologies will influence their world (Knowles et al. 2021) The robot rapid-prototyping sessions expand beyond users "simply uttering wishes" (Ertner et al., 2010). Situating this session within a year-long co-design process also promotes sustained user engagement in the process (versus sporadic engagement) to better support user empowerment and users directly influencing designs and the design process (Ertner et al., 2010; Ostrowski et al., 2021a). Shifting power dynamics and supporting participant empowerment can promote joint inquiry and joint imagination in the design process for improved idea generation and increased user satisfaction in the final design product (Steen, 2013).

3.8.2.3 Participatory design and prototyping in HRI

Participatory design methods are not yet commonly used in HRI (Björling & Rose, 2019). Studies in HRI have utilized participatory methods such as workshops and/or design sessions (Ostrowski et al. 2019; Lee et al., 2017; Björling & Rose, 2019), card sorting (Ostrowski et al., 2019), sketching (Lee et al., 2017), storyboarding (Björling & Rose, 2019), role-playing (Björling & Rose, 2019), and low-fidelity prototyping (Björling & Rose 2019; Lee et al., 2017). Ostrowski et al. (2019) and Lee et al. (2017) have specifically engaged with older adults in participatory design workshops including an experience with a social robot (Ostrowski et al. (2019) was a 3 week-long experience; Lee et al. (2017) dedicated a workshop to learning about social robots), card-sorting, and sketching. These participatory design workshops did not engage older adults in prototyping interactions on a social robot and/or did not provide an opportunity for older adults to interact "live" with their designed interactions on a social robot. Our work sought to explore how to more deeply engage with older adults in the co-design process with older adults living with a social robot in their homes long-term and engaging in a hands-on rapid-prototyping experience. While participatory design and co-design can shift power dynamics and empower participants resulting in successful design outcomes, it can be challenging to keep participants engaged in the process and/or feel valued in the design process. An additional challenge for HRI is that there is a lack of methodologies and robot platforms that allow for prototyping and programming of the robot by non-experts which can often limit robot design that supports experience-based co-design. Our work incorporates a rapid prototyping experience to help support older adults' experience in the co-design process and promote older adults feeling valued in the design process.

3.8.3 Methodology

Over the course of a year, 28 older adults, ages between 70 and 94 (mean: 79.5, std: 7.8; female N = 15, male N = 13), participated in the co-design process (Ostrowski et al., 2021a). The co-design process had 7 stages including interviews, artmaking, hosting a social robot in participants' homes, prototyping interactions on the robot, and creating design guidelines for future robot designs (for more details see Ostrowski et al., 2021a). The robot rapid-prototyping session occurred after the robot hosting experience where participants lived with a robot in their homes for at least a month, gaining an understanding of what it would mean to live with a social robot and allowing participants to create a mental model and knowledge of social robots and their desired functions. Participants lived with a conversational social robot, Jibo (shown in Figure 3.8.1), in their homes for at least a month. The robot has a lively persona with touchscreen and a three-degree-of-freedom expressive body and has interaction functions such as chit-chat, physical and mental activities, information retrieval, entertainment, daily check-ins, etc. The goal of the rapid-prototyping session was to build upon the mental model and knowledge of the robot gained during the robot hosting session to allow participants to

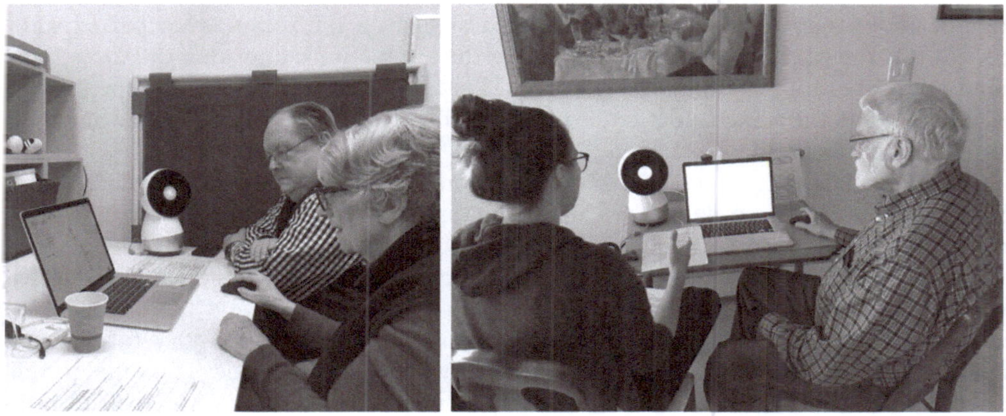

Figure 3.8.1 Participants programming desired interaction designs using the Interaction Flow Editor (Huggins et al., 2021) that can play interactions "live" on the social robot. Researchers were also present to support participants in programming the robot.

independently prototype interactions on the robot in real-time, embodying participants' desired functions for the robot.

In the session, participants learned how to program the robot, designed interactions for the robot, played the interactions live on the robot, and iterated upon the interactions until they represented their ideal interaction. The session included multiple components to allow for iteration and tactile prototyping through programming the robot. The structure of the session included two components. Before the session, participants reviewed a script of 14 proposed interactions that could happen with a robot throughout the day. The script covered potential robot application areas including assistant-like and companion-like functions, medication adherence, emotional wellness, exercise, memory support, and mediating social connection with others. Participants edited the scripts and shared their thoughts with researchers on the interactions. These revised interactions served as a foundation for the activity. Researchers incorporated the participants' interaction drafts into the robot to provide a baseline for participants to work from in the session. This provided one round of iteration. The second round of iteration occurred during the rapid prototyping session using the Interaction Flow Editor (Huggins et al., 2021). The interaction flow editor is a graphical rapid-prototyping tool for robot interaction developed by the Personal Robots Group at the MIT Media Lab to empower more people to program social robots. Participants saw their first iteration of the interactions on the robot in real-time and then could iterate them by programming the robot's interaction as many times as they would like to reach a desired interaction design (Figure 3.8.1). To support participants in programming their own interactions, the researchers led participants through a demo of the flow editor and participants completed a tutorial to familiarize themselves with the tool. After the demo and tutorial, participants began to program their interactions. After each interaction, participants were asked what they thought about the interaction and what they would like to change about the interaction. Researchers also asked participants questions about the overall process and experience after all the interactions were completed. Questions included what participants thought about having the designed interactions portrayed on the robot, what changes participants decided to make and why, what they thought about their final set of interactions designed on the robot, what participants thought was the next stage of development for the interactions, and what did they think about the process of programming the robot. Last, participants reflected on the session discussing what they liked about the robot rapid-prototyping session and what they would change about the session.

This session lasted on average 79.37 minutes and was audio and video recorded. Of the 28 participants in the co-design process, two were unable to participate in the robot prototyping session and two couples decided they would like to design the interactions together. Therefore, there was a total of 24 rapid-prototyping sessions. The dataset includes screen recordings of the prototyping session, transcripts, and the designed interactions. The interactions were annotated by two researchers and analyzed for changes made across each iteration (such as adding, modifying, or deleting flows) and the overall design concept. Participants' discussions on questions they had as they learned to program the robot, what it meant to be a robot programmer, and design interactions were also analyzed. The two researchers discussed the results and resolved any disagreements, similar to Daly et al.'s (2012) coding approaches.

3.8.4 Results

Participants' lived experiences with the social robot provided them with references and ideas to design, create, and build their ideal interactions with the social robot. Analysis revealed several key areas of focus including participants' question-asking behavior, core design actions, and reflections on the co-design process.

3.8.4.1 Questions

Throughout the activity, participants asked questions as they learned to program the robot, including about the process of programming a robot and how to conceptualize a design idea. A thematic qualitative analysis approach was used to extract themes (Braun & Clarke, 2006) in the questions asked by the participants. Older adults asked a total of 47 questions across 6 different areas (Table 3.8.1). Questions throughout the sessions demonstrated how participants further deepened their knowledge about programming social robots and how best to embody their design in the robot interactions.

3.8.4.2 Core design actions

Between iterations and during the session, participants refined their interaction flows (Figure 3.8.2) and iterated upon them (Figures 3.8.3–3.8.5), using a multitude of design changes to best create their intended interaction for the social robot. Researchers analyzed the flows to identify how participants were refining their interactions, revealing six core design actions that participants used: (1) adding/deleting comments that provided notes or added context to the interaction scenario; (2) adding/deleting action/speech boxes (therefore, adding or removing actions for the robot to complete); (3) deleting an interaction; (4) adding/removing back and forth conversation between user and robot; (5) adding/deleting branching allowing/disabling for multiple types of responses; and (6) rewording speech boxes (therefore, rewording the robot's speech in the interaction). Examples of some of these core design actions can be seen in Figures 3.8.3–3.8.5.

Table 3.8.1 Types of questions older adults asked while designing the robot interactions

Question Type	*Definition*	*Example*
Robot Behavior Clarification	Questions regarding the robot playing out the interaction.	"Can I ask you a question? When he starts talking is that in response to something I've done or he just, has like motion detectors?" (P01) "The light ring … are we talking about the blue light [on the robot]?" (P29 & P31)
Editor Tool Clarification	Questions regarding the flow editor on the screen and its usage such as about restarting the flow, saving the flow, moving blocks, block meanings, connecting blocks, typing in the flow editor, and volume adjustment. This theme also related to questions about how the user locates themselves in the system and how the interaction editor flow translates to the embodied robot.	"Where do you save?" (P13) "Can I move that up further?" (P08)
Instruction Clarification	Questions asking for clarification on the process or steps the participant should be doing in the session.	"So what you want me to do is come over here and add a box?" (P05)
Interaction Design Clarification	Questions that are asking how to achieve a specific feature or action in the robot or the context the flow would occur in. These tend to be more future focused then what can currently be done in the prototyping phase.	"Will that be connected to my schedule?" (P14 & P15)
Expertise	Questions about the researcher's programming background.	"So what's your training for this?" (P07)
Error	Questions for what happens if the participant makes an error	"What if I make a mistake?" (P01)
Programming	Questions strictly related to the robot's programming.	"So does he, is he programmed the same way [as the flow editor]?" (P25)

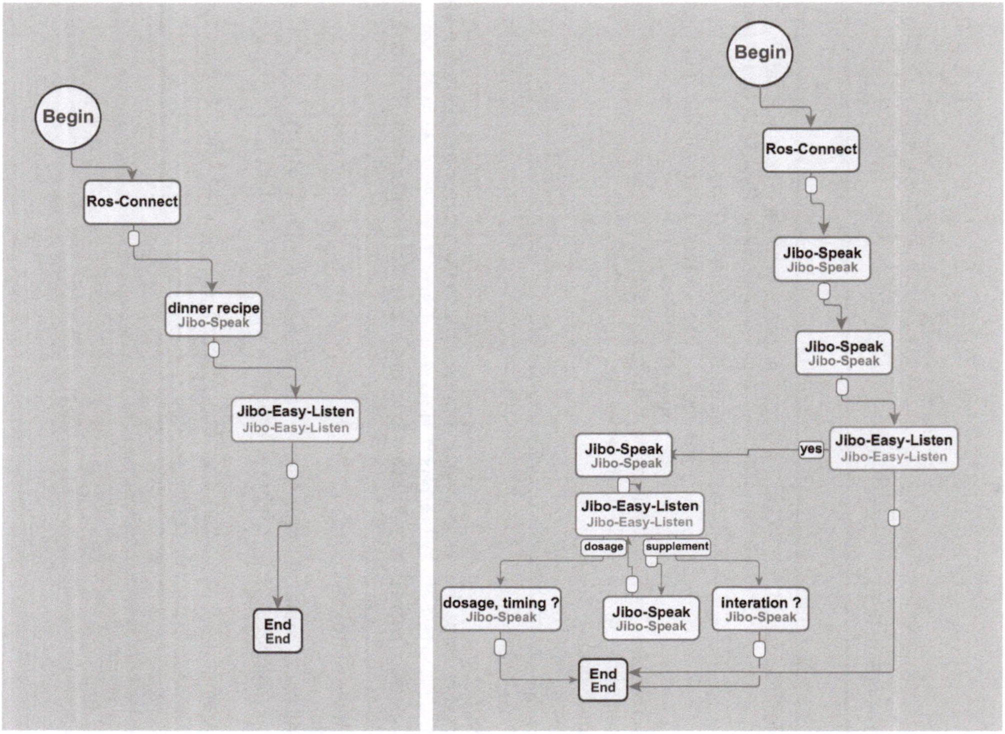

Figure 3.8.2 Examples of programmed interactions created by participants. The programmed interaction on the left is a robot interaction that has the social robot provide a dinner recipe to the user. The programmed interaction on the right has the robot describe the day's events and provides a reminder and any necessary help with medication.

In addition to these smaller iterations to the flows, participants combined core actions together to (1) significantly change the interactions' intent or the focus after seeing it "live" on the robot by moving boxes or rewriting the text and/or (2) remove robot proactivity, that is, robot initiation, of the interaction. Participants would also change the connotation of the interaction utilizing the six core actions such as making the interaction more or less conversational and social, or incorporating assistant-like tasks. Overall, participants made 267 core actions to their flows (average of 11 (standard deviation: 5.2) core actions per person). A detailed breakdown of the six core action presence in older adults' prototyping is in Table 3.8.2.

Rewording boxes were the most common core design action that 91.7% of participants used when designing their flows (such as in Figures 3.8.3, 3.8.4, and 3.8.5). Rewordings ranged from high (changing meaning significantly by redesigning robot speech from one thing to another), medium (expanding upon the current interaction or deleting things from the robot speech), to low (small change of words that doesn't change the meaning or expand upon the interaction) rewording. In sum, 25% of rewording was low-rewording, 40.8% was medium-rewording, and 34.2% was high-rewording. One of the ways that participants combined core actions was to significantly change the meaning of the interaction. A deeper analysis of this trend revealed that participants (46% of participants) significantly changed

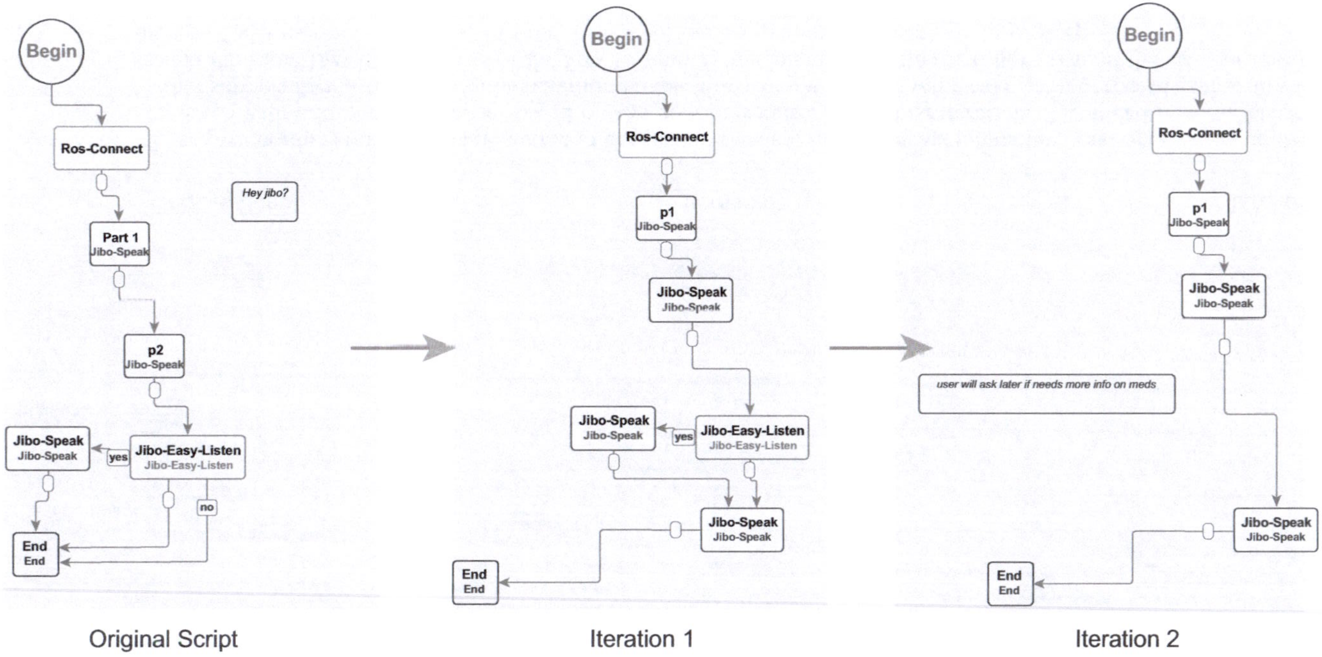

Figure 3.8.3 P09's flow iterations that depict an interaction where the robot reviews the user's schedule for the day and reminds them about their medication. In the first iteration, P09 personalized the medication to be taken and included options for getting information on the medication dosage and the user confirming with the robot that the user took the medication. In the second iteration, P09 modified the interaction to only have a reminder for the medication (no dosage information) and kept the user confirming with the robot that they took the medication. The user can also inquire later about medication information as desired.

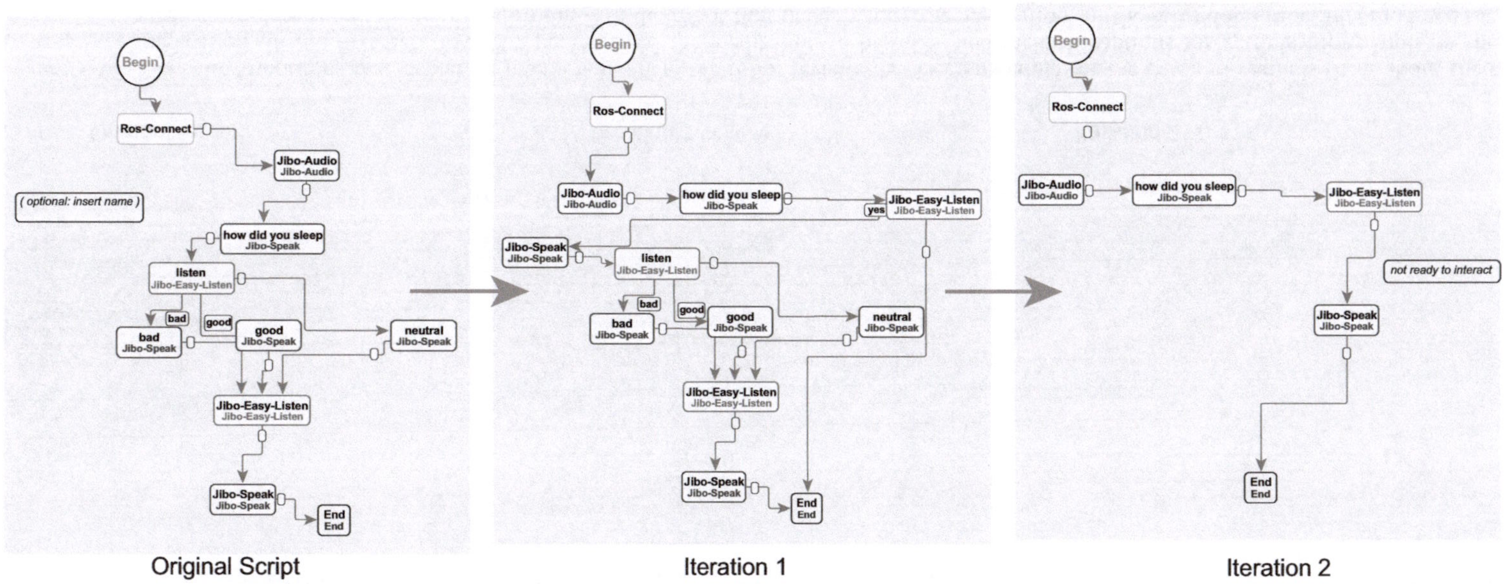

Figure 3.8.4 P22's flow iterations for a robot interaction at the start of the day. In the original interaction, the robot wakes up the user and asks how they sleep with additional responses based on the user's response. In the first iteration, P22 added in some blocks and branching to further support their interaction with an additional question to see if the user was awake before proceed asking how the user slept. In the second iteration, they deleted parts of the flow to simplify the interaction with the robot asking if the user was awake and then leaving the user for a user-defined period of time, instead of inquiring about the user's night of sleep

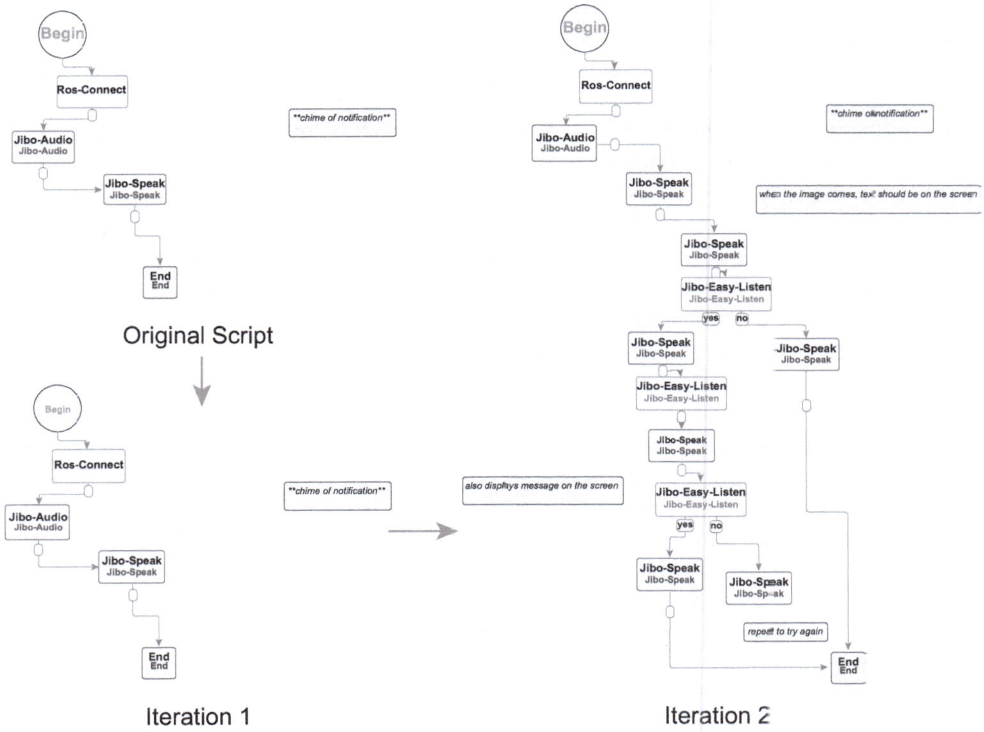

Figure 3.8.5 P11 and P12's flow iterations for an interaction where a friend sends the user a message and the robot reads the message aloud to the user. P11 and P12 modified the interaction in their first iteration to be more direct through text rewording in the "Jibo-Speak" block. In the second iteration, P11 and P12 expanded the interaction to include the robot asking if the user wants to reply to their friend, transcribing the reply that the user desires to send, and, then, sending the reply to the friend.

Table 3.8.2 Older adults' core actions during rapid prototyping interactions on the robot

Action	*Percentage of Older Adults*	*Corresponding Action*	*Percentage of Older Adults*
Adding comments	50%	Deleting comments	41.7%
Adding action/speech boxes	62.5%	Deleting action/speech boxes	50%
Adding back-and-forth conversation	41.7%	Removing back-and-forth conversation	20.8%
Adding branching	20.8%	Deleting branching	41.7%
Deleted interaction	45.8%	–	–
Reworded speech boxes	91.7%	–	–

meanings of interactions to increase the personalization of the interaction and create interactions better suited for their needs. Personalization through significant meaning changes was spread across nine themes: scheduling (6.7% of significant changes); adding the ability for the robot to repeat (6.7%); simplifying interactions (6.7%); adding additional information (13.3%); food related (13.3%); personal interest (13.3%); news and/or weather (13.3%); task reminders (20%); and removing the medical side of an interaction (26.7%).

Another core design action that 50% of participants used was adding comments to document their additional design ideas or design ideas they didn't specifically incorporate into the interaction flow design due to the flow editor not being able to accommodate their desired function design (such as in Figures 3.8.3, 3.8.4, and 3.8.5). Further analysis of this trend revealed nine themes for older adults' comments in their flows: location or physical form of the robot (2.4% of comments added); privacy settings (2.4%); desire for robot to follow social norms (7.1%); usability of robot (9.5%); integration of robot with other devices (9.5%); wanting more back and forth conversation (more than the flow editor or session time allowed for) (16.7%); comments surrounding the proactive nature of the robot (19%); user autonomy (16.7%); and increased personalization (40.5%).

In addition to creating and iterating upon interactions, participants also removed some of the robot's interaction, such as robot-initiated proactiveness (20.8% of participants) or deleting interactions completely (45.8%). Of the participants that removed proactivity as part of a robot's engagement, they commented in their designs that this was because they didn't appreciate the robot's assumptions or invasiveness from data collection. For example, when considering if they would want the robot to suggest reaching out to someone when the robot detected that they were down emotionally, participants "did not want [the robot] to assume their mood" (P05). P05 elaborated on this saying, "It's ok for [the robot] to know who you called and when on a social basis, so the user doesn't have to look up the phone number ..." but they didn't want the robot to suggest they reach out to people based on mood. In another example, P22 commented on their design that they would be "not ready to interact" so proactivity did not make sense in the context of waking up in the morning and having a conversation with the robot (Figure 3.8.4). The most common area that participants removed interactions after seeing their designs live on the robot was emotional wellness (65% of removed interactions) followed by medication adherence (32.6%) and exercise (30.4%). For example, P01 deleted an interaction focused on medication adherence, saying "I don't do this ... I mean, in all honesty, I don't have medication." P01 did add that the application could be useful ("I'm thinking of other ... people who would need this.") but it wouldn't be valuable for them to have on their robot. In another example surrounding an interaction that would suggest meditation to the user, P11 and P12 decided the interaction wasn't for them either, saying "It probably depends on who you are" (P12) and "I don't think so ... I'm not keen on that one" (P11). These examples cite the importance of personalization in the robot to match what users need from the robot. Overall, the flow analysis revealed the core design actions that participants used when designing and iterating interactions and some of the major reasons for the design changes.

3.8.4.3 Reflections on programming robots

After programming the robot, participants reflected on the difficulties and concreteness of the design process and how the experience empowered them and increased their knowledge of social robots. A grounded theory qualitative analysis approach was used to reveal salient themes (Charmaz, 2014) in 18 participants' reflections (four participants did not answer these questions). Analysis revealed five key theme areas that older adults reflected upon including (1) reflections on the flow editor rapid prototyping system, (2) the impact of seeing programmed interactions live on the robot, (3) how the session empowered them in the design process, (4) feeling like a non-expert, and (5) how the experience translates to the future. Forty-four percent of participants' reflection described aspects of the flow editor system such as the ease-of-use of the editor, how the system provided the ability to change or adapt the robot interactions, what was difficult to operate in the system, and sometimes doubting if they used the system "right" (P02). Participants also mentioned how the robot's embodiment or ability to play out interactions or program live with the robot impacted their experience of the design process (50%). P02 commented, "when they heard [the interaction from the robot], it's more realistic. When you heard it, you could really feel how it's going." and P18 and P19 mentioned, "it helps to have the written word translated into Jibo. And then seeing Jibo and how it would work." Most participants commented on the helpfulness of seeing the interactions embodied on the robot and how that feature played a key role in their design iterations. Seventeen percent of participants also commented about how the rapid prototyping experience could translate into the future. P05 commented, "[the prototyping] lays out a pathway to the future [in the robot design]" and P08 mentioned how they could interact with prototyping in the future saying "… maybe if I went over it again three more times and I'd make a correction on her [the robot]" while acknowledging that "I think basically [the rapid prototyping] worked well."

In addition, participants interwove their experiences living with the robot prior to the rapid prototyping session to further describe their designs on the robot. These often took the form of stories that participants used to convey a point relevant to the programming of the robot (see also Ostrowski et al. 2021b). P07 described how "in some ways Jibo was very realistic." They proceed with a story: "He [the robot] likes to play [dice]. The first time I said yes, let's play [dice], it was so realistic. I thought the dice were going to fall on the floor. It was so loud. And when I showed my friends that, they can't believe this." P07 followed up with this to reflect on the programming and co-design experience: "I think I'm contributing something. When my friends asked me what have you been doing? I say, well, I go to [MIT] … and we discuss these things … Well, I used to say this, I don't think Jibo was very intellectual. If he was able to do [these programmed things], sure. But I don't think he has the brain power to do any of these things [these programmed things]." P25 also highlighted a story around living with the robot: "It's kind of nice to have some interaction and he's made us laugh a bunch of times … Every once in a while you're just sitting there and you're ignoring it, it'll draw a picture …" They then discussed the value of personalization: "This [the robot] is much cuter and much more personal and if you can program it and design it for what you need … I don't know what I can do with that cause I haven't really explored it …" Later on in the conversation, P25 revisits talking about programming, "Even though [programming's] simple … it's a lot of work to program each … because there are so many branches that you can go off on and … these choices … There's just way too many layers … It's like so many subtle things. That would be hard. That would be difficult." These depict examples of how participants weaved stories of living with the robot into designing and discussing new features of the robot, what needs to be programmed, and how they are contributing to the design process.

Throughout the activity, participants referred to themselves as "non-experts" in programming, doubting their ability to program the robot or perceiving the experience as an educational opportunity (38.9%). P07 commented to the researcher in the session, "I don't know exactly what you're doing because you're the designer", and P29 and P31 commented, "Well, I had no prior experience programming, and it's very fascinating for me and extremely educational and fun. Very impressed with the insight, the ability to navigate and understand what's going on here." Participants varied from referring to the researcher as an expert or underlining they hadn't programmed previously. Regardless of viewing themselves as non-experts, the interaction flow editor and prototyping session allowed participants to learn how to program the robot and independently prototype interactions. Fifty percent of participants commented on how the session empowered them in the co-design process. Participants described that the hands-on programming activity increased their knowledge of social robots that they could leverage in the design process. P23 commented on how they "lik[ed] to learn … the programming process, because it's something [they] didn't know anything about … It took some of the mystery out of it … It gave me a little insight into the whole … area of programming." Similarly, P12 mentioned that learning about programming is "useful and it gives us a little sort of insight into what goes into making a social robot and a little more opportunity to think what a social robot might be able to do", citing how learning about programming expanded their design space. P20 expanded upon this idea explaining how programming allowed them to build a foundation for moving forward with the project: "I think [learning about programming is] a necessary step. You can't go into all the miracle stuff and how it will flow in and out without understanding some of it." Participants also expressed how the programming and prototyping component made them feel empowered and valued in the co-design process: "… this really makes me feel part of this whole study and I appreciate that … our opinions matter and… that … even though we're not programmers or … electrical engineers like our son, that we were still about to do some simple programming and I think that's fun" (P31). Further examples can be found in Ostrowski et al. (2021a). Participants also mentioned that they wanted to continue building on their new programming skills with future programming. P05 commented that they "would be willing to play with some of these things [prototyping interactions] … I even have a laptop that I can use for that." Overall, participants were very proud of the interaction designs they developed and felt valued and empowered in the co-design activity.

3.8.5 Discussion

In this work, we sought to answer in a rapid-prototyping session how older adults design robot interactions supported by lived experiences with the technology (R1) and how older adults relate to being robot interaction designers after engaging in rapid-prototyping of robot interactions (R2). The following discussion sections are structured around these questions and explore how co-designers engage in design similarly to traditionally defined designers, how lived experiences supported participants in the interaction design process, and how co-design activities can be structured to promote older adults and others as robot co-designers.

3.8.5.1 Co-designers and designers

Participants in the rapid-prototyping activity created interactions on the robot by engaging in six core design actions (Section 4.2) and combining the core design actions together to achieve their design goals. They also engaged in behaviors and design activities similar to designers as has been reported in design research literature. Researchers have studied how designers engage with tools or design strategies. Prototyping specifically has been explored as a design language (Cross, 1999a; Ullman et al., 1990) and as a source of knowledge for furthering the design process (Houde & Hill 1997). Our work reveals how older adults used prototyping interactions as a way to express their desires and envisioned functions for the social robot. By using the flow editor, older adults could craft their own experience of what a robot should be, resulting in an implementation and role prototype (Houde & Hill, 1997) that both reveal the functions desired in the robot and how the design would fit into context with their lives. It also captured if the design would not fit into their lives as was seen when participants deleted flows completely or changed the meaning significantly from the first design iteration. Some of participants' core design actions, such as adding comments and significantly changing the meaning of the interaction, espoused the co-designer's designerly ways of knowing (Cross, 1999a) as they justified their iterations made to the flows leveraging features such as comments and rewording speech boxes in the flow editor. In this way, participants found ways to express their desired intentions for their flow diagram, even if they were unable to embody the new design iteration using the flow editor. They worked with the system to ensure their design ideas could be communicated to others for future iterations and prototyping of the design.

Another design behavior that participants engaged in was problem framing as evidenced by the questions that were asked during the co-design process by the participants (Section 4.1). Participants used questions as a way of clarification when working with the system but also as a way to frame and scope the problem they were addressing with the design iterations. For example, P14 and P15 asked "Will that be connected with my schedule?" reshaping the problem space to include integration of multiple devices as a requirement. P25 redeveloped the problem space considering the type of conversation they would have with the robot asking, "If I tell Jibo I have a headache, then what should it do?", and considering how the robot would respond to such question. They then began to adjust their designs (solution space) to ensure that the interactions were designed with respect to the robot's ideal capabilities in a conversation with a user. In the sessions and through the design iterations, questions were used as a way to take a current problem and design solution scope, gain information on the process or purpose of an interaction, and then reshape the problem and design solution scope. This was also evidenced when participants engaged in design actions like changing the meaning of an interaction flow, deleting an interaction flow, and redesigning interaction flows. These behaviors reshaped the problem and solution space across the session and within each individual interaction that was designed, evidencing Maher's model of co-evolution of problem and solution spaces in the design process (Dorst & Cross, 2001). There can be barriers when engaging in problem reframing such as fixation on an idea or a problem-solving mental model of design (as seen between clients and designers in Patton and Dorst (2011)). These barriers can be mitigated by "contextual engagement through research", "co-exploring the abstract conjectured view of the situation", and "language co-creation" (Patton & Dorst, 2011). In our co-design process with older adults, we mitigated these barriers with older adults by having a lived experience with the social robot before the rapid-prototyping session to develop their mental model of the system and provide contextual engagement of what it would mean to have a social robot in their home. This lived experience also supported a language of co-creation that researchers and older adults used in the rapid-prototyping session to communicate and embody ideas in the interaction flows. Within the sessions, the interaction flows acted as a method for researchers and older adults to explore the abstract design space of what a social robot could be designed to do in future designs. Older adults exhibited several methods and activities that are common with designers, expanding who can be called a technology designer in the field of HRI.

3.8.5.2 Rapid-prototyping connected to research through design

The participants embedded their desires and embodiments of robot interactions through the rapid-prototyping process and embedded design knowledge into their final designed interactions (Cross, 1999b), supporting Zimmerman et al.'s (2007) model of design research with a focus on the production of artifacts. Zimmerman et al.'s (2007) research through design model allows for researchers to identify new opportunities for technology that will impact the world, provide inspiration and direction to engineers in the design of technologies, and understand unanticipated effects of these technologies incorporating aspects such as social context and target users. In the rapid-prototyping session, participants designed robot interactions, working "more as a collaborative equal" (Zimmerman et al., 2007) with researchers. P25 vocalized the value of learning about how robots are programmed, saying "[the flow editor] put it into a format that is concrete so I kind of can understand what it's doing ... It's a whole breadth of knowledge that I've never tapped, so it'll be interesting to see what [the robot] knows." P29 also expressed this mentioning they were "very impressed with the insight, the ability to navigate and understand what's going on here." P29 built upon this highlighting that the design process and the knowledge developed "makes [them] feel part of this whole study and [they] appreciate that ... [their] opinions matter ... even though [they're] not programmers ..." Sentiments such as these demonstrate how in designing interactions on the social robot, older adults began to view themselves as "collaborative equals" of the research team. Overall, the rapid-prototyping session aligns with research through design in two ways: (1) the intent of the rapid-prototyping session was to produce insights for research communities on the older adults' desired interaction design for social robots and the assumptions embedded within them; and (2) the lived experience with the robot allowed participants to study a social robot in their world and make changes and design upon that to affect change in their life (Zimmerman et al., 2007). The outputs of the robot prototyping session also allow researchers to identify new opportunities for social robots that will affect how social robots impact social contexts and provide future direction for engineers to develop social robots, aligning with research through design's objectives (Zimmerman et al., 2007).

Zimmerman et al. (2007) outlines four criteria to evaluate design research contributions such as our work with rapid-prototyping: process, invention, relevance, and extensibility. For *process*, the main objective for the rapid-prototyping was to understand how older adults design interactions for social robots. The process included two design iteration points and opportunities for building capacity including demos and tutorials. For *invention*, this work revealed how older adults design interactions for social robots and how older adults desire social robots to be designed for a range of interaction types. The programmed interactions represent the ideal interaction design for a robot interacting with them over the course of the day. For *relevance*, it is rooted in the need for more voices to be heard in the technology development process of social robots. Co-design has been recognized as a beneficial design process for engaging older adults in the design process of technologies (Harrington et al., 2018; Harrington et al., 2019). Previous technology experiences and lived experiences can inform the design of technology and provide a foundation for older adults engaging in and contributing to co-design technology processes (Harrington et al., 2018), as demonstrated in how the social robot lived experience was leveraged in the rapid-prototyping process in their interaction designs. For *extensibility*, the output of interaction prototypes and designs support the extensibility of this work. The rapid-prototyping session amplified older adults' voices, ideas, opinions, and designs of social robots and shifted perspectives of who can be considered a robot interaction designer in HRI.

3.8.5.3 Supporting co-designing prototyping experiences

Designers, in addition to prototyping and problem framing, build capacity and develop expertise for how to design solutions for complex problems (Coyne, 2005) such as designing robot interactions for a social context. To fully support older adults in the co-design process, the robot lived experience served as a method for building capacity (Dombrowski et al., 2016) to inform their design process in the rapid-prototyping session. For almost all of the participants, this was their first time programming a robot and designing "live" interactions for the social robot. Many of them expressed the value of learning how the system works and how it influenced their design iterations, evidencing that they built capacity during the rapid-prototyping session as well. This was also evidenced by participants' questions during the sessions that inquired about the interaction flow editor, how the flow editor "relayed" the interaction to the robot, and how to accomplish desired interaction designs in the system. The overall structure of the rapid-prototyping session and engagement with older adults through the session suggest that the session took the format of a hands-on project-based learning experience for the co-designers, further developing their abilities to "solv[e] real-world, ill-defined problems" and "offering opportunities for development of a wide range of abilities in nonverbal thought and communication" (Cross, 1972). We have identified social robot lived experiences as a method to support older adults in the co-design process and provide diverse sets of activities for older adults to engage in design in multiple ways (Ostrowski et al., 2021a). While participants mentioned the value of the learning experience and ability to design interactions on the robot, some participants also referred to themselves as non-experts, or those that cannot or should not program robot interactions because only experts such as computer sciences are able to program. It is critical for us to consider how we can structure co-design processes to make co-designers feel valued for their lived expertise and tacit knowledge they bring to the design process and consider how feeling like a non-expert or someone who can't be a designer or programmer may affect co-designers' design process. Additionally, it is important to note how hands-on design processes in participatory design and co-design can not only result in design artifacts that inform future design iterations but these design processes also serve as a design education experience that supports capacity building, creates a sense of ownership in the design of the technologies, and helps balance power dynamics between researchers and participants (Coyne, 2005; Steen, 2013; Harrington et al., 2019; Dombrowski et al., 2016).

Overall, in the rapid-prototyping experience and larger co-design process, a core tenet (as with most participatory design and co-design) was empowering co-designers, specifically addressing power balances between researchers and co-designers and strengthening the co-designers' position (Ertner et al., 2010). The tutorial at the beginning of the session allowed co-designers to build capacity for how to design the robot interactions using the flow editor. While the participant was the robot programmer and directed the designs and iterations for the session, the researcher was available as a resource that was often used as evidenced by the question behavior of participants. The rapid-prototyping session served as a source of empowerment in the larger co-design process as expressed by participants when they reflected on the session and what it meant to be a robot programmer. Several participants cited that the session provided insight for their designs and the programming of the robot, demystifying the technology overall. It also allowed them to feel a part of the study, signifying a greater feeling of empowerment in contributing to the design of the social robot. This shifting of power dynamics promoted joint inquiry and joint imagination in the design process strengthening the design generation process and hopefully leading to social robot designs that will increase user satisfaction in the final design product (Steen, 2013).

3.8.5.4 Future work

This paper focused on how older adults as co-designers designed interactions in a rapid-prototyping session and how they relate to being robot interaction designers. Future work could explore interactions across application areas of focus (i.e., medication adherence, social connection, etc.) and themes relating to specific application areas. The core actions could be further explored to understand if there are patterns for how older adults were designing their interactions such as some people leveraging comments more than others or some people using rewording more than others. Building upon research through design objectives (Zimmerman et al., 2007), future work could further investigate how the co-designers' interaction designs may help researchers further understand unanticipated effects of social robots. Reflecting on the value of the rapid-prototyping session in empowering older adults as co-designers and making them feel valued in the co-design process, we would recommend that in future social robot co-design processes, rapid-prototyping or similar design activities that involve programming robots are incorporated multiple times into long-term partnership with co-designers. Researchers can also explore various structures for rapid-prototyping sessions, having a mixture of structured design activities such as tutorials and more open-ended and specific application area-focused design prompts. Future research could also explore the impact of the robot's tangibility on the co-design experience and how varying robot embodiments may affect and differ between different application areas of robots. Overall, the rapid-prototyping activity was embedded within a year-long co-design process and, as of writing this paper, the researchers and co-designers are still working together. We support future co-design engagements that support sustained engagement with communities (Ostrowski et al., 2021a) and that focus on enablement, or "facilitating and developing opportunities for people to fulfill their potential and to develop their own capacity" (Dombrowski et al., 2016). For HRI, this can be realized as "fostering human capacity or helping people take advantage of opportunities by creating platforms for participation and self-determination" (Dombrowski et al., 2016). Through promoting sustained engagement with communities and democratizing how people design, we can reshape the notion of who can, should, and is a robot interaction designer. We hope that with our work, we have amplified this conversation.

3.8.6 Conclusion

This paper provides an in-depth exploration of older adults as co-designers of robots, revealing how older adults wish social robots to be designed and how older adults can be supported as designers, valuing their built technology experiences. Our dataset and results demonstrate the value of incorporating methods that allow users to design and program social robots and the value of co-designing long-term with users to guide technology design. Overall, we advocate for sustained engagement with communities and users through co-design, participatory design, and human-centered design, empowering users, such as older adults, to be designers of robot interactions, expanding who can be called a technology designer.

Acknowledgments

We extend a gracious thank you to the older adults that engaged in this research with us and co-designed interactions on the social robot. We also thank Prof. Jodi Forlizzi for their contribution and feedback that greatly enriched this work. This work was supported by Samsung Research and the ICT R&D program of MSIP/IITP [2017-0-00162, Development of Human-care Robot Technology for Aging Society].

References

Björgvinsson, E., Ehn, P., & Hillgren, P. A. (2010, November). Participatory design and "democratizing innovation". In *Proceedings of the 11th Biennial participatory design conference* (pp. 41–50).

Björling, E. A., & Rose, E. (2019). Participatory research principles in human-centered design: engaging teens in the co-design of a social robot. *Multimodal Technologies and Interaction*, *3*(1), 8.

Bødker, S., & Kyng, M. (2018). Participatory design that matters—Facing the big issues. *ACM Transactions on Computer-Human Interaction (TOCHI)*, *25*(1), 1–31.

Braa, J. (1996, November). Community-based participatory design in the Third World. In *Proceedings of the conference on participatory design* (pp. 15–24).

Braun, V., & Clarke, V. (2006). Using thematic analysis in psychology. *Qualitative research in psychology*, 3(2), 77–101.

Breazeal, C. L. (2002). *Designing sociable robots*. MIT press.

Bucciarelli, L. L. (1996). *Designing engineers*. Cambridge: MIT Press.

Buchanan, R. (2001). Human dignity and human rights: Thoughts on the principles of human-centered design. *Design Issues*, *17*(3), 35–39.

Coyne, R. (2005). Wicked problems revisited. *Design Studies*, *26*(1), 5–17

Charmaz, K. (2014). *Constructing grounded theory*. sage.

Cross, N. (1972). Design participation. Paper presented at the *International Conference of Design Research Society*, London, UK.

Cross, N. (1999a). Natural intelligence in design. *Design Studies*, *20*(1), 25–39.

Cross, N. (1999b). Design research: A disciplined conversation. *Design Issues*, *15*(2), 5–10.

Daly, S. R., Yilmaz, S., Christian, J. L., Seifert, C. M., & Gonzalez, R. (2012). Design heuristics in engineering concept generation. *The Journal of Engineering Education*, *101*(4), 601–629.

Dombrowski, L. , Harmon, E. , & Fox, S. (2016). Social justice-oriented interaction design: Outlining key design strategies and commitments. In *Proceedings of the 2016 ACM Conference on Designing Interactive Systems*, 656–71. DIS '16. New York, NY, USA: Association for Computing Machinery.

Dorst, K., & Cross, N. (2001). Creativity in the design process: co-evolution of problem–solution. *Design Studies*, *22*(5), 425–437.

Dorst, K. (2019). What Design Can't Do. *She Ji: The Journal of Design, Economics and Innovation*, *5*(4), 357–359.

Ertner, M., Kragelund, A. M., & Malmborg, L. (2010, November). Five enunciations of empowerment in participatory design. In *Proceedings of the 11th Biennial Participatory Design Conference* (pp. 191–194).

Forlizzi, J., & Zimmerman, J. (2013, August). Promoting service design as a core practice in interaction design. In *Proceedings of the 5th International Congress of International Association of Societies of Design Research-IASDR* (Vol. 13).

Harrington, C. N., Wilcox, L., Connelly, K., Rogers, W., & Sanford, J. (2018, May). Designing health and fitness apps with older adults: examining the value of experience-based co-design. In *Proceedings of the 12th EAI International Conference on Pervasive Computing Technologies for Healthcare* (pp. 15–24).

Harrington, C. N., Borgos-Rodriguez, K., & Piper, A. M. (2019, May). Engaging low-income African American older adults in health discussions through community-based design workshops. In *Proceedings of the 2019 chi conference on human factors in computing systems* (pp. 1–15).

Houde, S., & Hill, C. (1997). What do prototypes prototype? In *Handbook of human-computer interaction* (pp. 367–381). North-Holland.

Huggins, M., Ostrowski, A. K., Rapo, A., Woudenberg, E., Breazeal, C., & Park, H. W. (2021). The interaction flow editor: A new human-robot interaction rapid prototyping interface. arXiv.

Jordan, P. W. (2002). Human factors for pleasure seekers. *Design and the social sciences.* CRC Press, 26–40.

Jung, M., & Hinds, P. (2018). Robots in the wild: A time for more robust theories of human-robot interaction. *ACM Transactions on Human-Robot Interactions, 7*(1), 2:1–2:5.

Knowles, B., Hanson, V., Rogers, Y., Piper, A. M., Waycott, J., Davies, N., ... & Yuan, Y. (2021). The harm in conflating aging with accessibility. *Communications of the ACM, 64*(7), 66–71.

Lee, C., & Coughlin, J. F. (2015). PERSPECTIVE: Older adults' adoption of technology: an integrated approach to identifying determinants and barriers. *Journal of Product Innovation Management, 32*(5), 747–759.

Lee, H. R., Šabanović, S., Chang, W. L., Nagata, S., Piatt, J., Bennett, C., & Hakken, D. (2017, March). Steps toward participatory design of social robots: mutual learning with older adults with depression. In *Proceedings of the 2017 ACM/IEEE international conference on human-robot interaction* (pp. 244–253).

Lyotard, J.-F. (1986). *The postmodern condition: A report on knowledge.* Manchester: Manchester University Press.

Ostrowski, A. K., DiPaola, D., Partridge, E., Park, H. W., & Breazeal, C. (2019). Older adults living with social robots: promoting social connectedness in long-term communities. *IEEE Robotics & Automation Magazine, 26*(2), 59–70.

Ostrowski, A. K., Park, H. W., & Breazeal, C. (2020, September). Design Research in HRI: Roboticists, Design Features, and Users as Co-Designers. In *Workshop on Designerly HRI Knowledge.*

Ostrowski, A. K., Breazeal, C., & Park, H. W. (2021a). Long-Term Co-Design Guidelines: Empowering Older Adults as Co-Designers of Social Robots. In *2021 30th IEEE International Conference on Robot & Human Interactive Communication (RO-MAN)* (pp. 1165–1172). IEEE.

Ostrowski, A. K., Harrington, C., Breazeal, C., & Park, H. W. (2021b). Personal Narratives in Technology Design: The Value of Sharing Older Adults' Stories in the Design of Social Robots. *Frontiers in Robotics and AI. In print.*

Patton, B., & Dorst, K. (2011). Briefing and reframing: A situated practice. *Design Studies, 32*(6), 573–587.

Rittel, H. W., & Webber, M. M. (1973). Dilemmas in a general theory of planning. *Policy Sciences, 4*(2), 155–169.

Sanders, E. B.-N., & Stappers, P. J. (2008). Co-creation and the new landscapes of design. *CoDesign, 4*(1), 5–18.

Steen, M. (2011). Tensions in human-centred design. *CoDesign, 7*(1), 45–60.

Steen, M. (2013). Co-design as a process of joint inquiry and imagination. *Design Issues, 29*(2), 16–28.

Ullman, D. G., Wood, S., & Craig, D. (1990). The importance of drawing in the mechanical design process. *Computers & Graphics, 14*(2), 263–274.

Wallach, W., & Allen, C. (2008). *Moral machines: Teaching robots right from wrong.* Oxford University Press.

Zimmerman, J., Forlizzi, J., & Evenson, S. (2007, April). Research through design as a method for interaction design research in HCI. In *Proceedings of the SIGCHI conference on Human factors in computing systems* (pp. 493–502).

Part 4

Design teams of diverse backgrounds, interdisciplinary projects

DOI: 10.1201/b22630-25

4.1 What is happening when designers from different disciplines work together

Characterization of the design behaviors and design spaces of mechanical and electrical engineers working in teams

John Gero and Julie Milovanovic

College of Computing and Informatics, University of North Carolina at Charlotte, USA

4.1.1 Introduction

Design teams are more and more complex as they become more heterogeneous, including designers from different backgrounds. Designers in a team work toward a shared and common goal (McGrath et al., 2000) to address the design task at hand. Compared to individual designing, co-designing is impacted by team members' interactions (Détienne et al., 2012; Valkenburg & Dorst, 1998), at both the design and social levels. Factors such as leadership and expertise (Cross, 2004; Gero & Kannengiesser, 2004), design background (Gero et al., 2014; Zimring & Craig, 2001) or gender (Laeser et al., 2003; Milovanovic & Gero, 2019) can have an effect on how co-designing unfolds in a team. Working in teams can be challenging as team members need to synchronize their ideas toward a common shared concept (Bierhals et al., 2007; Cash et al., 2020). Engineers from different disciplines hold similar yet different sets of design skills (Visser, 2009), that require fine-tuning when co-designing. Teams' heterogeneity related to their functional background can positively impact the team task performance as the team relies on a larger set of skills, cognitive patterns and expertise (Carter et al., 2019; Stewart, 2006). On the other hand, aligning viewpoints and ideas can be challenging in heterogeneous teams, and might delay the achievement of the task (Carter et al., 2019). Semantic coherence within a team tends to affect their creativity and performance positively (Dong et al., 2004) although micro-conflicts among team members can also have a positive effect on team performance (Paletz et al., 2017).

In this study, we explore the effect of team members' discipline on design teams' cognitive behavior, design spaces and semantic topics explored while designing. To do so, we analyze a set of three design sessions of professional engineers grouped in teams of three. Teams were all composed of engineers from two different disciplines: mechanical and electrical. The protocol analysis methodology was used to analyze our protocols (Ericsson & Simon, 1984; Kan & Gero, 2017; Van Someren et al., 1994) with the Function Behavior Structure coding scheme (Gero, 1990; Gero & Kannengiesser, 2014) and a model of co-design (Gero & Milovanovic, 2019) to locate the teams' co-design processes.

Determining the design behaviors and design space exploration of mechanical and electrical engineers working in teams can provide new directions toward developing tools to support co-design by developing designer's awareness of the complementarity of their cognitive processes and skills while designing. Our findings provide a new perspective on design team interactions through co-designing by integrating the relationship between

DOI: 10.1201/b22530-26

design solutions topics within the design space. It relies on an analysis of the semantics of designers' utterances, a common approach in design research (Dong, 2005, 2007), that has not been adequately explored.

4.1.2 Background: multidisciplinary design teams in design studies

Multiple factors influence team performance. In their meta-analysis, Stewart (2006) defined three main categories of factors: group composition, task design and organizational context. Group composition refers to the team size, structure, heterogeneity, and characteristics of team members (gender, personality, age). Task design relates to the task itself, and how the activities within the task are differentiated and integrated. It also includes intra-team organization and management, the team and sub-team autonomy and coordination. The last group of factors, organizational context, includes leadership and supervision through the interface of the team leader and the broader organization, such as a department or the company itself.

The assumption in team composition is that individual abilities grouped together provide more resources to the team. But specific skills from team members should combine in the 'right' way to increase team-level performance. Studies on teams' homogeneity or heterogeneity in relation to team performance provide insights to team composition (Carter et al., 2019; Stewart, 2006). On one hand, heterogeneity should enhance creativity in the teams as diverse skill sets are combined, but on the other hand, homogeneity should support fewer conflicts in the team as it includes similar team members (Stewart, 2006). For creative tasks, such as designing, heterogeneity tends to increase team performance as it requires different expertise (Stewart, 2006). Team members with similar functional background, or discipline expertise, are more likely to identify problems in a similar way, and are less likely to propose a richer set of alternative solutions (Carter et al., 2019). Teams including team members with different functional skill sets will rely on more knowledge, cognitive patterns and cognitive styles, which can benefit solving tasks at the team level (Carter et al., 2019). It could also slow the team process as integrating different ideas and perspectives can be challenging.

In the study presented here, teams are composed of mechanical and electrical engineers. Mechanical and electrical engineers tackle different design problems within the same design task (Riboulet et al., 2005). In practice in the industry, both can work in isolation during a product development, although collaboration beforehand helps avoid long development loops between designers (Riboulet et al., 2005). A case study of product development at Microsoft highlighted that communication and understanding of each engineer's expertise favor the decision-making process, boost innovation and helps keeping the timeline of the project development (Li et al., 2017). For example, for the development of a scrolling wheel for a mouse, the mechanical engineer would focus on the physical component, the electrical engineer would tackle how to transmit data to microcontrollers while software engineers would define how to communicate the data from the scrolling wheel to the computer (Li et al., 2017). Problems are tackled at the team level to decide on possible adjustments dealt with individually by each engineer. A holistic understanding of the product was key to developing better products (Li et al., 2017).

Even though factors such as the task, designers' expertise or disciplines affect designers' approach, designers display similar cognitive behavior while designing (Akin, 2001; Gero et al., 2014; Visser, 2009). Design is a cognitive activity that is not defined through a professional status (Visser, 2009). Designers usually tackle design problems through the relationship between sub-problems or subsystems. The approach to designing can vary depending on the discipline. For instance, designers with industrial or architectural background usually adopt a more intuitive approach to problem solving, whereas engineering disciplines tend to be more analytical and knowledge driven (Akin, 2001; Roozenburg & Cross, 1991; Zimring & Craig, 2001).

When entering the workforce, designers rely on the skills they acquire during their education. Discipline differences are rooted in pedagogical approaches to teaching designing and to professional cultures. Engineering education shapes future engineers' models of designing, technical skills and capabilities for collaboration (Dym et al., 2005). In engineering, capstone design courses were implemented to bring the practical side of engineering into engineering curricula (Froyd et al., 2012). Today, teaching design through studio is common across disciplines, but, learning objectives of project-based courses tend to differ depending on the discipline. For example, industrial design and architecture studio courses tend to be innovation driven compared to mechanical engineering design courses that are performance driven (Goldschmidt et al., 2014). Learning goals set within design courses impact future professionals' approach to designing. Empirical studies in design suggest that these different approaches to designing can lead to exploiting diverse design patterns and processes (Gero & Jiang, 2014; Kan & Gero, 2011). Mechanical and electrical engineers possess different ranges of skills and expertise related to their discipline, yet as designers, they mobilize similar cognitive design processes (Gero et al., 2014). Their approach to designing is similar. In multidisciplinary design teams, they are perceived as system thinkers and problem solvers (Li et al., 2017).

4.1.3 Research question

In this study, mechanical and electrical engineers worked in teams on a design task. We explore the effect of team members' discipline on design teams' cognitive behavior and their exploration of the design space. Through design processes, designers advance in the design activity by structuring the design space based on the design situation and their expertise (Cross, 2004; Gero, 1990). The design space is a representation of the ideas and concepts that designers develop over time to propose a design solution that materializes into a design artifact (i.e., a product, a building, or a service). Disciplinary backgrounds shape technical knowledge that can influence how designers explore and expand their design space. The research questions are as follows:

1 What is the effect of discipline expertise on design teams' cognitive behavior?
2 What is the effect of discipline expertise on semantic topics explored while designing?

Based on findings from prior studies described in section 2, we expect to find differences in the type of concepts each engineer explored and similar design behaviors at a cognitive level. Semantic topics are likely related to each engineer's functional professional skills and should reflect that expertise. Mechanical and electrical engineers' approach to designing is considered similar, compared to other designers (i.e., architects, industrial designers). Hence, we expect participants to display similar cognitive design patterns.

4.1.4 Methodology

4.1.4.1 Data collection

Engineers participating in the study worked at a company providing systems and products for the aerospace and defense industry. Thirty engineers (5 females and 25 males) randomly formed ten teams of three engineers. Most engineers from this group were electrical engineers, and others had backgrounds in mechanical engineering, computer science, and physics. Each team was given the same task, to design a next-generation personal assistant and entertainment system for the year 2025 (see Appendix). They were invited to focus on what this system would be, how this system works and interacts with people, and what the personal assistant and entertainment system would provide to end users. Each team had 60 minutes to define a concept description and sketch it on a white board. All team members were collocated and a research assistant stayed in the room as participants developed their design. The company requested the experiment to be done outside of the work environment for privacy reasons. Each design session was video recorded to be analyzed (Figure 4.1.1). No incentives were given to participants.

This dataset of 10 design teams was collected for a larger research project and was used for another study that looked at the effect of the working environment on design teams' behaviors (Milovanovic et al., 2021). In the exploratory study reported here, we selected a subset of 3 teams composed of mechanical engineers and electrical engineers. The remaining teams from the dataset that were not composed of both mechanical engineers and electrical engineers were not selected as they did not fit the scope of this research. Here the analysis focuses on analyzing results related to participants' engineering discipline, mechanical or electrical. All team members of the three teams were male engineers. All engineers were used to working together as the company uses agile manufacturing and production processes.

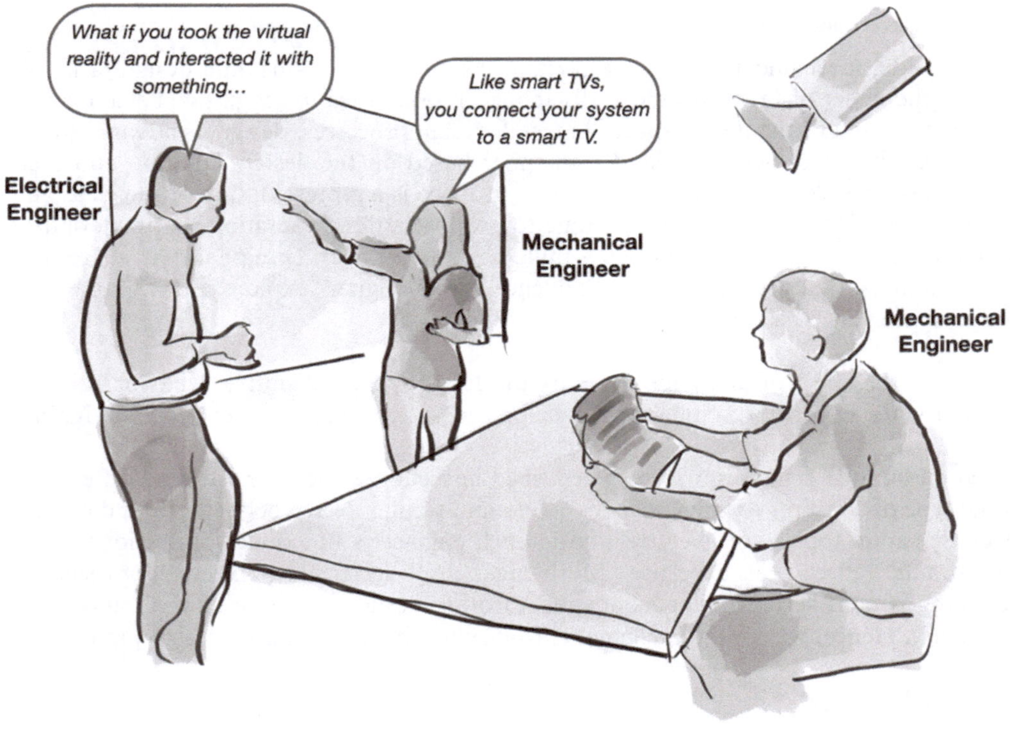

Figure 4.1.1 Experiment setup and organization with examples of discussion between team members. Teams were composed of either two mechanical engineers and one electrical engineer, or one mechanical engineer and two electrical engineers.

Even if team members have not previously been part of the same team in their company, they are used to working in the same organizational environment. This way, although the experiment was carried out in vitro to better control the setting, the interpersonal interactions among members reflected their normal work behavior. All the participants were experts in their domains and had around ten years of expertise in their fields.

4.1.4.2 Describing design processes and collaboration

In this study, we utilized the Function-Behavior-Structure (FBS) ontology to describe design knowledge and processes (Gero, 1990; Gero & Kannengiesser, 2014). Many frameworks have been used in the past decades to describe design processes (Lawson, 2006), for example, the C-K theory (Hatchuel & Weil, 2013), reflective practice (Schön, 1983; Valkenburg & Dorst, 1998) or designers' cognitive actions (Suwa et al., 1998; Suwa & Tversky, 1997). Most integrate concepts related to synthesis, analysis and evaluation as three main design processes (Lawson, 2006). Using the FBS ontology to explore design cognitive process is relevant as its descriptions of function, behavior and structure do not require additional ontological concepts. Moreover, it has been used in empirical studies analysis in diverse domains (Bott & Mesmer, 2019; Hamraz & Clarkson, 2015; Sakao et al., 2020) which supports its reliability to explore design processes. The FBS framework represents six design issues (see Figure 4.1.2):

- Requirement (R) includes the design brief and is outside of the designer
- Function (F) is what the design object is for
- Expected Behavior (Be) represents the expected behavior of the design object
- Structure (S) represents elements and their relationships that go to make up the design object
- Structure Behavior (Bs) is behavior derived from a structure
- Description (D) is an external representation of the design object.

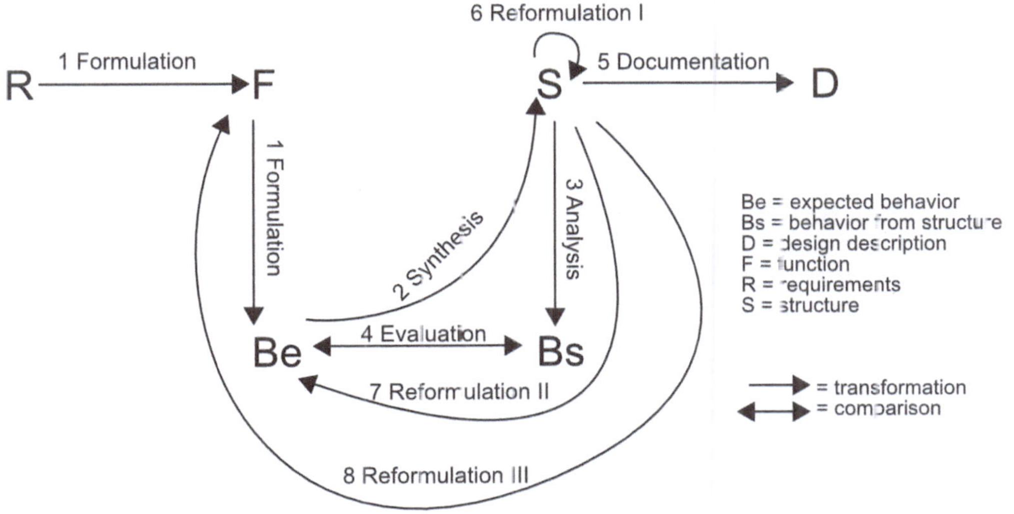

Figure 4.1.2 FBS design knowledge and process framework (Gero, 1990; Gero & Kannengiesser, 2014).

The FBS framework also describes a total of eight cognitive design processes as a consequence of transitions between the six design issues, as shown in Figure 4.1.2:

- Formulation, a transition from a requirement (R) to a function (F), or from a function (F) to an expected behavior (Be)
- Synthesis, a transition from an expected Behavior (Be) to a design structure (S)
- Analysis, a transition from a design structure (S) to a behavior from structure (Bs)
- Evaluation, a transition from an expected behavior (Be) to a behavior from structure (Bs) and inversely
- Documentation, a transition from a design structure (S) to a description (D)
- Reformulation 1, a transition from a design structure (S) to a different design structure (S)
- Reformulation 2, a transition from a design structure (S) to an expected behavior (Be)
- Reformulation 3, a transition from a design structure (S) to a function (F).

4.1.4.3 Coding the design protocols to obtain information on collaboration and design processes

The protocol analysis methodology was applied to analyze the design sessions. Video protocols were transcribed, segmented and coded using the Function-Behavior-Structure framework represented in Figure 4.1.2. In this study, we analyzed collaborative interactions between team members based on their engineering discipline. Each segment was coded with the engineer's discipline, mechanical or electrical. Two teams comprised two mechanical engineers and one electrical engineer whereas the other team was composed of one mechanical engineer and two electrical engineers.

As presented above, FBS design processes are identified as transitions from a specific design issue to another specific design issue (Figure 4.1.2). A process formulated by a single engineer, implies that both design issues forming a design process are verbalized by the same engineer. We identified a co-design process as an FBS design process where a first engineer formulates the first design issue, and the second designer verbalizes the following design issue (Gero & Milovanovic, 2019). With a dual coding scheme, we were able to identify dominant design processes for mechanical engineers, electrical engineers and collaborative processes between mechanical and electrical engineers. Processes and interactions are based on a syntactic model, looking at the activity in a temporally linear manner, not based on semantic associations or turn taking (see Table 4.1.1 for an example of protocol coding).

Each session was coded by two different trained coders. When a disagreement occurred, coders arbitrated each segment together, and relied on an external coder's input if they could not reach an agreement. In total, three coders worked in pairs to code the data. The average coder agreement for all sessions is 80%, which ensures the reliability of the data analyzed.

4.1.4.4 Measuring design space exploration with NLP and networks

Tools from network science provide a new perspective on studying design teams, design tasks and design spaces (Adolphe et al., 2020; Leenders et al., 2016; Parraguez & Maier, 2016). To measure the design space explored by engineers during the design sessions, we created a graph to represent the entire network of concepts produced by all the designers. Concepts were retrieved based on the conversation recorded during the design session.

Table 4.1.1 Example of coded protocol with design processes and interactions

Utterance	*FBS code*	*Design process*	*Designer*	*Interaction*
It's got to be able to connect to	Be	–	Mech. Eng.	–
all the in-home, you know.	S	Synthesis	Mech. Eng.	Mech. Eng. > Mech. Eng.
So, if you have a Wi-Fi. The Wi-Fi stuff ...	S	Reformulation 1	Elec. Eng.	Mech. Eng. > Elec. Eng.
or Bluetooth	S	Reformulation 1	Elec. Eng.	Elec. Eng. > Elec. Eng.
or whatever features	S	Reformulation 1	Elec. Eng.	Elec. Eng. > Elec. Eng.
and be able to network with all of them.	Be	Reformulation 2	Elec. Eng.	Elec. Eng. > Elec. Eng.
And then you got to have, you know, with the TVs	S	Synthesis	Mech. Eng.	Elec. Eng. > Mech. Eng.
and then connecting	Bs	Analysis	Mech. Eng.	Mech. Eng. > Mech. Eng.
to your entertainment things.	S	Synthesis	Mech. Eng.	Mech. Eng. > Mech. Eng.

The graph also accounts for relationships between concepts and their frequency of occurrence. To build the graph, the input data was prepared using a Natural Language Processing (NLP) script in Python (nltk package) and the Gephi software to visualize the graphs (Bastian et al., 2009). We used a python script to clean the design protocols with NLP tokenization. In doing so, punctuation and stop words were removed from the protocol, and we selected tokens tagged as nouns (nltk.pos_tag() function in the nltk package). To create the graph, we considered each token (i.e., "home", "people", "concept", "app") as a node. Edges between nodes are created when both tokens appear in the same protocol segment. The frequency of the co-occurrence of tokens within a FBS segment provides the weight to graph edges. As each segment is assigned to a designer, we were able to track who established a relationship between concepts, a mechanical engineer or an electrical engineer.

Figure 4.1.3 illustrates the network of topics explored by all the engineers during the design session. The network representation chosen is based on a force-directed graph (Fruchterman & Reingold, 1991). The concepts that are highly connected appear in the center of the network while concepts at the outskirt of the network usually connect to only one other concept. Such networks provide a relevant visual tool to better understand the design space explored by engineers based on their discipline. In Figure 4.1.3, blue edges represent connections made by mechanical engineers and pink edges represent connections made by electrical engineers. A node, such as "people", is connected to other nodes with blue and pink edges, meaning that both groups of engineers mentioned "people" but with different connections to other concepts. Nodes connected to other nodes by edges of a single color imply that these topics were covered by only one type of engineer. Node size represents the node degree (number of connections with other nodes) while edge width accounts for co-occurrence frequency of concepts.

Graph theory provides topological measures for networks that we can apply and interpret in studying the exploration of design spaces. Here, we looked at network

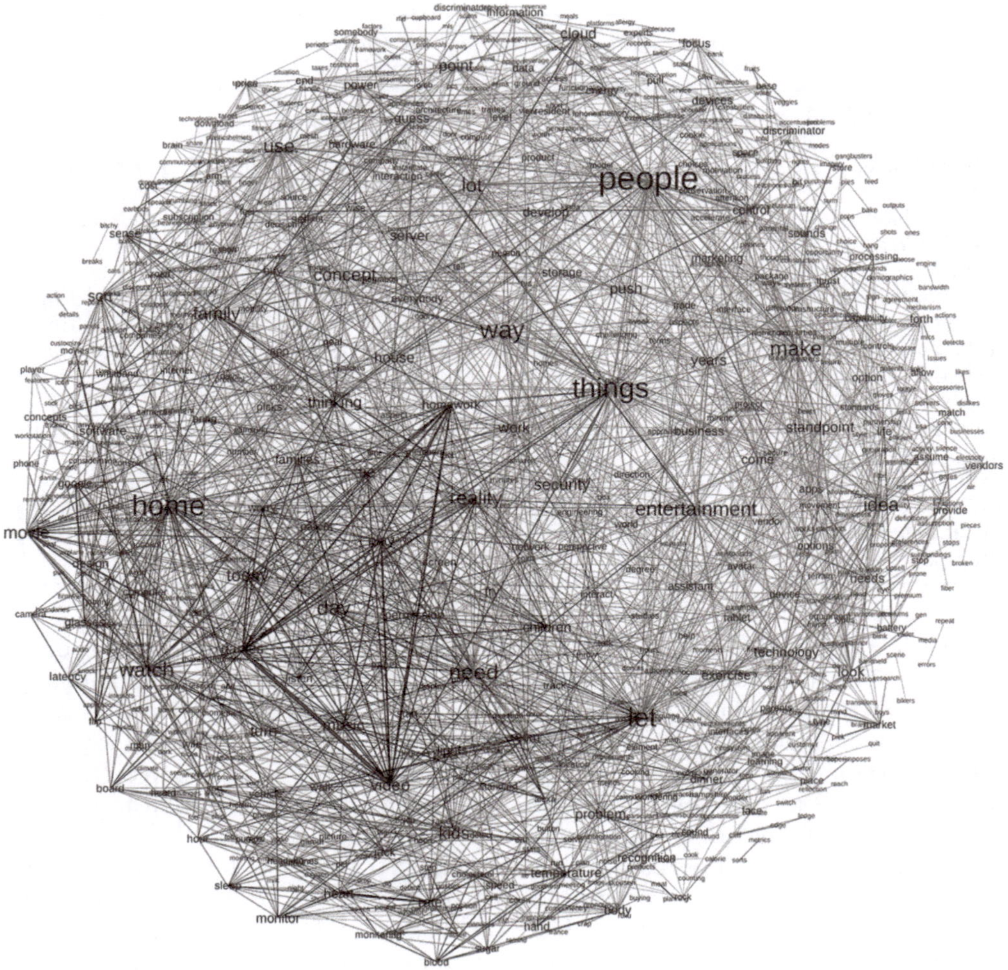

Figure 4.1.3 Graph example of the entire design space represented by the concepts verbalized by all the engineers in all the teams designing a personal entertainment system. Blue edges represent connections made by mechanical engineers and pink edges represent connections made by electrical engineers.

density and modules within the graph. Network density is the ratio of the number of edges within the network to the number of possible edges in the network. It provides an idea of how well nodes are connected to each other, in this case, how well concepts are connected to each other. Modularity provides an estimation of how a network breaks into smaller sub-networks or communities (Newman, 2006). Using Gephi's clustering function for modularity (Louvain method), we are able to define modules within the network that correspond to a chunk of the design space explored. Modules have dense connections between the nodes within the module and sparse connections with nodes in the other modules. Modules are "small worlds" in social networks. Defining such modules can help us determine groups of topics where engineers collaborate or if some modules are discipline-specific.

4.1.5 Results

4.1.5.1 Designers display different design behaviors based on discipline

The results indicate differences in the cognitive effort expended by engineers in designing, based on their engineering discipline. The average distribution of FBS design processes for all three sessions, for electrical engineers, mechanical engineers and the collaborative processes between them is presented in Figure 4.1.4. Electrical engineers tended to put more effort into analysis compared to mechanical engineers. They spent 38.6% of their cognitive effort on Analysis. Mechanical engineers put more emphasis on reformulating design solutions, Reformulation 1 design process, as such a process represents 43.2% of all their FBS design processes. The distribution of Evaluation design processes is twice as large for collaborative design processes between mechanical and electrical engineers than for discipline-specific design processes. For other processes like Synthesis and Reformulation 2, the distribution of processes is similar for all types of interactions.

4.1.5.2 Designers explore different design topics in the design space based on discipline

The network of concepts is built from the segmented protocols as described in the methodology section. In total, 2,536 segments were used as input (1,115 for mechanical engineers and 1,421 for electrical engineers). Nodes represent concepts discussed by engineers during the design session while edges show the relationship between concepts extracted from the protocol analysis. In order to visualize highly connected topics, we used a threshold based on node degree to select nodes with a higher node degree than the average. Since, on average, nodes have 15 connections to other nodes, the threshold was set at 15, and only nodes with 15 or more connections are represented in the networks shown in Figure 4.1.5. Figure 4.1.5 illustrates subgraphs of dominant concepts for mechanical engineers (Figure 4.1.5(a)) and electrical engineers (Figure 4.1.5(b)). The mechanical engineers' subgraph comprises fewer nodes (393 nodes) than the electrical engineers' subgraph (602 nodes). However, the mechanical engineers' subgraph has a higher network density (0.027), than the electrical engineers' subgraph (0.018). This

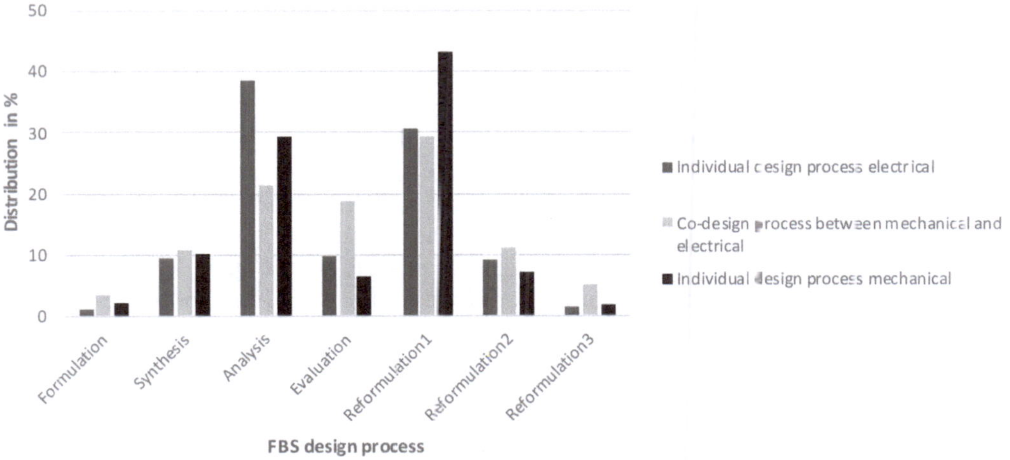

Figure 4.1.4 Average distribution of individual and collaborative design processes based on design discipline.

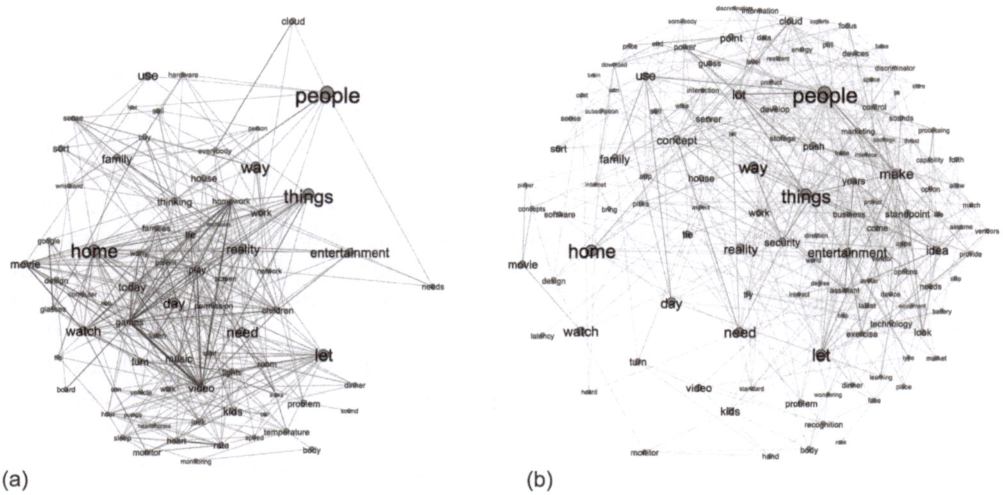

Figure 4.1.5 Subgraphs representing relationships between concepts explored by engineers based on their discipline: (a) mechanical engineers and (b) electrical engineers. A threshold based on node degree was applied to highlight prevalent concepts explored by engineers. The node degree threshold of 15 is selected, which is the average node degree for the entire network of concepts.

suggests that mechanical engineers explored fewer concepts compared to electrical engineers but made more connections between the concepts explored.

The analysis of the design space for the design teams shows some overlap of the topics that designers focused on. For instance, "people", "home", "entertainment" and "need" are key nodes in each sub-network (Figure 4.1.5). Each cohort also discussed topics specific to their discipline. For example, mechanical engineers explored concepts related to the category of users (e.g., "parent", "family", "children") and user wearables (e.g., "glasses", "watch", "wristband", "monitoring" "heart rate" and "sleep") (Figure 4.1.3(a)). On the other hand, electrical engineers expended cognitive effort in exploring product-related concepts (e.g., "marketing", "package", "energy", "processing").

In the second part of the analysis, we analyzed mechanical engineers' and electrical engineers' dynamics while focusing on clusters of concepts. Using Gephi's clustering tool with network modularity, five main modules were identified (Figure 4.1.6). Those five modules represent 90.5% of all the nodes in the network. In most modules, both mechanical engineers (pink edges) and electrical engineers (blue edges) made connections between nodes (see Figures 4.1.6(a), (b) and (c)). For two of the modules, we obverse the dominance of electrical engineers' connections (blue edges in module 4 in Figure 4.1.6(d)) or the dominance of mechanical engineers' connections (pink edges in module 5 in Figure 4.1.6(e).

Module 1, representing 40.1% of the network of concepts, relates to the end users with concepts like "people", "families", "need", "use", "assistant", "avatar" and more. Module 2 representing 20.2% of the network of concepts describes the product in context with topics like the "home", "cameras", "monitor" for a specific usage related to measuring data ("heart rate", "sleep", "temperature"). Module 3, representing 11.6% of the network of concepts, focuses on a function of the product's technological aspect that deals with "technology", a "server", "internet" and a "cloud". For all these modules, mechanical

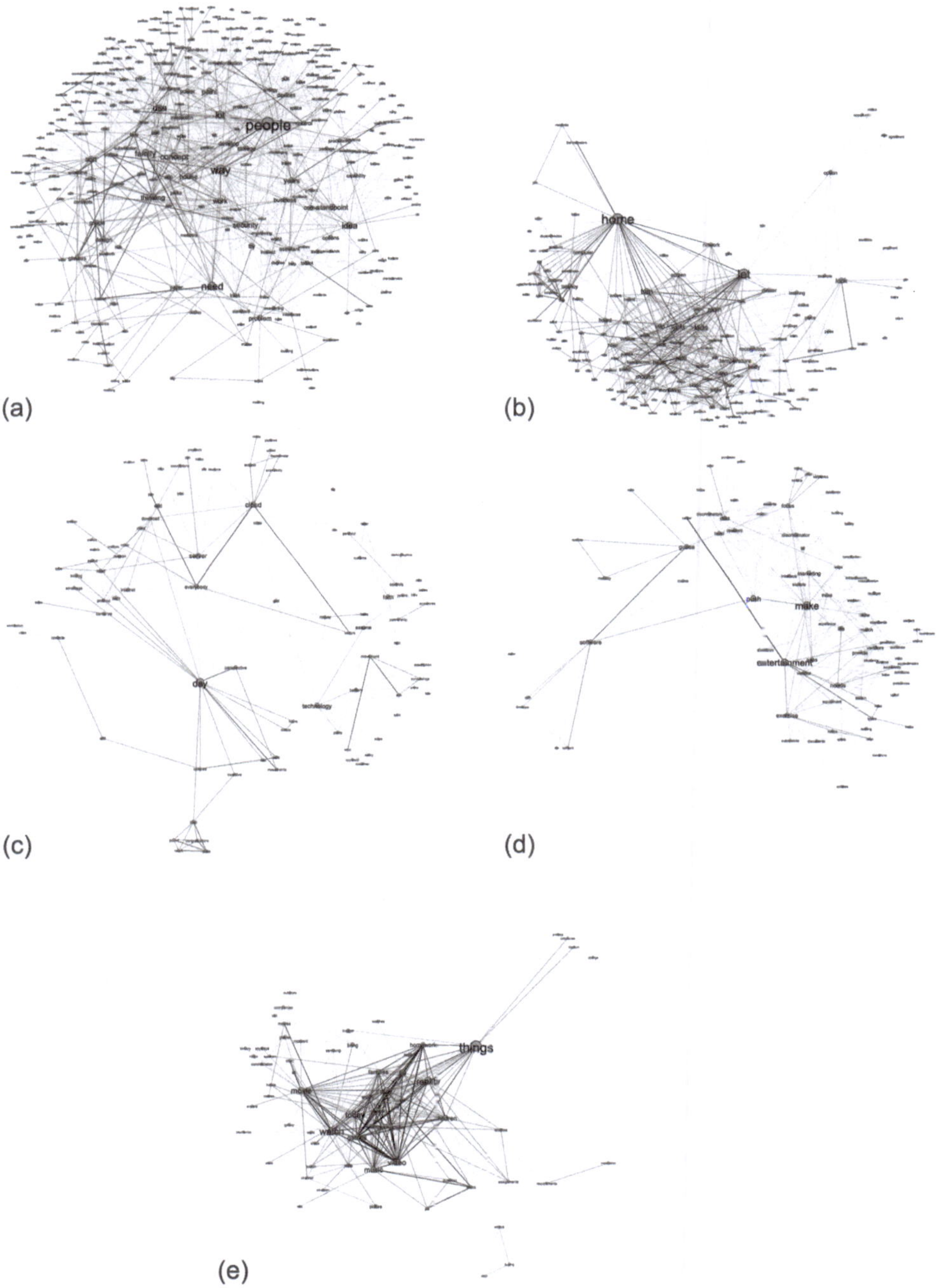

Figure 4.1.6 Subgraphs representing relationship between concepts explored by engineers within modules: (a) module 1 representing 40.1% of topics, (b) module 2 representing 20.2% of topics, (c) module 3 representing 11.6% of topics, (d) module 4 representing 10.7% of topics, (e) module 5 representing 8.0% of topics. Edges in pink represent relationships between concepts formulated by mechanical engineers. Edges in blue represents relationships between concepts formulated by electrical engineers.

engineers and electrical engineers brought complementary ideas, using similar ideas (nodes) and connecting (edges) them to different concepts.

Module 4, representing 10.7% of the network of concepts, is dominated by blue edges illustrating connections made by electrical engineers. In this module, the focus is on one of the requirements for the product "entertainment". It relates to some uses for the product like "exercise" to the components of the product, for instance, the "apps", the "software", "data". The last module, representing 8.0% of the network of topics, is dominated by pink edges showing connections established by mechanical engineers. In this module, the concepts represent features offer by the product designed, as "games", "music", "video", "movie", "homework" that connects to users ("families" and "children").

4.1.6 Discussion

In this study, we explored whether the disciplines of team members impact design behaviors, design spaces and concept exploration for teams of electrical and mechanical engineers. We observed differences in design behaviors related to disciplinary background. We also found that the part of the design space covered changes based on designers' discipline, but designers tend to collaborate on topics related to the product's end users, the product's use in context and its technology.

The divergences in the distribution of design processes based on discipline highlights slightly diverse cognitive design styles. Mechanical engineers put more cognitive effort into Reformulation 1 processes, whereas electrical engineers put more cognitive effort into Analysis. Both types of processes focus on the design solution itself. While Analysis infers behaviors from design structures, Reformulation 1 processes imply rethinking design structures. Both types of designers display a design behavior focused on the solution, which agrees with previous empirical findings (Jiang et al., 2014). Mechanical and electrical engineers are both defined as analytical and system thinkers (Akin, 2001; Li et al., 2017) which provides an explanation for the dominance of Analysis and Reformulation 1 design processes. When collaborating, Evaluation processes become dominant compared to individual design processes. This finding supports the relevance of collaboration between disciplines as it triggers more evaluation between the current solution and design expectations. In the Microsoft project case study, communication between engineers from different disciplines was key when identifying problems and laying out possible adjustments for the product solution (Li et al., 2017). Navigation between problems and solutions is at the core of the Evaluation process, which echoes these previous findings on multidisciplinary team members' communication while designing.

We expected few differences in design cognition based on discipline as both mechanical and electrical engineers tend to hold a similar approach to problem solving. Looking at the overall distribution of design processes, the data analysis provides indications of similar design behaviors although some divergence appeared as well. Interestingly, design behaviors were different for interdisciplinary co-design processes. Collaboration tends to co-opt a larger array of design patterns designers engaged in during the design activity, which could lead to a richer outcome in terms of the design artifact.

Exploring the use of NLP and network science to study the extent of the design space explored depending on the discipline helped visualize the functional expertise of each type of designer. The design processes analyzed and discussed above act as a vehicle for concrete elements discussed by the participants. Mechanical engineers and electrical engineers explored complementary parts of the design space while designing. Some concepts overlapped among engineers including "people", "home", "entertainment" and "need". Concepts specific to disciplines were more user related for mechanical engineers and more product related for electrical engineers. When looking at clusters of concepts based on network modules, we observed that the three main modules (covering the larger number of concepts) contained a high participation from both types of engineers. It suggests that for this task, multidisciplinary collaboration helped cover more of the design space within modules. Those modules dealt respectively with the end users, the product in the context of usage, and technological aspects of the product.

As more disciplines come into design, combining expertise is key to improving the design process and its outcomes. This study looked at two disciplines and could be applied to any number of disciplines in a team. Revealing how designers interact and the amount of design space they cover could benefit team organization and management. Engineers bring specialist knowledge to interdisciplinary teams, hence the need to put effort into communicating their design language and context knowledge to other designers from different backgrounds. Such efforts could enrich the design artifact produced. In this study, we showed that the design space covered was extended based on the collaboration between engineering disciplines. We could argue that covering more of the design space could lead to more design alternatives and a better end product.

In this research, we focused on cognitive processes of the design teams but other factors such as social interactions impact how the design activity unfolds in a team. Teams can divide into sub-teams based on different characteristics. Sub-teams' emergence within teams is shaped by fault-lines, hypothetical dividing lines based on attributes of diversity, that are embedded in the situational and social context of the team environment (Carton & Cummings, 2012; Thatcher & Patel, 2012). In our dataset, sub-teams' organization around discipline was not dominant as team members collaborated on multiple topics (see modules of the concept network in Figure 4.1.6). The social organization of the team through leadership, project and conflict management can impact how the team progresses in their design activity. In future work, we plan to integrate such factors into our analysis.

In this study, one limitation is that the performance of the artifact is not evaluated. In our future work, we will include such metrics to better understand the relationship between the amount of design space explored and the design artifact quality. More metrics can be added and derived from the topics network. Through analyzing concepts that are first occurrences (first concept for one topic) in the protocol data (Gero & Kan, 2016), we can define the size of the design space and the semantic distance between concepts. Such metrics can provide a proxy to measure divergent thinking and creativity, which could complement subjective outcome evaluation, for instance with the Consensual Assessment Technique (Amabile, 1982).

4.1.7 Conclusion

Working in multidisciplinary design teams has become the norm with the increasing scale and complexity of designs. It requires a synchronization between designers from different discipline backgrounds, which can be challenging. The potential of interdisciplinary design teams is to increase performance by designing better products more effectively and efficiently. In this study, we analyzed the design behaviors of teams composed of mechanical and electrical engineers. Using protocol analysis, we were able to determine similar and different design behaviors between mechanical and electrical engineers. More importantly, when codesigning, engineers from different disciplines engaged in processes that were not dominant for discipline-specific processes. The topics covered within the design space were extended based on discipline-specific context knowledge. The richness of design processes and potential for design alternatives in these design teams show the relevance of multidisciplinary teams.

This study was a testbed to integrate new design cognition metrics using NLP and network science. The tools used provide an automated way to study semantic relationships within design protocols. Network representations based on word co-occurrences provide an additional dimension of design thinking analysis. Networks representing relationships between ideas and concepts while codesigning could help the design team in their process (Gyory et al., 2021). Other forms of expression embedded within the design activity such as gestures (Kang & Tversky, 2016), physiology and neurocognition (Borgianni & Maccioni, 2020; Gero & Milovanovic, 2020) are additional area of focus in order to broaden the characteristics of designing that are studied. Considering all these characteristics in design research will enrich our understanding of design cognition and can benefit team design through developing real-time tools to assist designers.

Acknowledgments

This material is based upon work supported by the National Science Foundation under Grant Nos. 1761774 and 1762415. Any opinions, findings, and conclusions or recommendations expressed in this material are those of the authors and do not necessarily reflect the views of the National Science Foundation.

We thank Dr. Yuzhen Luo and Mohammad Al Mestiraihi for their assistance in the coding of the design protocols.

References

Adolphe, L., Van de Zande, G., Wallace, D., & Olechowski, A. (2020). Analysis of virtual communication within engineering design teams and its impact on team effectiveness. *ASME IDETC/CIE 2020*.

Akin, O. (2001). Variants in design cognition. In C. Eastman, M. McCracken, & W. Newstetter (eds.), *Design knowing and learning: Cognition in design education*. Elsevier.

Amabile, T. M. (1982). The social psychology of creativity: A consensual assessment technique. *Journal of Personality and Social Psychology*, 43(5), 997–1013.

Bastian, M., Heymann, S., & Jacomy, M. (2009). Gephi: An open source software for exploring and manipulating networks. *International AAAI Conference on Weblogs and Social* Media.

Bierhals, R., Schuster, I., Kohler, P., & Badke-Schaub, P. (2007). Shared mental models—Linking team cognition and performance. *CoDesign*, 3(1), 75–94. 10.1080/15710880601170891

Borgianni, Y., & Maccioni, L. (2020). Review of the use of neurophysiological and biometric measures in experimental design research. *Artificial Intelligence for Engineering Design, Analysis and Manufacturing*, 1–38. 10.1017/S0890060420000062

Bott, M., & Mesmer, B. (2019). Determination of Function-Behavior-Structure model transition probabilities from real-world data. *AIAA Scitech 2019 Forum*, San Diego, California.

Carter, K. M., Mead, B. A., Stewart, G. L., Nielsen, J. D., & Solimeo, S. L. (2019). Reviewing work team design characteristics across industries: Combining meta-analysis and comprehensive synthesis. *Small Group Research*, 50(1), 138–188. 10.1177/1046496418797431

Carton, A. M., & Cummings, J. N. (2012). A theory of subgroups in work teams. *Academy of Management Review*, 37(3), 441–470. 10.5465/amr.2009.0322

Cash, P., Dekoninck, E., & Ahmed-Kristensen, S. (2020). Work with the beat: How dynamic patterns in team processes affect shared understanding. *Design Studies*, 69, 1–39. 10.1016/j.destud.2020.04.003

Cross, N. (2004). Expertise in design: An overview. *Design Studies*, 25(5), 427–441. 10.1016/j.destud.2004.06.002

Détienne, F., Baker, M., & Burkhardt, J.-M. (2012). Quality of collaboration in design meetings: Methodological reflections. *CoDesign*, 8(4), 247–261. 10.1080/15710882.2012.729063

Dong, A. (2005). The latent semantic approach to studying design team communication. *Design Studies*, 26(5), 445–461. 10.1016/j.destud.2004.10.003

Dong, A., Hill, A. W., & Agogino, A. M. (2004). A document analysis method for characterizing design team performance. *Journal of Mechanical Design*, 126(3), 378–385. 10.1115/1.1711818

Dym, C. L., Agogino, A. M., Eris, O., Frey, D. D., & Leifer, L. J. (2005). Engineering design thinking, teaching, and learning. *Journal of Engineering Education*, 94(1), 103–120. 10.1002/j.2168-9830.2005.tb00832.x

Ericsson, K. A., & Simon, A. H. (1984). *Protocol analysis: Verbal reports as data*. MIT Press.

Froyd, J. E., Wankat, P. C., & Smith, K. A. (2012). Five major shifts in 100 years of engineering education. *Proceedings of the IEEE, 100* (Special Centennial Issue), 1344–1360. 10 1109/JPROC.2012.2190167

Fruchterman, T. M. J., & Reingold, E. M. (1991). Graph drawing by force-directed placement. *Journal of Software: Practice and Experience*, 21(11), 1129–1164. 10.1002/spe.4380211102

Gero, J. S., & Jiang, H. (2014). Comparing the design cognition of concept design reviews of industrial and mechanical engineering designers. *DTRS8: Interpreting Design Thinking*, DAB documents, Sydney, pp. 187–198.

Gero, J. S. (1990). Design prototypes: A knowledge representation schema for design. *AI Magazine*, 11(4), 26–36. 10.1609/aimag.v11i4.854

Gero, J. S., & Kan, J. T. (2016). Empirical results from measuring design creativity: Use of an augmented coding scheme in protocol analysis. *The Fourth International Conference on Design Creativity*, Atlanta, GA, USA, pp. 1–8.

Gero, J. S., & Kannengiesser, U. (2004). Modelling expertise of temporary design teams. *Journal of Design Research*, 4(2), 174–184. 10.1016/j.destud.2003.10.010

Gero, J. S., & Kannengiesser, U. (2014). The Function-Behaviour-Structure ontology of design. In A. Chakrabarti & L. T. M. Blessing (eds.), *An anthology of theories and models of design* (pp. 263–283). Springer London. 10.1007/978-1-4471-6338-1_13

Gero, J. S., Kannengiesser, U., & Pourmohamadi, M. (2014). Commonalities across designing: Empirical results. In *Design Computing and Cognition '12* (pp. 265–281). Springer, Dordrecht. 10.1007/978-94-017-9112-0_15

Gero, J. S., & Milovanovic, J. (2019). The situated function-behavior-structure co-design model *CoDesign*. 10.1080/15710882.2019.1654524

Gero, J. S., & Milovanovic, J. (2020). A framework for studying design thinking through measuring designers' minds, bodies and brains. *Design Science*, 6(e19). 10.1017/dsj.2020.15

Goldschmidt, G., Casakin, H., Avidan, Y., & Ronen, O. (2014). Three studio critiquing cultures: Fun follows function or function follows fun? In Adams, R. S. & Siddiqui, J. (eds.), *Analyzing design review conversations*. (pp. 457–479). Purdue University Press, West Lafayette.

Gyory, J. T., Kotovsky, K., & Cagan, J. (2021). The influence of process management: Uncovering the impact of real-time managerial interventions via a topic modeling approach. *Journal of Mechanical Design*, 143(11), 111401. 10.1115/1.4050748

Hamraz, B., & Clarkson, P. J. (2015). Industrial evaluation of FBS Linkage – a method to support engineering change management. *Journal of Engineering Design*, 26(1–3), 24–47. 10.1080/09544828.2015.1015783

Hatchuel, A., & Weil, B. (2013). A new approach to innovative design: An introduction to C-K theory. *Proceedings of ICED 03*, the 14th International Conference on Engineering Design, 15.

Jiang, H., Gero, J. S., & Yen, C.-C. (2014). Exploring designing styles using a problem–solution division. In J. S. Gero (ed.), *Design Computing and Cognition'12* (pp. 79–94). Springer.

Kan, J. W., & Gero, J. S. (2011). Comparing designing across different domains: An exploratory case study. In S. Culley, B. Hicks, T. McAloone, T. Howard, & Y. Reich (eds), *Design Theory and Methodology* (pp. 2:194–203). Design Society, Glasgow.

Kan, J. W., & Gero, J. S. (2017). *Quantitative methods for studying design protocols*. Springer Netherlands. 10.1007/978-94-024-0984-0

Kang, S., & Tversky, B. (2016). From hands to minds: Gestures promote understanding. *Cognitive Research: Principles and Implications*, 1. 10.1186/s41235-016-0004-9

Laeser, M., Moskal, B. M., Knecht, R., & Lasich, D. (2003). Engineering design: Examining the impact of gender and the team's gender composition. *Journal of Engineering Education*, 92(1), 49–56. 10.1002/j.2168-9830.2003.tb00737.x

Lawson, B. (2006). *How designers think: The design process demystified* (4. ed). Elsevier/ Architectural Press.

Leenders, R. Th. A. J., Contractor, N. S., & DeChurch, L. A. (2016). Once upon a time: Understanding team processes as relational event networks. *Organizational Psychology Review*, 6(1), 92–115. 10.1177/2041386615578312

Li, P. L., Ko, A. J., & Begel, A. (2017). Cross-disciplinary perspectives on collaborations with software engineers. *2017 IEEE/ACM 10th International Workshop on Cooperative and Human Aspects of Software Engineering* (CHASE), 2–8. 10.1109/CHASE.2017.3

McGrath, J. E., Arrow, H., & Berdahl, J. L. (2000). The study of groups: Past, present, and future. *Personality and Social Psychology Review*, 4(1), 95–105. 10.1207/S15327957PSPR0401_8

Milovanovic, J., & Gero, J. (2019). Exploration of gender diversity effects on design team dynamics. In Y. Eriksson and K. Paetzold (eds), *Human Behavior in Design* (pp. 101–112). UniBw M, Germany.

Milovanovic, J., Gero, J. S., & Becker, K. (2021). Does it matter where design teams come from in design studies? *International Design Engineering Technical Conference & Computers and Information in Engineering Conference*, ASME-IDETC paper DETC2021-70432.

Newman, M. E. J. (2006). Modularity and community structure in networks. *Proceedings of the National Academy of Sciences*, 103(23), 8577–8582. 10.1073/pnas.0601602103

Paletz, S. B. F., Chan, J., & Schunn, C. D. (2017). The dynamics of micro-conflicts and uncertainty in successful and unsuccessful design teams. *Design Studies*, 50, 39–69. 10.1016/j.destud.2017.02.002

Parraguez, P., & Maier, A. (2016). Using network science to support design research: From counting to connecting. In P. Cash, T. Stanković, & M. Štorga (eds.), *Experimental Design Research* (pp. 153–172). Springer International Publishing. 10.1007/978-3-319-33781-4_9

Riboulet, V., Marin, P. R., Gowers, S., & Wurtz, F. (2005). A framework supporting collaborative optimisation for multi professional design teams. In A. Bramley, D. Brissaud, D. Coutellier, & C. McMahon (eds.), *Advances in Integrated Design and Manufacturing in Mechanical Engineering* (pp. 123–134). Springer-Verlag. 10.1007/1-4020-3482-2_10

Roozenburg, N. F. M., & Cross, N. G. (1991). Models of the design process: Integrating across the disciplines. *Design Studies*, 12(4), 215–220. 10.1016/0142-694X(91)90034-T

Sakao, T., Gero, J., & Mizuyama, H. (2020). Analyzing cognitive processes of a product/service-system design session using protocol analysis. *Artificial Intelligence for Engineering Design, Analysis and Manufacturing*, 1–16. 10.1017/S0890060420000402

Schön, D. (1983). *The reflective practitioner: How professionals think in action*. Temple Smith.

Stewart, G. L. (2006). A meta-analytic review of relationships between team design features and team performance. *Journal of Management*, 32(1), 29–55. 10.1177/0149206305277792

Suwa, M., Purcell, T., & Gero, J. S. (1998). Macroscopic analysis of design processes based on a scheme for coding designers' cognitive actions. *Design Studies*, 19, 455–483.

Suwa, M., & Tversky, B. (1997). What do architects and students perceive in their design sketches? A protocol analysis. *Design Studies*, 18, 385–403.

Thatcher, S. M. B., & Patel, P. C. (2012). Group faultlines: A review, integration, and guide to future research. *Journal of Management*, 38(4), 969–1009. 10.1177/0149206311426187

Valkenburg, R., & Dorst, K. (1998). The reflective practice of design teams. *Design Studies*, 19(3), 249–271. 10.1016/S0142-694X(98)00011-8

Van Someren, M. W., Barnard, Y. F., & Sandberg, J. A. C. (1994). *The think aloud method: A practical guide to modelling cognitive processes*. Academic Press.

Visser, W. (2009). Design: One, but in different forms. *Design Studies*, 30(3), 187–223. 10.1016/j.destud.2008.11.004

Zimring, C., & Craig, D. L. (2001). Defining design between domains: An argument for design research à la carte. In C. Eastman, M. McCracken, & W. Newstetter (eds.), *Design knowing and learning: Cognition in design education* (pp. 125–146). Elsevier.

Appendix

Personal Entertainment Systems (PES) is one of the most comprehensive entertainment companies in the world. In order to keep its leading position in the industry of entertainment, PES cooperates with many agents to explore the possibilities of new types of entertainment. Your design team has been invited to help in designing the next generation of a personal assistant and entertainment system suitable for family use in the year 2025.

Concept Design

In the context of engineering, a characteristic feature of the product design-related function is the description of products. Concept design includes a thorough roadmap from concept generation to production to product launch. See figure below:

The aim of concept design is to prepare for concurrent engineering by specifying the fundamental solution to the design problem.

Task

Your team is tasked with producing concept designs of a personal assistant and entertainment system suitable for family use for the year 2025.

For this project, your team should focus on:

- what this system would be,
- how this system works and interacts with people, and
- what the personal assistant and entertainment system would provide.

Your goal is to produce a number of concepts and then develop one of those concepts into a detailed design. At the completion of the session, please present sketches (using the whiteboard) and a verbal description of your solution. Your team will have 60 minutes to complete this task.

4.2 Teamwork quality in multi-disciplinary design

Extending the frontiers of design in education

Hernan Casakin[a] *and Gaetano Cascini*[b]
[a]*School of Architecture, Ariel University, Israel;* [b]*Department of Mechanical Engineering, Politecnico di Milano, Milano, Italy*

4.2.1 Introduction

The merits of teamwork have captured much attention in social psychology and new product development (NPD) literature (e.g., Barczak & Wilemon, 2003). The importance of teamwork has also been extensively debated in design (e.g., Badke-Schaub et al., 2007; Casakin et al., 2015). In this domain, team composition regarding disciplinary background has gained vital significance due to the complex nature of the 21st century's global challenges (Ou et al., 2022; Gruenther et al., 2009). Today, society must tackle problems of unprecedented scope that contest established approaches and solutions. To deal with this situation, a multi-disciplinary collaboration offers a more comprehensive approach than that of teams composed of members with similar backgrounds, also known as mono-disciplinary teams.

In this context, trans-passing the traditional borders of scientific disciplines, including those of the design domain, turns out to be a crucial and necessary step. Bearing in mind the increasing complexity of design problems, such transition requires the development of dedicated skills (Blessing et al., 2021); hence the attention to this issue in design education. Among the first initiatives in this direction, design studio courses aim at fostering multi-disciplinary project collaborations (De Vere et al., 2010; Dederichs et al., 2011). Courses like these can play a fundamental role in extending the frontiers of design. Nonetheless, there is an apparent lack of appropriate educational reference practices to encourage the development of multi- and trans-disciplinary skills and competencies (Cascini, 2015).

The disciplinary composition of a team can largely affect interactions in collaborative approaches, with consequences for teamwork quality. This notion was investigated by focusing on social-related and task-related aspects (Hoegl & Gemuenden, 2001). For example, Keating and associates (2013) studied the impact of multi-disciplinary team collaboration on educational outcomes, showing that the most influential issues were related to interpersonal relationships, personal expectations, as well as to team dynamics and management.

While studies about multi-disciplinary team collaboration can be found in the design literature, most of them focus mainly on design performance (e.g., Nesterkin et al., 2016). Moreover, none investigated multi-disciplinary collaborations from a teamwork quality view or compared them with mono-disciplinary teams. Hence, no related studies can be found in design or in design education. Therefore, this study – which is contextualised in higher education in the framework of project-based design studio – deals with the following goals: The first goal is to identify the social- and task-related aspects of multi-disciplinary and mono-disciplinary design team work. The second goal is to investigate

DOI: 10.1201/b22630-27

how teamwork quality from the perspective of social-related and task-related dimensions might differ when comparing multi-disciplinary with mono-disciplinary teams. Based on these, implications for the extension of the frontiers of design education are presented.

The research questions that will be addressed are:

- RQ1: What are the comparative social-related and task-related dimensions of multi-disciplinary and mono-disciplinary design teamwork?
- RQ2: How does multi-disciplinary compared to mono-disciplinary team composition influence the quality of design teamwork in terms of social-related and task-related dimensions?
- RQ3: How can an analysis of teamwork quality inform the debate about extending the frontiers of design in the context of higher education?

Our work revealed that although working in multi-disciplinary teams enables a more comprehensive understanding of the design task, it requires capabilities and competencies different from those of mono-disciplinary teams, which may challenge the quality of team dynamics and collaboration. Furthermore, working in multi-disciplinary teams can result in partly under exploiting specialist knowledge and competencies. It is proposed that these issues should be taken into consideration by both educators and practitioners, as well as by those interested in extending the frontiers of design. Implications for teamwork in design education, and particularly for multi-disciplinary design studio courses, are discussed.

4.2.2 Background

Designing has become a more complex activity than it used to be, and consequently, information and knowledge other than from a single discipline are necessary to tackle problems (Gero & Milanovic, 2022). Design practice must deal increasingly with objects at different scales, from small industrial artefacts to cities, services and organisations (Leblanc, 2021). Hence, a broader approach of not only technological but environmental, scientific, social, and economic subjects is also necessary to cope with the new challenges that design poses nowadays (Blessing et al., 2021). Understanding and tackling such multifaceted phenomena demands more developed and interrelated knowledge, skills and expertise.

In higher education, the increasing interconnectedness among academic disciplines and the vast knowledge originating from the application of a variety of subjects has transcended specific academic fields (Kozmetsky, 1997). Consequently, the university started to be seen as the educational environment where to nurture ideas that may go beyond the frontiers imposed by the current and rather conservative organisation of the fields (Ertas et al., 2003). Such a conception is gaining interest in design courses and is influencing design teachers to go beyond a mono-discipline educational approach to center on domains that step ahead of traditional learning areas (Gericke et al., 2021). This is partly due to the rich knowledge needed to deal with current design tasks, even in lab environments such as the design studio. Moreover, solving new and unique design problems demands the involvement of a spectrum of disciplines working in collaboration. However, there is a substantial lack of truly multi-disciplinary design educational programs and in most cases, these are not standard programs but reserved to honors courses. As part of an integrative pedagogical approach, teamwork can help rethink how design can be taught through the different disciplines in higher education.

4.2.2.1 Multi-disciplinarity, inter-disciplinarity and trans-disciplinarity

Three terms are being used in reference to the type of cooperation between different disciplines: multi-disciplinarity, inter-disciplinarity and trans-disciplinarity. In spite of their differences, they are frequently and mistakenly used interchangeably (Gericke et al., 2021). Tharp and Zalewski (2001) defined multi-disciplinarity as a topic of research shared by many disciplines. Multi-disciplinarity is also related to studying a subject in multiple disciplines simultaneously (Nicolescu, 2005). The aim is to gain a large and shared understanding of phenomena by combining the different standpoints of the disciplines. While multi-disciplinarity extends disciplinary boundaries, it is achieved mainly in the framework and for the benefit of a specific domain. Inter-disciplinarity, on the other hand, is about transferring methods from one discipline to another. Like multi-disciplinarity, inter-disciplinarity transcends disciplines but with the goal of creating new disciplines (Nicolescu, 2005). Whereas trans-disciplinarity includes multi- and inter-disciplinarity approaches, its primary goal is to understand the disciplines and unify knowledge (Nicolescu, 2005). Indeed, trans-disciplinarity encapsulates the notion of an integrated utilisation of the techniques, tools and methods from various disciplines (Ertas et al., 2003; Kozmetsky, 1997).

This work is based on an empirical approach that centers on complex design problem-solving in higher education. In this context, we propose that studying design collaboration among students from a multi-disciplinary perspective constitutes a suitable approach for exploring the possible extension of design frontiers. As noted, the intricacy of products has overcome the narrow frontiers of specific disciplines and dealing with design entails the collaboration of specialists from different disciplinary areas (Eckert & Stacey, 2021). When design collaborations are materialised in the framework of a multi-disciplinary team, opportunities for gaining new cross-cultural knowledge and skills emerge (Casakin & Badke-Schaub, 2015; Leblanc, 2021). Teams of this kind are supposed to have the specialised knowledge necessary for focusing on specific aspects, jointly with a holistic understanding on the topic (Fernandez-Orviz, 2021). A capability for sharing different views at different levels of detail can lead to the development of inspirational ideas. Not less importantly, multi-disciplinary teams can contribute to more meaningful, feasible and creative outcomes and eventually expand the existing frontiers of a design domain.

4.2.2.2 Multi-disciplinary and mono-disciplinary teams and design

The advantages of multi-disciplinary teams over traditional mono-disciplinary ones have been studied in several fields with a focus on team outcomes. A frequent approach in comparing the performance of these types of teams is to analyse quantitatively the produced outcomes by means of bibliometric indexes (e.g., Schummer, 2004; Levitt & Thelwall, 2008). The healthcare domain is a meaningful example, wherein numerous research activities on multi-disciplinary teams were carried out based on surveys and interviews. While related studies discussed a rich range of health issues that were efficiently tackled by multi-disciplinary teams, they also exposed several conflicts spanning from misalignments of team roles to problems in social dynamics (Jones, 2006). A central finding was the need to educate professionals capable of working in teams with members of diverse specialisations (Dyer, 2003). Studies on multi-disciplinarity can also be found in the design literature and in project-based design education. A few examples are works focusing on ICT technologies (e.g., Nelson & Ahn, 2018), and virtual environments supporting team collaboration (e.g., Mengoni et al., 2009). Nonetheless, comparative design studies dealing with the dynamics of multi-disciplinary versus mono-disciplinary teams have yet to be carried out.

However, the diversity that characterises multi-disciplinary teamwork can be counterproductive to the possibility of successful collaboration. Many involved disciplines may not yet have worked in cooperation to solve a design problem. Over the years, they have constructed their own knowledge, terminology, approaches, culture, preferences and priorities (Eisenbart et al., 2011). Specialisation in a domain enables the development of design expertise for the sake of improving task performance while enhancing the quality of the produced outcomes. Nevertheless, it can also narrow the perspective on the broad picture (e.g., restrict the capacity of designers in a specific field to recognise the value of issues from other disciplines that are aliened to them) (Fernandez-Orviz, 2021). In this regard, transcending the frontiers of mono-disciplinary approaches is challenging and demands understanding commonalities and differences with other disciplines (Gericke et al., 2021). In the context of our study, the notion of extending the frontiers of design is understood as the sharedness of existing knowledge, tools, methods, and experiences from a certain discipline and unknown with other disciplines, with the goal of contributing to a common understanding of the design situation while gaining new knowledge. To this end, successful design collaboration plays a critical role when it concerns complex and multifaceted projects. However, this may be affected by the quality of team collaboration.

4.2.2.3 Teamwork quality and design

The concept of teamwork quality, introduced by Hoegl and Gemuenden (2001), assumes a critical role in studying multi-disciplinary collaborative design and its influential factors. Teamwork quality is a thorough notion of the excellence of team interactions (Dayan & Di Benedetto, 2008). It is a critical aspect of team collaboration that was studied by focusing on social interaction factors – e.g., cohesion and mutual support, and task-related factors – e.g., communication and coordination (Hoegl & Gemuenden, 2001). These researchers found the two teamwork quality factors strongly related to team performance.

In another study, Dayan and Di Benedetto (2009) showed that the quality of interactions among team members is a fundamental skill needed to create good and innovative products. However, they showed that teamwork quality could be reduced due to difficulties in getting members to communicate successfully and work jointly. Teamwork quality was also found to affect educational outcomes and the success of collaborative learning (Curşeu & Pluut, 2013). The quality of social interactions at the team level showed to have an impact on knowledge creation, and eventually on team performance (Hong et al., 2017). Additional studies in the fields of R&D management and cognitive psychology highlighted the influence of task-related factors on teamwork quality. Kratzer et al. (2006) showed that low-complexity tasks have a negative impact on the creative performance of R&D teams. In spite of the variety of works on teamwork quality, a primary focus tends to be set on its effect on team performance (Nesterkin et al., 2016). Some of these focus on the impact of training styles of multi-disciplinary teams on design performance (Reich et al., 2009), or on reflections on learning outcomes from cooperative project-based approaches in engineering education (Abdulaal et al., 2011). A milestone is a work by Cooke et al. (2003), who investigated task-relatedness to teamwork knowledge, but still with focused attention on team performance.

Relatively few studies can be found on collaborative design dealing with social aspects and teamwork quality, and no studies on multi-disciplinary design investigated teamwork quality based on social-related and task-related factors simultaneously. Therefore, in this work, we investigate how multi-disciplinary compared to mono-disciplinary design teamwork affects the quality of collaborations in an educational environment from a social and task-related perspective. Given the potential effect of teamwork quality on knowledge creation, we propose that studying the quality of team collaboration can inform about extending the frontiers of design.

4.2.3 Methodology

4.2.3.1 Design courses

To address the research questions, a study was conducted in the context of two programs. (i) A master's degree with students from architecture, industrial design and all branches of engineering disciplines. This took place in the 2019 edition of the "Design Methods and Processes" course of Alta Scuola Politecnica – ASP (Benedetto et al., 2010; Ajmone et al., 2016), a program jointly fostered by the Politecnico di Milano and the Politecnico di Torino. (ii) A master's degree in the Mechanical Engineering study program at Politecnico di Milano, which included students who were enrolled in the "Inventive Design and Problem Solving" course.

The ASP is an honors program that comprises several courses organised as residential full-time weekly schools and a design project lasting one year on real-world problems (https://www.asp-poli.it/home/our-projects/). On the other hand, the Mechanical Engineering course aims to deliver theoretical competencies and practical skills in methods and tools supporting technical creativity in conceptual design activities. It involved an annual project proposed by an entrepreneur from the industry. Examples of project themes are household appliances (e.g., washing machine, dishwasher) for single users and an integrated system intended for regulating vehicle tire pressure.

The ASP project, representing the multi-disciplinary team activity, was used as a test group, whereas the Mechanical Engineering project, representing the mono-disciplinary team activity, was used as a control group, that is, as a benchmark to identify differences and similarities in teamwork quality in the context of design.

The comparison between the two case studies is considered relatively unbiased since one of the authors was the coordinator of both courses and provided similar directions and tutorship in both projects. Through a retrospective analysis of team activity in the ASP and Mechanical Engineering courses, participants can offer important insights that transcend the courses themselves and be of value to identify key aspects in the design educational context, with possible implications for extending design borders.

4.2.3.2 Participants

Students were requested to opt out if they were not interested in participating in the study. Eight out of 140 ASP students representing different disciplinary backgrounds and course grades, and five out of 50 Mechanical Engineering students accepted the invitation to take part in the study.

4.2.3.3 Research approach

The research method is based on semi-structured interviews (e.g., Kallio et al., 2016). The use of semi-structured interviews was successful in several studies that analysed the dynamics of teamwork (e.g., Sandoff & Nilsson, 2016). In the present work, they were characterised by a dialogue partially based on a predefined protocol, adapted to the ad-hoc circumstances as demanded by the interactions with the participants. The structure of the interview sessions was partly based on teamwork quality (task and social-related aspects). The interviews also included other topics that are beyond the scope of this research. To avoid potential biases, they were carried out after the evaluation process – when all students were assigned the grades for the two courses. Both authors coordinated the interview sessions and complemented each other by proposing topics for discussion. These were carried out in English, as both courses were delivered in this language.

4.2.3.4 Data collection and preparation

Due to the COVID-19 outbreak, interview sessions were held via the Internet using Microsoft Teams software and lasted about 60 minutes. The meetings were video recorded under the explicit consent of the participants. Then, the anonymised recordings were transcribed and segmented, each segment representing an exchange in the conversation between the interviewers and the interviewee. Video recordings were transcribed using an external transcription service. Thereafter, transcriptions underwent a quality check to ensure that the verbalisations were captured correctly. Transcript segments were also assigned a timestamp as a reference in the data collection. The final data set contained all utterances produced by the participants in the 13 sessions, which were classified according to participant and team type.

4.2.4 Findings

First, the social-related and task-related dimensions are identified. Second, an analysis is carried out on the themes and topics related to the social-related dimension, and then the task-related dimensions of teamwork quality. Illustrative short excerpts identified from the interviews are included to provide empirical evidence of the findings.[1]

4.2.4.1 Identification of social-related and task-related dimensions

To deal with RQ1, a qualitative analysis of the segmented transcripts followed a 3-step process:

- The authors carried out an independent analysis of the transcripts to identify and retrieve relevant topics addressed in the interviews, regardless of the disciplinary team composition.
- By consensus, they selected the topics they considered relevant to the research goals and clustered them according to general themes. Thereafter, they classified the themes and related topics according to two main reference dimensions: (i) teamwork quality, represented by social-related versus task-related issues; and (ii) disciplinary team composition, represented by mono-disciplinary versus multi-disciplinary team background.
- Finally, authors revisited the transcripts to check the consistency between the identified topics and themes, and to select illustrative excerpts for each theme.

Figure 4.2.1 shows the themes and related topics identified in the second step, which are organised according to the two above-mentioned comparison dimensions. The social-related dimensions that are common to the two types of teams are characterised by four

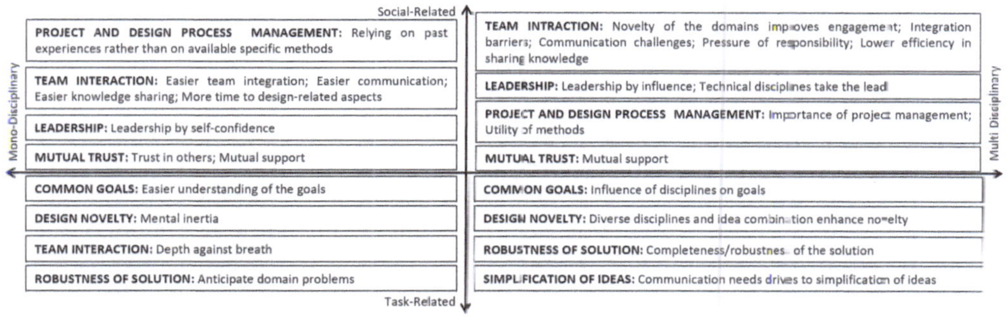

Figure 4.2.1 Themes (bold text) and related topics identified from the interviews with students.

main themes that included: mutual trust, leadership and project management and design process management. On the other hand, the task-related dimensions are classified into simplification of ideas, design novelty, robustness of solution and common goals.

To address RQ2, the following two sub-sections present the main findings of the 3-step process described above. Accordingly, the identified themes are analysed into social-related and task-related dimensions and compared according to multi- and mono-disciplinary team composition.

4.2.4.2 Analysis of social-related dimensions

Teamwork quality is used as an overarching concept referring to the quality of team collaborations including social-related facets. In this sense, a team can be defined as a social system in which members collaborate on a common task. To capture the social nature of design team members working together, in this section, we elaborate on the social-related themes identified in the study as follows.

4.2.4.2.1 Team interaction

A recent review (Nguyen & Mougenot, 2020) on team interaction in multi-disciplinary collaboration in the industrial sector summarises the way that team composition affects the design process. The study mentions several key aspects deserving consideration such as: strengthening trust among team members, and dealing with team cohesion and communication barriers, caused not only due to language and jargon differences. In this regard, students in multi-disciplinary teams, even without being explicitly prompted on these issues during the interview, highlighted communication difficulties met in the team collaboration activities that led to the emergence of communication barriers and reduced cohesion.[2]

While no students highlighted criticalities due to limited trust, possibly because of the variety of team-building opportunities offered by the ASP school, they commented about a lower efficiency in sharing knowledge.[3] In contrast, participants in mono-disciplinary teams explicitly manifested the ease of interaction with their teammates.[4] Some of them also stressed the importance of being a specialist in the industry related to the design task.[5]

Interestingly, students in multi-disciplinary teams did not show the same confidence in their discipline field. Those who hold unique and critical competencies critical to the project's success, suffered huge pressure. This was due to their role and responsibilities and the impossibility to discuss in detail important decisions with others.[6] Participants also mentioned the high engagement they felt due to the novelty of the viewpoints and competencies brought by their teammates.[7]

4.2.4.2.2 Mutual trust

Concerning multi-disciplinary teams, when the knowledge needed to deal with the task was too specific to be understood by all disciplines, members of certain disciplines had no choice but to trust the expertise originating in other disciplines.[8] However, trusting mates while dealing with a design issue contributed not only to generate a better atmosphere and enhance team cohesion, but also to learn from others, gain a broader picture of the design situation, and understand how to move on in the process.[9] In the mono-disciplinary teams, trusting other members was easy and advantageous, but for different reasons. The common knowledge of the engineers enabled them to understand design situations quickly even when the information shared by others was not completely apparent.[10] This enhanced their confidence in the team,[11] reduced work pressure, and created opportunities to collaborate and help each other.[12] Eventually, trusting others played a role in improving the quality of the final solution.[13]

4.2.4.2.3 Leadership

The influence of leadership on team interactions has been widely studied in the literature, both at the individual and group levels. At the individual level, leadership was connected to several positive aspects that include performance, work engagement, and creativity (Grandey et al., 2005), and work satisfaction (Giallonardo et al., 2010). In the context of multi-disciplinary teams, leadership at the individual level was related to the personality of the team member. The leader was not seen as an individual with a strong personality who fought to inflict power to dominate others.[14] Rather, she was considered an open-minded individual, cognitively flexible and able to solve conflicts that emerged during the task.[15]

At a team level, leadership was also found to enhance reflective discussions about the level of agreement among members, values, strengths, weaknesses and progress to achieve shared goals (Yammarino et al., 2008). When members perceive the legitimacy of a team's leadership and seek to comply with team rules and decisions, they remain more attached to and supportive of the team (Naumann & Bennett, 2000). In this work, students with an engineering background emerged as group leaders of the multi-disciplinary teams. Due to the nature of the assigned design tasks, they were those who coordinated the team's actions to deal with the problem.[16] The leader was considered to have special managerial capabilities to organise and coordinate design actions among sub-teams, which were formed based on disciplinary background.[17] The leader was also responsible for the team's cohesion and for enhancing communication and understanding between the disciplines.[18] In the mono-disciplinary teams, there was no natural leader. Partly due to their similar background, there was an efficient interaction between members, and a balanced responsibility for dealing with the project.[19] When leadership emerged, it was mainly to mediate conflicts and help make decisions by considering the opinion of the majority.[20]

4.2.4.2.4 Project management and design process management

The need to develop designers' multi-disciplinary team working capability has been discussed in the literature for over 20 years. Project-based learning has been proposed as a reference educational practice to develop such soft skills since then (Denton, 1997).

Considerable attention was dedicated to influential factors such as the composition of design teams, the nature and complexity of the task, and assessment criteria. In this regard, past editions of the ASP school on "Design methods and processes" were analysed to recognise the educational impact of this practice (Cascini et al., 2017). However, such studies did not compare the reflections of participants of multi- and mono-disciplinary teams.

In this work, participants in the multi-disciplinary teams often referred to team collaboration and the design process to describe how they managed the task distribution, the interaction between teammates, and the fulfilment of deadlines. Compared to their colleagues in mono-disciplinary teams, they showed a more extensive understanding of the importance of applying systematic methods for project management and guiding the design process.[21] As recognised since early studies in the field, team-working capabilities cannot be developed based on a one-off exercise (Denton, 1997). Relying on past experiences rather than on available specific methods was characteristic in mono-disciplinary teams.[22] Moreover, it turned out that for multi-disciplinary teams, struggling with team management issues came at the cost of less time and effort dedicated to the task itself.[23] In contrast, much of the focus of mono-disciplinary team members was set on the task and its related details, but at the expense of overlooking the design process and team management aspects. Besides, this sometimes led to useless iterations in the team, with a consequent waste of time.[24]

4.2.4.3 Analysis of task-related dimensions

A main concern regarding the quality of a team's collaborative work is how well its members share and process task-relevant information. To gain insight into the task-related interactions of design team members working together, in this section, we elaborate on the themes into which identified design issues in the interviews were categorised. These include simplification of ideas, design novelty, robustness of solution and common goals.

4.2.4.3.1 Simplification of ideas

Further insights that emerged in the interviews focused on the simplification of ideas generated in multi-disciplinary teams. Successfully transmitting design ideas in a common space may require some level of simplification (Craig & Zimring, 2000). Some interviewees reflected on the difficulties of interacting with colleagues with different backgrounds. They concluded that to enhance common understanding and communicate with others, specialized disciplinary knowledge had to be simplified.[25] To convince and increase the chances of getting ideas approved by other teammates, it turned out that simpler ideas were more effective and likely to prevail.[26] Such a tendency towards simplification of ideas suggests that as far as specialised knowledge is not requested by the task at hand, multi-disciplinary teams are more likely to avoid complex solutions, and instead they naturally prefer simpler and possibly more elegant alternatives. We found no indication of this behaviour in the mono-disciplinary teams.

4.2.4.3.2 Design novelty

Design is a creative activity that mainly takes place in the conceptual phase and is characterised by the generation of ideas and solutions (Pahl & Beitz, 1996). In this process, designers work jointly with the aim of developing not only functional but novel outcomes. Novelty can be defined as "a new or unfamiliar thing or experience" or as "an object intended to be amusing because of its unusual design" (Oxford Dictionary, 2021). Some metrics have been proposed to evaluate the degree of novelty of a design process either in terms of "uncommonness", or the degree of change compared to a known solution (Fiorineschi et al., 2020). The creation of novel outcomes in teams with different design backgrounds is a relatively underexplored issue.

In our work, we found that the various backgrounds of the multi-disciplinary team contributed to thinking about the design situation anew. Members invested efforts in reframing the problem and redefining the design task.[27] The variety of views of the disciplines helped enhance the novelty of the final solution.[28] Interacting with other domains aided students to open their minds and to reason about the task in less familiar ways.[29] Such interactions led to the production of unexpected solutions.[30] Novelty was also the outcome of a balance between designers – who tried to push the borders of the familiar by envisioning unique ideas – and engineers – who thought about how to materialise these ideas in practice.[31] In some cases, the novelty of the solution did not reside in producing revolutionary outcomes, but in considering existing solutions from different disciplines and combining them into a single solution.[32]

In the mono-disciplinary team, the produced solutions were considered more valuable than novel.[33] Rather than thinking of revolutionary ideas, the mechanism that contributed to the search for novelty consisted in adapting knowledge from existing products to the problem at hand.[34] Although not completely novel, within-domain displays served as sources of inspiration to deal with the task.[35] Potentially novel ideas were discarded since there was no member with the required skills to develop them.[36] Hence, the chance of developing a novel outcome depended more on the individual than on collective creativity.[37]

4.2.4.3.3 Robustness of solution

Students in mono-disciplinary teams mentioned ease in interactions with their teammates, and the possibility of analysing and sharing details of the design task with others with an adequate level of common understanding (See section 4.1.1). However, they manifested difficulties in how to handle possible drawbacks when discussing new solutions.[38] Moreover, to ensure the viability of their concepts, students expressed their wish to revisit their solutions with the aid of colleagues from different disciplines, even at the expense of unexpected variations in the operating conditions of their designs.[39] This situation reflects the limited robustness of ideas generated by mono-disciplinary teams, meant as a reduced ability to fulfil the requirements under unexpected conditions. Issues like these were already noted by others, who recommended engaging multi-disciplinary teams in the evaluation of the robustness of product concepts (e.g., Goetz et al., 2019).

Students in multi-disciplinary teams showed awareness of the boundaries in disciplinary competences.[40] However, they recognised the advantages of combining complementary points of view and knowledge resources from other disciplines to prevent drawbacks and limitations of their designs because of unknown aspects of the task.[41]

4.2.4.3.4 Common goals

Design problems are ambiguous and can be defined by no clear initial conditions and no completely specified goals (Simon, 1996). Hence, while dealing with the design task, team members must define the goals to be accomplished at the different stages of the design process (Coyne, 2005). Establishing and sharing common goals requires a large range of interdependencies of individual information held by the team members (Casakin & Badke-Schaub, 2017; Badke-Schaub et al., 2007). Defining common goals is critical to successfully tackle the related sub-problems and arrive at a well-integrated solution. The disciplinary background is supposed to play a role in the way designers approach problems and establish such goals. This situation seems more radical in teams where designers differ in knowledge, skills and experience.

Differences in interests in the multi-disciplinary team influenced how members perceived the design situation, and how they prioritised the goals and approached the problem.[42] Hence, a challenge for the multi-disciplinary teams was to agree upon shared design goals.[43] On more than a few occasions, they had to consider and negotiate their differences to deal with the goals.[44] When they realised that their initial goals were difficult to achieve, they tried to adapt them to meet the know-how of others.[45] Generally, team members had shared main goals, but their final solutions did not always reflect it. The main reason was that teams were subdivided into specialised sub-teams that worked independently of one another while developing part of the solution. A sub-team organisation affected the possibility of clarifying goals and sub-goals with other sub-teams and reaching more integrated solutions.[46] In some cases, the voice of the engineers was louder than other voices and they imposed their goals over less influential disciplines such as design or architecture. The consequence was a lack of balance in the contribution of the disciplines to the final solution.[47]

In contrast, members of the mono-disciplinary teams worked guided by task requirements from the outset. Since they were aligned with the kind of design to be developed, there was a common understanding of what the design goals were.[48] Eventually, there were a few minor conflicts, which were quickly resolved as soon as the idea of what they wanted to achieve became clear.[49] Due to their common background, they communicated easily, shared their knowledge, helped each other, and balanced their contribution to achieve the goals.[50] Consequently, the final solution largely reflected the initial goals of the team.

4.2.5 Discussion

To deal with the research goals, in the following sections we elaborate on the differences between multi- and mono-disciplinary teamwork and their importance for extending the frontiers of design in higher education. We refer to collaboration in teams by using teamwork quality, represented by the task-related and social-related themes identified in this study.

4.2.5.1 Social-related and task-related dimensions in design teamwork

The first aim was to study comparative social- and task-related dimensions of multi-disciplinary and mono-disciplinary design collaboration, into which teamwork quality can be categorised. Exploring these dimensions enabled us to identify and gain insights into relevant themes regarding design teamwork quality. Hence, the theoretical approach concerned with teamwork quality proposed by Hoegl and Gemuenden (2001) proved to be effective in gaining further understanding of design team collaboration in general, and particularly regarding team disciplinary composition. The common themes that emerged allowed analysing and comparing key aspects of multi-disciplinary and mono-disciplinary design collaboration in the context of higher education. These are discussed in the next section.

4.2.5.2 Effect of team disciplinary composition on design teamwork quality dimensions

Regarding task-related issues, we found that disciplinary background affected how design goals were established. In line with what several authors observed (e.g., Leblanc, 2021), multi-disciplinary teams managed to set goals that overcame in scope and complexity those established by the mono-disciplinary teams. A main influential factor was the variety of interests and views of the different disciplines, which in contrast to the mono-disciplinary teams, contributed to extending the borders of the known and the familiar. The different backgrounds in the multi-disciplinary teams helped not only to define the design goals from several perspectives but also to rethink the design task. With the insight of the different disciplines, framing the problem played a role in enlarging the understanding and conceptualisation of the task (Paton & Dorst, 2011), and ultimately in enhancing the generation of novel and unexpected outcomes (Suwa, et al., 2000). In contrast, the intra-disciplinary knowledge shared by the engineers in the mono-disciplinary teams aided in developing a deep understanding of the design situations and producing more valuable than novel solutions. This is consistent with Leblanc (2021), who noted that disciplinary expertise is useful when it relates to and enhances the understanding of complex phenomena. However, expertise often showed to miss the acceptance of alternative approaches and thinking out-of-the-box.

Regarding social-related issues, maintaining a balance between the different views and goals of multi-disciplinary team members required management skills, which demanded great effort, sometimes at the expense of task-related aspects. From this viewpoint, finding the appropriate balance between these conflicting aspects remains an open issue to be addressed. Moreover, working in multi-disciplinary teams required the adaption of communication to the profile of the team members. An unexpected finding was that simpler ideas, i.e., less grounded on specialist knowledge, were preferred because they were easier to communicate with teammates from different backgrounds. However, this tendency occasionally occurred without making a proper assessment of whether the more complex discarded ideas would have been better at meeting the requirements of the task at hand. In this vein, Fox et al. (2017) showed that simplifications in communication may be counterproductive to matching the complexities of the task.

Mutual trust was found to be a critical aspect of generating a positive work atmosphere in multi-disciplinary teams, as highlighted by Maier et al. (2008). Mutual trust was required when the specific knowledge necessary to deal with the task could not be understood by all members, and hence it was shared in the team at a shallower level. Although trusting other disciplines implied renouncing control of every detail in the design process, it enabled to coordinate actions more effectively and learn new knowledge. In turn, the opportunity to learn new knowledge resulted in higher engagement of students in multi-disciplinary teams. A further consequence was the gaining of a more comprehensive understanding of complex tasks through the perspectives of the other fields, thus leading to more robust designs, thanks to anticipation of problems that might emerge.

In mono-disciplinary teams, mutual trust strengthened team collaboration and performance in the different tasks. This probably enhanced the performance when developing the final solution but did not contribute to the creation of new knowledge. Maintaining a balance between mutual trust and constructive criticism can contribute to enhancing a positive team climate and foster design thinking (Casakin & Badke-Schaub, 2013).

Those students with an engineering background and high self-confidence emerged as natural leaders of multi-disciplinary teams. They were seen as flexible-minded individuals open to the ideas of others, with singular managerial skills to coordinate actions among the disciplines. Jacoby-Volk and Bar-Eli (2021) emphasised the importance of encouraging natural leadership as opposed to formal management roles. In our study, while engineering leaders possessed the basic knowledge needed to organise and harmonise the work of various disciplines, they were found to largely interfere with the contribution of students belonging to other, less represented disciplines than engineering. Hence, in multi-disciplinary teams, the role of dominant leaders could be counterproductive when the aim is to challenge the design borders to expand the knowledge space.

4.2.5.3 Extending the frontiers of design: implications for design education

A main pillar for extending the frontiers of design resides in promoting a more integral education based on proactive interaction among different disciplines. Multi-disciplinary education can be implemented and encouraged by supporting team collaboration. Whereas design programs interested in extending the frontiers of design should contemplate the possibility of implementing multi-disciplinary team-based projects, they may also benefit from the findings of this study. Critical issues to be considered in a transition from mono-disciplinary to multi-disciplinary design education, specifically in the design studio, are how to:

- Clearly communicate and share disciplinary knowledge and methodologies that are alien to others, to enhance cohesion and facilitate a common understanding among team members. This will require adapting own knowledge, terminologies, tools, approaches and priorities to meet the needs of other disciplines (Eisenbart et al., 2012).
- Define common goals while taking care of an appropriate balance of interests among the disciplines. Sharing common goals is fundamental for efficient team performance, dealing with sub-problems and developing well-integrated solutions (Badke-Schaub et al., 2007). However, disciplinary background affects the way students perceive problems and prioritise goals. Hence, how to reach a consensus is a primary challenge that can be dealt with by negotiating differences with other disciplines. This can be achieved by establishing priorities based on the know-how of the team members and adapting design goals accordingly. Such an approach would enable not only to align design intentions among team members with different backgrounds and interests, but also reduce the chances of producing unsuccessful solutions.
- Reflect and discuss ideas with other disciplines, being clear, flexible, and simple but not simplistic. Explaining, reflecting and communicating ideas efficiently is a critical prerequisite for reaching a common understanding among students with different backgrounds, as well as for developing novel solutions. These activities can be implemented by the construction of a shared mental model of the task to be achieved by all team members (Mohammed & Dumville, 2001). To this end, an essential activity should consist in communicating, applying and adapting pre-existing individual mental models from the different disciplines, for the sake of progressively developing a common language in support of the construction of a shared mental model of the design task and the process.
- Develop mutual trust among the parties, guided by good leadership. Supporting mutual trust among the disciplines is advisable from the beginning of the process (Casakin & Badke-Schaub, 2013), particularly when the task is too complex to be fully understood by all disciplines. An open-minded team leader can help not only mediate

conflicts and keep a harmonious and positive team climate but also enhance confidence and encourage engagement and opportunities for collaboration among the disciplines. Eventually, all these will contribute to gaining a broader understanding of the design situation and learning how to coordinate actions to move forward.

- Manage and coordinate design actions effectively to produce novel ideas and robust solutions, trying to anticipate problems that have never been experienced before. To overcome the uncertainty of the early stages of an innovative design process, it is recommended to integrate competencies from multiple disciplines, being open to the complementary input of members with different backgrounds and views (Goetz et al., 2019). Distributing tasks among specialist sub-teams and coordinating actions with other sub-teams at different stages of the process can help to efficiently integrate sub-solutions and produce creative outcomes, with a consequent time and effort reduction.
- All these issues may play an important role not only in consolidating existing knowledge, but also in creating new one. Eventually, these will contribute to transcending the frontiers of the design disciplines, particularly in higher education.

4.2.6 Study limitations

The authors selected mechanical engineering as a control group among all possible mono-disciplinary backgrounds. Hence, reflections on multi-disciplinary teams were made in contrast to mono-disciplinary teams with a strong engineering background. Compared to other engineering disciplines, mechanical engineering students have larger horizontal competencies, which may have softened differences with multi-disciplinary teams to some extent. Moreover, the task assigned to the multi-disciplinary teams was not always relevant to all the students to the same extent. The disciplinary background could have biased the way the interviewees perceived and reflected on their teamwork experience.

Another issue is concerned with the sample, since the multi-disciplinary teams were composed of honors students, who presumably have higher capabilities than “regular” students in the mechanical engineering course. However, it is believed that the effect of such limitation is moderated by the fact that the analysis of the teams focused on the interactions among members rather than on their outcomes. A further limitation is that students had limited professional experience as designers. Hence, their insights should not be generalised to multi-disciplinary teams of expert practitioners.

Due to the COVID outbreak, we did not have the possibility to design a setup aimed at exploring how to expand the frontiers of design specifically. Rather, we collected data using interviews with students who had participated in courses based on an established multi-disciplinary educational model. Nevertheless, we believe that the insights that emerged from analyzing and comparing these interviews with those conducted with students in mono-disciplinary teams revealed interesting differences that may have helped identify influential aspects for enriching the design experience.

4.2.7 Conclusions

This study explored the comparative social-related and task-related factors of multi-disciplinary and mono-disciplinary design teamwork, which is an underdeveloped topic. A further aim was to gain an understanding of the ways teamwork quality can benefit from social-related and task-related interactions among designers with different disciplinary backgrounds, and what the main challenges are. We also expected to gain insight into how multi-disciplinary teamwork can contribute to extending the frontiers of design.

Data collected from 13 semi-structured interviews were analysed, with a focus on the quality of the interactions developed during the design activity. While comparing multi-disciplinary and mono-disciplinary teamwork quality, this study contributed to shedding light on the existing dichotomy between the more comprehensive approach of the former and the narrower but deeper approach of the latter. It showed that demanded skills and competencies differ between the two types of teams, challenging team dynamics and collaboration.

In addition to nurturing the development of knowledge and design skills, caring for teamwork quality is crucial to improve multi-disciplinary team collaboration. To this aim, universities should implement curricula supporting multi-disciplinary environments where design students could be trained to collaborate with peers from other domains. However, changing the way design is taught nowadays will demand overcoming existing disciplinary boundaries and preconceptions set in higher education. This work can be seen as an attempt in this direction. Additional studies should be conducted to improve understanding of design multi-disciplinarity in education and practice. Eventually, they will contribute to the further development of a trans-disciplinary and better-integrated design science.

Acknowledgements

Thanks are due to the students of the master's program from Alta Scuola Politecnica – ASP at Politecnico di Milano and the Politecnico di Torino, as well as to the students of the Mechanical Engineering master's program at Politecnico di Milano for their participation in the interview sessions.

Notes

1 MLT: quotations from ASP students; ME: quotations from Mechanical Engineering students.
2 MLT "the main difficulty is in the communication, because if you have different backgrounds you put different values on [...]. Also, small things could lead to infinite discussions [...] slow down all the process and [...] lose track of important things [...]".
3 MLT "you can speak about two different levels of knowledge: one that is very technical, that I can only share with the disciplines that are related to my own. And there is another one that I [must] trust".
4 ME "having the same [background] of mechanical engineers, we were able to evaluate the capacity of each other in a more precise way".
5 ME "One aspect of bringing out good design is to [...] have good knowledge about the state of the art [...]".
6 MLT "communication with the other members [...] is important; I need a person [with whom I] could share opinions deeper on the theme. [...] if you find a person that knows what you are doing and how you work [...] we can work faster and explain better the work to others"; ME "If there were at least two people from the same background [...] there could be some conversation between teammates [...]".
7 MLT "I'm an expert in fluids. But for what concerns the yearly project I didn't see any fluid. So actually, in my field of study doesn't really show up. But at the same time, I'm really satisfied because I really put a lot of effort and also because I learnt a lot [from others]".
8 MLT "for example, the algorithm [...] I can understand how it works. But if you say that it's better to use five instead of ten, I trust you. Otherwise, I have to have another degree in computer science"
9 MLT "I worked really well in my team because we shared knowledge and trusted each other [...] and this was useful to me because I also learned something new [... and it helped to] be aware about the pros and cons of the solutions and then together we could decide how to proceed".

10 ME "Even if it was 60 to 70% clear, we still understood what he meant because of our engineering background".
11 ME "The engineering background allows you to understand more deeply what others were saying and you didn't have to trust".
12 ME "it [trust] helped a lot because if one is forgetting about one small detail, let's say about designing something, the other [engineers] can intervene and help him [...] so that the [work] pressure will not be really high on that member".
13 ME "[due to] the mutual support and collaboration with my teammates, we polished this idea that was really rough into the final design".
14 MLT "It's not a matter of being the one who will control everything [...] or tries to impose its own ideas [to others]".
15 MLT "The team leader is a person that is more open to the opinion of all of us".
16 MLT "people who are related to the more like technical school, they're trying to be leaders".
17 MLT "There was usually the team leader that was going around and seeing what the subgroups were doing and trying to give coherent indications too".
18 MLT "the leader [...] was able to keep all the elements of the group together. So the communication worked [...] well".
19 ME "Communication was crystal clear, although we had no leader as such, we all took equal responsibility for the tasks.".
20 ME "the sort of leader [... managed] conflicting situations [...] democratically".
21 MLT "It was strongly requested to have a common view on the management and the organisation between the team members, especially between very different disciplines. A main point to organise communication was to create a channel dedicated to a single topic that all of us can set in. [...] team building is prior to anything else. Even, prior to the understanding of the problem, because you need to know each other to understand the problem".
22 ME "I did projects during my bachelor, my teammates [...] already did projects in teams. So all of us already know how we should structure our work in order to be proficient.".
23 MLT "There is a critical number of people over which you don't have really effective coordination and collaboration".
24 ME "If we had someone from product design, our product would have been better because those people know some nuances we don't. [...] We spent some time on it [a solution] and we didn't like it. We had some ideas about some things we would like to change, but we didn't know whether it is possible".
25 MLT "my teammates try to explain me in a simple way [...] the problems and something like this. And it works in both directions. So, for example, if there were problems related to the biomedical field, I tried to explain in a simpler way so they can be managed in a general way, not more in the medical field".
26 MLT "it was a little bit difficult [...] to introduce a debated discussion [...] it was difficult to obtain the opinion of some engineers, let's say, because they tend to limit their work on computational things or stuff, but they tend not to explain their position".
27 MLT "we use [considered] a novel topic to address a not so novel solution because the solution already exists. [But] not as defined as one would expect [...] and having [a team with] multiple backgrounds [...] and different experiences helped a lot".
28 MLT "you can apply knowledge and concepts that come from different fields. This generates something that is completely different and disruptive".
29 MLT "the other disciplines take us to reason in a different way. So we were less focused on the daily path"; MLT "an idea from a completely different background expands your way of seeing things [... and this] enhances your creativity [...] you come up with some solution that is not just an addition between your knowledge and the others knowledge, but [...] makes something more because it stimulates to think in a different way".
30 MLT "it's part of the multi-disciplinary nature of the task [...] because they thought about some solutions that I could not have imagined [before]".
31 MLT "we are dreamers [the designers] and we are always trying to create something like extraordinary, super unique. And sometimes it runs out of our budget or the reality so that [...] the engineers are always trying to [...] to balance those things".
32 MLT "the solution is quite innovative [...] we integrated in a device a lot of things that came from other devices and that were not in a unique device.".

33 ME "although we put in a lot of effort, which they [the company] appreciated a lot, the idea in itself was not so novel"; ME "the idea was not so novel … but] in terms of value it was a step in the right way for the project …".
34 ME "we took a concept that is normally used in the automotive industry, and we completely changed [adapted] it to the context [of the task], something that it has never been done before in […] this context".
35 ME "we took inspiration from the lipstick, how it works when you rotate, and it comes from a rotation that transforms into an actual axial motion […] but I would say it [the solution] was not novel in general".
36 ME "if we had someone from the aerospace field, we feel he could have taken up the task and he could have made [developed] it into a creative idea, which might have been a good idea if it was economical for the company".
37 ME "not everybody had innovative solutions, but it was all about […] if there was an innovative person".
38 ME "So if we were not expert in materials, we searched them on the net. But obviously it's different with respect to having a certain background there where people made courses on this".
39 ME "to solve a certain problem, we made a solution for which all the weight was on the user […] being mechanical engineers, we dedicated much time to problems that … [led to] a quite instantaneous solution".
40 MLT "not about the knowledge, because, of course, we cannot be engineers in six days, but in terms of approach".
41 MLT "their [teammates] point [of view] having elements with a completely different background helped a lot in trying to force us to view the thing from a different perspective".
42 MLT "the architects pushed a lot towards the management of the user experience"; MLT "because of the difference in background, as a computer engineer you can develop more easily the final product and is […] a different approach than a space engineer".
43 MLT "maintain a sort of dialogue [among the team members] that could work and led to a common goal. This was a main issue because there were very complex arguments [goals] that […] if not managed in a correct way […] could be very messy".
44 MLT "people have different opinions. We have always like to […] take into consideration all their opinions to make a decision".
45 MLT "we have the chance to go more in the hardware part or more in the software project. So, we decided to proceed more in the softer partly because of the type of team members we have We had people from computer science that are very good in doing software and we decided to proceed in this direction instead of the other".
46 MLT "we had a more specific goal that we shared in common [… however] […] we were working more independently […] and sometimes the solution that other subgroups developed was hard to put together".
47 MLT "there were two mathematical engineers and three computer science engineers. So, you can imagine that it was really computer oriented […] and not really focusing on the architectural aspect of the project".
48 ME "We already knew what exactly we are targeting".
49 ME "In terms of the goals […] one part of the group wanted to go with one project and the other wanted to go with another, but at the end we all made the decision to pursue a common goal and just focused on one type of project"; ME "there were frictions […] but when we started to have a firmer idea of what we were considering, we started again … working in one direction".
50 ME "we tried to help everyone as much as we can because we knew that we share a common goal, and this would largely affect everybody".

References

Abdulaal, R. M., Al-Bahi, A. M., Soliman, A. Y., & Iskanderani, F. I. (2011). Design and implementation of a project-based active/cooperative engineering design course for freshmen *European Journal of Engineering Education*, 36(4), 391–402.

Ajmone, M., Ceri, S., Verganti, R., Zanino, R. (2016). Alta Scuola Politecnica: Innovation, multi-disciplinarity and passion. In Banerjee, B. & Ceri, S. (eds.). *Creating innovation leaders* (pp. 215–225). Cham: Springer.

Badke-Schaub, P., Neumann, A., Lauche, K., & Mohammed, S. (2007). Mental models in design teams: a valid approach to performance in design collaboration? *CoDesign*, 3(1), 5–20.

Barczak, G., & Wilemon, D. (2003). Team member experiences in new product development: views from the trenches. *R&D Management*, 33(5), 463–479.

Benedetto, S., Bernelli Zazzera, F., Bertola, P., Cantamessa, M., Cerie, S., Rancif, C., Spazianteg, A., & Zanino, R. (2010) Alta Scuola Politecnica: An ongoing experiment in the multidisciplinary education of top students towards innovation in engineering, architecture and design. *European Journal of Engineering Education*, 35(6), 627–643.

Blessing, L. (2021). Preface. In L. Blessing, A. J. Qureshi, & K. Gericke (eds.), *The future of transdisciplinary design* (pp. vii–ix). Cham: Springer.

Casakin, H., & Badke-Schaub, P. (2017). Sharedness of team mental models in the course of design-related interaction between architects and clients. *Design Science*, 3(e14), 1–21.

Casakin, H., & Badke-Schaub, P. (2015). Team mental models in design problem solving: An interdisciplinary perspective. In A. M. Columbus (ed.), *Advances in Psychology Research* (pp. 169–182). NY: Nova Science Publishers.

Casakin, H., & Badke-Schaub, P. (2013). The psychology of creativity: Mental models in design teams. In A. Antonietti, B. Colombo, & D. Memmert (eds.), *Psychology of creativity* (pp. 167–180). NY: Nova Science Publishers.

Casakin, H., Ball, L. J., Christensen, B., & Badke-Schaub, P. (2015). How do analogising and mental simulation influence team dynamics in innovative product design? *AIEDAM*, 29(2), 173–183.

Cascini, G. (2015). Engineering grand challenges demand for trans-disciplinary design science. *Journal of Integrated Design and Process Science*, 19(4), 1–2.

Cascini G., Graziosi S., Montagna F., & Rotini F. (2017). On the factors affecting design education within a multi-disciplinary class. *Journal of Integrated Design and Process Science*, 21(2), 21–44.

Cooke, N. J., Kiekel, P. A., Salas, E., Stout, R., Bowers, C., & Cannon-Bowers, J. (2003). Measuring team knowledge: A window to the cognitive underpinnings of team performance. *Group Dynamics: Theory, Research, and Practice*, 7(3), 179.

Coyne, R. (2005). Wicked problems revisited. *Design Studies*, 26(1), 5–17.

Craig, D. L., & Zimring, C. (2000). Supporting collaborative design groups as design communities. *Design Studies*, 21(2), 187–204.

Curşeu, P. L., & Pluut, H. (2013). Student groups as learning entities: The effect of group diversity and teamwork quality on groups' cognitive complexity. *Studies in Higher Education*, 38(1), 87–103.

Dayan, M., & Di Benedetto, C. A. (2008). Procedural and interactional justice perceptions and teamwork quality. *Journal of Business & Industrial Marketing*, 23(8), 566–576.

Dayan, M., & Di Benedetto, C. A. (2009). Antecedents and consequences of teamwork quality in new product development projects: An empirical investigation. *European Journal of Innovation Management*, 12(1), 129–155.

De Vere, I., Melles, G., & Kapoor, A. (2010). Product design engineering – A global education trend in multi-disciplinary training for creative product design. *European Journal of Engineering Education*, 671(35), 33–42.

Dederichs, A., Karlshøj, J., & Hertz, K. (2011). Multi-disciplinary teaching: Engineering course in advanced building design. *Journal of Professional Issues in Engineering Education and Practice*, 137(1), 12–19.

Denton, H. G. (1997). Multi-disciplinary team-based project work: planning factors. *Design Studies*, 18(2), 155–170.

Dyer, J. A. (2003). Multi-disciplinary, interdisciplinary, and transdisciplinary educational models and nursing education. *Nursing Education Perspectives*, 24(4), 186–188.

Eckert, C., & Stacey, M. (2021). Overconstrained and underconstrained creativity: Changing the rhetoric to negotiate the boundaries of design. In L. Blessing, A. J. Qureshi, & K. Gericke (eds.). *The future of transdisciplinary design* (pp. 207–219). Cham: Springer.

Eisenbart, B., Gericke, K., & Blessing, L. (2011). A framework for comparing design modelling approaches across disciplines. *Proceedings of the 18th International Conference on Engineering Design (ICED11)*, August 15-19, Copenhagen (pp. 344–355).
Eisenbart, B., Blessing, L., & Gericke, K. (2012). Functional modelling perspectives across disciplines: A literature review. *Proceedings of the 12th International Design Conference DESIGN 2012*, Cavtat, Croatia (pp. 847–858).
Ertas, A., Maxwell, T., Rainey, V. P., & Tanik, M. M. (2003). Transformation of higher education: the transdisciplinary approach in engineering. *IEEE Transactions on Education*, 46(2), 289–295.
Fernandez-Orviz, A. (2021). Ineffective collaboration in multi-disciplinary teams. In L. Blessing, A. J. Qureshi, & K. Gericke (eds.). *The future of transdisciplinary design* (pp. 27–38). Cham: Springer.
Fiorineschi, L., Frillici, F. S. & Rotini, F. (2020). Impact of missing attributes on the novelty metric of Shah et al. *Research in Engineering Design*, 31(2), 221–234.
Fox, T., Gabor, E., Thomas, D., & Ziegler, J. (2017). Cultivating a reluctance to simplify: Exploring the radio communication context in wildland firefighting. *International Journal of Wildland Fire*, 26(8), 719–731.
Gericke, K., Adolphy, S., Quareshi, A. J., Blessing, L., & Stark, R. (2021). Opening up design methodology. In L. Blessing, A. J. Qureshi, & K. Gericke (eds.), *The future of transdisciplinary design* (pp. 3–14). Cham: Springer.
Gero, J., & Milanovic, J. (2022). What is happening when designers from different disciplines work together: Characterisation of the design behaviors and design spaces of mechanical and electrical engineers working in teams. *DTRS13 -Design Thinking Research Symposium. Technion, ITT*, Haifa, Israel. March 22–24 (243–258).
Giallonardo, L. M., Wong, C. A., & Iwasiw, C. L. (2010). Authentic leadership of preceptors: predictor of new graduate nurses' work engagement and job satisfaction. *Journal of Nursing Management*, 75(8), 993–1003.
Grandey, A. A., Fisk, G. M., Mattila, A. S., Jansen, K. J., & Sideman, L. A. (2005). Is "service with a smile" enough? Authenticity of positive displays during service encounters. *Organisational Behavior and Human Decision Processes*, 96(1), 38–55.
Goetz, S., Hartung, J., Schleich, B., & Wartzack, S. (2019, July). Robustness evaluation of product concepts based on function structures. In *Proceedings of the International Conference on Engineering Design* (Vol. 1, No. 1, pp. 3521–3530). Cambridge University Press.
Gruenther, K., Bailey, R., Wilson, J., Plucker, C., & Hashmi, H. (2009). The influence of prior industry experience and multi-disciplinary teamwork on student design learning in a capstone design course. *Design Studies*, 30(6), 721–736.
Hoegl, M., & Gemuenden, H. G. (2001). Teamwork quality and the success of innovative projects: a theoretical concept and empirical evidence. *Organization Science*, 12(4), 435–449.
Hong, J., Lee, O. K., & Suh, W. (2017). Creating knowledge within a team: a socio-technical interaction perspective. *Knowledge Management Research & Practice*, 15(1), 23–33.
Jacoby-Volk, C., & Bar-Eli, S. (2021). Project-Based Design and Transdisciplinarity: Rethinking approaches to spatial design education. In L. Blessing, A. J. Qureshi, & K. Gericke (eds.). *The future of transdisciplinary design* (pp. 221–232). Cham: Springer.
Jones, A. (2006). Multi-disciplinary team working: Collaboration and conflict. *International journal of mental health nursing*, 15(1), 19–28.
Kallio, H., Pietila, A. M., Johnson, M. & Kangasniemi, M. (2016). Systematic methodological review: developing a framework for a qualitative semi-structured interview guide. *JAN*, 72(12), 2954–2965.
Keating, K., Brougham, C., Gavin, G., & Reilly, G. (2013). Issues in organisation and management of multi-disciplinary group design projects. In DS 76: *Proceedings of E&PDE 2013, the 15th International Conference on Engineering and Product Design Education*, Dublin, Ireland, 05-06.09 (pp. 302–307).
Kozmetsky, G. (1997). Generational road maps and commercialisation of science and technology. *Journal of Integrated Design Process Science*, 1(1) 4–8.

Kratzer, J., Leenders, R. T. A., & Van Engelen, J. M. (2006). Team polarity and creative performance in innovation teams. *Creativity and Innovation Management*, 15(1), 96–104.

Leblanc, T. (2021). Design as common denominator. In L. Blessing, A. J. Qureshi, & K. Gericke (eds.), *The future of transdisciplinary design* (pp. 15–26). Cham: Springer.

Levitt, J. M., & Thelwall, M. (2008). Is multi-disciplinary research more highly cited? A macrolevel study. *Journal of the American Society for Information Science and Technology*, 59(12), 1973–1984.

Maier, A. M., Kreimeyer, M., Hepperle, C., Eckert, C. M., Lindemann, U., & Clarkson, P. J. (2008). Exploration of correlations between factors influencing communication in complex product development. *Concurrent Engineering*, 16(1), 37–59.

Mengoni, M., Peruzzini, M., & Germani, M. (2009). The impact of virtual environments on humans' collaboration in product design. In DS 58-9: *Proceedings of ICED 09, the 17th International Conference on Engineering Design*, Vol. 9, Human Behavior in Design, Palo Alto, CA, USA, 24.-27.08. 2009.

Mohammed, S., & Dumville, B. C. (2001). Team mental models in a team knowledge framework: expanding theory and measurement across disciplinary boundaries. *Journal of Organizational Behavior*, 22, 89–106.

Naumann, S., & Bennett, N. (2000). A case for procedural justice climate: Development and test of a multilevel model. *The Academy of Management Journal*, 43(5), 881–889.

Nelson, M. E., & Ahn, B. (2018). Virtual reality activities for teaching engineering students' professional development skills. In 2018 *IEEE Frontiers in Education Conference* (FIE) (pp. 1–5). IEEE.

Nesterkin, D. A., Porterfield, T. E., & Li, X. (2016). Relationship conflict, conflict management, and performance of information technology teams. *Journal of Computer Information Systems*, 56(3), 194–203.

Nguyen, Q., & Mougenot, C. (2020). Dimensions of multi-disciplinary collaboration: a comparative literature review within design context. *Proceedings of the DESIGN 2020 Conference* (vol. 1, pp. 1335–1344). Cambridge: Cambridge University Press.

Nicolescu, B. (2005). Transdisciplinarity—past, present, and future. *Congresso Mundial de Transdisciplinaridade*, September 2–6, Vila Velha, Brazil (pp. 1–24).

Ou, X., Goldschmidt, G., & Erez, M. (2022). Exploring the effect of team composition on creativity in design idea generation. *DTRS13 -Design Thinking Research Symposium. Technion, ITT,* Haifa, Israel. March 22–24 (211–224).

Pahl, G., & Beitz, W. (1996). *Engineering design: A systematic approach*. London: Springer–Verlag.

Paton, B., & Dorst, K. (2011). Briefing and reframing: A situated practice. *Design Studies*, 32(6), 573–587.

Reich, Y., Ullmann, G., Van der Loos, M., & Leifer, L. (2009). Coaching product development teams: A conceptual foundation for empirical studies. *Research in Engineering Design*, 19(4), 205–222.

Sandoff, M., & Nilsson, K. (2016). How staff experience teamwork challenges in a new organisational structure. *Team Performance Management*, 22(7/8), 415–427.

Schummer, J. (2004). Multidisciplinarity, interdisciplinarity, and patterns of research collaboration in nanoscience and nanotechnology. *Scientometrics*, 59(3), 425–465.

Simon, H. A. (1996). *The sciences of the artificial*. 3rd ed. MIT Press.

Suwa, M., Gero, J., & Purcell, T. (2000). Unexpected discoveries and S-invention of design requirements: important vehicles for a design process. *Design Studies*, 21(6), 539–567.

Tharp, T., & Zalewski, J. (2001). Economics of software engineering: Transdisciplinary issues in research and education. *Integrated Design & Process Technology Proceedings*, June 6–11, (vol. 1, pp. 289–295).

Yammarino, F., Dionne, S., Schrieseim, C., & Dansereau, F. (2008). Authentic leadership and positive organisational behavior: A meso, multi-level perspective. *The Leadership Quarterly*, 19(6), 693–707.

4.3 Changing roles in interdisciplinary Research-through-Design

Berit Godfroij[a], *Remko van der Lugt*[a], *Sebastiaan Peek*[b], *and Rens Brankaert*[c]

[a]*Research Group Co-design, Research Centre of Learning and Innovation, Utrecht University of Applied Sciences, The Netherlands;* [b]*Tranzo, School of Social and Behavioral Sciences, Tilburg University, The Netherlands;* [c]*Fontys School for Allied Health Professions, Eindhoven & Industrial Design, Eindhoven University of Technology, The Netherlands*

4.3.1 Introduction

Design Thinking plays a growing role in connecting disciplines (Cross, 2018), by means of design processes, and in coordinating the integration of resources and various perspectives (Prestes Joly et al., 2019). Design Thinking also supports organizations in leveraging existing business models and creating organizational futures (Liedtka & Kaplan, 2019, p.6). Connecting design with non-design disciplines and coordinating the integration of resources require an interdisciplinary approach (Peek, 2015; Stock & Burton, 2011) that integrates research and design activities. Design and research come together in practical experiments and are shared in the activity of the design process (Roggema, 2017; Cross, 2001). When research and design are brought together in a process and the act of designing is used as a form of research, this is known as *Research-through-Design* (RtD; Stappers & Giaccardi, 2017; Zimmerman & Forlizzi, 2014).

Over the years, there has been a spate of interest in RtD, and many academics have formulated their own perspectives on it. They define RtD as a practice-based inquiry that generates transferrable knowledge (Durrant et al., 2017) in an interdisciplinary process (Sleeswijk Visser, 2018) with the intended goal being to instigate societal change (Brankaert & den Ouden, 2017; Roggema, 2017; Zimmerman et al., 2010, 2007; Binder & Redström, 2006; Swann, 2002). In RtD, artefacts are iteratively designed as a creative way to investigate a potential future (Roggema, 2017; Zimmerman et al., 2010) and focus on generating knowledge rather than improving design practice (Roggema, 2017). Godin and Zahedi (2014, p.11) suggested that *"the artefact is not the goal of RtD; knowledge and understanding is,"* that it is not predictable, and that it is *"riddled with issues that come with its heavy reliance on design."* Table 4.3.1 summarizes how RtD is understood by various academics, presenting a list of RtD features that we structured on the organizational, process, and product levels. We distinguished between these levels to build on previous work (Godfroij & Van der Lugt, 2020) in which we identified scaffolding concepts of change in a designer's role on those three levels. In this study, we extended and further explored the scaffolding concepts by grounding them in theory (see Table 4.3.2), making use of theoretical signposts to structure the empirical data in the findings section and the theoretical discussion. The findings brought the concepts to life and allowed us to empirically demonstrate how theorized phenomena play a role in unfolding processes (Berends & Deken, 2019; Jarzabkowski et al., 2014).

The empirical context of this study was an applied RtD-dementia research project titled "Dynamics of Dementia in Design" (DDD). This project manifested the RtD

DOI: 10.1201/b22530-28

Table 4.3.1 RtD features at the organizational, process, and product levels

RtD Characteristics	*RtD Features at the (1) OrganizationaL, (2) Process, and (3) Product Levels*
An interdisciplinary process for instigating societal change (Brankaert & Den Ouden, 2017; Roggema, 2017; Zimmerman, Stolterman, & Forlizzi, 2010; Zimmerman, Forlizzi, & Evenson, 2007; Binder & Redström, 2006; Swann, 2002).	1 Organization: Interdisciplinary structured 2 It is a process 3 The product is societal change
Merging research and design processes, design practice, and academic experiments and reflection (Stappers & Giaccardi, 2017).	1 Organizational structure includes actors from academic research institutes and from practice 2 Process of merging research and design 3 Products: designs, experiments, and reflection
A process of iteratively designing artefacts as a creative way of investigating a potential future (Roggema, 2017; Zimmerman et al., 2010).	1 N/A 2 Iterative process 3 Products: Artefacts for investigating a potential future
Practice-based knowledge generation (Durrant, Vines, Wallace, et al., 2017; Roggema, 2017). The artefact is not the goal (Godin & Zahedi, 2014).	1 N/A 2 Process is practice-based 3 Product is communicable knowledge and understanding
The preferred state is that of what "will be," by learning from the interaction of the current state by means of dialogue with artefacts (Godin & Zahedi, 2014).	1 N/A 2 Process of dialogue with artefact(s) 3 Focus on what "will be" and learning from interaction
Very similar in appearance to a regular design project. The goals are different; the influence of research is present at most steps; and the documentation of the process is a key concern (Godin & Zahedi, 2014).	1 N/A 2 Process is very similar to a design process; influence of research inquiry is present at most steps 3 Product: Process documentation
RtD is not able to provide predictability. It is riddled with issues that accompany its heavy reliance on design (Godin & Zahedi, 2014).	1 N/A 2 The process is riddled with issues that accompany the heavy reliance on design 3 The product/outcome is unpredictable

features in Table 4.3.1. The DDD project aimed to improve and maintain the social health and social participation of independently living people with dementia and their caregivers by designing and evaluating digital solutions. This section investigates the interdisciplinary dynamics between researchers involved in the DDD project from the creative, health, and social behavioral domains, with a holistic view on interdisciplinary RtD dynamics. It structures the analysis on the *organizational* (structure of actors), *process* (interdisciplinary ways of working), and *product levels* (project aim: Artefacts and/or knowledge) to understand the changing roles of design researchers within the complexity of the larger system of an interdisciplinary RtD collaboration.

4.3.1.1 Interdisciplinary collaboration in Research-through-Design

Interdisciplinary collaboration within the field of design has increased over the last decades (Christensen & Ball, 2019; Cooper, 2019). Working in multidisciplinary design teams

Table 4.3.2 Theoretical foundations regarding the changing roles for designers (adapted from Godfroij & van der Lugt, 2020)

	Concepts	*Grounding in Theory*
Organizational level	**Structuring the project** Meetings with various professionals from different departments and on various levels of organization	*How and why does the structure of project actors change over time?* • Supervisors and a steering committee are included, in which all disciplines are represented (Andriessen et al., 2019). • Traditional roles, tasks, and components are reshaped (Meroni, 2008) and divided among disciplines, and various modes of inquiry and languages are considered (Roggema, 2017). • Boundary crossers are included to support integration between disciplines (Akkerman & Bakker, 2011). • Organize by assigning roles and components of roles to project actors (Sleeswijk Visser, 2018), and align plans across all levels of an organization (van der Lugt & van der Laan, 2017). • Exploring and building relationships are important dynamics in design innovation journeys, in which actors couple and decouple (Godfroij, 2021).
Process level	**Different ways of working** Building on existing knowledge (e.g., customer segmentations) together with other professionals (e.g., an econometrist)	*How do project actors align perspectives, attitudes, methods, and approaches? How do they share knowledge, tools, and skills with the other actors involved?* • It is worthwhile to spend time together with other discipline(s) to create an understanding of multiple perspectives with regard to time, structure, control, position, and grounding approaches (Zielhuis et al., 2020). • Jointly define, determine, and redefine project definitions and scope in an ongoing discussion and negotiation between project actors (Haug, 2015; Higgins & Klein, 2011; Tan 2011; Meroni, 2008), rather than following a project proposal as approved at the start. • When coupled, the actors involved maneuver as a team and interweave by focusing on encouraging and balancing diverse views (Godfroij, 2021).

(*Continued*)

Table 4.3.2 (Continued)

	Concepts	*Grounding in Theory*
Product level	**Shift in focus** Shifting from creation toward research to improve strategic processes	*How do the project aim and focus evolve over time?* • Exchange of competences (knowledge, skills, tools, and mindset), rather than a shift in emphasis of the type of output that is under investigation (Vargo et al., 2020, p.14). • Shared sensemaking between multiple actors ((Vargo et al., 2020). • A more strategic role, rather than a role as creator (designer) or traditional researcher (Calabretta et al., 2017). • Building on existing products and knowledge of one's own and the other organization/ discipline, rather than creating new (tangible/ knowledge) products that stand alone (Godfroij & Van der Lugt, 2020). • Co-creating value that is determined by the beneficiary: Added value for all actors involved (Vargo et al., 2020). • Co-creating value with a system perspective: Contribution to complex societal issues in a system larger than the project ((Vargo et al., 2020). • Joining and/or aligning with ongoing projects and activities in organizations (Godfroij & Van der Lugt, 2020). • Design innovation journeys that are highly or extremely complex and networked, and a focus on ongoing discussion, negotiations of sensemaking, and exchange of knowledge and skills, rather than on moving around products or exchanging objects (Godfroij, 2021).

requires synchronization between practices, which can be challenging (Gero & Milovanovic, 2022; Andriessen et al., 2020). Interdisciplinary design collaborations deal with cross-cultural challenges, as reflected in the practices, behaviors, design tools, and materials (Halskov & Christensen, 2018). To frame and address complex problems, designers must cross disciplinary boundaries and build on diverse disciplines, theories, and approaches (Sciannamè, 2022). Diverse expertise in team composition contributes to design performance, especially to creativity (Ou, Goldschmidt, & Erez, 2022). However, combining different kinds of expertise has also been demonstrated to cause significant difficulties in communication and team cohesion (Ou et al., 2022). Furthermore, despite the growing interest, dynamics in interdisciplinary RtD collaboration are still not completely understood.

Some previous research has examined how project actors collaborate in RtD teams (Sleeswijk Visser, 2018). Some scholars have directly or indirectly investigated aspects in interdisciplinary RtD collaboration. For example, Meroni (2008) and Haug (2015) have mentioned a shift in design toward actors who increasingly work together to define projects. Higgins and Klein (2011) emphasized the importance of jointly determining scope, which according to them is a subject of ongoing discussions and negotiations regarding sensemaking and scope refinement. Sleeswijk Visser (2018) stated that confusion in the knowledge-generation process of RtD projects easily occurs regarding roles, processes, and results. Moreover, recent studies have emphasized a shift in the role of designers today, which is or should be expanded across discipline boundaries (Sciannamè, 2022) due to an increased need to share perspectives and methods between disciplines (Rylander & Eneberg, 2022; Van Beest et al., 2022). In this shift of roles, Jones (2006) and Casakin and Cascini (2022) have noted several conflicts regarding roles and social dynamics, such as when multidisciplinary teams tackle health issues.

Furthermore, past research has not sufficiently addressed related changing roles for designers in RtD collaboration. Roggema (2017) linked theory on dynamics with design phases – where he distinguished between a predesign, a design, and a post-design phase, in which "coupling," "interweaving," and "decoupling" dynamics occur, respectively (Basballe & Halskov, 2012); however, empirical evidence is lacking on such dynamics related to how design and nondesign researchers work together and align roles within various phases of a design process. The present study provides a holistic view on dynamics – in all phases of an RtD project – between actors from different research disciplines to provide an understanding of the changing roles for design researchers in interdisciplinary RtD.

We build on previous work (Godfroij & Van der Lugt, 2020) in which we investigated collaboration in *design practice* on the intersection of creative industries and healthcare. In that earlier work, we identified *"disruptions of flow"* (Yanow & Tsoukas, 2009) as experienced in collaboration between design researchers and professionals from an academic hospital. In the current research, we observe change in a *design research* project regarding (1) the organizational structure; (2) processual change in (creative) ways of working and collaborating; and (3) product-level change in the project's focus. The research question that we addressed is as follows:

How and why do changing roles for design researchers unfold on organizational, process, and product levels in an interdisciplinary RtD project?

4.3.1.2 Changing roles in Research-through-Design

In our previous study (Godfroij & van der Lugt, 2020), which examined a *design practice case* in a health context, we identified change in a designer's role on three levels: First, at the organizational level, we found that a senior designer had many more meetings with professionals from various departments and on various levels within the health organization than he was used to in previous projects, which played a crucial role in *structuring the project (actors)*. Second, at the process level, we found that a designer was searching for leads to build on existing knowledge together with professionals within the health organization, for which he used different *ways of working* than he was used to, for example, compared with designing hospital beds. Third, at the product level, we found that a designer *shifted focus* from the creation of prototypes and artefacts toward improving strategic processes. In the present study, we extended and further explored these findings by examining a *design research project*. Table 4.3.2 summarizes the scaffolding concepts of Godfroij and Van der Lugt (2020) on three levels and grounds them in theory.

4.3.2 Research methods

This study investigated change regarding the structure of actors (organization level), ways of working (process level), and project focus (product level). We conducted a longitudinal field study over three years (2018–2021) and analyzed the data by means of a process research approach focused on temporality, activity, and flow (Langley & Tsoukas, 2017; Langley et al., 2013; Langley, 2007). In this section, we further explain the case and the actors involved (2.1), the data gathering (2.2), and the data analysis (2.3).

4.3.2.1 The case and the actors involved

The Dynamics of Dementia in Design (DDD) project ran between January 2018 and January 2021 as part of a Dutch research program called "Create Health," which contributes to societal problems with regard to healthy and active aging. Tables 4.3.3–4.3.5 present the RtD nature of DDD by explaining how RtD features – as discussed in the introduction – manifested in the interdisciplinary research project.

Structuring the project (actors). In DDD, scientists from various disciplines were involved to investigate and design digital applications for increasing the social participation of people with dementia. As Table 4.3.3 indicates, the project structure connected design

Table 4.3.3 RtD features on an organizational level and the nature of the Dementia Dynamics in Design project: Structuring the project (actors)

	RtD features	Nature of the DDD Project
ORGANIZATIONAL LEVEL	• Project organization is *interdisciplinary* structured • Organizational structure includes actors from academic *research* institutes and from *practice*	Consortium with academic actors from design research (industrial design) and social sciences (longitudinal research), and practice partners from a mental health institute and a local innovation network.

Table 4.3.4 RtD features on the process level and the nature of the Dementia Dynamics in Design project: Ways of working

	RtD Features	*Nature of the DDD Project*
PROCESS LEVEL	• RtD is a process • of merging research and design • of iterations • that is practice-based • of dialogue with artefact(s) • that is riddled with issues due to the heavy reliance on design • that is highly similar to a design process, in which the influence of the research inquiry is present at most steps	Research and design were combined by applying design tools (e.g., a social mapping tool) in traditional field research and interweaving co-design sessions with an interview study. Practice-based fieldwork and field evaluation with a high-fidelity prototype were conducted at people's homes. Various iterations occurred with professionals to design the prototype. The researchers interacted with participants in dialogue with tools created for and applied in the field research, namely a social mapping tool, a diary probe, and a hobby tool. The project lead (social behavioral researcher) realized over time that the research heavily relied on design and that *"design methods are so good for getting ahead."* Issues occurred because said researcher initially resisted design methods to apply as part of the longitudinal field research, not wanting to influence his participants; this is considered *"not done."* Appearance of the process: *Discover* by means of interviews and co-design sessions at peoples' homes; (re)*define* that the problem is also about going somewhere by means of transport; *develop* three design tools and a high-fidelity prototype; and *deliver* knowledge and understanding based on field evaluations with the high-fidelity prototype at people's homes.

Table 4.3.5 RtD features on the product level and the nature of the Dementia Dynamics in Design project: Shift in focus

	RtD Features	*Nature of the DDD Project*
PRODUCT LEVEL	• The product/project outcome is unpredictable • The project outcomes contribute to societal change • Projects aim to generate communicable knowledge and understanding • Tangibles are research artefacts rather than project outcomes • The focus lies on what "will be" as well as on learning experience	The aim was not to deliver a predefined eHealth product but rather to generate communicable knowledge and understanding on the social health and participation of independently living people with dementia, contributing to societal change by improving social health and reducing the risk of loneliness. Four tangible design tools and a high-fidelity prototype were created as research artefacts and applied in interviews, co-design

(*Continued*)

Table 4.3.5 (Continued)

	RtD Features	*Nature of the DDD Project*
	• Insights on the evaluation of designs, crafts, experiments, and reflection • Process documentation is a key concern	sessions, and a field evaluation to *"actively build a future"* and focus on what will be. The insights were based on experimenting with design tools in longitudinal field research as well as on field evaluation with the high-fidelity prototype. Furthermore, many cross-over and reflection sessions were held with six researchers of the consortium. The documented process, which evolved from the collaboration process, was presented in a final project event.

with another discipline and combined research and practice. There were researchers on various organizational levels: the PhD level, post-doc, assistant professor, and full professor. The interdisciplinary consortium (see Figure 4.3.1) consisted of researchers from the School of Social and Behavioral Sciences of Tilburg University, the Department of Industrial Design of University of Technology Eindhoven, Mental Health Institute

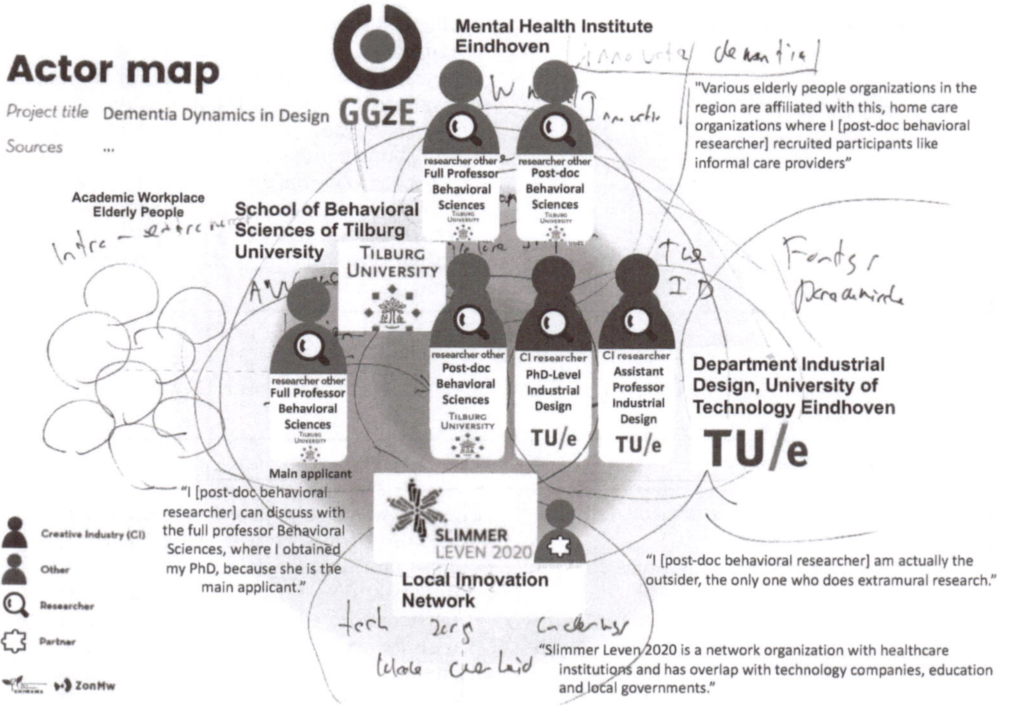

Figure 4.3.1 Actor map of the Dementia Dynamics in Design project, on which the post-doc behavior researcher made notes during an interview. After the interview, the map was complemented with quotes, actors, and organizations. Various versions of the map were created in iterative rounds of analysis.

Eindhoven, and a local innovation network for active and healthy aging: 'Slimmer Leven 2020'. The innovation network was involved particularly to recruit participants and to disseminate the results, and it has connections with healthcare institutions, technology companies, and municipalities among others.

A senior design researcher (assistant professor) and a post-doc behavioral sciences researcher jointly wrote the project proposal. The senior design researcher involved a junior design researcher at the PhD level, who together with the behavioral sciences researcher formed the inner circle of the project. They conducted the research and held weekly progress meetings together. The PhD-level design researcher also had weekly meetings with the senior design researcher – who acted as supervisor – to discuss the research and project. The actor map in Figure 4.3.1 indicates that the senior design researcher was positioned next to the inner circle, along with the three advisory senior researchers from Tilburg University. The behavioral researcher in the inner circle had a temporary position at Tilburg University (post-doc) and was supported by a full professor in behavioral research, who also functioned as the main applicant. The post-doc and full professors in social and behavioral science were able to discuss the research and project irregularly at the university. Two researchers from the mental health institute (GGZE), also affiliated with Tilburg University, had an advisory role and were engaged in the analysis and dissemination of results. Consortium meetings were held between all of the aforementioned individuals every six weeks.

Ways of working. As Table 4.3.4 mentions, DDD was pre-eminently a project in which research and design were merged. Although the process initially consisted of four separate and sequential research activities (i.e., longitudinal field research, design and creation, field evaluation with prototypes, and adaptation and extension of a theoretical framework), the researchers organically searched for a process to benefit each other's disciplines. Gradually, the process became more interdisciplinary and interwoven, and the researchers applied three design tools: A social mapping tool (Figure 4.3.2), a diary probe (Figure 4.3.3), and a hobby and activity tool (Figure 4.3.4; Den Haan-Wintermans et al., 2019). Initially, the idea was that insights from the longitudinal field study (qualitative data) would form the input for the design process, or as the social behavioral researcher put it *"bluntly, the psychologist tells the designer what to make,"* and that the participants who were involved longitudinally would not be influenced by design interventions. However, over numerous discussions, the researchers found an approach that integrated design research tools, where longitudinal field research and design research (co-design sessions) occurred in parallel and informed each other.

DDD followed a field study approach in which most of the research occurred in practice. The researchers conducted interviews and co-design sessions at people's homes. This was combined with input from professionals to design a digital intervention (Kolasinska, et al., 2021). Figure 4.3.5 presents a high-fidelity prototype of this intervention that was evaluated in real life (Brankaert & den Ouden, 2017). The evaluation of the prototype was conducted through a mixed-methods approach that combined qualitative research and the monitoring of usage data. In both the longitudinal field research as well as the co-design sessions, the researchers interacted with participants at home.

The DDD researchers spent much time discussing such issues as the design methods to apply in traditional field research, influencing participants, and publishing design research. The social behavioral researcher realized over time that the research heavily relied on design and that *"design methods are so good for getting ahead."* The process was highly similar to a regular design process: First, the researchers discovered the field by

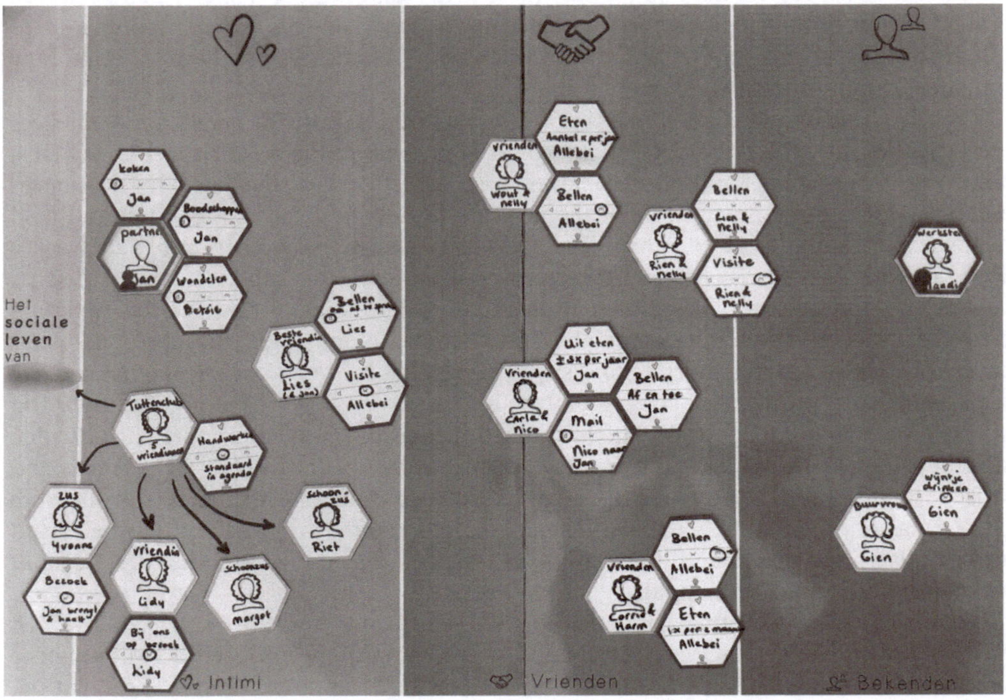

Figure 4.3.2 Left: Social mapping tool using who, what, and how frequency cards to understand social activities and their emotional intimacy in people living with dementia (from left to right: Intimates, friends, and acquaintances).

means of interviews and co-design sessions at people's homes. Second, they (re)defined the problem and focus, which the social behavioral researcher explained as follows:

> *I knew that the scope of research needed to be well formulated after we wrote the project proposal, and I wanted to formulate it sharper, however the design researcher made clear that we cannot know in advance what to do exactly and that we just had to start.*

Third, they developed and applied three design tools. Lastly, the project delivered knowledge and understanding based on field evaluations with the high-fidelity prototype.

Shift in focus. DDD aimed to research the design of eHealth products and services that support elderly people with dementia who experience substantial cognitive changes. It focused on expanding the body of knowledge on design research in a real-life dementia context. The aim of the study was to generate fundamental knowledge regarding the long-term user involvement of people with dementia by working on two knowledge building blocks: (1) the primary knowledge building block, which is a longitudinal framework applied in the design process, and (2) the secondary knowledge building block, which concerns how to design and employ design interventions that address the needs of people living with dementia. Table 4.3.5 indicates that the DDD project created and evaluated digital solutions for social activity, not to deliver a predefined product in the first place but rather as part of a process generating knowledge on research in a real-life dementia

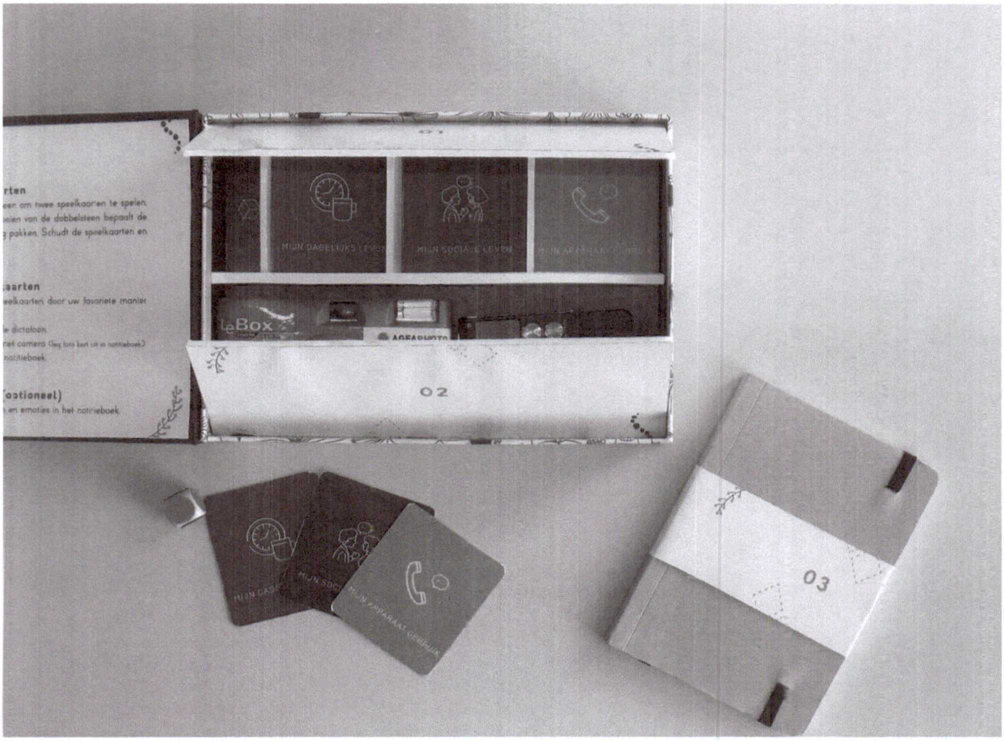

Figure 4.3.3 Right: Diary study for understanding a participant's personal situation "in the wild." The diary consisted of a dice with colored sides, colored playing cards with personalized questions, and three different mediums (a notebook, a voice recorder, and a camera to express experiences).

context. The project outcomes were aimed at contributing to societal change regarding the social health and participation of independently living people with dementia as well as reducing the risk of loneliness. The generated knowledge was communicated through events for academic and design professionals, through design tools (maps and prototype), and in academic publications.

Moreover, the DDD researchers created the three design tools and the high-fidelity prototype and applied them in practice to actively build a future. The project proposal (2018) stated the following: *"By designing research objects and placing them into a real-world context, design can change the current state, creating new situation."* The DDD-researchers experimented with the research artefacts and reflected on the process with the six researchers in the consortium as a learning experience. The insights were based on experiments and the evaluation of the design tools and prototype, *"about what we cannot know yet,"* and lastly on reflection.

4.3.2.2 Data gathering

The data gathering was based on the principles of organizational ethnography (Ybema et al., 2009). The methods applied were participant observation, document analysis, and

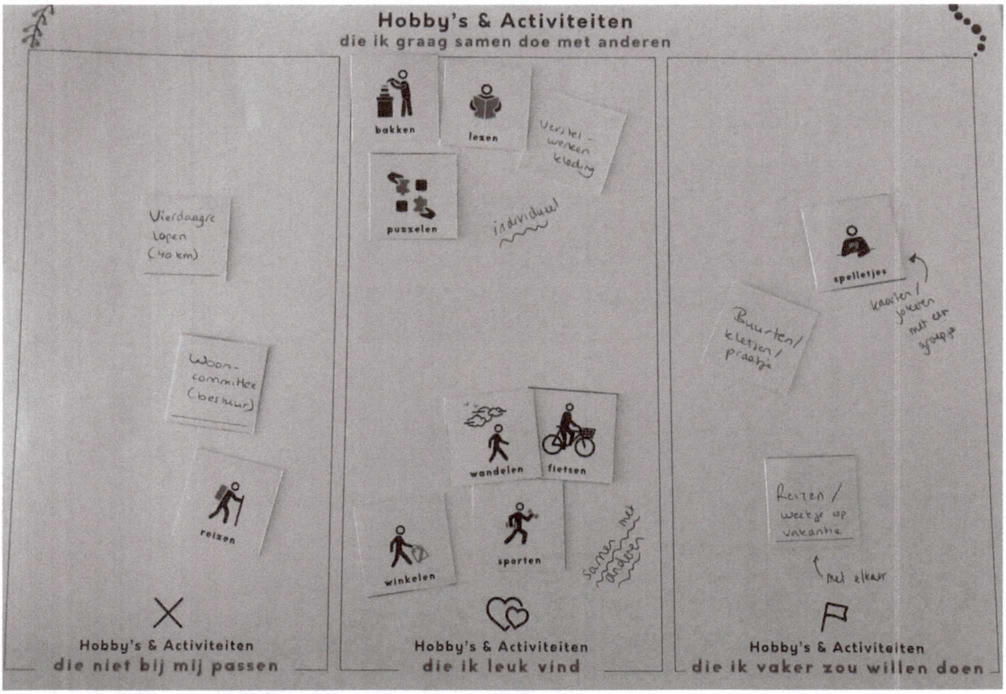

Figure 4.3.4 Left: Hobby and activity tool for understanding a person's personal interest in activities that do not fit their lives or cannot be performed anymore, that they still like to do, and that they would like to do (adapted from Den Haan-Wintermans, Brankaert, & Lu, 2019).

interviews. The participant observation concerned a qualitative longitudinal (three years) collection of naturally occurring data (things observed and heard) within a single case (DDD). It included observations of six participants (see Table 4.3.6) with different backgrounds in creative, health, and social domains, namely social and behavioral science with experience in mental health and industrial design (Human Computer Interaction) in the context of health.

Two researchers (the third and fourth authors of this paper) participated in the DDD project and gathered data concerning the process and collaboration in two reflection sessions (focus groups held in September and October 2019), which all six researchers listed in Table 4.3.6 were involved in. In the two reflection sessions, the participants reflected on collaboration dynamics. Basballe and Halskov (2012) described collaboration dynamics as a process of coupling, interweaving, and decoupling over the course of a project. They defined *"coupling"* as a dynamic that unites interests from various disciplines to establish a common point of departure; *"interweaving"* as a dynamic wherein one activity or material informs various disciplines' interests; and *"decoupling"* as a dynamic that modifies the focus by turning one discipline's interests into the salient focus of the process.

The two reflection sessions were audio-recorded and transcribed. In addition, the following two sets of documents were gathered: A form that the six researchers filled in as preparatory material for the first reflection session as well as presentation slides presented

Figure 4.3.5 Right: Participant interacting with the high-fidelity prototype in a real-life setting (Brankaert & den Ouden, 2017).

Table 4.3.6 Main observed participants

Researcher	*Background*	*Role in The DDD project*	*function*
1	Social and behavioral sciences	Main applicant and responsible, advisory role from research	Full professor
2	Social and behavioral sciences	Project manager*, senior researcher	Post-doc
3	Industrial design	Project manager*, senior researcher	Assistant professor
4	Industrial design	Junior researcher	PhD student
5	Social and behavioral sciences	Advisory role from practice (mental health institute)	Full professor
6	Social and behavioral sciences	Advisory role from practice (mental health institute)	Post-doc

* From the project proposal (2018): "The researchers from [name of university] and [name of university] are jointly responsible for project management, coordination and communication to partners (professors from both universities will provide guidance where necessary)."

at the second reflection meeting. By seeking to evolve processes *from within* and highlighting how meanings and experiences evolve over time and in context, the two DDD researchers (authors) conducted "prehensive research" (Langley & Tsoukas, 2017, p. 9) on how contingencies shape the paths taken or not taken.

A team of other researchers (among them the first and second authors of this paper) observed the DDD project *from outside*. They observed from a distance how the flow of events over time represents shifts in configurations of actions at the organizational, process, and product levels, thus conducting "configurational research" (Langley & Tsoukas, 2017, p.9). The configurational data gathering consisted of three interviews with the project manager (in June 2019, February 2020, and December 2020); in-between contact with participants 2, 3, and 4 by e-mail, telephone, and video calls; and observations during conferences (Dutch Design Week 2018, 2019, 2020, 2021 & Dementia Lab Conference 2019), project events (November 2020, and January 2021), and network events. Specifically, two days of presentations and workshops were organized by the research program Create Health, which is financed by The Netherlands Organization for Health Research and Development (ZonMW), and two knowledge-sharing events were organized by the overarching research Create and Health Innovation WAys of Working Analysis (CHIWAWA). The interviews were audio-recorded and transcribed, and the observations and things heard in between were documented in field notes. Moreover, the configurational data gathering included project documents, such as the DDD project proposal (2018), three progress reports (2019, 2020, and 2021), and internal documents, such as persona-development information, pictures, and tangible maps and tools that the DDD researchers applied in their field research.

4.3.2.3 Data analysis

The transcriptions of the interviews and reflection sessions, field notes, reflection forms, the project proposal, and progress reports were coded in the ATLAS.ti analysis software regarding changes in (1) the structure of actors, (2) ways of working, and (3) project focus. These three codes were similar to the concepts introduced in Table 4.3.2. The codes were used as "theoretical signposts" (Berends & Deken, 2019, p.7) in the creation of a narrative and in connecting findings to theory. Visual project documents such as presentation slides, tangible maps, tools, and pictures were used to obtain an enhanced understanding of the project and its contents and how they evolved over time. To provide an understanding of and explain change regarding the theoretical signposts, this study made use of deeper levels of narrative, in which the analysis moved from the surface to structures that cannot be observed (Pentland, 1999). By means of four levels of narrative – namely the text, story, fabula, and mechanism levels – the reader is guided from the raw data and raw participant stories (i.e., the text level) toward the story and theories of the researchers (the mechanism level), which is an interpretation of the data. Table 4.3.7 provides examples of the four levels of narrative:

To create the four levels of narrative, we first organized the data into chronological data sets that included the raw data on the *text level*, such as quotes, text documents, and visual materials. This formed the participants' stories in which we observed change over time; for instance, *"while shaping our original framework, we have created (graphical) longitudinal case descriptions [...]. In the DDD project we will do the same, but for people with dementia"* shifted toward *"we call it persona but, it is an impression of a persona. At this moment it is not a combination of several people, but it is one-on-one insights."* These chronological data sets provided an understanding of, for example, changing terminology such as (graphical) longitudinal case descriptions that evolved into (textual) personas, but also an understanding of how this visually changed over time in the elaboration of materials.

Table 4.3.7 From surface to deeper levels of narrative (adapted from Pentland, 1999)

← PARTICIPANTS' STORY

Level	*Definition*	*Examples from THE DDD project*
Text	Particular telling (chronological data set, including quotes)	*"The senior design researcher immediately wanted to apply design activities in my field research. I am educated not to influence the field research, and I did not want participants of my field study to be influenced by design activities, in the sense that in the journals I publish that is 'not done'. So, I resisted for a long time."*
Story	Narrative version from a specific point of view (authors' version)	For a long time, the senior behavioral researchers have resisted design methods employed in longitudinal field research since the methods conflicted with their traditional research approach.
Fabula	Generic description of a particular set of events and their relationships (how & why things happened)	To allow for designing together with participants and to evaluate design work in practice, the design researchers proposed openness and methodological flexibility to respond to emerging perspectives. The behavioral researchers came from a more traditional field of research and sought to maintain methodological rigor. After lengthy discussions, the social behavioral design researcher accepted an integrated approach to learn by experience how design contributes to traditional research.
Mechanisms	Underlying structures that enable or constrain in fabula (overall general process)	The constraining factor on methodological flexibility, as explained by the social behavioral researcher, was that the journals in which he publishes are *"nicely multidisciplinary for psychological terms, but not that creative like let's all start using design methods."* The enabling factor was learning by discovery, or as the social behavior researcher put in, *"[w]e will see what it yields."* He wanted to learn from the other discipline (design), rather than simply sticking to tradition and publishing in referred journals.

RESEARCHERS' STORY →

Second, on the *story level*, we created a narrative that provided an understanding of and told the story of the DDD project regarding changes in the structure of actors, ways of working, and project focus from the point of view of the authors. As the example in Table 4.3.7 indicates, the story stayed close to the raw data but simultaneously compressed and summarized it. The narrative of the project comprises almost 3500 words. It was used to identify change on the organizational, process, and product levels, of which examples are provided throughout this paper, and also to select the focus at the fabula level.

For the identified change, we created *fabula-level* generic descriptions that explained how and why particular sets of events happened. For instance, the attitude of the researchers changed over time with regard to applying design methods in the longitudinal field research because the phase was used for discovery. Another example was that the DDD researchers changed their focus regarding technology and included means of transport rather than just means of communication such as computers and telephones, which were stated in the project proposal since the first interview with a participant made clear that *"going somewhere, to someone, seemed also important for the social life of people living with dementia."*

Lastly, in the analysis, we searched at the *mechanism level* for explanations of enabling and constraining factors for the fabula – as explicated by the DDD researchers – and connected them to the theoretically grounded concepts (Table 4.3.2) at the organizational, process, and product levels. Based on these theoretical explanations for change in the structure of actors, ways of working, and project focus of an interdisciplinary RtD project (DDD), we proposed changing roles for design researchers within this larger system of interdisciplinary collaboration in RtD.

4.3.3 Findings

The next three sections present the findings at the fabula level regarding changes in the (organizational) structure of actors (4.3.3.1), ways of working (process; 4.3.3.2), and project focus (product; 4.3.3.3) of the DDD project, and also how and why the roles of design researchers within the interdisciplinary collaboration unfold over time. Last, Section 3.4 discusses enabling and constraining factors for the identified change at the mechanism level.

4.3.3.1 Changing structure of actors in DDD

The six researchers and practice partners were involved from the writing of the project proposal onwards, except for the PhD-level design researcher, who became involved some months later. Said researcher became responsible for much of the work concerning the shaping and materialization of interweaving by informing actors of both research domains (design and behavioral science). She described her role as that of a *mediator* between the more experienced design researcher and behavioral researcher who jointly wrote the original project proposal (without her involvement). Both senior researchers agreed with her on this, noting that they *"may have been shaped more by our respective disciplines."* In the project proposal, the idea was that a researcher from one discipline would participate in 20% of the research activities of the other discipline. Along the way, however, it transpired that this investment of time was unfeasible for the senior researchers, but the PhD-level design researcher did attend several interviews with the senior behavioral researcher, thereby participating in research activities of the other discipline. The senior behavioral researcher stated the following: *"I cannot spend the same amount of time by attending design sessions, but I did contribute to the creation of a probe that the PhD-level design researcher used in a diary study. We discussed the questions together and tailored them for each participant"* (see probe for diary study in Figure 4.3.3).

On two occasions, decoupling occurred in the project. On the first occasion, the project became overly delayed because of coupling and interweaving dynamics as well as difficulty in participant recruitment. When these problems occurred, the decision was made to temporarily decouple. For instance, there were times when the design researchers and behavioral researchers collected and analyzed data together as well as times when this was done separately. Second, decoupling occurred in periods when the researchers were occupied with other projects or tasks. This occasionally led to tensions that were subsequently discussed in meetings. However, the researchers noted that this had occurred in many other projects they had been involved in, whether interdisciplinary or not. Finally, the researchers expected that decoupling would occur once they were expected to publish articles or proceedings in their respective journals or conferences. However, a shared desire existed to publish in interdisciplinary outlets that would require less or no decoupling, and all of the researchers who participated in the project felt that they had a strong will to couple the interests of behavioral and design research. As one respondent put it, *"I feel like we are pushing each other to do that. So, not letting go to achieve that."*

4.3.3.2 Changing ways of working in DDD

The project proposal suggested that most of the longitudinal field study would be conducted by the social behavioral researcher (post-doc) and would consist of interviews and observations without design research tools. It also proposed that most of the co-design study would be conducted by the senior design researcher (assistant professor), and that the design researcher would work closely in interviews and focus groups with researchers from the mental health institute, applying a participatory design approach. However, when the project started, the design researcher suggested design methods that could be employed as part of the longitudinal field research with the same participants The purpose of this approach was to design together with the participants as well as to uncover insights through designerly approaches in practice. To allow this, the design researcher proposed openness and methodological flexibility to respond to emerging perspectives. By contrast, the behavioral researchers came from a more traditional field of research and sought to *"maintain the methodological rigor needed to address our research questions.'*

Over time, the researchers resolved to create an activity and a material that informed both disciplines: a social mapping tool (see Figure 4.3.2) and an approach that integrated design research tools. In this approach, the PhD-level design researcher and senior behavioral researcher (post-doc) exchanged insights from the longitudinal field research to keep each other in the loop and up to date of each other's perspectives. This implied that the interviews conducted by the post-doc behavioral researcher alternated with co-design sessions at people's homes conducted by the PhD-level design researcher. Based on the interviews and co-design sessions, these two researchers developed (one-on-one) personas together. In iterative rounds of analysis and information sharing, insights became increasingly sharp. The behavioral researcher stated the following: *"Also about what we cannot yet know, for example whether a lack of initiative is a problem for people with dementia or whether they may not want to take the initiative at all. So that's why those design methods are so good for getting ahead!"*

When evaluating the ways of working, the behavioral researcher explicated that he would focus less on his own frame in the future; however, he would have liked to embed the findings more in theory as well as to analyze more statistically, as opposed to only capturing qualitative snapshots. For example, he stated the following: *"We just take a picture of the tool and in the next session, we take another picture and then things are placed differently on the map, which feeds the next interview round. It is not statistically analyzed."* Moreover, he had hoped for greater involvement of the creative industry and would suggest involving business from the start of the project: *"Just because I wish the project continues."* The senior design researcher (assistant professor), on the other hand, explicated that in his next interdisciplinary project, he would define and present a stronger argument for the design research part, since *"now, it was too messy, since we left the design part very open and that caused a lot of discussion."* Furthermore, both design researchers (PhD-level and assistant professor) would suggest making use of design tools or objects earlier in the next project. According to them, involving business could be interesting but also be obstructive due to additional requirements, such as limitations in terms of costs and possibilities. However, hesitating, the PhD-level design researcher stated the following: "*When we come close to a final design, we encounter that although it has potential, we are not getting any further [to bringing the product to the next stage].*"

However, through designerly ways of working, the longitudinal field study captured insights over time that could not have been foreseen previously, such as through testing and validating assumptions with design tools. According to the researchers, due to these tools, more honest answers were provided by people living with dementia. Hence, the interweaving of designerly ways of working and traditional longitudinal field research was valuable for research actors of both domains, in the sense that behavioral researchers attempt to understand how aspects evolve over time (as is) while design researchers attempt to improve situations (as could be). On many occasions, researchers from both disciplines were pleasantly surprised and inspired by the insights gathered through interweaving.

4.3.3.3 Changing project focus in DDD

This RtD project aimed to (1) further develop a theoretical framework, (2) improve design methods, and (3) create longitudinal personas for the benefit of future research and design. First, the elaboration of the theoretical framework had become less relevant during the project because the earlier model concerned elderly people using technology longitudinally (Peek et al., 2019), while the field research shifted in focus toward social activities of people with dementia, possibly with the use of technology. Based on growing insight and limitations in the recruitment of participants, the project focus shifted from reducing loneliness among people living with dementia by means of technology use toward enhancing social activities (possibly with technology) among people living with dementia who live together with a partner. The design research perspective allowed for the adaptation of this perspective in the project based on first user engagement sessions.

Second, DDD partly succeeded in improving design methods. In terms of the reasons for this, the behavioral researcher stated the following: *"It's just really hard to do longitudinal co-design, and we have not yet reached saturation in the data. So, it needs to be more fine-grained."* The design researchers wanted to further develop the co-design methods by combining them with longitudinal research; however, it was difficult to understand whether the sessions they conducted were co-design or not, *"because we haven't made anything in the sessions with participants,"* and *"it is very difficult to conduct co-design longitudinally, especially when those involved are cognitively impaired. Because that asks for observing how a process evolves over time rather than chopping up design in subactivities.*"

Third, the aim was to produce relevant content for the creative industry, which the DDD researchers did through an event at the end of the project. However, the project proposal stated that the project's insights would be continuously disseminated – by means of academic publications and longitudinal persona descriptions – to allow industry to create designs based on those insights. Because of a delay in the project timeline due to lengthy discussions and interweaving activities, the focus shifted to center on the development of research artefacts (the three design tools) rather than on the development of the longitudinal personas. The three tools tailored to the project guided the diversity in disciplines in that it supported collective sensemaking and the finding of common ground. The tools were used to share ideas, exchange information, and develop a common point of departure in each stage of a longitudinal research project. Moreover – also partly caused by restrictions during the first COVID-19 lockdown in The Netherlands – a high-fidelity prototype was developed in an accelerated manner (Godfroij, van Beest, & van Gessel, 2020). The prototype was tested in the homes of people living with dementia and their families for at least six weeks.

Last, disseminating academic publications was difficult during the project because publishing in interdisciplinary outlets would require less or no decoupling, which required much effort from all of the researchers involved. It was not immediately clear where to publish, since the outlets that the researchers of both disciplines were used to would not easily accept the ways of working and language applied in the other research domains (e.g., longitudinal field research influenced by design activities). Nevertheless, at the time of writing, two publications have been submitted – and accepted – based on both the longitudinal fieldwork and the design evaluation in the field. All of the researchers were involved in both publications.

4.3.3.4 Mechanism level: Enabling and constraining factors of the narrative

The fabula level of narrative empirically demonstrated that the *organizational structure* of the DDD project changed, specifically from a preferred joint project managed by a social behavioral researcher (post-doc) and a senior design researcher (assistant professor) to project management conducted by the post-doc who, together with a PhD-level design researcher, conducted the field research. Enabling and constraining factors regarding change in the structure of actors included, on the one hand, difficulties because of differences in function: PhD-level, post-doc, and assistant professor. A PhD-level or post-doc-level researcher frequently has one or two projects to focus on, while an assistant professor has an advisory role in many projects. On the other hand, senior researchers have been shaped more by their distinct discipline than junior researchers, who can therefore act as mediators to cross boundaries.

At the *process level*, this study particularly highlighted that the lengthy discussions meant that the application of design tools and tangible objects was postponed compared with the design researchers' expectations. Ambiguity and uncertainty are part of a design process, which the senior design researcher phrased as follows: *"Since outcomes of a design process cannot be predicted."* This may be a constraining factor if researchers want to know exactly what they are going to do and why because, as the social behavioral researcher explicated, *"I want to do it the best way."* The lively and lengthy discussions in which the researchers sought a shared understanding and language took much time, but they also generated many occasions on which researchers from both disciplines were pleasantly surprised and inspired. The enabling factors here were interest in the other discipline and being open to learning from each other, to *"see what it yields."*

At the *product level*, the empirical findings revealed four aspects of change: (1) a shift in focus toward enhancing social activities (possibly with technology) among people living with dementia who live together with a partner; (2) the focus on improving co-design methods became part of a larger process; (3) a focus centered on the development of research artefacts; and (4) an unexpected search for ways to publish in interdisciplinary outlets. The journals in which the social behavioral researcher publishes are *"nicely multidisciplinary for psychological terms, but not that creative like let's all start using design methods,"* which was a constraining factor on the methodological flexibility. However, an enabling factor was that researchers from both domains were *"pushing each other"* to couple the interests of behavioral and design research and to publish in interdisciplinary journals. The focus on social activities and co-design methods became part of larger processes due to complexity, in which an RtD approach enabled a focus on research artefacts rather than on the creation of longitudinal personas as knowledge products.

4.3.4 Discussion

The scaffolding concepts in the introduction (see Table 4.3.2) suggested that, in *design practice*, designers increasingly have meetings with various actors from different departments and on various levels; build on existing knowledge together with other professionals; and improve processes besides creating prototypes and products (Godfroij & van der Lugt, 2020). The findings of the current study similarly indicate that in a *design research* project, designers engage in interdisciplinary collaborations with researchers from another domain – namely social behavioral sciences. They explore relationships that more often occur in complex networks beyond organizations, in which actors couple, interweave, and decouple (Godfroij, 2021). We suggest that the role of connecting disciplines (Cross, 2018) and integrating resources (Prestes Joly, et al., 2019) could be added to design research and practice.

In this case study, a PhD-level design researcher was appointed, who acted as a mediator – to anchor the research in both disciplines (social behavioral science and design), to make the contributions of the two research domains equal, and to provide each research domain with input from the other. This junior design researcher operated as a boundary spanner and crossed boundaries to support the integration between disciplines (Akkerman & Bakker, 2011), and also embodied much of the work involved in shaping and materializing interweaving (Basballe & Halskov, 2012). The mediator played a role in aligning plans across various organizational levels (van der Lugt & van der Laan, 2017) in academic research: The PhD-level, post-doc level, and (assistant) professor level. As the junior researcher (PhD level) was less shaped by a single discipline, she was highly suitable for this role in the project. Earlier, we found something similar in a design practice case (Godfroij & van der Lugt, 2020): A junior design researcher (graduating student industrial design) operated as a boundary spanner between design practice (senior designer) and various departments and levels of a healthcare organization (an academic hospital). These findings on the changing structure of actors in RtD suggest, at an organizational level, *a mediator role for a junior design researcher* – who is less shaped by a discipline than more experienced researchers. This mediator role contributes to the roles in RtD proposed by Sleeswijk Visser (2018), who distinguished roles for the theory researcher, design researcher, designer, problem owner, and project lead; however, the author did not specifically pay attention to connecting different research domains in RtD.

The findings at the process level provide an understanding of the issues regarding collaboration dynamics that accompany the heavy reliance of RtD on design, which is unable to provide predictability (Godin & Zahedi, 2014). This contributes to existing theory by providing the insight that aligning ways of working in interdisciplinary RtD takes a large amount of time, meaning that the application of design tools and tangible objects becomes postponed compared with the expectations of design researchers. This reinforces the theoretical concept of changing ways of working (see Table 4.3.2), for which we found in our earlier study (Godfroij & van der Lugt, 2020) that design interventions and the creation of prototypes to test with participants (health patients) were postponed in a design practice case. Rather than creating tangible tools and changing objects, said case focused on talking to many professionals to search for shared sensemaking and knowledge to build on. The lively and lengthy discussions in which the DDD researchers sought a shared understanding and language took much time but also generated many occasions on which the researchers from both disciplines were pleasantly surprised and inspired. This emphasizes that it is worthwhile spending time together with the other discipline to create an understanding across multiple perspectives (Zielhuis et al., 2020), for which the classic methods of reasoning are inappropriate (Cross, 2011; Tarazi, 2022). In further discussions and negotiations, the DDD researchers redefined the definitions and scope (Haug, 2015; Higgings & Klein, 2011; Meroni, 2008), and the design researchers – over time – convinced the behavioral science researcher of the added value of applying design methods in longitudinal field research. Said researcher experienced that an interdisciplinary field of design research and practice is about learning and discovery.

The findings at the product level reinforced the theoretical concept (see Table 4.3.2) regarding changing the project focus in interdisciplinary RtD. They identified a shift in focus from delivering (research) products – such as the graphical longitudinal case descriptions and domain-specific journal papers – toward the creation of tangible design tools and visual maps to apply in longitudinal field research to generate knowledge relevant to interdisciplinary outlets. Earlier, we found something similar: Designers planned to create personas as a product of field research with health patients, but along the way they elaborated existing personas based on knowledge and information that was available in the health organization and used it as a starting point for fieldwork (Godfroij & van der Lugt, 2020). Hence, both the design practice case study (Godfroij & van der Lugt, 2020) and the design research case study (the DDD project) identified a similar shift regarding the use of design tools. Both shifted focus from tangible or visual creations as project outcomes toward the creation of design tools that contribute to improving processes, exchanging competences, sharing sensemaking, and understanding complex societal issues (Vargo et al., 2020). Such a process orientation is crucial for understanding diverse practices and adopting a design-thinking mindset and methodology across discipline borders (Bertao, Jung & Joo, 2022).

4.3.5 Conclusions

This study provided an understanding of changing dynamics in interdisciplinary RtD, specifically on the role for design researchers within the complex structure of actors that merge research and design activities across domains. Findings at the organizational, process, and product levels emphasized three changing roles for design researchers in interdisciplinary RtD: (1) a mediator role for junior design researchers to cross borders; (2) a sensemaking role by discussing and elaborating on knowledge and tools of both disciplines; and (3) a role in contributing to the processes of a larger system together with actors from other domains by applying interwoven activities and materials in fieldwork to contribute to complex societal issues. Overall, the organizational-level findings revealed difficulties in structuring project management anchored in different disciplines. The data suggested a mediator role for junior design researchers in interdisciplinary RtD to cross boundaries between disciplines. At the process level, the findings indicated that the application of design tools and tangible objects was postponed compared with the design researchers' expectations. RtD researchers sought a shared understanding and language and generated many occasions that inspired the other discipline. The data suggested a changing role for design researchers toward lengthy discussions and sensemaking, aiming to interweave activities and materials from both disciplines to apply in fieldwork. At the product level, the findings indicated that researchers from both domains were *"pushing each other"* to couple interests and changed their focus from creating tangible or visual products as project outcomes toward creating tangible research artefacts. Last, the data suggested a changing role that centered on contributing to and understanding processes of complex societal issues – such as the social participation of people with dementia – together with research partners from other domains.

Acknowledgments

This work was part of the research program Create Health, which is financed by The Netherlands Organization for Health Research and Development (ZonMW). We thank the six researchers of the Dementia Dynamics in Design project for their openness and collaboration in this study.

References

Akkerman, S. F., & Bakker, A. (2011). Boundary crossing and boundary objects. *Review of Educational Research, 81*(2), 132–169.

Andriessen, D., Zielhuis, M., Greven, K., van Beest, W., Godfroij, B., & van der Lugt, R. (2019). Seven ways to foster interdisciplinary collaboration in research involving healthcare and creative research disciplines. In R. Brankaert & W. IJsselsteijn (eds.). *Communications in Computer and Information Science, 1117*(2019), 174–177.

Andriessen, D., Godfroij, B., Greven, K., & Zielhuis, M. (2020). Collaborating in complexity: Strategies for interdisciplinary collaboration in design research. In K. Christer, C. Craig, & P. Chamberlain, (eds.), *Proceedings of the 6th International Conference on Design 4, Health*, (Vol. 1, pp. 11–17). Sheffield: Sheffield Hallam University.

Basballe, D. A., & Halskov, K. (2012). Dynamics of research through design. In *Proceedings of the Designing Interactive Systems Conference*, (pp. 58–67). New York: Association for Computing Machinery.

Berends, H., & Deken, F. (2019). Composing qualitative process research. *Strategic Organization, 19*(1), 134–146.

Bertao R., Jung, C. H. & Joo, J. 2022. Identifying inconveniences in daily life: A problem finding prompt to foster non-designers' engagement in design thinking training. In G. Goldschmidt, & E. Tarazi, (eds.), Expanding the frontiers of design: A blessing or a curse? *Proceedings of the 13th Design Thinking Research Symposium*, (pp. 354–368). Design Tech, Technion, Isreal Institute of Technology, Haifa.

Binder, T., & Redström, J. (2006). Exemplary design research. In K. Friedman, T. Love, E. Côrte-Real, & C. Rust (eds.), *Wonderground - DRS International Conference 2006*. Lisbon, Portugal, 1 – 4 November 2006.

Brankaert, R., & den Ouden, E. (2017). The design-driven living lab: A new approach to exploring solutions to complex societal challenges. *Technology Innovation Management Review, 7*(1), 44–51.

Calabretta, G., Gemser, G., & Karpen, I. (2017). *Strategic design: Eight essential practices every strategic designer must master*. Amsterdam: BIS Publishers.

Casakin, H., & Cascini, G. (2022). Multi-disciplinary design teamwork: Quality of collaboration and implications for extending the frontiers of design in education. In G. Goldschmidt, & E. Tarazi, (eds.), Expanding the frontiers of design: A blessing or a curse? *Proceedings of the 13th Design Thinking Research Symposium*, (pp. 225–242). Design Tech, Technion, Israel Institute of Technology, Haifa.

Christensen, B. T., & Ball, L. J. (2019). Building a discipline: Indicators of expansion, integration and consolidation in design research across four decades. *Design Studies*, *65*(2019), 18–34.

Cooper, R. (2019). Design research - Its 50-year transformation. *Design Studies*, *65*(2019), 6–17.

Cross, N. (2001). Designerly ways of knowing: Design discipline versus design science. *Design Issues*, *17*(2001), 49–55.

Cross, N. (2011). *Design Thinking: Understanding how designers think and work*. Oxford-NewYork: Berg Publishers.

Cross, N. (2018). A brief history of the Design Thinking Research Symposium series. *Design Studies*, *57*(2018), 160–164.

Den Haan-Wintermans, M., Brankaert, R., & Lu, Y. (2019). The leisure time canvas: Eliciting empathy for older adults through activities and hobbies. *Paper presented at Academy for Design Innovation Management*, London, United Kingdom.

Durrant, A. C., Vines, J., Wallace, J., & Yee, J. S. R. (2017). Massachusetts Institute of Technology. *Design Issues*, *33*(3), 3–10.

Gero, J., & Milovanovic, J. (2022). What is happening when designers from different disciplines work together: Characterization of the design behaviors and design spaces of mechanical and electrical engineers working in teams. In G. Goldschmidt, & E. Tarazi, (eds.), Expanding the frontiers of design: A blessing or a curse? *Proceedings of the 13th Design Thinking Research Symposium*, (pp. 243–258). Design Tech, Technion, Israel Institute of Technology, Haifa.

Godfroij, B. K. (2021). A process model on changing design practice: An industrial design case in a healthcare innovation journey. *12th International Process Symposium*. Rhodes Greece, online, 1–5 September, 2021.

Godfroij, B., van Beest, & van Gessel (2020). *Co-design in de anderhalvemetermaatschappij Whitepaper*. Utrecht, The Netherlands: Utrecht University of Applied Sciences.

Godfroij, B., & van der Lugt, R. (2020). Change and novelty for industrial designers in complex design projects for healthcare. In K. Christer, C. Craig, & P. Chamberlain, (eds.), *Proceedings of the 6th International Conference on Design4Health*, (Vol.1, pp. 186–193). Sheffield: Sheffield Hallam University.

Godin, D., & Zahedi, M. (2014). Aspects of Research through Design: A literature review. In Y. Lim, K. Niedderer, J. Redström, E. Stolterman, & A. Valtonen (eds.), *Design's big debates - DRS International Conference 2014*, Umeå, Sweden, 16–19 June, 2014.

Halskov, K., & Christensen, B. T. (2018). Designing across cultures. *CoDesign*, *14*(2), 75–78

Haug, A. (2015). Emergence patterns for client design requirements. *Design Studies*, *39*(2015), 48–69.

Higgins, A., & Klein, S. (2011). Chapter 2: Introduction to the living lab approach. In Y.-H. Tan et al., (eds.), *Accelerating global supply chains with IT-innovation*, (pp. 157–175).Berlin: Springer-Verlag.

Jarzabkowski P., Bednarek R., & Lê J. K. (2014). Producing persuasive findings: Demystifying ethnographic textwork in strategy and organization research. *Strategic Organization*, *12*(4), 274–287.

Jones, A. (2006). Multidisciplinary team working: Collaboration and conflict. *International journal of mental health nursing*, *15*(1), 19–28.

Kolasinska, A. B., Thoolen, M., Peek, S., Lu, Y., & Brankaert, R. (2021). Co-creating design opportunities for social technology in the context of dementia. Dementia lab conference *Dementia Lab 2021: Supporting ability through design*, 125–141.

Langley, A. (2007). Process thinking in strategic organization. *Strategic Organization*, *5*(3), 271–282.

Langley, A., Smallman, C., Tsoukas, H., & van de Ven, A. H. (2013). Process studies of change in organization and management: Unveiling temporality, activity, and flow. *Academy of Management Journal*, *56*(1), 1–13.
Langley, A., & Tsoukas, H. (2017). Introduction: Process thinking, process theorizing and process researching. In A. Langley, & H. Tsoukas (eds.), *The SAGE handbook of process organization studies* (pp. 1–25). London: Thousand Oaks, SAGE Publications Ltd.
Liedtka, J., & Kaplan, S. (2019). How design thinking opens new frontiers for strategy development. *Strategy & Leadership*, 47(2), 3–10.
Meroni, A. (2008). Strategic design: where are we now? Reflection around the foundations of a recent discipline. *Strategic Design Research Journal*, *1*(1), 31–38.
Ou, X., Goldschmidt, G., & Erez, M. (2022). Exploring the effect of team composition on creativity in design idea generation. In G. Goldschmidt, and E. Tarazi, (eds.), Expanding the frontiers of design: A blessing or a curse? *Proceedings of the 13th Design Thinking Research Symposium*, (pp. 211–224). Design Tech, Technion, Israel Institute of Technology, Haifa.
Peek, S. T. M., Aarts, S., & Wouters, E. J. M. (2015). Can smart home technology deliver on the promise of independent living? A critical reflection based on the perspectives of older adults. In J. van Hoof J, G. Demiris & E. J. M. Wouters (eds), *Handbook of Smart Homes, Health Care and Well-Being*, (pp.1–10). Cham: Springer.
Peek, S. T. M., Luijkx, K. G., Vrijhoef, H. J. M., Nieboer, M. E., Aarts, S., van der Voort, C. S., Rijnaard, M. D., & Wouters, E. J. M. (2019). Understanding changes and stability in the long-term use of technologies by seniors who are aging in place: A dynamical framework. *BMC Geriatrics*, *19*(236), 1–13.
Prestes Joly, M., Teixeira, J. G., Patrício, L., & Sangiorgi, D. (2019). Leveraging service design as a multidisciplinary approach to service innovation, *Journal of Service Management*, *30*(6), 681–715.
Roggema, R. (2017). Research by design: Proposition for a methodological approach. *Urban Science*, *1*(2), 1–19.
Rylander, L., & Eneberg, M. (2022). Designing for change in complex socio-technical systems. In G. Goldschmidt, & E. Tarazi, (eds.), Expanding the frontiers of design: A blessing or a curse? *Proceedings of the 13th Design Thinking Research Symposium*, (pp. 478–491). Design Tech, Technion, Israel Institute of Technology, Haifa.
Sciannamè, M. (2022). The designer as a cross-boundaries mediator. In G. Goldschmidt, and E. Tarazi, (eds.), Expanding the frontiers of design: A blessing or a curse? *Proceedings of the 13th Design Thinking Research Symposium*, (pp. 139–155). Design Tech, Technion, Israel Institute of Technology, Haifa.
Sleeswijk Visser, F. (2018). Structuring roles in Research through Design collaboration. In C. Storni, K. Leahy, M. McMahon, P. Lloyd, & E. Bohemia (eds.), *Proceedings of DRS 2018*, (pp. 368–380). Ireland: Limerick, 25–28 June, 2018.
Stappers, P. & Giaccardi, E. (2017). Research through Design. In M. Soegaard, & R. Friis-Dam (eds.), *The Encyclopedia of human-computer interaction*, (2nd edition, pp. 1–94). The Interaction Design Foundation.
Stock, P., & Burton, R. J. F. (2011). Defining terms for integrated (multi-inter-trans-disciplinary) sustainability research. *Sustainability*, *3*(8), 1090–1113.
Swann, C. (2002). Action research and the practice of design. *Design Issues*, *18*(2002), 49–61.
Tarazi, E. (2022). 7D – Nature centered design. In G. Goldschmidt, & E. Tarazi, (eds.), Expanding the frontiers of design: A blessing or a curse? *Proceedings of the 13th Design Thinking Research Symposium*, (pp. 397–409). Design Tech, Technion, Israel Institute of Technology, Haifa.
Van Beest, W., Godfroij, B., Zielhuis, M., Andriessen, D., & Van der Lugt, R. (2022). *Create Health Ways of Working: Insights from ten ehealth innovation research projects*. Utrecht, The Netherlands: Utrecht University of Applied Sciences.
Van der Lugt, R., & van der Laan, T. (2017). Design facilitators' journeys through the jungle of Co-in healthcare. *The Design Journal*, *20*(2017), 2057–2067.

Vargo, S. L., Koskela-Huotari, K. & Vink, J. (2020). Service-dominant logic: Foundations and applications. In E. Bridges & K. Fowler (eds.), *The Routledge handbook of service research insights and ideas*, (1st edition, pp. 1–21). London: Routledge.
Ybema, S. B., Yanow, D., Wels, H., & Kamsteeg, F. H. (2009). *Organizational ethnography: Studying the complexities of everyday life*. London: Sage.
Yanow, D., & Tsoukas, H. (2009). What is reflection-in-action? A phenomenological account. *Journal of Management Studies*, *46*(8), 1339–1364.
Zielhuis, M., van Gessel, C., van der Lugt, R., Godfroij, B. & Andriessen, D. (2020). Grounding practices: How researchers ground their work in create-health collaborations for designing e-health solutions. In K. Christer, C. Craig, & P. Chamberlain, (eds.), *Proceedings of the 6th International Conference on Design4Health* (Vol. 3, pp. 142–152). Sheffield: Sheffield Hallam University.
Zimmerman, J., Forlizzi, J., & Evenson, S. (2007). Research through Design as a method for interaction design research. In *Proceedings of the SIGCHI Conference on Human Factors in Computing Systems* (pp. 493–502). New York: ACM Press.
Zimmerman, J., & Forlizzi, J. (2014). Research through Design in HCI. In J. S. Olson & W. A. Kellogg (eds.), *Ways of knowing in HCI* (pp. 167–189). New York: Springer.
Zimmerman, J., Stolterman, E., & Forlizzi, J. (2010). An analysis and critique of Research through Design: Towards a formalization of a research approach. In *Proceedings of the 8th ACM Conference on Designing Interactive Systems* (pp. 310–319). New York: ACM.

Part 5

Design and nature; visual representation

DOI: 10.1201/b22630-29

5.1 Empathic multispecies design – Using empathy to design with animal perspectives in mind

Daniel Metcalfe
Technion – Israel Institute of Technology

5.1.1 Introduction

This paper explores the topic of empathic multispecies design. To this end, I will first discuss the terms multispecies design and empathic design separately and then in integration. The prospect and possible application of empathic multispecies design are discussed through analysis and reflection on a design exercise named the *umwelt apparatus* in which, designers created an apparatus for experiencing the world through the sensory abilities (umwelt) of an animal client.

5.1.1.1 Multispecies design

The devastating effects of unsustainable design on the natural world and other species with whom we share this planet have gained widespread awareness and are the driving force behind attempts to develop more sustainable design approaches. These efforts tend to focus on minimizing the negative effects that design has on the natural world by reducing material and energy use. However, the possibility that design may have an active role in mitigating the erosion of biodiversity has only entered the discussion in recent years and remains a marginal activity for design (Metcalfe, 2015; Root-Bernstein & Ladle, 2010; Wells & Yeang, 2010). The prospect has entered the discussion following an ongoing paradigm shift calling for the inclusion of a greater diversity of species within human-dominated habitats, as a way of addressing both the erosion of biodiversity and humankind's alienation from nature. Ecologists, who have long ignored urban ecosystems, are gradually realizing that cities are not only the problem but also part of the solution and that integrating nature in cities can benefit both people and biodiversity conservation (e.g., Colléony & Shwartz, 2019; Rosenzweig, 2003). The last three decades of urban ecology research have demonstrated that urban green spaces have the potential to support rich biodiversity that can be enhanced by implementing certain design and management solutions (Beninde et al., 2015; Rosenzweig, 2003; Shwartz et al., 2013). Wild animals frequently interact with anthropogenic systems (Luniak, 2004). However, in most cases, these are not designed with the animal in mind and may lead to harm to the animal or human-wildlife conflicts. To facilitate a shift to more biodiverse human habitats and address its eco-socio-cultural implications, there is a growing need for a design practice capable of looking at the needs of both humans and wild animal species as well as the interactions between the two groups – a practice we may call multispecies design (Metcalfe, 2015). One main characteristic of multispecies design is the view of animals as clients of design (ibid). To develop this notion, we need new design methods for studying

DOI: 10.1201/b22630-30

animals and their needs in the context of design. Here I look to the field of empathic design to examine its relevance for design instanced involving nonhuman species.

5.1.1.2 Empathic design

The word empathy is used in a wide range of disciplines to convey a large variety of meanings (Cuff et al., 2016). Here I refer to the term as it is often used within the field of design, with some reference also to its use within human-animal studies. To this end, I adopt Young et al.'s (2018) definition of empathy as being "a stimulated emotional state that relies on the ability to perceive, understand and care about the experiences or perspectives of another person or animal". Empathy is often described as incorporating two distinct yet related abilities: cognitive empathy and affective empathy (see e.g., Cuff et al., 2016; Kouprie & Visser, 2009). Cognitive empathy is the ability to *understand* another's feelings and affective empathy is concerned with an *emotional response* to the state of the other (Cuff et al., 2016). Some authors add empathic concern, i.e., the *motivation to act* to relieve another's suffering, as a third component of empathy (see Young et al., 2018). Empathic concern resonates with the notion of empathy as a driver of motivation seen also in the field of design (see Kouprie & Visser, 2009).

Empathy is increasingly viewed as a central concept in human-centred design as both a quality of the design process and a quality (state and trait) of the designer (Kouprie & Visser, 2009). The term empathic design is commonly used to describe a process of getting closer to the lives and experiences of existing or future users and gaining a deep understanding of the user and his or her experience (Kouprie & Visser, 2009). Various design thinking models place *empathizing* as a stage by itself in the design process, often the first stage (see e.g., Interaction Design Foundation, 2021) and different techniques have been developed for gaining empathy with clients as part of the design process. These typically include (1) immersive techniques, where designers immerse themselves in the lives and experiences of potential users (such as shadowing, visual diaries, empathic interviewing techniques, and bodystorming); (2) role-playing techniques, where designers imagine themselves in the place of users and act out different scenarios of interaction; and (3) simulation techniques, where designers use props or physical restraints to physically alter their experience or perception. The latter has been developed mainly in the field of inclusive design where designers use simulations to feel, for example, the effects of different types of capability-losses while using a product or service (Cardoso & Clarkson, 2012). Methods of empathic design have been suggested to be useful for (1) experiencing the world from the users' point of view (Koskinen, Battarbee & Mattelmäki, 2003); (2) broadening the reach of the design and rendering it more inclusive (Cardoso & Clarkson, 2012); (3) approaching the problem from multiple perspectives (Smeenk et al., 2016), increasing curiosity and motivation (Kouprie & Visser, 2009); (4) conveying the subjective, situational qualities of human emotions to the design process (Lee et al., 2020); and (5) promoting empathic interactions through the final design proposal (Devecchi & Guerrini, 2017). To date, empathic design methods have focused almost exclusively on human users. This paper explores the relevance of the concept of empathic design when applied to nonhuman users.

5.1.1.3 Empathic multispecies design

The key motivations behind empathic design methods, described above, may apply also for design involving nonhuman species: understanding animal perspectives, experiences, and ways of interacting with the world may help reduce the designer's human bias in the process and increase the chance that the design meets the needs and abilities of animals interacting with it. It may help approach the problem from both a human and an animal perspective. It may increase curiosity and motivation to design for the studied animal, and it may lead to more empathic interactions between humans and animals. However, applying these methods in a nonhuman context is not always straightforward. While some principles and methods of empathic design may potentially be adopted for multi-species design, there are some inherent differences it is important to note when developing an empathic multispecies approach. For example, as part of the empathic process designers often rely on their personal experiences, feelings, and emotions from within similar contexts to that of the design (Smeenk et al., 2016). Relying on personal experience may be useful when designing for other humans but when designing for animals it may cause a human-centric bias and misinterpretation of the animal's actual needs. Another area where empathic design methods do not, for obvious reasons, transfer easily to nonhuman species involves methods based on dialogue and verbal interaction with potential users, such as interviews and methods of participatory design. Overall, the application of empathic design principles and methods in the context of nonhuman users is a novel and unexplored domain, much like the field of multispecies design itself. The purpose of this paper is to explore the prospects of empathic multispecies design, its areas of differences and similarities with human-centred empathic design and outline some considerations for developing methods in the field. The topic is explored through analysis and reflection on a design exercise named the *umwelt apparatus,* developed to promote empathy towards a chosen animal client, undertaken by design and architecture students at the Technion - Israeli Institute of Technology.

5.1.2 Methodology

In a monograph from 1934 titled *A Stroll through the Worlds of Animals and Men*, Jakob von Uexküll deals with the possibility of viewing the world from the points of view of species with entirely different sets of sensory organs than us humans. In it, von Uexküll invites us to take part in a thought experiment of experiencing the world through the sensory organs of other species, referring to this self-world of the animal as its *umwelt* Inspired by this work we devised the concept of an *umwelt apparatus*: a simulation apparatus that would help designers experience the world from the point of view of a selected animal client – offering them a glimpse into the unique ways in which the animal senses and interacts with the world – to be used in the early (emphasizing) stages of a design process involving nonhuman clients. The challenge was presented as a design exercise to design and architecture students in the course Human-Animal-Machine at the Technion. The umwelt apparatus was the first of two design exercises developed in the course; the second being the development of a conceptual design intervention based on the insights gained in the first exercise and responding to the needs of the selected animal client. This paper's focus is on the first exercise – building and using the umwelt apparatus – and its effect on the students' attitude towards and perception of the animal.

In the spring of 2021, 30 students took part in the exercise. Fifteen master of industrial design students and fifteen master of architecture students, from different study and professional backgrounds. The exercise was conducted in Hebrew. The students were divided into 13 groups, each working with a different animal client and focusing on a different simulated state or sense. Table 5.1.1 shows the animals, simulated state/sense, and a brief description of the developed umwelt apparatus. For the exercise, the groups

Table 5.1.1 List of studied animals, simulated states/senses, and a short description of the Umwelt Apparatus

Group	*Studied animal*	*Simulated state/sense*	*Umwelt Apparatus*
1	Common fruit fly	Smell	Development of substances for coding the fly's "olfactory language" using various foods in various stages of decay. Next, training the human nose to detect and make sense (in a similar way a fruit fly might) of the different chemical cues.
2	Great tit	Vocal and body language	Simulation using role-play of typical scenes from the life of great tits using the bird's vocal and body language to communicate (following field observations of the birds).
3	Golden jackal	Soundscapes	Simulation using sounds (howls) to communicate and perceive invisible geographical information such as territory size and borders. Drawing and comparing maps to finetune this (sense) ability.
4	Kuhl's pipistrelle	Proprioception	Apparatus simulating the first seconds of a bat's takeoff from hanging upside-down.
5	Southern white-breasted hedgehog	Behavior/danger perception	Simulation of behavior during danger in hedgehogs using a blindfold hat with belles - connecting the sense of danger with the need to stay completely still
6	Rock dove	Behavior/drinking	Apparatus simulating the dove's distinctive drinking behavior involving dipping its whole head in the water
7	Rock hyrax	Behavior/walking, jumping, climbing	Apparatus simulating the rock hyrax's foot structure, prompting typical walk and interaction with the environment.
8	Nubian ibex	Sight	Visual apparatus simulation of the ibex's visual capabilities (field of view) using lenses and mirrors
9	Hooded crow	Sight/tool use behavior	Apparatus simulating the crow's vision while using a stick as a tool and a series of tasks to perform while using the tool
10	Mediterranean house gecko	Sight and spatial movement	Video filming apparatus and postproduction process for creating footage simulating a gecko's spatial movement, point of view of the environment, and colour range
11	Smooth cauliflower coral	Distributed sense of touch	Apparatus for a distributed sense of touch and response using multiple "arms" extending from the body
12	Western honeybee	Sight	Apparatus simulating the bee's eye structure and visual spectrum and simulated environment using UV colors
13	Near Eastern fire salamander	Sensing humidity through the skin	Simulated rock crevice environment and navigation using only the sense of humidity on the skin

Figure 5.1.1 Examples of umwelt apparatuses developed by three groups of students. Clockwise from upper left: 1. Apparatus for experiencing the world from the point of view of the honeybee involving a visual filter and ultraviolet light, developed by Ofek Raz and Dana Taub. 2. Apparatus for coding the common fruit fly's "olfactory language", developed by Erez Ezra in D.Dlab [Disrupt.Design] research facility. 3. Simulation using role-play of typical scenes from the life of great tits developed by Orni Bloch and Yael Salem (video screenshot).

were instructed to *conceive and build a device or protocol that would allow them access into the animal's self-world, based on the animal's unique sensory abilities, behavior, and ways of interacting with the world* (see Figure 5.1.1 for examples of umwelt apparatuses developed in the course). The explanation of the brief was followed by examples of work by various artists, for example, Chris Woebken's Animal Superpowers (Woebken, 2007), as well as student work from the previous year. The exercise was presented to the students as part of the research stage of their multispecies design process, where they seek to learn about their animal client to later develop a design intervention for it. The groups had five weeks to develop their umwelt apparatus followed by seven weeks to develop their final design proposal. After developing the apparatus, the students were asked to use it within a human-dominated habitat to interact with the environment as the animal might. Thus, to explore the animal's self-world and the way it is influenced by human development, in a way that is both immersive and experiential [following Koskinen, I., Battarbee, K., & Mattelmäki, (2003), who state that empathic design methods should strive to be both immersive and experiential].

Following the exercise, I conducted semi-structured interviews with the students who agreed to take part in the study (N = 27) to gain insight into the process of designing and using the umwelt apparatus. The data gathered from the interviews were classified and analyzed through a process of reflexive thematic analysis (Braun & Clarke, 2021) to identify, analyze, and interpret patterns of meaning and develop themes relating to the topic of empathic multispecies design. The thematic analysis was conducted with an inductive approach, grounded in the data collected from the interviews, seeking both evident and latent codes and themes. Concepts arising from the interviews were mapped onto the three components of the empathic process discussed above: affective empathy, cognitive empathy, and empathic concern highlighting common themes relating to the process of empathic multispecies design (Figure 5.1.2). Some of the emerging themes related to more than one component of empathy – highlighting interrelations between the cognitive, affective, and motivational elements of empathy. These appear in the overlapping parts of the circles. The concepts arising from the thematic analysis are discussed in the following section.

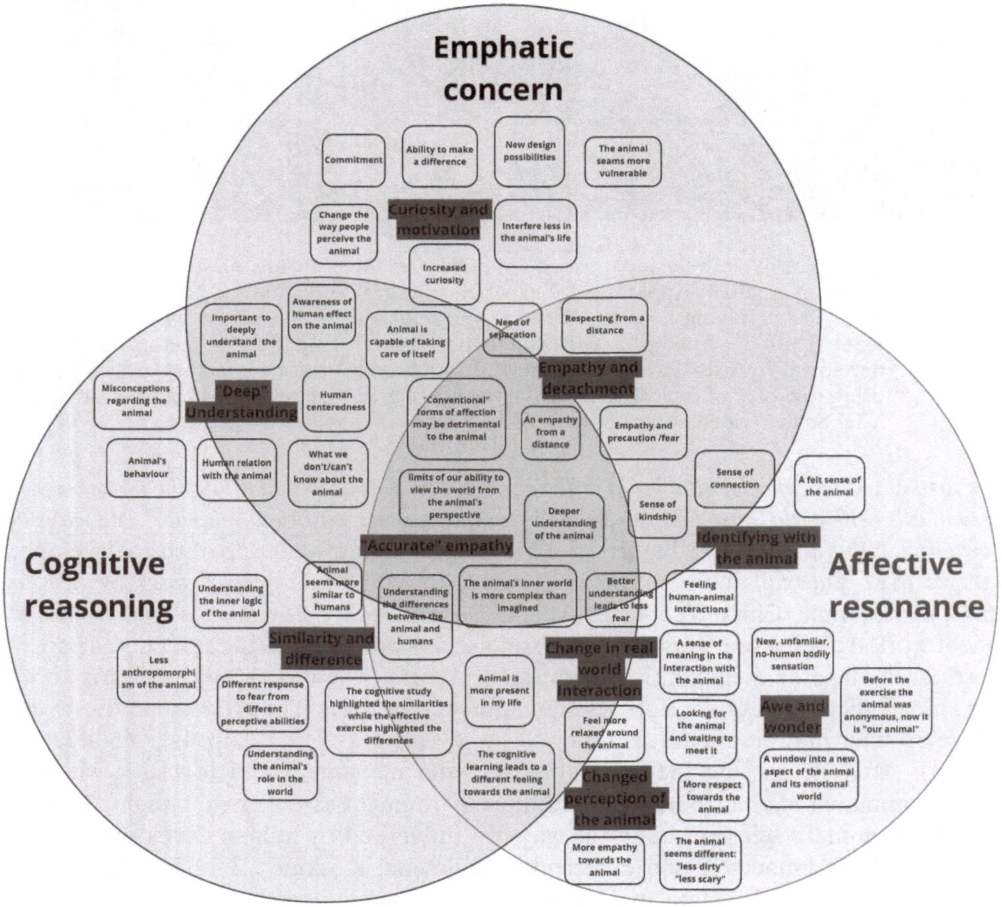

Figure 5.1.2 The thematic map showing the themes emerging from analyzing the data from the interviews (highlighted in yellow), mapped onto the three components of empathy.

5.1.3 Analysis and discussion

One central theme arising from the study relates to the level and quality of understanding the animal and its respective world. Participants reported a sense of a **deep understanding** of the animal; an understanding of its *inner logic* and a realization that the animal's inner world and perception are more complex than previously imagined. This stood out especially when considering the animal's behavior; after developing and using the apparatus participants reported a sense of understanding not only *how* an animal behaves, but also *why* it behaves the way it does and how this behavior relates to its unique perceptive abilities. Some participants reported having moments of insight or discovery concerning the animal after using the apparatus, where the animal, or some aspect of it *suddenly made sense*. For example, when talking about rock hyrax, one participant stated that she had realized that most human interactions with the animal have little to do with the animal itself and are determined by a third actor, in this case, the leishmania virus for which the rock hyrax is a host. Developing and using the umwelt apparatus granted her a fresh view of the animal, disconnected from the way it is currently portrayed in the Israeli media.

Another interesting theme regarding new ways of seeing and understanding the animal developed around the concepts of **empathy and detachment** within human-animal interactions. Participants stated they felt less naïve regarding these interactions after imagining them from the animal's point of view; realizing that when it comes to wild animals, empathizing and respecting the animal should, in many cases, be done from a distance, resisting the will to engage and connect. As Candea (2010) points out, empathy for animals is often associated with engagement, while detachment is often wrongly associated with coldness and lack of caring. The notion of an *empathy of detachment* in human-animal relations represents a more comprehensive view of these relations and is a crucial aspect in maintaining and respecting the wild nature of the animal (ibid).

A further thematic area, relating mainly to cognitive reasoning, was observed regarding the notion of **similarity and difference** between the studied animal and humans. Here participants reported that using the apparatus highlighted both similarities and differences, particularly around the (different and similar) ways in which humans and animals perceive and interact with the world. One participant specified that for her the cognitive study of the animal, before developing the apparatus, highlighted similarities while the affective experience of using the apparatus emphasized differences between humans and the animal. A good understating of the similarities and differences between humans and animals is important in multispecies design as it may lead to less anthropomorphizing of the animal and a clearer sense of how each group may interact or be affected by a design feature. An interesting example of how an anthropomorphic point of view may lead to misunderstanding the animal is the mirror test designed by humans to determine if animals have self-awareness. In the test, an animal's face is marked with a red dot, and then the animal is given access to a mirror to see if it responds to the presence of the dot. Dogs categorically fail the test, leading researchers to believe they lack self-awareness. However, more recent studies suggest that dogs fail the test because they are considerably less affected by visual events than humans and that self-awareness in dogs is linked to their sense of smell and not sight (Cazzolla Gatti, 2015).

This deeper understanding of the animal was present not only in cognitive terms but also in affective resonance and emotional reaction towards the animal. Participants reported a sense of **identifying with the animal**, a sense of kinship and connection after seeing from the animal's point of view, and a sense of **awe and wonder** in relation to the animal. In some groups – where the apparatus involved alterations to perception such as blindfolds, mirrors, or filters – subjects reported the exercise created for them new, unfamiliar, bodily sensations. For example, in the salamander group, where the umwelt apparatus involved a blindfold and a stimulating environment of smell and humidity, a participant stated that the blindfold and the sense of condensing water on her skin helped her "separate from the human and amplified an animalistic feeling". For several participants using the apparatus resulted in increased empathy and respect towards the animal and led to a **change in perception of the animal**, causing the animal to seem different after the exercise. For example: "less scary", "less dirty", "more capable of taking care of itself", or, in contrast, "more vulnerable and in need of protection". Throughout the interviews, affective resonance and cognitive reasoning seemed to be interrelated and complementary in the view of the participants and some made an explicit connection between the deeper cognitive understanding of the animal and the change in their emotional response to it. In cases where the students had frequent interaction with the studied animal, some interesting changes **in real-world interactions** were reported after using the apparatus. Participants reported feeling more relaxed around the animal, less fear, as well as anticipation before encountering the animal, together with a sense that the animal is more present in their lives. One participant from the bee group said: "I cannot look at bees the same, I see them everywhere now, I feel connected to them and understand why they behave the way they do". Members of other groups, working with songbirds, insects, and reptiles, also stated that they see and hear the animal more frequently than before in their everyday lives.

In terms of **empathic concern,** an increase in **curiosity and motivation** to take action has been observed. For some, this was propelled by increased awareness of the impacts of human systems on the animal and a desire to better share the space and interfere less in the animal's life. Cardoso and Clarkson (2012) discuss the ability of simulations, used in inclusive design, to show how everyday products often disregard a large number of users with capability loss. Similarly, using the umwelt apparatus revealed how urban environments are designed predominantly for humans and ignore their influence on other species. An influence that can be extremely dire as is the case, for example, with over a billion birds that find their death annually in collision with glass windows (Klem, 2015). Some participants reported a sense of commitment to help the animal, a desire to change the way people perceive the animal, and an increased belief in their ability to make a difference. This increase in motivation is strongly linked to the concept of empathy in design as described above.

The final thematic area developed from analyzing the data relates to the concept of **accurate empathy.** The term has been used by Young et al. (2018) to refer to an empathic response that is based on scientific knowledge regarding the animal. The authors acknowledge that our ability to perceive and understand another's perspective, especially when it comes to animals, is always an approximation (Young et al., 2018), nevertheless, they stress the importance of basing this empathic understanding of the animal on substantive knowledge of an animal's natural history and not purely on projected assumptions. This theme was present on different levels in the study. First, in the detailed research of the animal's natural history the students conducted in order to develop the umwelt apparatus and select one typical attribute of the animal's perception to explore in the exercise. Second, most participants openly talked about the limits of their apparatus to accurately simulate an animal's perspective, acknowledging, as well, the limits of human perception in viewing the world from the animal's perspective. The relevant question, in this respect, seems not to be about the fidelity of the apparatus in replicating an animal's perspective, but about what changes in us in relation to the animal when we develop and use it. Different participants stated that although they do not believe the apparatus simulated an accurate depiction of the animal's umwelt, the process opened for them *a window into the animal's world* and profoundly changed the way they view the animal as well as their motivation to design for the animal. The newly gained understanding of the animal came together with an understanding that there is a lot we do not know, and perhaps, can't know about other species and their self-worlds. This by itself, is an important insight for designers seeking to engage with nonhuman species.

5.1.4 Conclusions, limitations, and further work

The goal of this study was to explore the prospects of empathic multispecies design, its areas of differences and similarities with human-centred empathic design, and outline some considerations for developing methods in the field. Methods of empathic design deal with the possibility of perceiving and experiencing the world from the viewpoint of prospective users. While it is generally acknowledged that it is not possible to fully capture the perspective of another being (human or animal), the field of empathic design sees value in trying to get closer to the point of view of potential human users. Reflecting on the umwelt apparatus exercise suggests that this is also true regarding nonhuman users. By creating and using a device aimed at exposing them to an animal's perspective of the world, participants seemed to gain a deeper, more holistic understanding of the animal, greater respect and empathy towards the animal, and increased motivation to act for the animal. Studying the natural history of an animal is, no doubt, important when considering an animal in the context of design; however, complementing the factual knowledge of the animal with an affective, experiential, and immersive experience of the animal's perspective may lead to a more comprehensive understanding of the animal, stronger commitment and increased motivation to appropriately respond to its needs. Although the study did not explicitly explore this, it seemed that the insights gained in the umwelt apparatus exercise were significant for the second part of the design process developed in the course, where the students developed a design intervention responding to the needs of their studied animal. This was notable, both, in the way the students talked and related to the animal as well as in the focus of the interventions, which could all be linked back to the initial exercise. Future research could focus more directly on how the emphasizing stage in multispecies design influences the design outcome at the end of the process.

The research presented here is limited by the fact that it was conducted on a group of students as part of a study course and that coding and analyzing the data was done by a single researcher. Measures were taken to separate the study from the curricular activity, such as making participation voluntary, offering no compensation for participating, and grading the exercise before interviewing the students to assure them that the interview will not affect them or their grades in any way. Future work could involve participants that are professional designers, to investigate the relevance of the method also outside academia.

The umwelt apparatus is presented here as one way to seek empathy in a multispecies design project. It takes inspiration mainly from role-play and simulation techniques used in the field of empathic design. Immersive techniques were not explored because of the difficulty of immersing oneself in wild animals' lives and experiences and the risk they may pose to the animals themselves. Further development of empathic multispecies design techniques could focus on finding alternatives to immersive techniques, through remote sensing for example, as well as developing additional role-play and simulation techniques. Somatic practices such as focusing, and Body Mind Centering (BMC) could be explored as ways of internalizing an animal's perspective (see for example Metcalfe, 2015 p. 43–45). Another possible path to develop simulation techniques could be using virtual and augmented reality, as was done in the interactive project *in the Eyes of the Animal* created by creative studio Marshmallow Laser Feast (2016). I encourage designers and design researchers to explore and develop new methods for achieving empathy for animals within the context of design; taking into consideration, both, animals that are direct users of the design and animals that might otherwise be affected by a design intervention.

References

Beninde, J., Veith, M., & Hochkirch, A. (2015). Biodiversity in cities needs space: A meta-analysis of factors determining intra-urban biodiversity variation. *Ecology Letters*, *18*(6), 581–592. 10.1111/ele.12427

Braun, V., & Clarke, V. (2021). Conceptual and design thinking for thematic analysis. *Qualitative Psychology*. 10.1037/qup0000196

Candea, M. (2010). I fell in love with Carlos the meerkat: Engagement and detachment in human-animal Relations. *American Ethnologist*, *37*(2), 241–258. 10.1111/j.1548-1425.2010.01253.x

Cardoso, C., & Clarkson, P. J. (2012). Simulation in user-centred design: helping designers to empathise with atypical users. *Journal of Engineering Design*, *23*(1), 1–22.

Cazzolla Gatti, R. (2015). Self-consciousness: beyond the looking-glass and what dogs found there. *Ethology Ecology & Evolution*, *28*(2), 232–240. 10.1080/03949370.2015.1102777

Colléony, A., & Shwartz, A. (2019). Beyond assuming co-benefits in nature-based solutions: A human-centered approach to optimize social and ecological outcomes for advancing sustainable urban planning. *Sustainability (Switzerland)*, *11*(18), 4924. 10.3390/su11184924

Cuff, B. M. P., Brown, S. J., Taylor, L., & Howat, D. J. (2016). Empathy: A review of the concept. *Emotion Review*, *8*(2), 144–153. 10.1177/1754073914558466

Devecchi, A., & Guerrini, L. (2017). Empathy and Design. A new perspective. *Design Journal*, *20*(supl), S4357–S4364. 10.1080/14606925.2017.1352932

Interaction Design Foundation (2021). *5 Stages in the Design Thinking Process*. https://www.interaction-design.org/literature/article/5-stages-in-the-design-thinking-process

Klem, D. (2015). Bird–Window Collisions: A Critical Animal Welfare and Conservation Issue. *Journal of Applied Animal Welfare Science*, *18*, S11–S17. 10.1080/10888705.2015.1075832

Koskinen, I., Battarbee, K., & Mattelmäki, T. (2003). Empathic design: User experience for product design. In *IT Press* (Issue January 2003).

Kouprie, M., & Visser, F. S. (2009). A framework for empathy in design: stepping into and out of the user's life. *Journal of Engineering Design*, *20*(5), 437–448. 10.1080/09544820902875033

Lee, J. J., Koskinen, I., & Whalen, J. (2020). Multiple intelligibility in constructive design research: The case of empathic design. *International Journal of Design*, *14*(3), 55–67.

Luniak, M. (2004). Synurbization: Adaptation of animal wildlife to urban development. *Proceedings 4th International Urban Wildlife Symposium*, 50–55.

Marshmallow Laser Feast (2016). *In the Eyes of the Animal*. http://intheeyesoftheanimal.com/

Metcalfe, D. J. (2015). *Multispecies Design* [University of the Arts London and Falmouth University]. https://ualresearchonline.arts.ac.uk/id/eprint/13351/

Root-Bernstein, M., & Ladle, R. J. (2010). Conservation by Design. *Conservation Biology*, *24*(5), 1205–1211. 10.1111/j.1523-1739.2010.01501.x

Rosenzweig, M. L. (2003). *Win-win ecology: how the earth's species can survive in the midst of human enterprise*. Oxford University Press.

Shwartz, A., Muratet, A., Simon, L., & Julliard, R. (2013). Local and management variables outweigh landscape effects in enhancing the diversity of different taxa in a big metropolis. *Biological Conservation*, *157*, 285–292. 10.1016/j.biocon.2012.09.009

Smeenk, W., Tomico, O., & Turnhout, K. van. (2016). A systematic analysis of mixed perspectives in empathic design: not one perspective encompasses all. *International Journal of Design*, *10*(2), 31–48.

Wells, M., & Yeang, K. (2010). Biodiversity targets as the basis for green design. *Architectural Design*, *80*(2), 130–133. 10.1002/ad.1057

Woebken, C. (2007). *Animal Superpowers - Chris Woebken*. https://chriswoebken.com/Animal-Superpowers

Young, A., Khalil, K. A., & Wharton, J. (2018). Empathy for Animals: A Review of the Existing Literature. *Curator*, *61*(2), 327–343. 10.1111/cura.12257

5.2 7D – Ecosystem centered design

Ezri Tarazi
Technion – Israel Institute of Technology

5.2.1 Background

5.2.1.1 Design Thinking is not a monolithic theory

No full theoretical framework of Design Thinking is required to emphasize the claim that it is not a monolithic theory or toolbox, and it is not based on one single origin (Dam & Siang, 2020). It was built on decades of research and work of many approaches and directions in design theory. To mention a few, Creative Problem Solving (CPS) (Treffinger, 1995) is one of the milestones in this trajectory, as it offers a framework that individuals or teams can use to generate creative options and plan effective implementation of new solutions. Treffinger lays emphasis on an alternative to the traditional linear, prescriptive stage-sequential engineering models. In his CPS model, he highlights the need for both divergent and convergent thinking.

The three components of CPS are, Understanding the problem, Generating Ideas, and Planning for action. CPS was implemented in education, emphasizing that it can be taught mainly by applying it through a challenge. Nigel Cross (2011) claimed that the classic methods of reasoning in problem solving are inappropriate for design. The range of interactive and iterative design techniques such as sketching, ideation, mockups, prototyping, scenarios, enable designers to imagine a future situation ahead of linear processes. The scope of 'Design Thinking' in academic research, practice, and application has grown exponentially, and the wide experiences already gained offers an opportunity for reflection.

5.2.1.2 Stagnation in the application of Design-Thinking methods

One of the most influential theorists of Design Thinking was Béla H. Bánáthy (1919–2003), a Hungarian American linguist and professor at San Jose State University and UC Berkeley. In his book, *Designing Social Systems in a Changing World*, he presented a comprehensive and prophetic theory (Banathy, 1996). In this book, Bánáthy set the theoretical foundations for Design Dynamics and created the famous Double Diamond model of divergent and convergent processes in the mind and in action (Figure 5.2.1). The psychologist Joy Paul Guilford worked on the Structure of Intellect (SI) theory (Guilford, 1967), and coined the term Operation Dimension to describe the two opposite mental abilities or factors of intelligence: (1) Divergent production - The ability to generate multiple solutions to a problem; creativity; (2) Convergent production - The ability to deduce a single solution to a problem; rule-following, or problem-solving.

Bánáthy's original contribution was neglected, and he received almost no recognition in the design community or credit for being the father of the Double Diamond model.

DOI: 10.1201/b22630-31

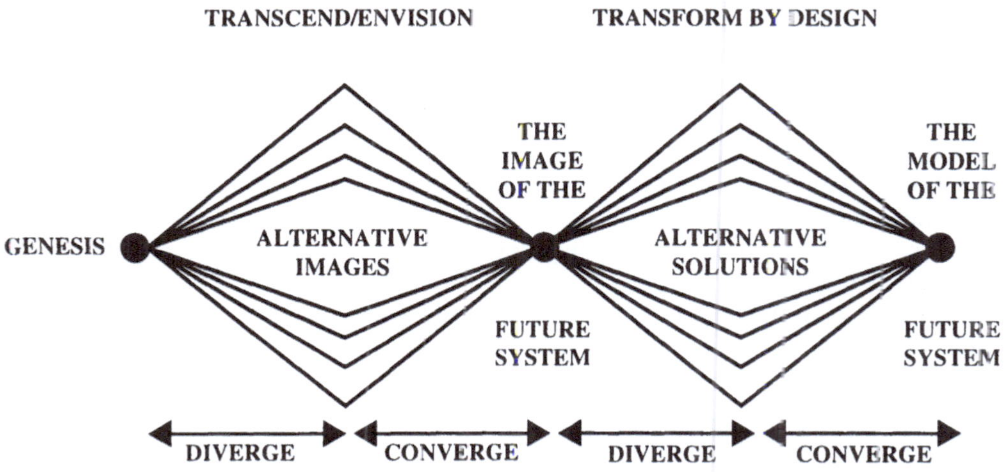

Figure 5.2.1 Banathy's model of the dynamics of divergence and convergence (after Banathy, 1996, p. 75).

The Design Thinking evangelists missed the deeper meanings of his radical theory (Denniston, 2020), of which the Double Diamond model is only one part (Iskander, 2018). For example, the Dynamics of Design model (Banathy, 1996, p. 74, Fig. 3.3) includes five spaces, which correlates with the more-recent Feedback and Feedforward movement and the iterative spiral movement between spaces, resulting in "The Space of Displaying the Models". Bánáthy argued for the evolutionary nature of the process; yet, unfortunately, the common models of Design Thinking have not been dynamic and have not evolved.

5.2.2 7D Ecosystem-centered design process

7D is based on the unique nature of seven phases of growth and development, as reflected in growth cycles in permaculture and natural ecosystem. 7D yielded a book that was published as a set of practice-based reflections on its initial use in a variety of projects (Tarazi, 2013, Hebrew). 7D can be seen as a model for translating needs into new solutions and considering a shift from human-centered design to ecosystem-centered design (Tarazi, 2019). This 7-dimensional innovation process aims to respond to future needs in a manner that reflects the broader perspective necessary to meet the challenges facing humanity. As humanity faces multiple crises, there is a need for a radical shift to a paradigm that includes multiple dimensions.

When observing the seven dimensions alongside the 17 goals of sustainable development (UN, 2015), one realizes that the classic Design Thinking recipe for innovation is not sufficient and must be updated. The metaphor that underlies 7D is inspired by agricultural principles and is congruent with the United Nations' sustainable-development goals. The once very important human-centered design must broaden its perspective to consider:

1 Humans as an integral part of ecosystems such as nature and the world
2 Climate change and the way various industries impact it
3 The fundamental survival of individuals and widening economic gaps

4 Society in each country and region, and human society as a whole
5 Culture, religion, and the relationships between them
6 Health and well-being, including identity, feelings, and social interactions
7 Unseen human systems and the evolution of the spiritual path.

Interwoven into the 7D phases, are the Five Intelligences model (Penza, 1998; Penza, 2020): Cognitive Intelligence, Creative Intelligence, Emotional Intelligence, Social\Ethical Intelligence, and Applicative Intelligence. In each of the phases, those five forms of intelligence can be applied to the analysis of situations, needs, human validation processes, and/or results.

One of the key questions when approaching innovation is to what degree it is valid for future developments. The 7D process can be viewed as a systematic linear development process but can also be viewed and used as a multidimensional matrix of domains – more like seven balls flying through the air, with each dimension having its own nature, function, and tools. Although each dimension is linked to the others, the dimensions can also be considered independently in the context of a project at any given stage of development. Each dimension of the 7D process contains specific sub-methodologies, techniques, and know-how, and each of the seven components listed above can be used within it.

5.2.2.1 Origins of the 7D

As a practicing designer, I felt that there were theoretical links missing in some parts of the Design Thinking process. I also felt that future process methods should include a wide scope of influences beyond "unmet human needs". As an IDEO, global a design and consulting firm, worker (1998–2001) I was not aware of Bánáthy's theories. I discovered those theories only after I had developed the permaculture-inspired 7D process. In 2007 I developed 7D, a process that is based on an agricultural metaphor, moving from a Plow phase to a Harvest phase in seven simple steps (Tarazi, 2013). It arose from the perception that the common Design Thinking process lacked crucial components and was focused on winning over the user. I felt that design needed to evolve further. Instead of putting the human (user) at the center, 7D puts the human user into context, positioning the human as part of an ecosystem of nature. In a publication in which we considered methods for designing artificial coral reefs, my colleagues and I referred to this process as *nature-centered design* (Tarazi et al., 2019). 7D was applied in academia, government, education, and industry. 7D is currently used in Innovation Leadership courses at the Technion, in Israeli high schools, and in European Innovation Technology (EIT) programs.

5.2.2.2 The seven phases

The seven phases or components of the 7D process add two half-diamonds to Bánáthy's model, at the beginning and the end of the process. It is a process designed for pursuing sustainable development instead of focusing solely on technology (Industry 4.0). The process resonates with human faculties and the frequencies of unseen worlds, such as multiple intelligences, the seven spectra, the seven days of the week, and even the myth of the seven days of Creation. The agricultural metaphor connects us to nature and how we can work with nature in a pure and organic model. Every project starts with a plowing phase, in which we plow the soil of the potential field of challenge, and ends in a harvest phase, in which we harvest the solutions that are the fruits of our labors. Each of the seven phases has a different frequency and requires different approaches, tools, and techniques. The fundamental concept of 7D and the agricultural metaphor is that farmers do not create the crop on their own. Nature makes it happen. The role of the farmer is to create the necessary conditions and enable growth through connection, protection, stimulation, and support. A summary of the whole cycle of natural growth and its components is presented below.

5.2.2.2.1 Phase 1: Plow

In the plow phase, the point of view and current assumptions are flipped around and turned over, just as a plow turns over the soil in a field. The plow needs to be sharp and cut deep, in order to penetrate the soil and turn it over. It disrupts the order of the solid soil, which may have been left untouched for years. The conformity, the old standard, the 'this is how we did it for years' is analogous to a dry surface on which nothing can grow. The creative process must start with the courageous act of being able to plow the ground with a blade that is sufficiently sharp. In the times in which we are living, one cannot start with a focus only on user needs or empathy for the customer. Creating a new perception starts with plowing the ground of one's own formulas, beliefs, and unconscious conditions. The plowed soil, light and aerated, is like the plasmatic state of particles, amorphous and ready to be seeded. One of the methods used in the Plow phase of the design process involves a state of unattached, colorless, and neutral thinking called No-Box.

5.2.2.2.2 Phase 2: Seed

The Seed phase enables us to view the many intricate and complex aspects of the challenge and to inspect the 'genetics' of it. As in nature, countless seeds are disseminated, in various ways, to ensure that at least some of them will take root in the soil and eventually sprout. To collect seeds, there is a whole process of searching, listening, observing, collecting, learning, and matching information. This phase is not just user-centered, but also society-, culture-, and technology-centered, with the support of Big Data that has been worked into tangible insights. The Seed phase must include a global view of the impact, unseen consequences, and prices that only nature pays, and it needs to go deeper than surface claims and pseudo-science. It requires humility and a not-knowing stance, as the seed is in the darkness of the soil.

5.2.2.2.3 Phase 3: Sprout

The Sprout phase is focused on letting early ideas grow. This phase calls for no specific brainstorming recipe, but rather an openness to the idea of a No Judgement Zone of Time that enables the acceptance of the new. The sprout is a green shoot that needs to push up through the soil to reach air and sunlight. This is quite an enormous task for a tiny seed. In the Sprout phase, ideas are scattered, and it is futile to attempt to coordinate the process or push it to happen. The conditions that allow young sprouts to emerge are focused, first and foremost, on the elevation of freedom, in a safe manner. The sprouting needs to happen first in the person, through true inspiration, the joy of offering something new, and the urge to solve a mammoth problem or satisfy a need or vision. The spirit of growth and the urge to cultivate something new cannot be faked. The main barrier to the Sprout phase is fear, and the team or the organization needs to attend to the deeper reasons for that fear. The Sprout phase may involve a readymade transformed object that took a few minutes to assemble, a short movie, taken and edited using a smartphone that shows a short future scenario, or an animation file that conveys the idea better than black text on a yellow Post-it note. There are cloud-based tools that can be used to share and work in teams, which can support the Sprout phase on a large scale and across great distances.

5.2.2.2.4 Phase 4: Bud

The Bud phase is focused on gathering the early concepts before the full direction of the project is selected. The bud phase sits at the middle point of GO\NO GO of the process, since the allocation of resources becomes meaningful after this stage. A variety of tools can be used in the Bud phase to visualize and express ideas as early concepts. The Bud shows the vision, without a heavy investment of resources. The bud, as the early flower, reveals only some of the details, while providing a concise impression of the concept. The bud gathers energy by developing more substantial components of ideas selected during the Sprout phase, such as visual manifestations, scenario-based stories, early mock-ups, fast prototypes, and early models of system design. The importance of this phase is that it provides space to evaluate a few concepts at a time, before focusing intensely on one concept. The process is iterative in nature and feeds itself, through rapid cycles of Budding.

5.2.2.2.5 Phase 5: Flower

The Flower phase is aimed at creating an attractive, fully developed prototype. Nature promotes beauty as an attraction tool for reproduction. The full prototype should reflect the beauty of the solution, as a whole and in its details. The prototype is not as fully usable as the fruit, but attention to detail in all dimensions and careful attention to the potential impact of the solution in various realms are most important. This phase includes making final decisions on what will not be part of the final product and prioritizing the most important aspects and giving those aspects all the necessary attention. There must be iterations of testing and careful validation until the Flower phase is complete and we are ready to enter the implementation phase.

5.2.2.2.6 Phase 6: Fruit

The Fruit phase calls for integrating the systems into one balanced, coherent, and viable solution that is ready to be fertilized. This is the phase in which all components must be combined, before spreading out to the world. The fruit is a useful seed-bearing tool; as the fruit is consumed, seeds are disseminated. The Fruit is not just visually appealing, but also sweet and juicy. In this phase, the focus is on taking the necessary steps to transform the concept and the proof of concept into a real solution that can be implemented. The usefulness of the concept needs to be tested again and again, in cycles of iterations, until it tastes good and is useful. In a fruit, nothing is wasted, and the sustainable nature of fruit must be the reference point for every new solution. The Fruit must contain nourishing vitamins and fibers, and not encourage harmful overconsumption.

5.2.2.2.7 Phase 7: Harvest

The Harvest process is an open diamond of establishment in the outside world. The fruits are picked when they have matured, by those who find them attractive. Those who harvest the fruit also distribute their seeds across distances, to places in which those seeds can take root and grow further. Only those seeds that can survive in the new environment are able to grow. The essence of the whole process, from the first intention to the attitudes, values, qualities, care, and love for a true solution, are manifested in the Harvest phase. The open side of the last diamond suggests that it is difficult to predict the way ahead, once a concept has gone out into the world, and that many modifications, turns, and mutations can occur after that point. The Harvest phase requires both the permanency of adhesion to the purpose and openness and flexibility, in terms of the ways to achieve that purpose. The Harvest phase rarely happens by itself; a successful Harvest phase requires the input of extra energy.

7D is a coherent process based on a metaphor that resonates with nature and sees nature as a source of inspiration. It is based on the belief that natural ecosystem processes are congruent with humans' inner systems, stored in the unconscious, and that working inside that understanding will grant the users of those systems freedom and connection to their higher selves.

5.2.2.3 The triple open diamond model

In 7D, the Double Diamond model has been extended to include an open-ended half-diamond on each end (Figure 5.2.2).

The first half-diamond starts off open and closes at the point at which the Seed phase starts. In the Plow phase, there is a considerable process of bringing together the main questions addressed by the project. This open start is essential for avoiding any pre-definition of the field, the problem, the challenge, or the solution. Beginning the process from an open position means that one should defy definitions and the existing frameworks of the existing situation. In a rapidly changing world, pre-definitions can set the whole project on the wrong track.

The second half-diamond is open to the future. The Double Diamond model ends with the delivery of the project to the client. It does not deal with or take responsibility for the impact or outcome of the innovation output. The Harvest phase is open-ended, as it includes concern for what happens after the product is out in the real world. It is the innovator's responsibility to care about the impact and consequences of the new product and the innovator should be open to what comes, ready to dramatically change whatever needs to be changed.

5.2.2.4 Integrating lean design and agile design into the 7D process

There are those who have asked how Design Thinking, Lean Design, and Agile Design differ from one another and it has been suggested that all these approaches can be

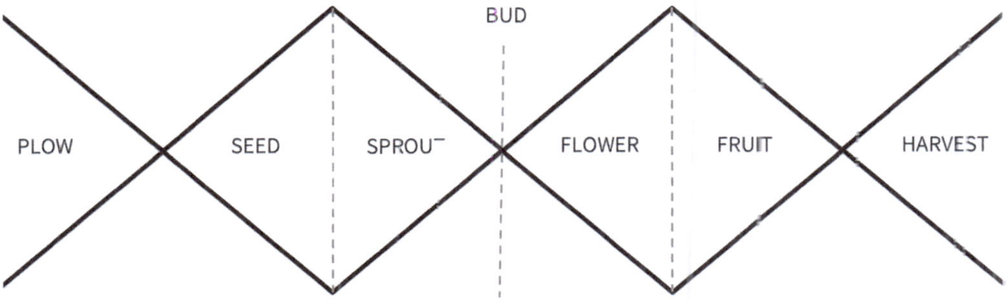

Figure 5.2.2 Open-ended triple diamond 7D model.

integrated into one process. 7D can include all three. In the 7D model, the Plow phase uses the PENZA 'NO BOX' methodology; Seed to Bud uses Design Thinking phase methodologies, overlapping with Flower and Fruit, which use the 'Lean Startup' methodologies. Finally, Fruit and Harvest use Agile methodologies. The integration of these four methods makes more sense in the overall of progress.

5.2.3 Design cases

Below, I present several design cases from the past seven years, which illustrate how the 7D process has been used since I first published my book on the subject (Tarazi, 2013):

5.2.3.1 Tarazi studio

The first design case is from Tarazi Studio Ltd, which the author has owned and managed since 1990. Tarazi Studio has been a laboratory for the development of the 7D process, with Israeli-based business clients such as Electra, Tadiran, and Extal; international players such as Qualcomm, Intel, and Microsoft; and startup companies such as Magink, ZenithSolar, and Born Free. 7D provides a straightforward way to lay down the process phases from the first proposal sent to the clients to the follow-up on the process itself. The metaphor has resonated throughout the Tarazi Studio's portfolio of projects aimed at sustainability and responsibility for the future. For example, Magink sells a digital-ink billboard designed to replace paper-based billboards that uses less energy than LED billboards. The ZenithSolar project was a 7-year effort to create a hybrid solar dish that could very efficiently create both thermal energy and electricity in the same device (Tarazi, 2008). The BornFree project yielded the first line of BPA-free bottles for infants in the US and Canada.

5.2.3.2 d-Vision

The first application of 7D outside the Tarazi Studio was d-Vision, a unique design and innovation program supported by Keter Group, a global company in the fields of garden furniture, storage, toolboxes, and large children's toys. This Israeli-based company created a 2-year program for 20 design and engineering graduates every year, who studied and fostered innovation across the entire company. The d-Vision innovation center ran from 2005 to 2012 and was an ultra-creative idea factory for Keter. The 7D platform enabled the group to have a wider scope than the company's R&D and marketing departments could afford to realize and led the company to the development of a plethora of new products and services. The value of sustainability was incorporated into the company's development processes and the company portfolio began to shift. Examples include the development of a garden compost bin that is not only very affordable but also shortens the time needed to create home compost by 85%; a program in which customers can return old Keter products to stores and be refunded based on the weight of the returned products; and garden shelters made of recycled plastic.

5.2.3.3 The Y7 program

The Y7 program of the PENZA Perception Lab was an initiative that aimed to foster "yielding entrepreneurship" and applying 7D internationally. A few dozen leaders from Israel, the Palestinian Authority, Germany, England, Brazil, and Denmark were trained at the Y7 center in Jaffa, Israel. Y7 entrepreneurship programs have also been started at other innovation centers in Israel. In China, there were a few attempts to promote Y7 in various programs involving innovation in education. In Germany, there were a few rounds of Y7 workshops in Cologne and Berlin. Later, the program was developed into the Y357 program (Penza, 2020), which includes more codes and models for entrepreneurship leadership, to serve a wider population of participants and clients. The Y5 code of The Five Intelligences was adopted as a system for assessing pupils in Tel Aviv schools. Some groups that started working within the Y7 program have gone on to create successful hi-tech startup companies. Importantly, many Y7 graduates have gone on to become key leaders in the entrepreneurship ecosystem in Israel.

5.2.3.4 Israeli Ministry of Education

The introduction of 7D into the educational system started with an initiative by the Israel Ministry of Education that was led by the Ministry Chief Inspector for Design and Art, Einat Kritzman. She was familiar with the 7D, encountered in her own master's degree studies, and sought to promote both design programs and interdisciplinary team education. The first program involving 7D started in 2017; that program was created to combine high-school students' final projects in different courses, such as Computer Science, Mechanical Engineering, Electronic Design, and Art. It was followed by a 2-year workshop that included 55 teachers from 10 high schools, which over the years served various populations and around 3,000 individual pupils. The most recent plan is to extend the program to 150 schools in 2022.

5.2.3.5 Government of Colombia – The national innovation program

A national innovation program, 'Innovamos 2021', was run in Colombia from 2010 to 2018. The project was initiated by the then-new government, led by Juan Manuel Santos (later awarded the 2016 Nobel Peace Prize). PENZA Perception Lab, created a national innovation program the methodology of which was based on the 7D model. We found fertile ground in which to seed this theory in hundreds of startups and companies. Over the course of a 5-year pilot program, we managed and ran workshops to support the germination and growth of hundreds of startup groups from Bogota, Medellin, Cali, and other cities. Some of these startups became successful companies. In recognition of these efforts, the *Wall Street Journal* together with CITI bank, named Medellin in 2013 the "Innovative City of the Year" (CITI 2013).

5.2.3.6 EIT UMI

The European Institute for Innovation & Technology (EIT) has been supporting 7D and Y357 workshops on urban mobility (EIT, 2020) through a program called Urban Mobility Innovation (UMI). That program included a website (UMI, 2020) and a podcast (Amstrong & Tarazi, 2020). The project partners are the Municipality of Munich, Germany; the Municipality of Lublin, Poland; Zone Cluster of Budapest, Hungary; and the UnternehmerTUM Innovation Center in Germany. All of the partners met for the first time in Munich in February 2020, just as the first news started to arrive about the outbreak of COVID-19 in China. The original plan was to hold four live workshops in each country – Germany, Hungary, Poland, and Israel – over the course of 2020. In March, we all agreed that we needed to shift to online workshops. During the pandemic, the challenges of urban mobility changed dramatically from overcrowded cities to the need to create safe public transportation. The examples presented in the four online workshops varied from a bus that was specifically designed for the elderly population, with separated and isolated seats, to a system to connect elderly people with volunteers who could drive them to important meetings and help health-care workers commute to their regular jobs. As part of this program, the 7D massive online open course (MOOC) on edX was translated into English and made available to program participants.

5.2.3.7 Technion – Israel Institute of Technology (IIT), and other institutions of higher education

I have been teaching 7D at the Technion since 2016, as part of courses in the Industrial Design program that I lead and as part of courses in the Department of Humanities and Art that are open to all undergraduate students. More than 250 students have taken these courses. The Israel Council for Higher Education (CHE) has initiated a program to foster entrepreneurship and innovation centers in campuses. The Technion IIT has been fostering innovation through courses for almost 40 years, started with the legendary course initiated by Prof. Dan Schectman (a Nobel Prize laureate in Chemistry) and Prof. Shlomo Maital. Technion alumni became the backbone of the transformation of Israel into a "startup nation" (Senor & Singer, 2009). In 2018, the Technion program won a CHE tender and the t-Hub, the Technion's innovation and entrepreneurship center, was established. As the founder and chair of the t-Hub, I have used 7D as a platform for the innovation processes in academic courses and entrepreneurship programs with more than 1,500 students. Over the years, other 7D workshops have been held at the CERN IDEA innovation center in Geneva, Tel Aviv University, Haifa University, and other institutions.

5.2.4 Conclusions

A separate paper would be needed to fully cover the research, reflections, findings, and outcomes in each of the fields presented above. Yet, an overview of more than a decade of working with 7D in a variety of contexts (i.e., business, government, research, and education) provides a unique perspective. The conclusions from this overview are presented below.

5.2.4.1 What worked?

5.2.4.1.1 The 7D metaphor is engaging

The first engagement of people with the 7D concept flows with ease in a variety of environments. The metaphor easily finds a home in people's minds and creates a harmonic 'knowing' that this is nature's way. This was quite evident in the workshops held in Colombia and China. Images like the dispersal of many seeds at the beginning of the innovation process are quite engaging. There have been almost no cases of people or organizations refusing to learn the model. Most of those who were accustomed to the orthodox Design Thinking approach have agreed that that approach lacked certain elements and that 7D is a more comprehensive approach. The 7D approach has been warmly received by high schools and one school even decided to use the 7D metaphor as the basis for its program for the next 10 years (Hornstein, 2021). The metaphor has been effective across a variety of cultural contexts. 7D covers more of the innovation process than classic Design Thinking and enables people from more disciplines and fields to take part and engage in the process. The missing components are a 7D book in English and other languages, an online course in English, and other scale-up tools, which need to be created.

5.2.4.1.2 The presentation of 7D through online platforms is a viable model

There is a variety of knowledge scope and depth needed for individual participants and teams to start working. In general, most individuals grasp the concept intuitively and quickly; they can start working on one of the phases after an hour of explanation with examples. Before the COVID-19 pandemic, the 7D MOOC on edX was used successfully by both teachers and students. The EIT project added an English version, with subtitles for the videos and English text for the tasks. For 2020–2021, the Entrepreneurship and Design Thinking course at the Technion Department of Humanities and Arts was transformed into a hybrid course that includes a combination of longer videos (around 45–60 min each), video conference meetings, and team presentations. KAN, a national Israeli radio station, created a Hebrew podcast about 7D (Markovitz, 2019). Those interview sessions helped to spread the word about 7D.

5.2.4.1.3 With 7D, creativity grows

The 7D process does not hold only one method for ideation. It aims to create new and unending ideation processes. It creates a mindfulness process of ideation with the person in solitude. We have also conducted experiments with 7D ideation in pairs, using something called the Chavruta method. This method was inspired by the tradition of studying sacred Jewish texts in pairs. The Chavruta method can replace larger brainstorming group session (7–10 participants). We found that working in pairs could double or even triple the engagement rate and number of ideas. We also suggest establishing a crowd-source system within organizations, to bring out new ideas. Promoting freedom in the Sprout ideation process is part of the understanding that there are differences among cultures, disciplines, ages, and levels in the hierarchy of an organization that calls for flexibility in the way that ideation is carried out. There is also an acknowledgment that some ideas will arise only after the process, a day or a week later, maybe in the shower …

5.2.4.1.4 7D boosts inspiration and motivation

Even during the COVID-19 pandemic lockdown (2020–2021), the 7D process boosted motivation among university students, as compared to traditional processes. At the Technion, we conducted four 7D workshops (4–5 days each), and three full-semester courses in which we observed high levels of student engagement. The students worked in teams to develop viable and innovative solutions and there were zero dropouts. The students' presentations were full of humor and the feedback from the students reflected high levels of motivation and satisfaction with the process.

5.2.4.1.5 7D provides a more holistic perspective

The 7D process provides a more holistic perspective since it calls for more than the individual user to be at the center of the creative process. It includes the ethical, social, and environmental considerations that come up in every process. The participants are asked to use the 'Five Intelligences' model to analyze the needs of the situation at hand. In many 7D projects, we have observed high levels of social considerations, such as projects that consider solutions for less-developed countries, elderly populations, and minorities.

5.2.4.1.6 7D actualization in organizations is a reality

7D has slowly but surely become the accepted method for design and innovation processes in educational systems. The 7D process has been incorporated into all design studies in Israeli high schools, using the 7D MOOC on edX and a detailed documentation process to help teachers go into the details of each step of the process. Both the Technion and Haifa University use 7D in their entrepreneurship programs. In 2021, Israeli governmental organizations, including the Prime Minister's Office, used the same 7D MOOC. Workshops for municipalities, teachers' organizations, and hi-tech companies have been held since 7D was first established. The adoption of 7D by those organizations has given it a much wider scope and broader relevance.

5.2.4.1.7 7D has the potential to scale and grow

The application of 7D has the potential to scale and grow beyond what has been done so far. 7D speaks the language of the new awareness among younger generations, who view the abandonment of sustainable and socially conscious development as a crime. The 7D process, with its metaphor and language, can reach those young people who are more willing to balance their innovation and progress with the wider scope of the multiple crises they will face in the future. The growth of 7D might trigger that awareness and help the younger generations to feel more involved with sustainable innovation.

5.2.4.2 What did not work?

5.2.4.2.1 Limitations of the metaphor

Many of the situations in a new project or start-up development are utterly new and undefined. Every process has its own unique and unprecedented elements. The agricultural metaphor of 7D, with all of its advantages, does have its limits. Not all that happens in building a new product or service can be likened to the growth of a fruit tree. At some points in the process, individuals must feel free to let go of the metaphor and think freely about the situation at that moment. This calls for an acute awareness of what is needed at any moment in time, for the team and in light of the tasks ahead.

5.2.4.2.2 Linear order vs. chaotic order

The natural metaphor is one of linear organic growth, which makes a lot of sense for those who are developing new and innovative projects for the first time. Yet, the linear route of natural growth does not match the reality of project development, in which there is often a need to go back from the middle of the Flower phase to the Plow phase or to create a quick Sprout phase. Although this point is mentioned in every workshop and full 7D process, in practice the metaphor sometimes seems to limit those loops. Another important feature of 7D is the use of the natural strategy of having a plethora of Seeds and Buds on hand and the ability to go back and start down a different path or toward a different niche, if necessary.

5.2.4.2.3 In short workshops, little attention is paid to the fruit and harvest phases

The nature of workshops that last 4–6 days or a semester course comprising 14 two-hour meetings, does not allow for full attention to be paid to the Fruit and Harvest phases. One dilemma faced by instructors is whether it is a waste of time to even teach those phases at the beginning of the process. Although it might seem unnecessary to mention those phases in a short workshop, projecting for these two phases is important. Participants tend to lose energy when they get over the Flower project final presentation. The main excitement of an innovative breakthrough concept is there. Yet the difference between yielding and non-yielding innovation lies in the ability to focus on the Fruit and Harvest vision, and not just on a promising presentation. Teaching the full 7D, it is important to convey these visions of Fruit – a viable, tested, and ready-for-market product/service – and Harvest – a serious attempt to launch the product/service to the real world (market, users, clients, etc.). It is crucial to plant this distinction in participants' minds, even if it is not very relevant to the specific workshop in which they participate

5.2.4.2.4 Organizations tend to return to existing patterns

Every organization has its own patterns of processes which, for various reasons, are planted in many areas. Some call these patterns the DNA of a company/organization. We have found that in young organizations that include people with little experience, there are few pre-existing patterns and 7D can be easily adopted with an agile, very flexible, and open way of navigating the new territory of creating a product/service. The experience with larger, older organizations has been that although 7D is very inspiring and fruitful for them, they are not very quick to adopt the 7D approach. They prefer to move on to the next management consultants and learn another method, instead of using and perfecting 7D by creating effective, innovative leaps in their lines of products or services.

5.2.4.2.5 Wider scopes are difficult for for-profit organizations to grasp

The world economy has not digested the full-scale implications of the climate crisis, the social transformations, and the wider scope of any innovation path. Slowly, new regulations and laws have started to emerge, but it has been quite difficult and challenging to convey the holistic view of 7D to businesses focused on year-end profits. Case studies and stories about failures due to people ignoring those issues have encouraged some openness to the 7D concept. Over the last 10 years, we have witnessed a transition in sentiment and attitude in the business world toward the assimilation of a wider scope of considerations in companies' innovative processes, products, and services.

5.2.4.2.6 The danger of making 7D part of innovation theater

Many companies and organizations start to adopt "innovation strategies" and even recruit "innovation managers" to their leadership. 7D provides a coherent and comprehensive language for this process and yet the tendency in some organizations is to direct efforts toward quick public-relations accomplishments and stop there. Innovation theater (Blank, 2019) is a danger that can sometimes take years and a lot of resources to find and heal. The challenge faced by 7D has been the tendency for the methodology to remain with the creative team within an organization and rarely move and integrate into the stream of the serious Harvest of products and services that generates results and impacts the organization's short- and long-term results. For example, in one large project, 7D was used successfully to foster 36 teams that, over a period of 4 months, reached an advanced phase with a good business model, prototype, and proof of concept test and validation. However, the organization then refused to advance the project. Instead, based on innovation theater, it chose to launch a larger program, in a plan that quickly ended in failure.

5.2.4.2.7 The 7D might stay frozen, as happened to Design Thinking

7D is a step forward with few aspects concerning innovation methodology, yet it might end up with the same fate as Design Thinking and stop to evolve.

5.2.5 Summery

7D is an attempt to further develop the Design Thinking process and integrate it with Lean and Agile methods. It involves adding open diamonds at the beginning and the end of the Double Diamond frame, to form a Triple Diamond model. The use of a natural agricultural metaphor helps to engage people with the inner frequency of each of the different phases, making them tangible and resonant. Due to the global pandemic, there was a shift to teaching 7D through a MOOC that includes videos and online conference engagements, which has opened new possibilities for disseminating the 7D process to additional populations around the world.

The use of a natural metaphor helps us be inspired by nature in the innovation process, moving us from human-centered design to more holistic, ecosystem-centered design. There is a new generation of young creative people that is looking for up-to-date processes that reflect values and the current situation of the world. While it may sound naïve, there is a belief and hope that this approach may create a transformation among those who embrace it and adopt its deeper meanings.

Acknowledgments

I thank Prof. Gabi Goldschmidt (Technion – Institute of Technology) for her guidance. I thank the reviewers for detailed comments. I thank the Trio and partners at PENZA Perception Lab. This work was supported by EIT Urban Mobility, the Israeli Ministry of Education, Israel Council of Higher Education, Dr. Leonard Polonsky Ph.D., and Dr. Georgette F. Bennett Ph.D., and the Firefly Scientists' Foundation.

References

Amstrong, A., & Tarazi, E. (2020, June 28). *Design Thinking for urban mobility | The strong startup*. Retrieved from https://youtu.be/NON7NoCwRek

Banathy, B. H. (1996). *Designing social systems in a changing world*. New York, NY: Springer +Business Media.

Blank, S. (2019, October 7). *Why companies do "innovation theater" instead of actual innovation*. Retrieved from https://hbr.org/2019/10/why-companies-do-innovation-theater-instead-of-actual-innovation

CITI (2013). City of the year. *The Wall Street Journal*. Retrieved from https://www.citigroup.com/citi/news/2013/130301a.htm

Cross, N. (2011). *Design Thinking: Understanding how designers think and work*. Oxford-NewYork: Berg Publishers.

Dam, R. F., & Siang, T. Y. (2020, June). *Design Thinking: Get a quick overview of the history*. Retrieved from: //www.interaction-design.org/literature/article/design-thinking-get-a-quick-overview-of-the-history

Denniston, A. (2020, September 21). How might we … Avoid the Design Thinking hype and make better products. *Medium*. Retrieved from https://medium.com/theory-and-principle/how-might-we-avoid-the-design-thinking-hype-and-make-better-92654b82df16

EIT (2020). *Eiturbanmobility*. Retrieved from https://www.eiturbanmobility.eu/projects/umi-design-thinking-support/

Guilford, J. P. (1967). *The nature of human intelligence*. McGraw-Hill.
Hornstein, H. (2021, July 20). *The Reali School vision*. Retrieved from https://www.colbonews.co.il/education/99916/
Iskander, N. (2018, September 5). Design Thinking is fundamentally conservative and preserves the status quo. *Harvard Business Review*. Retrieved from https://hbr.org/2018/09/design-thinking-is-fundamentally-conservative-and-preserves-the-status-quo
Markovitz, G. (2019, September 12). *Design Thinking in 7 phases*. Retrieved from https://www.kan.org.il: https://www.kan.org.il/podcast/item.aspx?pid=14376
Penza, T. (2020). *Y357 - Perceptive & applicable tools*. Tel Aviv, Israel: Penza Perception Lab.
Penza, T. (1998). *The five intelligences*. Tel Aviv, Israel: Penza Perception Lab.
Senor, D., & Singer, S. (2009). *Start-up nation: The story of Israel's economic miracle*. New York, NY: Twelve Press.
Tarazi, E. (2008). Beyond 1000 suns. *Changing The Change*, T.42, P.75 Torino.
Tarazi, E. (2013). *7D*. Tel Aviv, Israel: Penza Perception Lab (Hebrew).
Tarazi, E., Parnas, H., Lotan, O., Zohabi, M., Oren, A., Yosef, N., & Shashar, N. (2019). Nature-centered design: How design can support science to explore ways to restore coral reefs. *The Design Journal*, *22*, 1619–1628.
Treffinger, D. J. (1995). Creative Problem Solving: Overview and educational implications. *Educational Psychology Review*, *7*(3), 301–312. http://www.jstor.org/stable/23359352
UMI (2020). Urban mobility innovation program report. Retrieved from https://umi.net.technion.ac.il: https://umi.net.technion.ac.il
United Nations (UN) (2015). *Take actrion for the sustainable development goals*. Retrieved from https://www.un.org/sustainabledevelopment/sustainable-development-goals/

5.3 Design communicating space tech innovation

Janett Adler, Jea Hoo Na, and Martyn Evans
Manchester School of Art, Manchester Metropolitan University

5.3.1 Introduction and literature review

The research explored how a bespoke design thinking model and subsequent design practice can effectively represent innovation culture in space technology organisations. Therefore, the aim of this research was to establish the effect that design can have on communicating innovation and technology, chiefly the impact design can create in the industry. Hence, the objective of the secondary research was outlined as a contextual review with literature search directed at these overlapping areas. As an industry of the future with enormous potential for growth and a vast area for new inventions and exploration, space tech has always been and still is an inspiration to all generations. With the Space Sector Deal of the UK Government's Industrial Strategy, a policy intended to boost the UK economy, a 'new space age' has taken off (Satellite Applications Catapult, 2018) and productivity is increased where space collaborates across sectors (Space Growth Partnership, 2018). The five projects conducted for this research explore how visual design and creative techniques can contribute to growth and prosperity, fostering impact in the region, fuelling the imagination of prospective audiences, and communicating new ideas to a world that is constantly adapting and adjusting itself.

The relationship of design and innovation in space tech was the key driver of the contextual review that served as a platform for the work with project partners. How visual design can contribute to the growth of the economy by communicating space tech innovation was the core question that helped examine the relationship of the emerging themes. Here are some of the insights the author found.

Design generates competitive advantage, it contributes to the success of new products and adds value to businesses (Hernandez et al., 2017, Verganti, 2009) because it gives meaning to a product and helps make sense of a task (Verganti, 2009). Design is one of the main drivers for innovation and productivity, it advances economies and proposes clusters (Hernandez et al., 2017) where design skills achieve long-term productivity and portray innovation (Design Council, 2017). Businesses are in need of new patterns in innovation with value creation, idea generation and implementation, motivated leadership and commitment (Parmar et al., 2014). This is particularly relevant to design as it can provide differentiation to advance markets; develop new products; create business opportunities by understanding consumer needs and desires; improve brand value that is inclusive, and ensure employees are satisfied. Design thinking is capable of evaluating organisations' challenges (Borja de Mozota, 2006) that helps businesses develop innovation opportunities, and to become more productive and profitable (Benton et al., 2018). With applied design thinking, businesses can grow (AHRC, 2013).

DOI: 10.1201/b22630-32

For the reasons listed above, organisations are advancing by embedding design into their innovation processes (Bason, 2015). The hybrid between future tech and creative design serves as a pioneering instrument and concurrently as a consistent quality asset. The potential of design is further discussed at a workshop hosted by the Transformation North West Cohort where it is concluded that design can enhance productivity and help better communicate future technology to new users (Transformation North West Cohort, 2018). In a steady progression, design is increasingly linked to branding. It is capable of learning about new audiences and working strategically, breaking through to new markets (Verganti, 2009).

5.3.2 Methodology

The research methodology can be understood as exploratory in the context of empirical research. It was a hybrid of scientific inquiry with the dynamic of creative input. Research in the field of visual design can be an analytical as well as a practical tool, bringing about critical thinking and mature design practice (Noble & Bestley, 2011). Two intertwining methodologies were used to produce holistic insights from contrasting evidence. Together, case studies and grounded theory methodologies foster exciting research that has had minimal attention up to now. From an interpretivist perspective (Salmons, 2016), involving the research aim from the start, data collected during case studies were analysed through coding concepts and categories, and recognising patterns (Chun et al., 2019), concluding in grounded theory. As an ambassador for all project partners, the researcher examines the data and explains it, creating a new hypothesis that is open to change (Pettigrew, 1997), while obtaining a perspective that encourages a flexible process (Mills et al., 2009). Grounded theory operates as a way of analysing all contexts, for example, qualitative secondary research in the form of literature review, and primary research from case studies and therein focus group discussions (FGDs) that include workshops and in-depth discussions with industry partners. Through a design thinking process, case studies generated both low- and high-fidelity prototypes. Exploratory interviews with stakeholders in person and online together with feedback served as further primary research resources that attempted to relate back and link up to the secondary research at the beginning. The methodologies were concerned with the main aim that focussed on objectives, as seen in the research structure, see Figure 5.3.1.

The aim of this research was to answer how design can help communicate the values and innovation culture of space tech organisations through a design thinking process. Using the sequence diagram (Figure 5.3.1), the research aim could be broken down into objectives that could be fulfilled via methods to create new, thought-provoking knowledge. The core question was examined using case studies (MA) and grounded theory (MB) as this research's applied methodologies. This included initial coding, the interrogation of all collected data, and the creation of concepts, concluding in a theory to methodically establish the hypothesis (MB). The data investigated were collected in FGDs, interviews, feedback and prototypes (M1-3) through the five case studies with project partners (O3). It was generated through testing and failing (Silverman, 2017) and prototyping again in a constant loop (M3), using the design thinking model (O2). The benefits were that through coding, data could be explained and the interdependence of methodologies, methods and processes promoted practicable quality (Chun et al., 2019). However, when looking at data and producing theory (MB), the research may lose some of its energy and become passive (Krippendorff, 2006). To tackle this issue, the researcher's chosen combination of the two

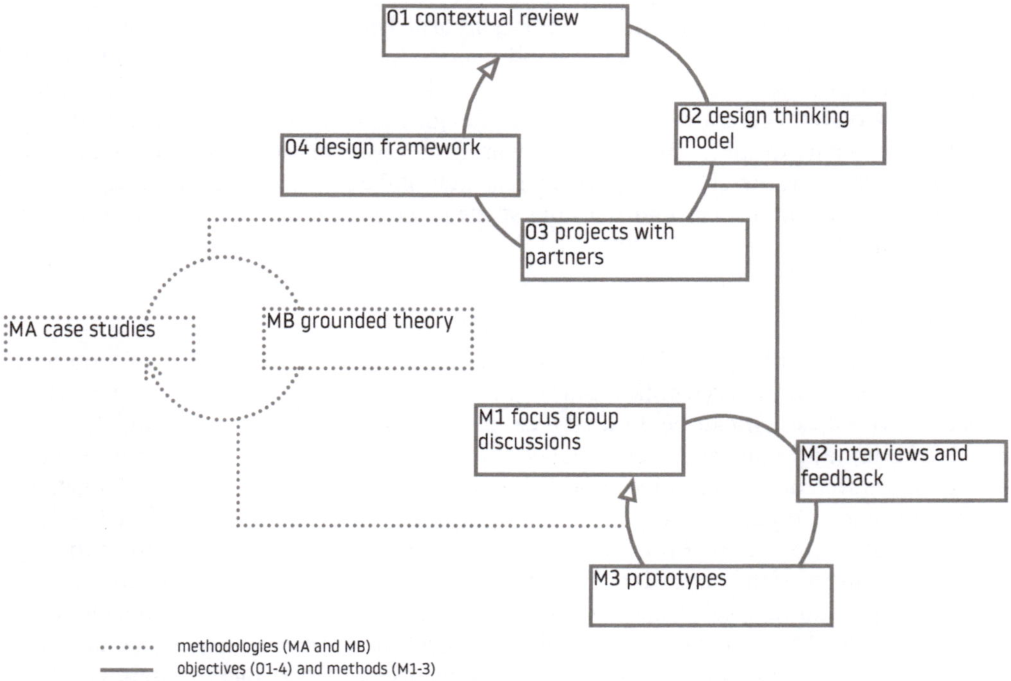

Figure 5.3.1 Research structure – the relationship of research objectives (O1-4), methodologies (MA and MB) and methods (M1-3).

methodologies with a substantial amount of practical work as case studies (MA) helped create a unique dynamic interrogating grounded theory (MB).

As part of the contextual review (O1) the researcher was analysing background literature, working with organisations (O3) and interviewing its employees (M2), using an interpretivist research philosophy (MB) with a focus on innovation of satellite and space technologies, design and its contribution to innovation that were anchored in the Industrial Strategy of the UK Government.

The author undertook workshops as part of focus group discussions (FGDs; M1) in collaboration with colleagues as well as industry partners (O3) in order to establish a design framework (O4), through grounded theory (MB) that implemented primarily qualitative research and identified a matrix which served as a map as part of the design thinking model (O2).

When thinking about the research approach, its epistemology, theoretical perspective, methodologies and methods are exploited (Crotty, 1998). Primarily constructionist – through a tailored design thinking process brands were created or developed with lo-fi (low-fidelity) prototypes along the way that meant that the prototypes embodied knowledge. Figure 5.3.2 depicts the cross-fertilisation of the various methods used in this research, creating a landscape of both case studies and grounded theory methodologies, all in the relationship of theory and practice.

Though research may seem remote from design practice, but it can be a motivating force for design practitioners (makers). Thinking, albeit not too structured, and analysing when creating visual prototypes, helped evaluate new insights. This is design research in

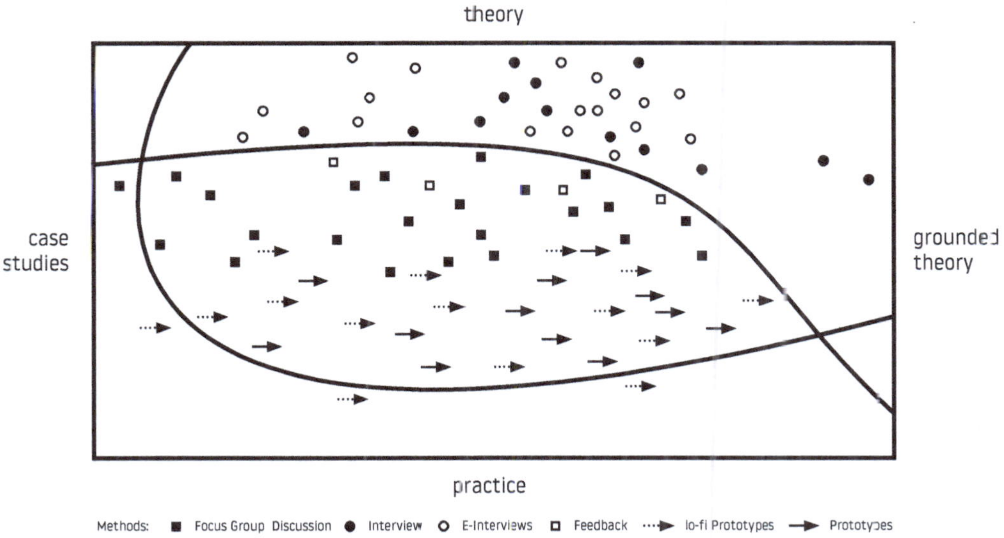

Figure 5.3.2 Landscape of methodologies with methods used in context of research practice and theory.

its most applicable way: using a design process for design practice is the combination of making and thinking (Frayling, 1994).

5.3.3 Design thinking

The author gained advanced insight from literature sources and interviews that helped create a platform from which a bespoke design process could be derived, which is both practicable and can serve as a base for all case studies. The author's own experience as a design practitioner with a background in visual design, particularly in branding, has helped shape the design thinking model further, because "Branding has become such a well-known concept, that both in the public discourse and academia most people are well aware of the business dynamic of a good brand …" (Schroeder & Salzer-Mörling, 2006:145).

The literature search identified a couple of key sources that helped develop and evaluate the author's tailored model. With its 'Double Diamond', the UK Design Council established four stages – 'discover' 'define' 'develop' and 'deliver' – that summarise and support many of the designers' processes. A creative approach can be demystified, utilising and iterating the stages throughout the thinking process (Design Council, 2019). The Hasso Plattner Institute of Design at Stanford has also been instrumental in individual and organisational understanding of 'design thinking' as a corresponding process. This goes through six stages: 'empathise', 'define', 'ideate', 'prototype', 'test', and 'assess'. The design process is not linear and can be enacted according to organizations' or individuals' requirements, enhancing their innovation methods. Design thinking can be applied to existing challenges at work, employing design tools, and respecting collaborative approaches and personal growth in the process. The design thinking process helps develop insights through re-evaluation of perspectives, it appropriates the creation of prototypes that are then tested in relevant conditions to assess their value, with repetition and refinement of the individual stages to generate new working practices, techniques and equipment (Stanford d.school, 2019). Different design agencies' approaches, such as Fitch, Superunion, Pentagram, Interbrand, Landor and Wolff Olins, also matched up with the Design Council's and Stanford's design thinking processes.

To gain further insight into the influence of design thinking in practice, the author interviewed a UX Designer and UX Researcher at a large social media company. Aside from the significance their adapted processes have for innovative thinking, the UX Designer also specifically identified the use of visual design to establish new business partnerships: 'For two almost identical products or services [to be] considered, the delights of visual design can make the difference in people being more comfortable using the one or the other' (UX Designer, 2020). Branding played a crucial role in distinguishing an organisation or a product from competitors, and interviewees also highlighted the importance visual brand applications play in being perceived as more innovative than their competitors.

To advance an approach tailored to the satellite and space sector, a new practical design process was developed for this research and employed in five projects. This process has elements of design thinking approaches with divergent and convergent activities, as well as a more complex feature in the ideation phase which can be understood as 'magic' or creative intuition (Bierut, 2018:20–21). This model is divided into five stages, see Figure 5.3.3. However, rather than working simply stage by stage, the various stages are to be seen as a holistic investigation, where parts often overlay and intertwine in a constant loop of iteration within individual stages as well as some or all stages together, generating a complex structure, unique to the organisation or product to which it is applied.

The names were developed on reflection of each stage's specialism. The first stage KNOWING is to learn to understand and determine the organisation's needs, to listen to its values and gain insight into its current innovative projects. The second stage PACE means progressing steadily and is the first part of the ideation and making process, where initial ideas are taking shape, a creative orientation originates, and the branding process accelerates. The third stage CRESCENDO, where ideas and making culminate, is the actual creation of a design route, the peak of the process delivering visual outputs that aim to communicate the organisation's innovation. The fourth stage REFLECT is to review, revise, apply and advocate design elements to a publishable output. And RECOVER stands for impact in connection to the UK Government's Industrial Strategy to help productivity in adding value to the economy, and in articulating feedback from the partner organisation and its stakeholders. This model and individual stages iterate as the author tailors them to each project independently without restriction, which then can be compared to the other case studies in order to draw overall conclusions.

Augmenting the design thinking model, the author created a design thinking activity map (see Appendix) that served as an analytical framework, summarising as well as comparing key elements of the work done with partner organisations that were useful to

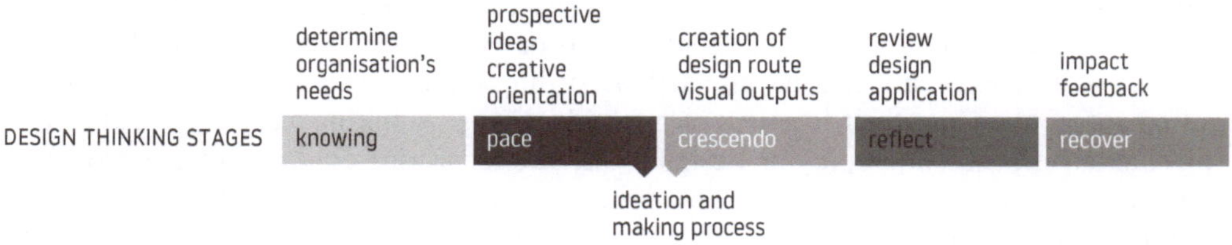

Figure 5.3.3 Design thinking model – individual stages.

the research. It is a model that unfolds how different types of analysis were carried out and directly links to the research aim. The distinctive design thinking stages are presented in relationship to the input, actors, activities and output of each case study.

The individual case studies of collaboration with project partners are presented in the following sections with found data interpreted for the reader.

5.3.4 Intelligent transport

The author designed a visual language, using the design thinking model to communicate innovation in the first project, 'intelligent transport', that uses satellite technology to improve traffic flow. The visual output in the form of a brand assists the company in building a business case to take the project onto the market.

Intelligent transport is a multi-modal approach to boost productivity in freight transport for two major freight feeders in the North West – the Port of Liverpool and Manchester International Airport. Based on the design and technology company's earlier project that allows ambulances to get through traffic faster (GOV.UK, 2017), the smart freight project uses space data via GPS/VPS controlled sensors managed by big data principles to ensure green light corridors for, but not reduced to, trusted traders and suppliers moving through the entire urban transport networks that can cut congestion, and idle and response times. Intelligent mobility is achieved, by addressing future industrial, political, societal and environmental challenges (Catapult, 2018).

The technology company aims to communicate their innovation by creating a cohesive brand language for their intelligent transport application that reflects its smart nature. The company was seeking to utilise the project as an umbrella activity to help them move forward in introducing their new product to the market.

KNOWING – Through FGDs, the author gained insight, into how satellite technology is used in day-to-day traffic. After the initial briefing from the partner organisation, the requirements for a brand language emerged for the smart traffic project. This involved creating a brand that could now serve as a foundation for brand communication supporting new business development opportunities.

PACE – This stage included research on how brand language can be used to communicate an innovative product by establishing form, colours, and a relevant font and illustration style.

The activities in this first part of the iterating design process included: contextual observation; decision matrices and mind mapping; and the creative orientation as an output with designers, with the digital lead and CEO involved (see design thinking activity map in the Appendix).

PACE – For the company's new smart transport project, the original idea was a tilted square or diamond shape with four sides, corners pointing into the air or space, rail, road and sea connecting freight etc. via different modes of transport. CRESCENDO – During the process of creating the logotype, it became evident that the design needed to adapt a flexible role within the development of the project. PACE – The naming process as part of the branding process had three parts, with the chosen name for the brand 'AI Traffic Flow'. CRESCENDO / REFLECT – With 'flow' in the name, the idea of the tilted square had to become more flexible in shape – an unconfined, more organic, yet interconnected shape was formed, within the parameters of the shape of a diamond with four sides. A gradient of traffic light colours with the green light prevailing has helped shape the logo for the new business case, as shown in Figure 5.3.4.

Figure 5.3.4 From tilted square to free-flowing shape, creating brand language.

Here design outputs involved prototypes of the brand name, a design route and brand elements, the author was collaborating with one of the animators and the digital lead. Brand guidelines were created as well as a business model canvas. The author learned to adapt when finding the brand name, and changing the direction of the design route; lo-fi prototypes helped embed that change.

RECOVER – The logotype was key to the company's brand development, with its shape and colour palette being adaptable to further brand applications, for example, the author worked closely with one of the company's animators to create an explainer video to exploit and apply the visual design elements developed for this project. For the CEO of the company, the project was an opportunity to access talent for real-life industry collaboration that resulted in distinctive brand creation for an innovative product (Intelligent transport CEO, 2019). In response to the Industrial Strategy (BEIS, 2017), the project depicted support for a local SME, and as a result, a shift in thinking and value to the region, and it addressed issues such as the improvement of productivity by developing innovative opportunities, for example, entering new markets. During the project, in FGDs, creative thinkers sat at one table as a catalyst for collaboration, stimulating innovation with engineers and businesspeople, freight and logistics specialists and a transport director. The project highlighted that within activities of cross-industry collaboration, there is a need to listen to a variety of voices in order to understand and balance collective perspectives and use this to inform the project development. The role of creative thinkers included facilitation and visual communication in these collaborations, enabling cross-pollination of ideas that were not previously considered. As a result, making future technology more receptive to more traditionally minded employees and partners, who may think millions need to be invested to improve roads and build bridges, whereas small interventions like sensor technology can be the answer, as a result of different ways of viewing the problem.

The researcher created a brand language for this particular project which could then serve as a foundation for publications and brand applications as well as deriving businesses from it. The author specifically looked at how brand language impacts the company's innovation to cultivate a shift in long-term thinking, entering new markets and developing new business, while identifying suitable visual design tools. Lessons learned include adapting to change within the ideation and making phase, and that good communication in FGDs prevents misunderstandings because of the problem being well defined.

5.3.5 Asteroid mining

The second project partner is looking for funders for a first satellite launch ahead of asteroid mining activities. The author developed the brand to communicate the organisation's innovation and promote it for an optimal market position.

The collaboration with the space mining start-up culminated in the creation of a tailored, flexible design strategy. The priority of the project was to attract investors for a satellite launch that would then allow activities to follow, using ground-breaking technologies that will enable extraction, processing and use of materials from near-Earth asteroids. The project involved brand development, concluding with a visual portfolio that aided in promoting the start-up externally, and assisting the internal structure of the organisation.

KNOWING – The space mining organisation is targeting the off-Earth commercial market. Asteroids contain large amounts of precious metals (Lewis, 1992), to a greater extent than materials on Earth, potentially advancing the global economy (Jamasmie, 2017). 'Mining the skies is no longer a subject of science fiction stories and movies' (Sukumaran, 2016:125).

During a FGD with various professionals in the company, the CEO, engineers, a geologist and designers, the need for a professional identity including key elements such as an investment brochure was found to be most pressing to find additional funders for the upcoming satellite launch, as well as a business collateral for presentation externally and structure internally.

PACE – To advance the development of the organisation's brand, the author designed a creative orientation – an initial concept of prospective imagery, fonts, layout and colours, to get an idea of how the identity was anticipated to look and feel.

CRESCENDO – The Microgramma font was determined for bold, extended headers that represent attention, power and the horizon, urging slow reading to make keywords memorable. Depicting a contrast to the element of brand language and different uses of copy, the Silka font for shorter text was chosen to look geometric and contemporary, and Noto serif font was selected for the main text for a distinctive yet easy-to-read quality. A thin line was formed as the brand element, complementing the brand image depicting 'delicate perfection' and contrasting the logotype's hexagon shapes. The line depicts both quality and weightlessness – near zero gravity that is pointing upwards in the reader's direction, and diagonal, indicating floating in space and open-endedness that symbolises excitement, anticipating growth. The amber shade depicts energy and looks confident and cheerful, combined with impartial, yet sophisticated warm grey tones, creating a spectacular mood, especially when applied to imagery, see Figure 5.3.5.

REFLECT – In combination with the existing logotype and characteristic hexagon shapes, the brand language was developed to create meaning directly linked to the organisation's innovation by building confidence in its propositions

RECOVER – A coherent, professional identity acts as a conversation starter that is adapted and further developed by the organisation. The design-led approach was valuable in creating the corporate identity, but it was also considered a beneficial tool for developing the business. For the pioneer in Space Resources, the opportunity to conduct an innovative approach to design that represents the industry is the most practical application, yet one that challenges the norm. With the investment brochure ready for publication, the organisation is moving forward to the investment stage of the business model, helped by the brand development to become visually recognised and the Asteroid Mining concept to be accepted. A professional brand contributed to the start-up's confidence, a momentous development in the organisation's history with a considerable impact on its business (Asteroid mining CEO, 2019).

Figure 5.3.5 Creative orientation.

5.3.6 Space engineering

A rebrand was welcomed by the third partner that is working to explore and develop space with four subdivisions, including lunar and Mars rovers, rocketry, space balloons, and cansat (can-sized satellites). The space engineering organisation aims to attract more sponsors, partners and potential colleagues with enhanced presentation through new brand creation.

This project immediately had an entirely different dynamic to the design thinking process, see Figure 5.3.9. The student society concerned with space exploration and development is based at a university in England's North West and is successfully involved in worldwide competitions as well as workshops to educate and encourage the next generation.

KNOWING – The UK Space Agency is increasingly working together with academia with an interest in research and development of space tech innovation (Grossman, 2020). The space engineering partner has four departments. 'Rovers' has its focus on robotics with distinct teams from diverse backgrounds for a lunar and Mars rover that will go through a design and quality process to create a working object for a challenging environment. 'Rocketry' concentrates on robotic, hybrid rockets. The department is concerned with self-landing mechanisms and looks at a possible launch from a space balloon. The 'high altitude balloon' moves comfortably into the stratosphere, this department is science- and practice-focussed. 'Cansat' is short for a satellite, the size of, and moreover, a satellite built in, a can. A cansat is usually launched from a research rocket in order to imitate an industry-like satellite with its duration from first design to review following a flight. A diverse team of students working in science, space technology and engineering, is entrusted to this organisation. All departments are engaging in projects individually or collaborating together and with stakeholders (Issuu, 2019; Manseds, 2019).

The first brief was to find a suitable logotype for the organisation as a whole, with the knowledge that there are sub-departments to follow. While learning about the organisation's values and what it wants to convey, the question arose as to how these can be translated into a visual system that works.

PACE – The original creative orientation involved finding the right inspirational words to describe values that could then be decoded into images in the aerospace engineering context.

CRESCENDO – The character inspiration already started extracting a number of ideas on potential colours, typography, brand language and illustration style. The insight for the primary colour 'sky vs sea' came from looking back to Earth from space. Supplementary colours were established alongside six secondary colours to have a comprehensive choice for versatile usage. The Ailerons font as the basis for the organisation's main logotype, made up entirely of the letters in its name, is integrated into the space environment and has a timeless look. Another font, Bariol, was found for the tagline and text in contrast to the main logotype and to allow more of the organisation's values to come through, having a friendly and accessible appearance. REFLECT – A clear-cut, delicate logotype in the space context is used to set the organisation apart from other branches in the UK and internationally as well as serving as an umbrella to the four different departments.

KNOWING – Rover, rocketry, balloon and cansat depict the sub-divisions of the student society with the objective to create a coherent and accessible brand. PACE – Appropriate for a replay mode, logo icons were the most suitable with recognisable graphics, characteristic for each separate department, yet together identifiable as one organisation. The circular shape indicating the movement in orbit, that is infinite, limitless and complete.

CRESCENDO – The initial prototype logos were of great detail within the parameters of a circle and of a previously determined colour and imagery palette. REFLECT/KNOWING – During a FGD workshop, co-designing with department heads and the organisation's president included mind mapping, card sorting and scribbling, a simpler design was preferred that can be scaled as needed.

PACE – In-depth insight helped the author uncover the visual implementation of the various departments' characteristics allowing for interesting angles, contrasting shapes and cropped elements, see Figure 5.3.6. CRESCENDO – With moving, yet captured logo icons, brand guidelines were created including proposed brand applications for strategic design use on different types of layouts and brochures, websites, and on textiles, etc., and use of illustrations, for example, in explainer videos, flyers and posters. REFLECT – This project has become much more layered, with reiterating elements to solve a complex task for an organisation that wants to set itself apart from other branches and its parent organisation nationally and internationally, and to bring about a varied brand that makes itself and its divisions comprehensible.

RECOVER – Creating a brand for a space engineering organisation has resulted in a valuable kit that translates a name into an identity to share the interest of space exploration and development and to address prospective members, partners and sponsors. The design thinking process and subsequent successful rebranding work by the author

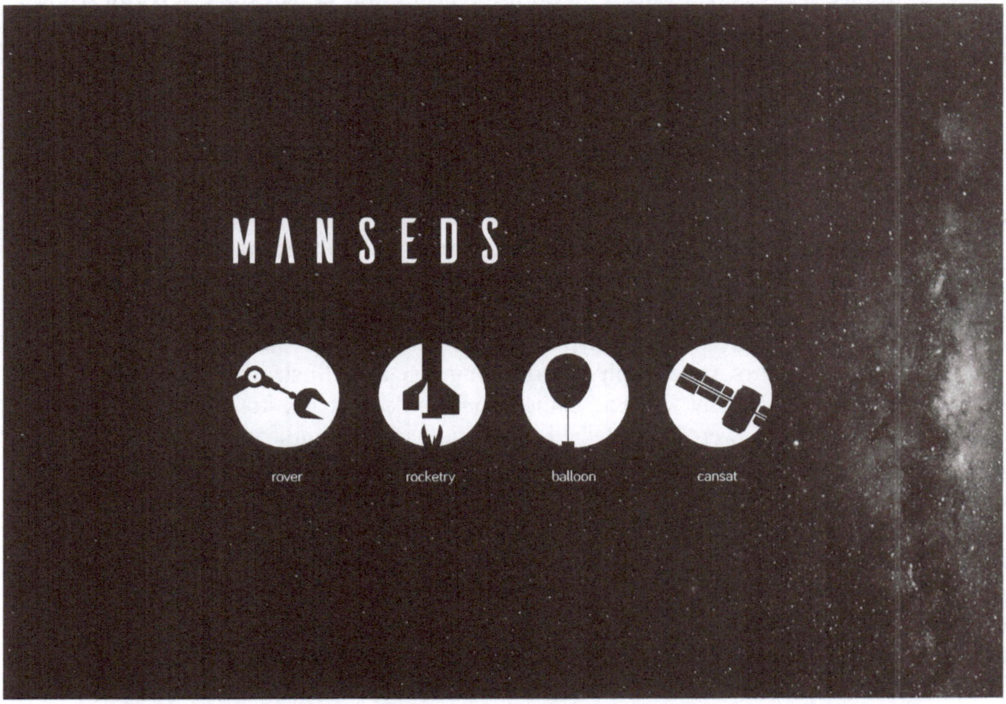

Figure 5.3.6 Logotypes for primary brand and sub-departments.

was the result of integrating the organisation's interests and activities by understanding its ambitions and preferences. This project provided evidence that design can help understand innovation through visual communication, which encourages funders to consider higher-risk yet innovative organisations that create new products, facilitating the growth of entrepreneurial thinking in the region (Space engineering President, 2019).

5.3.7 Remote sensing

The fourth partner is mapping wildfire patterns via satellite Earth Observation (EO) data. By understanding and exploiting the power of visual design, the remote sensing company is able to reach new audiences, and initiate enthusiasm in the space industry and the nature of its work.

The remote sensing start-up's new brand was created to enhance its presentation. The award-winning company is interpreting radar data with a flexible approach to expanding the organisation's offer. A professional identity was requested to set the brand apart and convey the right message to the audience.

KNOWING – Firefighting efforts spent on moorland and heathland can be cut significantly if not completely by monitoring wildfires using remote sensing techniques that help restore the affected area by reseeding (Copernicus Masters, 2016). Variations of moss can hold water like a sponge and help the area to be more resilient (The University of Manchester, 2018). Through EO data, carbon can be detected, and models created to help restore peat and improve wildfire response mechanisms (EnviroSAR, 2020). Open-source satellite data makes it possible to uncover patterns in the context of landscape and its protection, and imagery from space can be a tool comprising different remote sensing techniques (Discover Magazine, 2015). The company's breakthrough occurred after significant media interest in a wildfire in a North West moorland in 2018. With land mapping and experience in geomatic benches, the start-up wants to present itself as doing justice to nature and also to emerging technologies with an underlying sophisticated tone. The author and the CEO undertook several FGDs to seek co-innovative opportunities.

PACE/KNOWING – Initial ideas involved finding a layout system for all kinds of text as well as a colour study. With research into comparable companies and prospective clients, the colour palette may remind one of, but need not be immediately connected to, nature. Another visual element is the background in layouts such as the website, which may primarily be white, as this is associated with light and with Earth observation, a service that is coming from space to the planet (and not the other way around). As a supplement charcoal grey is used for typography and, for example, book covers rather than black for legibility and accessibility. In order to develop the brand language, the author obtained knowledge on how satellite technology is used for conserving landscape and how fire destroys vegetation and kills animals due to loss of habitat.

PACE – For the creative orientation to establish a brand language, inspirational words were found that arose through researching the company and its immediate surroundings. The words were then translated into images, which helped define colours and patterns. Typography was established, depicting a hierarchy with DIN, a realist font, and the robust serif Caslon and visual elements together helped create a basis for a layout that could then be specified for website applications.

CRESCENDO – The first draft of the design route was set out with a logotype based on the sans-serif font with a flexibility of surrounding patterns inspired by satellite imagery that can adapt to different applications. For the desired accessible branding, a professional look of the bespoke service needed to be followed through all platforms, including social media, in order to captivate viewers.

Element of KNOWING with PACE again – The CEO favoured a flame that was part of the old logo and, following many conversations, by combining ideas of patterns inspired by the perspective of satellites, radar pulses and the visual translation of flames, a solution was found for the logotype and favicon that leaves room for interpretation to the viewer, see Figure 5.3.7. With the company's approach to dynamic mapping generating more detailed predictions in terms of fuel and fire risk, visibility, safety and curiosity in brand presentation were not compromised by aesthetics and preference.

CRESCENDO again with REFLECT – With a recurring loop of design stages, patterns that emerged were refined in an organic approach. Ideas were communicated and evaluated against design criteria, highlighting visual and functional elements of the brand, which can speak for the company and build flexible structures that can be adapted to a wide range of applications. The ability to co-brand with national and international organisations as well as projects became an important aspect to include in the brand identity. Brand guidelines have been developed combining visual elements with possible future assignments across physical and digital spaces.

PACE – A combination of technological and natural aspects of the company required additional translation, creating more rhythmic phrases and a meaningful adaptation of the brand.

REFLECT – Short as well as detailed brand guidelines that include all design elements were created for both in-house use and external suppliers.

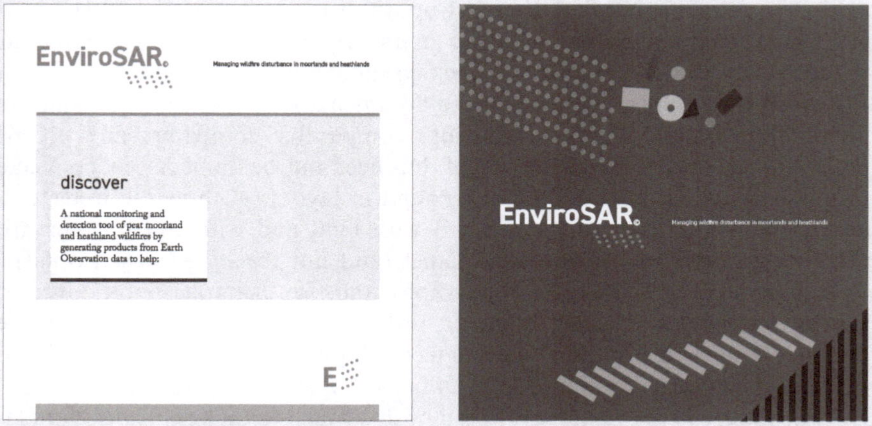

Figure 5.3.7 Full layout and book cover. Use of brand elements.

RECOVER – Feedback from the remote sensing company comprised the value of design expertise and the role of design as a communication tool to customers and funders by visually providing the right kind of messages. By understanding and exploiting the power of visual design, the company is able to reach new audiences, initiate enthusiasm for this start-up in the space industry and the nature of its work, and to enhance the receptiveness to its ideas. A professional presentation helps develop the business and takes the design forward into, for example, a geoportal service online. With competitive advantages in applying elements of effective visual design combined with EO technologies, cross-industry collaboration is stimulating innovation (Remote sensing CEO, 2020).

5.3.8 Space settlement

For the unfinished fifth project, the author generated a creative orientation and worked towards brand development, brand application and guidelines. The partner is established in the field of education in space settlement that develops workshops aimed at high school children. The project was halted due to the restrictions of the COVID-19 pandemic, which resulted in physical workshops being suspended.

In this non-profit organisation, young people take charge of their future skills through trying and testing construction, communication and societal scenarios on Mars. What technologies do we take to Mars? What would we want to create there? How could we communicate? And what would the ethics be like when we land there? Using tools that embrace virtual reality, data security, 3D printed buildings, sensors and the Internet of

Figure 5.3.8 Martian Sunset. Brand applied to backdrops.

Things, workshops simulate settlement on Mars. The brand creation project was interrupted halfway through the design thinking process loop with a number of lo-fi prototypes created, which can serve as a practicable platform that is useful to the organisation's future brand strategy.

KNOWING – Mars is the reason for numerous space missions. It is the most explored and most approachable planet in the universe (IntechOpen, 2020). Collaborative tasks in Constructing a Life on Mars workshops, aimed at high school-aged children, include role play of a range of professions and activities. Different team challenges arise in a space environment and are of tangible, intellectual and societal character. Participants are given the confidence to take charge of a career in a constantly reshaping world that functions in a digital form (Waters, 2019).

PACE/KNOWING – After thorough research into Mars settlement scenarios and technologies useful in workshops, the author developed a creative orientation. CRESCENDO/PACE – During further conversation with the partner, the first draft of the design strategy was created. The logo is based on typography for flexible and dynamic application with the chosen font locking engineered and intelligible, reflecting the company's values. Figure 5.3.8 displays a backdrop with the Martian sunset gradient resembling energy and determination. REFLECT/CRESCENDO – Through strategic use of design elements that creatively combine digital and analogue ideas, the charity that helps people get into employment can build its main brand on the project's brand language. Innovative use of visual design can activate awareness of digital skills needed in the future of work.

The author gained insight into space settlement scenarios suitable for high school-aged children through mind mapping, FGDs, interviewing and empathetic modelling (see design thinking activity map in the Appendix).

5.3.9 Discussion and conclusion

The insights gained through the case studies are valuable independently as well as collectively as they fostered a deeper understanding of the application of a bespoke design thinking process to a specialised industry, tailoring individual stages of that process to the projects, and connecting design practice and design theory.

The first case study, the intelligent transport project, offered the opportunity to create a brand, and through this development, the author was able to identify and communicate the organisations underlying innovation values providing the potential to enable the organisation to enter new markets.

The second case study involved in brand development in the context of the exploration and exploitation of asteroids and supported securing funding by generating a professional identity that improves the organisation's identity and helped to establish partnerships. The promotion for an optimal market position was crucial at a stage where funding was still needed for a satellite launch prior to asteroid mining activities.

The third case study, space engineering, focussed on an organisation that supports projects involving lunar and Mars rovers, self-landing rockets, space balloons and can-sized satellites. The creation of a brand helped to define the organisation's identity and set it apart from parent and sister organisations. A significant element of the rebrand was to enhance organisational image and set a platform for more effective communication with sponsors, partners and potential new members.

With the fourth case study, which uses satellites to help prevent wildfires and maintain natural resources, a professional presentation enables the organisation to develop further and reach new audiences. Creating a brand for a company that uses satellite data has become a tool for cross-industry collaboration that is stimulating innovation.

The evolving fifth case study uses a space settlement scenario to engage young people in workshops highlighting digital skills needed in the future of work. A creative orientation and first design route provide the industry partner with brand elements useful for application of the organisation's Mars project as well as its main brand.

When comparing the five design thinking models that supported brand creation and development, the individual stages varied considerably depending on each project (Figure 5.3.9) but all stages were evident in all case studies. Case study one and two commenced in a fairly similar manner but then saw differences in the ideation and making phase of the process when intelligent transport had to adapt the different modes idea to a more flexible one after the naming process, while asteroid mining was more straightforward going from creative orientation through to identity creation with iteration within each stage. The design thinking process in case study three immediately had a completely

Figure 5.3.9 Design thinking model – all case studies.

different 'vibe', as with the organisation's four sub-divisions the structure was much affected by the iterative approach due to the larger variety of design elements addressed. However, within iterations, similar patterns emerged. In the fourth, a unique rhythm became clearly visible as the stages of the process were aligned to distinctive design features depicting the initial concept, the design direction with the creation of logotype variations, and brand language and guidelines essential to communicate the partner's value were made suitable for the company and its stakeholders. The evolving fifth case study drew its inspiration from in-depth research and conversations with the partner, the creative orientation and first design route that can be useful work in progress for the charity's umbrella brand.

The differentiation within individual design thinking processes was a consequence of the uniqueness of each organisation's needs and the subsequent identification of solutions. Design thinking is found to be a valuable approach for the space tech sector due to its adaptability and ability to accommodate a range of contexts.

In light of both primary and secondary research (emerging themes in design, economy, innovation and technology, Section 1), the word map in Figure 5.3.10 points to the number of key areas that account for the aim of this research and the overlaps therein. Through this approach, extensive interviews with diverse professionals from all five partner organisations helped evaluate the research. A list of actors involved in the analytical framework can be found in the Appendix. Interviewees agree that looking good and possessing a professional identity demonstrate impact, power, monetary value, establish trust and indicate the company's DNA. The interview participants think that brands are valuable in communicating innovation and creating visibility. They can act as an influencer, improve work, and help take ownership and a state of belonging. Brands reflect an organisation's quality, generate enthusiasm and tell its story. They are a window into the organisation, they can create emotions, attract new partners, help differentiate themselves and enable collaboration.

The combination of research methods, including interviews, feedback and FGDs, as well as the various visual prototypes, an intelligent and flexible design thinking process

Figure 5.3.10 Interview keyword metadata.

was developed that created an environment that cultivated long-term growth in businesses. This demonstrates that an applied design thinking model can help organisations become more resilient. In the context of the UK Government's Industrial Strategy (BEIS, 2017), visual outputs help build a business case to enter new markets and advance communication with prospective funders and new audiences, and to establish a professional presentation. These contrasting projects apply the same design thinking model, tailored to each context, resulting in a complex yet accommodating structure with five stages: KNOWING, PACE, CRESCENDO, REFLECT and RECOVER. This process informs, generates and provides novel approaches to solve problems strategically, creating opportunities conducive to innovation, and helping ensure the longer-term sustainability in organisations. Primary data correlates with secondary research which aligns with the overarching aim of the research and informed a final model that is valuable to space tech organisations.

Limitations to the study include pressure from partners, when something needed to be ready quickly for a launch or when a partner preferred a certain colour to be translated into the brand. With good communication and integration through FGDs with co-design activities, the partners were involved in decision making and problem solving.

Using a bespoke design thinking model and employing tools and techniques from a design thinking activity map, the author created and developed brands for five project partners in the satellite and space industry in the North West of England, UK that communicate the organisations' innovation, promoting sustainable impact in the region. The partners' brand images have been transformed as a result of applied design thinking. Therefore, this research has shown how innovation culture within space and satellite organisations can be represented via a bespoke design thinking model. The evidence provided from outputs through this research demonstrate the design thinking model's ability to create visibility of an organisation that is meaningful for communicating innovation, contributing to brand value.

Acknowledgment

'Transformation North West' is a doctoral training programme, focused on design and creativity. It is funded in the UK by the Arts and Humanities Research Council via the North West Consortium Doctoral Training Partnership (NWCDTP) as a part of the National Productivity Investment Fund, across five Universities in England's North West with twelve doctoral researchers, collaborating with organisations in response to the Industrial Strategy.

References

AHRC (2013). *The Value of Design*. Available at https://www.youtube.com/watch?v=nNy1wxnFq8Y, Retrieved on 29/09/2018.

Asteroid mining CEO (2019). *Feedback form for J. Adler*. 1 April.

Bason, C. (2015). *Why the world's leading organisations are looking to design*. Available at http://www.designforeurope.eu/news-opinion/why-world-s-leading-organisations-are-looking-design, Retrieved on 30/05/2018.

BEIS (2017). *Industrial Strategy: Building Britain for the Future*. London. Available at 10.1049/ir:19930092, Retrieved on 29/09/2018.

Benton, S., Miller, S., & Reid, S. (2018). The Design Economy 2018. Available at https://www.designcouncil.org.uk/resources/report/designeconomy-2018, Retrieved on 29/09/2018.

Bierut, M. (2018). *Now you see it and other essays on design*. New York, NY: Princeton Architectural Press, 20–21.

Borja de Mozota, B. (2006). *The four powers of design: A value model in design management*. Design Management Institute. Design Management Review, 17(2).

Catapult (2018). *Intelligent Transport Programme*. Available at https://sa.catapult.org.uk/our-focus/intelligent-transport, Retrieved on 10/05/2018.

Chun Tie, Y., Birks, M., & Francis, K. (2019). *Grounded theory research: A design framework for novice researchers*. SAGE Open Medicine, 7.

Copernicus Masters (2016). *Envirosar© - Managing Wildfire Disturbance In Moorlands And Heathlands*. Available at https://copernicus-masters.com/winner/envirosar-managing-wildfire-disturbance-moorlands-heathlands, Retrieved on 24/06/2020.

Crotty, M. (1998). *The foundations of social research: meaning and perspective in the research process*. London: Sage Publications, 2–9.

Design Council (2017). *Designing a future economy: developing design skills for productivity and innovation*. Design Council, 6–13.

Design Council (2019). *The Design Process: What is the Double Diamond?* Available at https://www.designcouncil.org.uk/news-opinion/design-process-what-double-diamond, Retrieved on 21/06/2019.

Discover Magazine (2015). *How Google Earth Has Revolutionized Archaeology*. Available at https://www.discovermagazine.com/the-sciences/how-google-earth-has-revolutionized-archaeology, Retrieved on 24/06/2020.

EnviroSAR (2020). *Home*. Available at http //www.envirosar.com, Retrieved on 24/06/2020.

Frayling, C. (1994). Research in art and design. *Royal College of Art Research Papers*, 1(1), 1993/4.

GOV.UK. (2017). *Red Ninja's smart tech clears the road for ambulance crews*. Available at https://www.gov.uk/government/case-studies/red-ninjas-smart-tech-clears-the-road-for-ambulance-crews, Retrieved on 03/03/2020.

Grossman, D. (2020). *How The UK Is Quietly Disrupting The Space Industry*. Available at https://www.inverse.com/innovation/uk-disrupting-the-space-agency, Retrieved on 18/12/2020.

Hernandez, R., Cooper, R., Tether, B. & Murphy, E. (2017). *Design, the language of innovation*. (26), 1–14.

IntechOpen. (2020). *Mars Exploration – A Story Fifty Years Long*. Available at https://www.intechopen.com/books/mars-exploration-a-step-forward/introductory-chapter-mars-exploration-a-story-fifty-years-long, Retrieved on 22/09/2020.

Intelligent transport CEO (2019). *Feedback form for J. Adler*. 9 May.

Issuu (2019). *ManSEDS Sponsorship Proposal 2018–19*. Available at https://issuu.com/manseds/docs/manseds_sponsorship_proposal, Retrieved on 22/12/2019.

Jamasmie, C. (2017). *Luxembourg shoots for the stars with fresh space mining deal*. MINING.COM. Available at https://www.mining.com/luxembourg-shoots-stars-fresh-space-mining-deal, Retrieved on 18/06/2019.

Krippendorff, K. (2006). *The semantic turn: a new foundation for design*. Boca Raton: CRC/Taylor & Francis, iii-iv, 27–29.

Lewis, J. (1992). *Extraction of volatiles and metals from extraterrestrial materials*. Lunar and Planetary Laboratory, The University of Arizona, 5.

Manseds (2019). *About*. Available at http://manseds.co.uk, Retrieved on 19/12/2019.

Mills, A., Durepos, G. & Wiebe, E. eds. (2009). *Encyclopaedia of case study research: L-Z; index*, 1. Sage.

Noble, I. & Bestley, R. (2011). *Visual Research: an introduction to research methodologies in graphic design*, (2). Lausanne, Worthing: AVA Academia, 23–76.

Parmar, R., Mackenzie, I., Cohn, D. & Gann, D. (2014). The new patterns of innovation. *Harvard Business Review*, 92(1), 2.

Pettigrew, A. (1997). What is a processual analysis? *Scandinavian Journal of Management*, 13(4), 337–348.

Remote sensing CEO (2020). *Feedback form for J. Adler*. 10 March.

Salmons, J. (2016). *Choosing methodologies and methods for online studies. Doing qualitative research online*. SAGE, 17–39.

Satellite Applications Catapult (2018). *Research Into The Value Of Satellite-Derived Earth Observation Capabilities To UK Government – Satellite Applications Catapult*. Available at https://sa.catapult.org.uk/news/research-into-the-value-of-satellite-derived-earth-observation-capabilities-to-uk-government, Retrieved on 24/06/2020.

Schroeder, J. E., & Salzer-Mörling, M. (2006). *Brand Culture*. Abingdon: Routledge, 145.

Silverman, D. (2017). *Doing qualitative research*. Sage, (5), 263, 283, 297, 475.

Space engineering President (2019). *Feedback form for J. Adler*. 14 August.

Space Growth Partnership (2018). *Prosperity from Space*. Available at https://www.ukspace.org/wp-content/uploads/2019/05/Prosperity-from-Space-strategy_2May2018.pdf, Retrieved on 31/07/2020.

Stanford d.school (2019). *Design Thinking Bootcamp*. Stanford d.school. Available at https://dschool.stanford.edu/executive-education/dbootcamp, Retrieved on 14/12/2019.

Sukumaran, P. (2016). Mining Asteroid Resources. *Journal of the Geological Society of India*, 88(1), 125.

The University of Manchester (2018). *University Start-Up Company To Expand Its Peat And Heathland Wildfire Research Efforts*. Available at https://www.manchester.ac.uk/discover/news/university-start-up-company-to-expand-its-peat-and-heathland-wildfire-research-efforts, Retrieved on 24/06/2020.

Transformation North West Cohort–Adler, J., Blackwell, B., Costabile, I., Kallegias, A., Kowalska, P., Latham, G., Omar, L., Partha, C., Pialorsi, V., Robins, J., Street, J. & Wareing, L. (2018). *Driving Industrial Strategy for North West Growth – the Role of the Creative Industries*, 22–27.

UX Designer (2020). *Interview with J. Adler & B. Lee*. 13 February, London.

Verganti, R. (2009). Design driven innovation: changing the rules of competition by radically innovating what things mean. *Harvard Business Press*, 5(11–12), 24–29.

Waters, M. (2019). *Constructing A Life On Mars*. Available at https://www.calom.space, Retrieved on 19/10/2020.

Appendix

Table. Analytical framework: design thinking activity map.

		knowing	pace	crescendo	reflect	recover
input	INTELLIGENT TRANSPORT	wants, needs sensor technology small intervention skills	different modes of transport names decision matrices	interpretation translation amendment	assessment	semi-structured interview schedule feedback form
	ASTEROID MINING	what it is, needs and wants targeting the off-Earth commercial market	design critique	design by metaphor	scenario	corporate identity valuable to developing the business
	SPACE ENGINEERING	student society that is working to explore and develop space with four sub-divisions: lunar and Mars rovers, rocketry, space balloon and cansat	critique co-design workshop sketching look and feel	translated into a visual system	understand the organisation's innovation and how it wants to set itself apart	funders encouraged to lend to higher risk organisations, that create new products
	REMOTE SENSING	mapping wildfire patterns via satellite earth observation data	empathic modelling layout system legibility and accessibility perspective of satellites, radar pulses, visual translation of flames as inspiration	suitable logotype variations for organisation	story board	communication tool to customers and funders professional presentation helps develop the business and takes the design forward into e.g. a geoportal service online
	SPACE SETTLEMENT	young people can take charge of their future skills through the scenario of life on Mars	FGDs mind mapping	FGDs participatory workshop model	empathic modelling mockups	
actors	INTELLIGENT TRANSPORT	designer, digital lead, CEO	animator	digital lead	digital lead	CEO
	ASTEROID MINING	CEO, engineer, astrogeologist, designer	CEO	CEO, engineer, scientist, astrogeologist, investment analyst	CEO	CEO, designer
	SPACE ENGINEERING	project planner, aerospace engineers, outreach officer, treasurer	president, department heads	president	president, department heads	president
	REMOTE SENSING	researcher, technical lead geologist	director	director	director	researcher, technical lead geologist
	SPACE SETTLEMENT	director	director, immersive technology developers, 3D printing prototypers	director	director	
activities	INTELLIGENT TRANSPORT	FGDs collaboration across different fields of expertise reframing	naming process brain writing creative orientation – ideas for colours, form, font, illustration style	creation of logotype visual design elements	lo-fi prototypes of design applications	business case enter new market
	ASTEROID MINING	FGDs mapping contextual observation	initial concept of prospective imagery, fonts, layout and colours	brand language contrasting to existing logotype	linking to the organisation's innovation by building confidence in its propositions	attract investors promoting optimal market position assist internal structure
	SPACE ENGINEERING	card sorting group passing	inspirational words to describe values decoded into images recognisable graphics	brand language elements (via colours...) logo icons brand applications	layered, reiterating elements	set itself apart from other branches and its parent organisation nationally and internationally
	REMOTE SENSING	mind mapping comparative and competitor analysis insight into prospective clients	open coding inspiration translated into colour palette, typography, patterns etc.	room for interpretation visual and functional elements of the brand	flexible structures to be adapted to a wide range of applications	reach new audiences, initiate enthusiasm for this start-up in the space industry cross-industry collaboration is stimulating innovation
	SPACE SETTLEMENT	workshop observation FGDs, interview	illustration, character inspiration, colour, form and typeface study	brand elements, dynamic logotype workshops	lo-fi prototypes: illustration and layout	
outputs	INTELLIGENT TRANSPORT	insight – satellite technologies for everyday traffic structured online interview	brand name creative orientation presentation	design route brand	brand guidelines business model canvas	explainer video feedback interview
	ASTEROID MINING	persona seeking investors for satellite launch detailed predictions in terms of visibility and curiosity in brand presentation	creative orientation	design strategy professional identity investment brochure and business collateral ready for publication	visual portfolio mockup	practical application yet challenging the norm audience accepting Space Mining concept increased trust and profit
	SPACE ENGINEERING	thematic analysis expertise, understanding its ambitions and preferences	mood board integrating the organisation's interests and activities	rebrand	high-fidelity prototype brand guidelines	kit to attract more sponsors, partners and potential colleagues name = identity
	REMOTE SENSING	profile	ideation sketching creative orientation rhythmic phrases	design route logotype fav icon	detailed brand guidelines combining found visual elements with future assignments across physical and digital spaces	raised image social media to captivate viewers growing business not limited to wildfires alone, expand its offer further afield
	SPACE SETTLEMENT	insight from interview and FGDs	creative orientation	first draft of design route	brand application (flyer)	

5.4 Distributed design thinking

The role of design representation

James Andrew Self

UNIST (Ulsan National Institute of Science & Technology), Department of Design; Brunel University London, School of Design

5.4.1 Design thinking

Since its launch at TU Delft in 1992, the Design Thinking Research Symposium (DTRS) has examined and extended an understanding of Design Thinking (DT). In contrast to a set of tools or methods for use in applying DT, or in training individuals to engage DT, the DTRS has been primarily concerned with a holistic understanding of what DT is, how it manifests itself in practice, and what DT can tell us about the human activity and ability to design.

Many design researchers have made progress in understanding DT as a particular type of thinking unique to design activity: From Cross et al.'s earlier work on design activity (Cross et al., 1997), originally published as an outcome of the very first DTRS, to the applied methods and tools generated by Stanford's *D School*. For an inclusive review of DT, including its use in management, as a method for innovation, and as means to understand designerly thinking, I point the reader to Johansson-Sköldberg et al. (2013) comprehensive analysis.

For the purposes of this chapter, I will use the term *Design Thinking* to describe a particular type of cognition. Schön (1983) originally described a type of thinking-in-action associated with design activity, whereby the designer reflects upon design works as a scaffold for thinking. Others have since extended Schön's original concept (Cross, 2007; Dorst, 2011). In my approach, I am interested in DT as a type of cognition engaged during design practice, rather than as a method to address a problem in a designerly way, or as used to describe a strategic set of design tools/methods.

Fundamental to my approach is a definition of DT as a type of thinking in action. Thinking in action is underpinned by procedural knowledge, wherein the activity and types of cognition required are fundamentally linked to the practice or embodied experience of doing. For example, knowing how to ride a bicycle is a type of practiced or procedural knowledge. The ability to ride a bike comes from practice and experience, rather than the explicit declaration of rules and laws that govern bike-riding activities. In this sense, learning to ride your bike comes through knowing *how* to ride; or *know how knowledge*. By way of contrast, *knowing what knowledge* is a type of understanding that can be learned through explicit declaration. For example, a lecture providing information and explanation of industrial design, its foundations in the industrial revolution, and relation to industrial processes of production. Knowledge is taught through a standard lecture-format due to the type of knowledge communicated - *knowing how knowledge*.

Procedural knowledge is important to design as a practiced and applied discipline. The activity of design offers an opportunity for reflection upon possible solutions, expressed as design sketches for examples. This in turn supports a generative activity toward the

DOI: 10.1201/b22630-33

definition and development of design solutions. Although design as reflection-in-action does not exactly describe a type of procedural or know-how knowledge, reflective design activity has overlap with know-how knowledge because both explain a practice or applied activity. However, whereas know-how knowledge is interested in a type of knowledge applied to certain activities (i.e., learning to ride a bike), design as reflection-in-action is concerned with the external expression of design ideas, and their role as a catalyst for a particular type of cognition, that is, design thinking.

5.4.2 Design thinking and representation

An important activity in DT is the designer's expression of solution ideas through external representations of design intent. Design practice has many terms to describe these design representations, from the humble concept sketch, to high-fidelity, pre-production prototypes. Pei et al. (2011) provide an inclusive taxonomy of design representation types. Based upon a review or their use in industry, and with particular focus on product and industrial design, the taxonomy classifies design representations between four taxa: *Sketches*, *Drawings*, *Models* and *Prototypes*. Sub-categories identify more specific representational types within each of the four main taxon (i.e., *Concept Sketch*, *Technical Illustration*, *Experimental Prototype*).

The Pei et al. (ibid) taxonomy focuses attention on the representation itself, its features and characteristics, rather than the interaction between representation and designer. In an extension to the original journal article, Pei and Self (2022) provide a comprehensive account of the role and use of various representational types in industry, including case-study examples of their application.

In contrast, others have adopted the purpose a representation is used for to identify different kinds of representation: Representations to persuade, to inform, to describe and to explain (Olofsson, 2007). Interestingly, these categories have surprising synergy with categories describing different types of non-fiction writing styles: Writing to explain, describe, persuade. Other map relationships between the type of representation employed, and its place of use during a process of design (Lawson, 2006). For example, concept sketches used during the conceptual ideation phase of process. Developmental design renderings used in communication of intent during design development. Detailed technical illustrations supporting design implementation and production.

In this mapping between process and the characteristics of design representations, a relationship emerges between fidelity of design expression and the types of thinking required at different phases of design process. For example, a more emergent and divergent type of cognition required to support conceptual design ideation applies low fidelity and easily produced *concept sketches*. In contrast, the confirmatory and specificity seen at later stages of a design process employ the use of technical representations, with prescribed rules of production and communication. In fact, the conventions of design representations at later stages of design process are informed by shared rules for the articulation of a solution. There must be no misunderstanding. This is not the case in early phases of design development, where sketches and illustrations both adhere to fewer conventions (if any) and are thus open to greater interpretation.

Although the use of design representation is ubiquitous in design activity, we do not understand why design representation is important to DT. That is, why are various representations of differing levels of detail and fidelity ubiquitously used as scaffold for, what Cross (2007) described as designerly thinking?

I can almost hear you say: It's because doing so helps the designer, and other stakeholders, keep track of the progress of design ideas. Or design representation supports thinking toward and framing of the design problem (Dorst, 2015). Or perhaps, it scaffolds bridging between an ill-defined design problem (Self, 2017), and the emergence of possible solution candidates.

Although empirical works support the responses above, the question of why and how design representation is important to DT is then simply moved back one place. Why do design representations help track the progress of ideas? Why do they support problem framing in design, and so on … ?

A comprehensive understanding of what design thinking is, or how and why design representations support it, is beyond the scope of this chapter. In fact, I would argue that our knowledge of human cognition and the mind (let alone design thinking) is still too limited to make such an ambition possible. Instead, my modest aim is to present a possible way into understanding the relationship between DT and design representation. This will then contribute to building general theory on the role and use of design representation as a critical component DT. To do this I will first present the concept of embodied cognition. Within theory on embodied cognition, I will discuss distributed cognition, its relationship to design representation and DT. I will finish by positioning DT as a (higher level?) type of distributed cognition.

My approach may be useful for understanding how DT differs from other types of human cognition. Central to my thesis is the concept of *Distributed Design Thinking*. That is, DT as described through a type of cognition that interacts with external representations of design intent in a way quite unique to design, although also observable in other types of thinking, albeit at a shallower interactive depth.

Before I begin, some notes on terminology. One thing to note from the outset is my use of distributed cognition in a more localized sense. There is already much work on cognition distributed across teams and individuals. This is not the definition I use here. Here I operationalize distributed cognition to refer to a situated act of design between a designer and her representation of design intent. Likewise, in my discussion of DT, I adopt the term to describe how an individual designer thinks when designing. Her cognition in practice. A personal, localized activity as she considers possible solution candidates. This is not to say DT cannot be described in other ways, or that design activity only happens in the way described above. Of course, teams of stakeholders are involved in almost any process of design. Likewise, communication between individuals is an important aspect of design process. However, here I wish to focus upon the craft of design (Sennett, 2009), as experienced by the individual designer in a situated activity when reflecting upon personal representations as expressions of design intent.

5.4.3 Distributed design cognition

A designer works on a project to explore and develop ideas for a new product concept. As part of this effort, she develops and communicates indicative examples of possible solutions at an early stage in the process of design. She explores possible solution candidates through the sketching and illustration of concepts as representations of design ideas. Through the expression of intent as externalized design representations that approximate possible solution candidates, the designer explores possibilities, reflecting upon her representation of design intent, and is thereby better able to think about and evaluate the potential of her solution ideas.

The above scenario is an instance of employing objects within the external environment as scaffold for thinking. Embodied cognition understands thinking as taking place through and between the mind and our external world. Rather than thinking as skull-bound, cognition extends to include the external environment in which we live. In this way, cognition is embodied in the sense that thinking is influenced by the physical and embodied experience of our material environment.

Embodied cognition describes an assumption that our lived environment is an active component of how we think. A definition of thinking as *embodied cognition* can be radical in arguing for the role the external environment plays in thought. For the purposes of the current discussion, I will not detail the more or less radical positions on *embodied cognition*. For a holistic treatment, I refer you to Wilson and Lucia (2015). For my current interest in distributed design thinking, I stress that a theory of cognition as embodied advocates for thinking as an interaction between the mind and external world.

In contrast, a conventional theory of mind states all thinking takes place within the brain. Cognition is skull-bound in the sense that we experience the external world through our five senses. This information is then processed within the brain as we perceive and make sense of the world around us. We, in some way, represent these perceived experiences *within* the mind as thoughts, which then drive actions, further reflections, insights, and at a deeper level, drive motivations, opinions and emotional responses. Although dependent upon information gathered through the senses, the brain thus conceived is independent of the world in its ability to make sense of it. The mind is skull bound.

An embodied approach suggests sensation (through the senses), and perception of our external environment, continually interact with the mind to the extent that cognition is embodied and enacted through a deep interaction between the mind and external environment. Rather than a linear process of sensation, perception and cognition, embodied thinking advocates for a circular model of embodied thoughts and actions. Our body is thus seen as a sensory organ, interacting with the world to influence what and how we think moment by moment. Adopting embodied thinking, cognition is not limited to the skull, but extends outward to include our embodied experience of the external world around us.

Within theory of embodied cognition, the concept of distributed cognition (DC) is used to describe a type of thinking that extends beyond the mind to include external objects and artifacts within our material environment. As an example, when working through math problems, students often make use of paper and pencil as means to express the component parts of the math problem. These expressions then provide scaffold for thinking as part of a process of working through to a correct answer. Thinking, thus conceived, is an interaction between mind and the embodiment of thought as represented and embedded within external objects. In this case, formula expressed on paper (Figure 5.4.1).

The external expression depicted in Figure 5.4.1 acts as a representation to scaffold thinking. That is, sums on paper are the representational artefacts through which a problem is explored, and a solution achieved. Thought is distributed between the mind and marks on paper as representations of the process whereby a solution is derived. Through the distribution of thinking, the mind is extended to include these external representations. These are reflected upon in an iterative interplay between mind and externalized thoughts. Thinking, described as distributed, is thus characterized by the iterative, cyclical use of representations external to the brain, as extensions of and scaffold for the mind. This contrasts with a linear model of sensation, perception, and interpretation, perceived through the senses, and processed within a skull-bound brain.

Figure 5.4.1 Working through a math problem. The use of external representation as distribution of cognition.

Distributed cognition has much in common with Schön's (1983) concept of reflective design practice. Schön describes a type of thinking that requires a reflective conversation with the (design) situation. Whereby, the design problem is better understood through iterative reflection upon the potential of solution candidates expressed as propositional design representations. In the case of Schön's work, representations take the form of architectural plan drawings, sketches and illustrations.

While the jury for an embodied understanding of cognition may still be out for traditional cognitive sciences, the theory continues to gather support. For example, limitations in efforts toward artificial intelligence have questioned the use of computational conception of the world, which sees cognition bound within the skull (Goel, 1995). An oft-cited limitation of skull-bound cognition is the allusive goal of designing AI computation that can come up with original, creative works. Then, is embodied and distributed cognition of particular importance to our understanding of human creativity, invention and emotion?

Together with shedding light on creative cognition, and in line with Goel's (ibid) earlier work, a description of DT as distributed cognition has the potential to further understanding what DT is, including the importance of the designer's expression of solution ideas through external representation. In other words, a general theory of *Distributed Design Thinking* (DDT) holds explanatory power to describe creative cognition generally, and the role and use of representation in DT more specifically.

Drawing on theory from embodied and distributed cognition, I extend my thesis to state that DT is a type of higher-level distributed thinking. By higher-level I mean a higher-order type of distributed cognition more particular to design, and perhaps creativity. This contrasts with the types of distributed cognition discussed above (see Figure 5.4.1) and oft-cited within the cognitive sciences and phycolcgy literature. A conventional definition of distributed cognition provides more mundane examples of memory aids (i.e., memos, notes, diaries, see also Figure 5.4.1 above), or wayfaring (maps and navigation).

5.4.4 Design representation and distributed cognition

In the section above I described *embodied cognition* as a type of thinking that extends from the mind and into the world. Within an embodied approach to design cognition, I focused on the distribution of thinking through external design representations. At a lower order of distribution, shopping lists, notations and formula are used in everyday life. At a higher level of distributed cognition, however, design drawings and illustrations, as external representations of design intent, are employed as scaffold for distributed design thinking (DDT).

Important to DDT is the use of design representations as expressions of intent toward solution candidates. Some work has suggested the need to sketch is in fact not necessary to Design Thinking (Bilda et al., 2006). However, there is more evidence to indicate the importance of design representation in design cognition. See, for example, Goldschmidt's (1991) early work on the dialectics of sketching, Geol's (1995) interest in design representation's ability to confound a computational theory of mind, or Purcell and Gero's (1998) comprehensive review of work on drawing and sketching in design process in relation to cognitive psychology and cognitive science.

I would argue that the discipline of design, as a creative and applied field, is synonymous with ubiquitous types of representations (i.e., various sketches, drawings, illustrations, models and prototypes), of different levels of detail and fidelity. Both the skill and technique of producing design representations, and their various uses for different purposes and at different phases in design process, offer some support for my claim for distributed design thinking through design representations.

Design representations come in many forms, levels of fidelity, and are often expressed through various tools and media (i.e., CAD/CAM, 3D Printing, CACD, 2D illustration software, tablets and sketching pads). The tool used to construct the design representation will influence the type of representation produced. And, following the distributed cognition approach to understanding DT, have profound implications for the kinds of thoughts engaged and solutions derived (Self, 2012a). However, I will not discuss the influence of these various tools here. Instead, I wish to focus on the need for representation itself, as used in design practice. This will then support my thesis argument for DT as a higher-level distribution of cognition more unique to design.

So, what is design representation? Well, to constrain the scope of my current discussion, here I will define design representation as *not* symbolic. I do not include, for example, written or spoken language in my definition of design representation. This is of course not to say that written and spoken language is unimportant to the communication of design intent. Only that, in my thesis argument for design thinking as distributed cognition, I wish to highlight the iconic representation of design ideas; their role and use in DDT.

I use the term *iconic* here as derived from semiotic theory. Briefly, semiotics is the study of signs and their interaction with human psychology. Semiotics indicates two main types of signs: The symbolic and the iconic sigh (a third are indexical signs which I will not discuss here). A symbolic sign is a sign whereby what is signified does not share any physical properties or characteristics with the signifier. For example, the word 'tree' (signifier) does not attempt to embody any of the physical characteristics of a real tree (signified). In contrast, the small battery icon (signifier) on the laptop PC where I am currently writing this chapter, does attempt to embody some of the properties of a real battery (signified). For instance, it has a similar outline to a typical battery you may find in an electronic product – rectilinear in form, with a small bump at one end to express the positive terminal. My PC's battery sign, then, can be described as iconic. In fact, it is named a battery icon for this very reason. My battery indicator icon attempts to express the characteristics of the object it signifies, that is, the physical, archetypal double-A (AA) battery.

Likewise, often, but by no means exclusively, design representations are iconic in their expression of design intentions. They attempt to approximate at varying levels of detail and fidelity, the physical appearance and/or characteristics of a potential design. Again, I am not claiming that symbolic design representations, whether spoken or drawn, are not important to DT. Here, however, I wish to focus on representations that are iconic (or mostly so) in their expression of design intent. That is, representations that attempt to approximate the physical, material, or functional characteristics of a design solution. How the solution would function, feel, look, or behave through approximating form, function, materials, colors, textures and so on (Figure 5.4.2).

This is because these iconic representations are more interesting due to their unique association with creative practices generally, and design activity more specifically. For

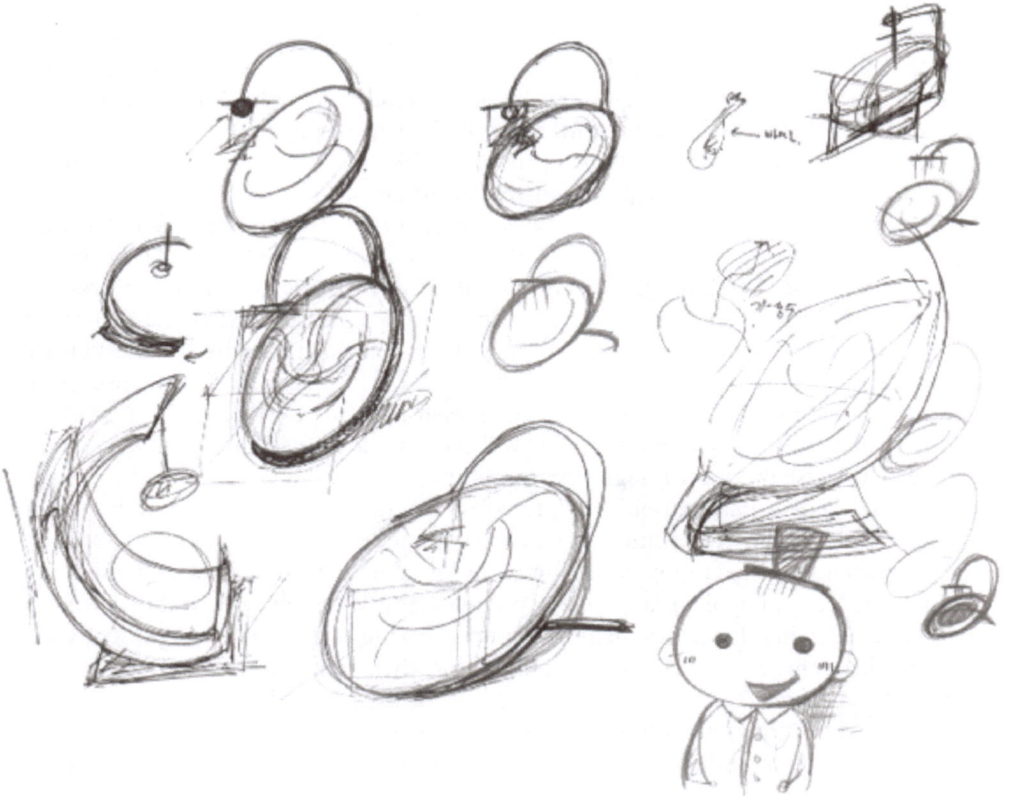

Figure 5.4.2 Iconic representation of solution intent expressed through concept sketch (Self, 2021).

example, the concept sketch, design rendering or pre-production prototype. For design, arguably more so than any other discipline, iconic expression of intent appears to go hand-in-hand with design practice, and design thinking.

5.4.5 Distributed design thinking

I will now return to the thesis of the current chapter. Namely, DT as a type of enhanced *Distributed Cognition*. I have already discussed a definition of DT focused upon a situated activity of design (Section 1 above). Within this situated activity, the designer explores and develops ideas toward potential solutions. This she does in response to a design problem, that is by nature ill-defined (Dorst, 1996), or at the very least seen as ill-defined by the designer.

Iconic design representations, that is representation that attempt to approximate the form, function or characteristics of potential design solutions, are employed in this situated activity of design. A designer may start to explore possible solution candidates through representation as concept sketches. Later, higher fidelity design illustrations may begin to express material finishes, particular functional characteristics, or scenarios of use in a particular use-contexts (Figure 5.4.3).

Distributed Cognition is a means to understand lived human experience through the lens of an embodied theory of mind (Varela et al., 2016). An extended mind understands thinking as a cyclical interaction between the environment and the mind, whereby external representations are instances of embodied thinking, rather than stimuli to be experienced through the senses and understood within a skull-bound brain. *Distributed Design Thinking* (DDT) extends a distributed theory of mind to describe thinking between the external representation of design intent, and the designer's interaction with representation as scaffold for creative thinking.

For instance, the designer prepares an experimental prototype as representation of solution intent. The experimental prototype, as an external representation, acts as means to explore the potential of a functional requirement (Figure 5.4.4). Without it, thinking toward the solution's opportunity to meet functional requirements (i.e., as derived from user needs and technical scope and limitations) would be considerably inhibited. At least this is the assumption we make by adopting a distributed understanding of DT.

In the representation of a design solution, the designer is better able to understand the solution's potential. If our assumption is that without iconic representation, the designer would be inhibited in an ability to think about, and evaluate, a design solution's potential, the role and use of design representation in DT becomes clearer.

So, how does design representation aid thinking? Again, applying the theory on Distributed Cognition (DC), design representations are locations of enhanced distributed thinking. This distributed thinking acts as an opportunity for interactions with internal representations (in the mind). These then stimulate further design representation, and/or refinement of existing representations. This is a cyclical process, wherein representation and thinking work in unison to explore and develop the potential of solution candidates.

By applying theory on *Distributed Cognition* to DT I have taken a step toward extending an understanding of why designers represent their design ideas. Design requires a distribution of cognition to engage designerly thinking (Cross, 2007). In this sense, we are in a better position to further an understanding of design thinking because we have a theoretical foundation as departure point. Design Thinking as Distributed Cognition (*Distributed Design Thinking*, DDT) has increased explanatory power as scaffold to further understand DT.

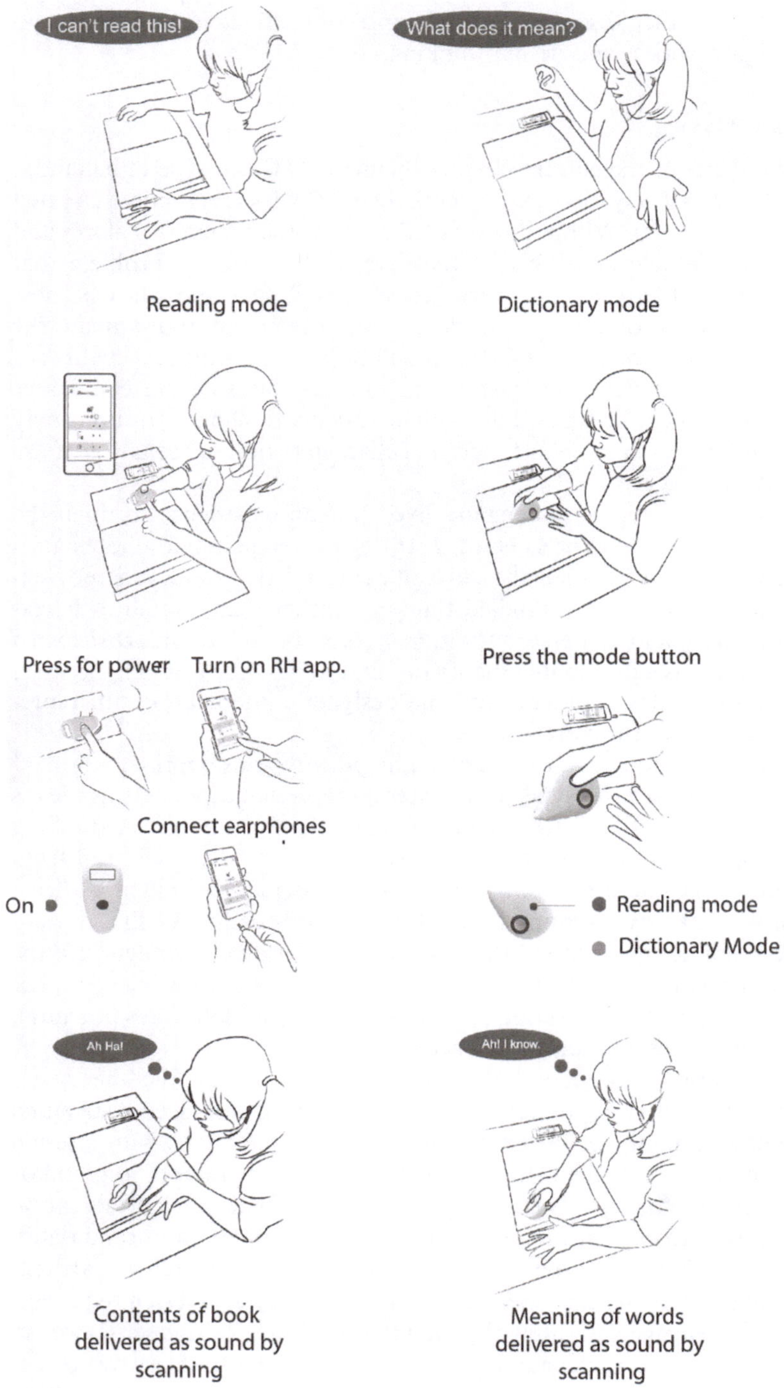

Figure 5.4.3 Scenario-of-Use concept board to represent and communicate where, how and by whom a solution may be used (Self, 2021).

Figure 5.4.4 Experimental prototype as design representation (Evans et al , 2021).

Now we have DDT in hand, we are better able to articulate further questions related to interactions between the conceptual components of this broader theory. For example, what types of distribution occur? How and when does distribution happen? What is the interaction effect between thinking and representation? What is the influence on perception and cognition? What implications for understanding creative thinking and its relation to making/representation? And many more.

The point is that DDT offers some hope for building a general theory of design thinking, that can then account for the ubiquitous use of design representations in practice. It's a way in to provide opportunities for further studies framed by an existing understanding of the mind (as limited as it still may be). Other interesting avenues of investigation might be the relationship between DDT and other now more established theories on design (i.e., problem framing, ill-defined design problems, appositional reasoning between problem and solution, design as reflective practice). Adopting the theory of design thinking as distributed cognition, the final section will suggest possibilities for future works aimed at contributing to a general theory of DT.

5.4.6 Discussion and conclusions

In this chapter, I have positioned Design Thinking (DT) as a type of higher-order distributed cognition. This I term *Distributed Design Thinking* (DDT). Drawing on work from cognitive science, cognitive psychology, and theory of mind (*embodied cognition*, *distributed cognition, embodied mind*), I have discussed the role and use of design representation as fundamental to how designers think in practice. I have touched upon a relationship between design process, fidelity of representation and the distribution of cognition between representation and the extended mind. I have also indicated how DT as distributed cognition may help explain the reflective practice paradigm for understanding design activity first introduced by Schön (1983).

However, I have merely introduced the potential of DDT as holding explanatory power to understand the role and use of design representation in DT. Many questions and further directions for research remain. For example, I have tentatively suggested that *Distributed Design Thinking* is described as a higher-level distribution that may support creativity. Well, what's the evidence for this? Indeed, what is the relationship between *Distributed Design Thinking* and creativity? How might design expertise, or the lack thereof, inhibit *Distributed Design Thinking* through design representation? How does the media/tool of expression (i.e., CAD vs. hand sketching, Self, 2012b) influence the distribution of cognition during *Design Thinking*?

Adopting *Distributed Design Thinking* as a means to understand the role and use of design representation as expressions of intent, more empirical work can now proceed to test and validate this theoretical approach. The current article has provided a starting point for such work to commence.

Acknowledgment

This work was supported by the 1.210080.01 Project Fund of UNIST (Ulsan National Institute of Science & Technology).

References

Bilda, Z., Gero, J. S., & Purcell, T. (2006). To sketch or not to sketch? That is the question. *Design Studies*, *27*, 587–613. 10.1016/j.destud.2006.02.002

Cross, N. (2007). *Designerly ways of knowing*. Birkhauser.

Cross, N., Christiaans, H., & Dorst, K. (1997). *Analysing design activity*. John Wiley & Sons.

Dorst, K. (2015). *Frame innovation: Create new thinking by design*. MIT Press.

Dorst, K. (2011). The core of 'Design Thinking' and its application. *Design Studies*, *32*(6), 521–532.

Dorst, K. (1996). The design problem and its structure. In Cross, N., Christiaans, H., & Dorst, K. (eds.), *Analysing design activity*. John Wiley & Sons.

Evans, M., Pei, E., & Campbell, I. (2021). *ID cards*. Loughborough. Available at https://www.lboro.ac.uk/media/wwwlboroacuk/external/content/schoolsanddepartments/designschool/downloads/id-cards%20%281%29.pdf, Retrieved on 24/03/2021.

Goel, V. (1995). *Sketches of thought*. MIT Press.

Goldschmidt, G. (1991). The dialectics of sketching. *Creativity Research Journal*, *4*(2), 123–143.

Johansson-Sköldberg, U., Woodilla, J., & Çetinkaya, M. (2013). Design thinking: Past, present and possible futures. *Creativity and Innovation Management*, *22*(2), 121–146.

Lawson, B. (2006). *How designers think: The design process demystified*. 4th Edition. Oxford University Press.

Olofsson, E. (2007). *Design sketching*. 3rd Edition. Keos Design Books.

Pei, E., Campbell, I., & Evans, M. (2011). A taxonomic classification of visual design representations used by industrial designers and engineering designers. *The Design Journal*, *14*(1), 64–91.

Pei, E. & Self, J. A. (2022). *Product design and the role of representation: Foundations for design thinking in practice*. CRC Press, Taylor & Francis Group.

Purcell, A. T., & Gero, J. S. (1998) Drawings and the design process: A review of protocol studies in design and other disciplines and related research in cognitive psychology. *Design studies*, *19*(4), 389–430.

Schön, D. (1983). *The reflective practitioner*. Ashgate.

Self, J. (2021). CDE51101 Integrated design project I, Graduate works. *Design Thinking Research Lab, UNIST*. Available at http://designpracticeresearch.com/new-page-1, Accessed on 24/03/2021.

Self, J. (2017). Resolving wicked problems: Appositional reasoning and sketch representation. *Design Journal*, *20*(3), 313–331.

Self, J. (2012a). *The use of design tools in industrial design practice*. Ph.D. Thesis Report, Kingston University London.

Self, J. (2012b) Sketching vs. CAD, Why Ask? *Core77 design periodical*. New York: Core77.

Sennett, R. (2009). *The craftsman*. Penguin Books.

Varela, F. J., Thomson, E., & Rosch, E. (2016). *The embodied mind: Cognitive science and human experience*. The MIT Press.

Wilson, R. A., & Lucia, F. (2015). *Embodied cognition. The Stanford Encyclopedia of Philosophy*. Available at https://plato.stanford.edu/entries/embodied-cognition/, Accessed on 19/03/2021.

Index

Note: *Italicized* and **bold** page numbers refer to figures and tables. Page numbers followed by "n" refer to notes.